ORAL CONTROLLED RELEASE FORMULATION DESIGN AND DRUG DELIVERY

ORAL CONTROLLED RELEASE FORMULATION DESIGN AND DRUG DELIVERY

Theory to Practice

Edited by

HONG WEN

Novartis Pharmaceuticals Corporation
East Hanover, New Jersey, USA

KINAM PARK

Purdue University
West Lafayette, Indiana, USA

A JOHN WILEY & SONS, INC., PUBLICATION

Published by John Wiley & Sons, Inc., Hoboken, New Jersey
Published simultaneously in Canada

For general information on our other products and services or for technical support, please contact our Customer Care Department within the United States at (800) 762-2974, outside the United States at (317) 572-3993 or fax (317) 572-4002.

Wiley also publishes its books in a variety of electronic formats. Some content that appears in print may not be available in electronic formats. For more information about Wiley products, visit our web site at www.wiley.com.

Library of Congress Cataloging-in-Publication Data:

Oral controlled release formulation design and drug delivery : theory to practice / edited by Hong Wen, Kinam Park.
　　p. cm.
　Includes bibliographical references and index.
　Summary: "This book describes the theories, applications, and challenges for different oral controlled release formulations. This book differs from most in its focus on oral controlled release formulation design and process development. It also covers the related areas like preformulation, biopharmaceutics, in vitro-in vivo correlations (IVIVC), quality by design (QbD), and regulatory issues"–Provided by publisher.
　　ISBN 978-0-470-25317-5 (cloth)
　1. Drugs–Controlled release. 2. Oral medication. I. Wen, Hong, 1967- II. Park, Kinam.
　[DNLM: 1. Delayed-Action Preparations. 2. Administration, Oral. 3. Drug Design. QV 785 O6248 2010]
　RS201.C64O725 2010
　615'.6–dc22

2010013930

Printed in Singapore

10　9　8　7　6　5　4　3　2　1

CONTENTS

PREFACE

As researchers in the fields of pharmaceutical sciences and drug development, we all understand the importance and challenges related to the design of oral controlled release formulations. The oral administration has been the first choice of drug delivery when a new drug is developed because of its easiness of administration and high acceptance by patients. Naturally, the oral formulations occupy the majority of all dosage forms. For this reason, there have been numerous research articles, patents, and books describing various aspects of development and clinical applications of oral formulations. A number of books dealing with various topics in oral drug delivery have been available, but they are either not comprehensive in topics or were published a while ago, necessitating update. We thought that it would be highly useful to prepare a new comprehensive book, covering all the major topics of oral controlled release formulation ranging from basics to practice that can serve as a useful reference book to the scientists in academia, industry, and government. Many practical examples in the book will be especially useful to graduate students and those scientists who do not have pharmaceutical background and training.

Since we initiated this book, we received many valuable suggestions from scientists active in the field on the structure and contents of the book. The book covers not only the fundamentals of preformulation, biopharmaceutics, and polymers, but also practical aspects in formulation designs, all of which are critical to achieving successful formulation development. The book also includes the most updated topics, such as new drug delivery technologies, quality by design (QbD), regulatory consideration for drug development, and competition between brand drugs using life cycle management (LCM) and generic drugs. In each chapter of the book, both theory and practical examples have been introduced to help readers understand the topic and apply the knowledge gained from the book directly to their own work. Since each chapter contains not only updated scientific information, but also authors' own experiences on a specific topic, the book as a whole provides many practical tips in formulation development, serving well as a reference book.

We want to thank all the authors for their hard work in writing their chapters, sharing knowledge, and making this book successful. Our thanks also go to all the reviewers who provided invaluable inputs for the chapters. We are also grateful to John Wiley & Sons, Inc. for agreeing to publish this book, and our special appreciation goes to Jonathan Rose who has been infinitely patient and supportive during the preparation of this book.

HONG WEN
Kinam Park

CONTRIBUTORS

Salah U. Ahmed, Abon Pharmaceuticals, Northvale, NJ, USA

Xiaoming Chen, Pharmaceutical Operations, OSI Pharmaceuticals, Cedar Knolls, NJ, USA

Nipun Davar, Transcept Pharmaceuticals, Inc., Point Richmond, CA, USA

Wendy Dulin, Pharmaceutical Development, Wyeth Research, Pearl River, NJ, USA

Linda A. Felton, College of Pharmacy, University of New Mexico, Albuquerque, NM, USA

Sangita Ghosh, Transcept Pharmaceuticals, Inc., Point Richmond, CA, USA

Michele (Xuemei) Guo, Analytical and Quality Sciences, Wyeth Research, Pearl River, NJ, USA

Bhaskara Jasti, Department of Pharmaceutics and Medicinal Chemistry, Thomas J. Long School of Pharmacy and Health Sciences, University of the Pacific, Stockton, CA, USA

Seong Hoon Jeong, College of Pharmacy, Pusan National University, Busan, South Korea

Joseph Kam, Faculty of Medicine, School of Pharmacy, The Hebrew University of Jerusalem, Jerusalem, Israel

Gemma Keegan, Vectura Group plc, Chippenham, Wiltshire, UK

Mansoor A. Khan, Division of Product Quality Research, Office of Testing and Research, Office of Pharmaceutical Sciences, Center for Drug Evaluation and Research, Food and Drug Administration, Silver Spring, MD, USA

Mannching Sherry Ku, Pharmaceutical Development, Wyeth Research, Pearl River, NJ, USA

Jaehwi Lee, College of Pharmacy, Chung-Ang University, Seoul, South Korea

Ping I. Lee, Department of Pharmaceutical Sciences, Faculty of Pharmacy, University of Toronto, Toronto, Ontario, Canada

Jian-Xin Li, Evonik Degussa Corporation, Piscataway, NJ, USA

Xiaoling Li, Department of Pharmaceutics and Medicinal Chemistry, Thomas J. Long School of Pharmacy and Health Sciences, University of the Pacific, Stockton, CA, USA

Ravichandran Mahalingam, Formurex, Inc., Stockton, CA, USA

Mirela Nadler Milabuer, Faculty of Medicine, School of Pharmacy, The Hebrew University of Jerusalem, Jerusalem, Israel

Venkatesh Naini, Teva Pharmaceuticals, Pomona, NY, USA

Ali Nokhodchi, Medway School of Pharmacy, Universities of Kent and Greenwich, Kent, UK

Hossein Omidian, Department of Pharmaceutical Sciences, Health Professions Division, Nova Southeastern University, Fort Lauderdale, FL, USA

Kinam Park, Departments of Biomedical Engineering and Pharmaceutics, Purdue University School of Pharmacy, West Lafayette, IN, USA

Zhihui Qiu, Pharmaceutical Development Unit, Novartis Pharmaceuticals Corporation, East Hanover, NJ, USA

Ali Rajabi-Siahboomi, Colorcon Ltd., West Point, PA, USA

Abraham Rubinstein, Faculty of Medicine, School of Pharmacy, The Hebrew University of Jerusalem, Jerusalem, Israel

Rakhi B. Shah, Division of Product Quality Research, Office of Testing and Research, Office of Pharmaceutical Sciences, Center for Drug Evaluation and Research, Food and Drug Administration, Silver Spring, MD, USA

Sheri L. Shamblin, Pfizer Global Research & Development, Groton, CT, USA

Yan Shen, Department of Pharmaceutics, School of Pharmacy, China Pharmaceutical University, Nanjing, China

Stephen P. Simmons, Pharmaceutical Development, Wyeth Research, Pearl River, NJ, USA

Shailesh K. Singh, Pharmaceutical Development, Wyeth Research, Pearl River, NJ, USA

John D. Smart, School of Pharmacy and Biomolecular Sciences, University of Brighton, Brighton, UK

Jean M. Surian, AstraZeneca Pharmaceuticals LP, Wilmington, DE, USA

Yue Teng, Pharmaceutical Development Unit, Novartis Pharmaceuticals Corporation, East Hanover, NJ, USA

Jiasheng Tu, Department of Pharmaceutics, School of Pharmacy, China Pharmaceutical University, Nanjing, China

Thirunellai. G. Venkateshwaran, Pharmaceutical Development, Wyeth Research, Pearl River, NJ, USA

Hong Wen, Pharmaceutical Development, Novartis Pharmaceuticals Corporation, East Hanover, NJ, USA

Xiaoguang Wen, WuXi AppTec Inc., Shanghai, China

Jong Soo Woo, College of Pharmacy, Yeungnam University, Gyongsan, Gyeongbuk, South Korea

Hua Zhang, AstraZeneca Pharmaceuticals LP, Wilmington, DE, USA

1

INTRODUCTION AND OVERVIEW OF ORAL CONTROLLED RELEASE FORMULATION DESIGN

HONG WEN

Pharmaceutical Development, Novartis Pharmaceuticals Corporation, East Hanover, NJ, USA

KINAM PARK

Departments of Pharmaceutics & Biomedical Engineering, Purdue University School of Pharmacy, West Lafayette, IN, USA

1.1 FUNDAMENTALS OF ORAL CONTROLLED RELEASE FORMULATION DESIGN AND DRUG DELIVERY

1.1.1 Overview

Due to the difficulty in developing new drugs, more and more emphasis has been given to developing new drug delivery systems for existing drugs as well as new chemical entities. Drugs can be delivered to patients by more than one route and by more than one type of dosage form. Even though "dosage form" and "drug delivery system" are used interchangeably, "drug delivery system" implies that a technology has been used to deliver a drug to the desired body site for drug release with a predetermined rate. Among various drug delivery systems, oral controlled release (CR) formulation is the most commonly used in pharmaceutical industry.

Delayed release, sustained release, and repeat action formulations are the three most common controlled release formulations [1, 2]. The most widely used example of delayed release form is enteric coated tablets [3, 4] or capsules, in which drug will not release in gastric fluid, that is, acidic environment, until it reaches the intestine, that is, neutral environment. In sustained release formulations, a portion of drug is released immediately, and the remaining drug is released slowly over an extended period of time, normally over 12–18 h. In fixed dosage combination (FDC), immediate release (IR) formulation for one drug and sustained release (SR) for another drug [5] or the same drug in both IR and SR formulation parts are popular approaches [6, 7]. For example, metoprolol succinate extended release and hydrochlorothiazide immediate release combination tablets have additive antihypertensive effects [8]. In Sanofi-Aventis' Ambien CR, there is a biphasic profile of dissolution, where the first phase is an IR phase and the second phase is a prolonged release phase [7].

For those drugs where prolonging blood levels of the drugs have no therapeutic advantages, there is no need to develop their controlled release formulations [9]. For example, drugs with a long half-life ($t_{1/2}$) (e.g., diazepam [10, 11] and amitriptyline [12]), drugs whose maintained effect is undesirable (e.g., β-lactamase antibiotic (amoxicillin) may induce emergence of resistant bacteria [13]), and drugs that require immediate effect (e.g., nitroglycerin for heart attack) [14] are not suitable in controlled release formulations.

1.1.2 Advantages and Disadvantages

In addition to extending the patent life of those drugs whose patent protection are expiring, there are many other benefits for patients by using an oral controlled release formulation [15–17]. They include maintenance of optimum drug concentration and increased duration of therapeutic effect [18], improved efficiency of treatment with less amount of drug [19], minimized side effects [20–23], less frequent administration [24], and increased patient convenience and compliance [18, 25]. The controlled release formulations are

Oral Controlled Release Formulation Design and Drug Delivery: Theory to Practice, Edited by Hong Wen and Kinam Park
Copyright © 2010 John Wiley & Sons, Inc.

also beneficial for the study of pharmacokinetic (PK) and pharmacodynamic (PD) properties of the drug [26, 27].

Like any other formulation, there are some disadvantages of oral controlled release formulations. In most cases, the amount of drug contained in the dosage form is higher than a single dose of conventional dosage forms. If the drug reservoir of a controlled release formulation is damaged and release the drug all at once, the drug concentration may go above the toxic level. Therefore, the potential of dose dumping has to be taken into consideration in controlled release formulation design. Furthermore, once the drug release begins, it is difficult to stop the release even if it is necessary. In addition, the cost of producing the controlled release formulation is higher than that of the conventional dosage forms. The relative higher production cost can be compensated if the benefit of the controlled release formulations is immediate and obvious to the patients.

1.1.3 Fundamental Release Theories

Based on different drug release mechanisms, quite a few drug release theories have been developed, which will be elaborated in the corresponding chapters. For all different types of controlled release systems except osmosis-based systems, the drug concentration difference between formulation and dissolution medium plays a very important role in drug release rate. The drug concentration can be affected by its solubility, drug loading, and/or excipients used. Besides drug concentration difference, the dissolution rate of polymer carriers can affect drug release rate in dissolution-controlled systems, and the diffusion speeds of both drug and dissolution medium inside polymer(s) can affect drug release rate in diffusion-controlled systems. For osmosis-based and ion exchange-based systems, the drug release can be affected by other factors as well. Overall, for most CR formulations, drug release can be affected by one or more mechanisms. Here, a few fundamental theories will be briefly discussed.

Fick's first law of diffusion is used in steady-state diffusion, in which the concentration within the diffusion volume does not change with time. The drug release rate is determined by drug release surface area (S), thickness (h) of transport barrier (such as polymer membrane or stagnant water layer), and the concentration difference (ΔC) between drug donor (C_d) and receptor (C_r), that is, between drug dosage surface and bulk medium.

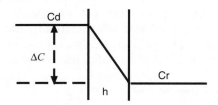

Fick's first law states that

$$M = JSt = \left(D \frac{\Delta C}{h} \right) St$$

where M is the total amount of solute crossing surface area S in time t, J is the flux rate, and D is the diffusion coefficient of the drug molecule in the unit of cm^2/s.

Fick's first law did not take into account the drug concentration changes with time in each diffusion volume, which have been taken into consideration by Fick's second law of diffusion. Based on Fick's second law, drug accumulation speed (dC/dt) is determined by drug diffusivity (D) and the curvature of drug concentration:

$$\frac{\partial C}{\partial t} = D \frac{\partial^2 C}{\partial^2 t}$$

Most commonly seen drug release rate for oral controlled release formulation is first-order release and/or zero-order release. Most oral controlled release formulations based on matrix and coating approaches are close to first-order release. Alza's osmotic pump and Egalet's erosion tube can release drugs at zero order. Based on the shape of release profile, there are five major release profiles: zero-order release (constant release rate); first-order release (decreasing release rate); bimodal release (two release modes, which can be either two separate immediate release modes or one immediate release mode followed by one sustained release mode [28–30]); pulsatile release (multiple release modes and multiple peaks of release rate [31]); and delayed release (e.g., enteric coated tablets [32–34]).

The two important phenomena in controlled release formulations are the lag time effect and the burst effect. In diffusion control system, if fresh membrane is used, it takes time for drug molecules on the donor side to appear on the receptor side. Under the sink condition, drug molecules will be released at constant rate into the receptor side and the steady state is reached. The time to reach the steady state is known as the "lag time." However, if the membrane saturated with a drug is used, a "burst effect" will be observed at the beginning of drug release. Gradually, the drug concentration inside the polymer membrane will decrease until the steady state is reached. Actually, for matrix approach controlled release formulation, because it takes time for polymer molecules to form hydrogel, "burst effect" is also a common phenomenon.

TIMERx™ is very versatile hydrogel-based controlled release technology, which can provide different release kinetics for a wide range of drugs by manipulating molecular interactions. The release profiles range from zero order to chronotherapeutic release. This technology does not need complex processing or novel excipients, but still achieves desired drug release profiles using a simple formulation development process. TIMERx™ is a pregranulated blend composed of synergistic heterodisperse polysaccharides (usually xanthan gum and locust bean gum) together with a saccharide component (generally dextrose). Different drug release kinetics can be achieved based on the synergism between the homo- and heteropolysaccharide components in the system. Finally, the drug release rate is controlled by the speed of water penetrating into the matrix [35, 36]. The material has good compressibility and can be mixed or granulated with drug and other necessary excipients to be compressed into tablets.

1.1.4   Limiting Factors for Oral CR Formulations

There are a few unique properties of the gastrointestinal (GI) tract that make development of oral CR formulations rather difficult. Figure 1.1 shows schematic description of the GI tract. Based on histology and function, the small intestine is divided into the duodenum, jejunum, and ileum, and the large intestine is divided into the cecum, colon, rectum, and anal canal. W. A. Ritschel reported the average length, diameter, and absorbing surface area of different segments of the GI tract, and the data clearly show that jejunum and ileum (small intestine) have similar surface absorbing areas that are significantly larger than those of other segments [37]. For

most drugs, there is better drug absorption in the upper GI tract, which is also consistent with the significant higher surface absorbing area in the upper GI tract.

1.1.4.1   *Relatively Short Gastric Emptying and Intestinal Transit Time and Varying pH Values*   Because oral dosage forms will be removed from the GI tract after a day or so, most oral CR formulations are designed to release all drugs within 12–18 h. The values in Table 1.1 show the approximate transit time in different GI segments. The presence of food in the stomach tends to delay the gastric emptying. Among different foods, carbohydrates and proteins tend to be emptied from the stomach in less than 1 h, while lipids can stay in the stomach for more than 1 h [37–41]. As a convenient resource, Gastroplus™ can provide rough estimation on the transit times and pH values of the GI tract under different situations and help to calculate corresponding drug PK profiles.

Table 1.1 shows that the small intestinal transit time is more reproducible and is typically about 3–4 h [38, 39]. Thus, the transit time from mouth to cecum (the first part of large intestine) ranges from 3 to 7 h. Colonic transit is highly variable and is typically 10–20 h [42–44]. Since most drugs are absorbed from the small intestine, the time interval from mouth to cecum for oral controlled release dosage forms is too short, unless the drug can be equally well absorbed from the large intestine. Thus, the release profiles of most oral controlled release dosage forms can be effective for only about 8 h. If the drug can be absorbed from the large intestine, the time interval for drug absorption can be increased to 1 day. Thus, certain drugs can be delivered for 24 h by a single administration of an oral controlled release dosage form. But many drugs require more than one administration if they have the upper GI tract absorption window and short half-life, unless the release of those drugs can be controlled at the upper GI tract with special design. The study on the GI transit time of once-a-day OROS® tablets of both oxprenolol and metoprolol showed that the median total transit time was 27.4 h with a range of 5.1–58.3 h [45, 46].

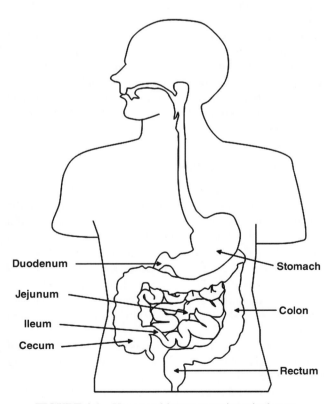

FIGURE 1.1   Upper and lower gastrointestinal tract.

TABLE 1.1   The pH Values and the Transit Time at Different Segments of the Human GI Tract [37–41]

Anatomical Site	Fasting Condition pH	Fasting Condition Transit Time (h)	Fed Condition pH	Fed Condition Transit Time (h)
Stomach	1–3.5	0.25	4.3–5.4	1
Duodenum	5–7	0.26	5.4	0.26
Jejunum	6–7	1.7	5.4–6	1.7
Ileum	6.6–7.4	1.3	6.6–7.4	1.3
Cecum	6.4	4.5	6.4	4.5
Colon	6.8	13.5	6.8	13.5

1.1.4.2 Nonuniform Absorption Abilities of Different Segments of GI Tract Drug transport across the intestinal epithelium in each segment is not uniform, and in general, it tends to decrease as the drug moves along the GI tract. Because drug absorption from different regions of the GI tract is different, the residence time of drug within each segment of the GI tract can profoundly affect the performance of the oral controlled dosage form, that is, the absorption of drug.

If a drug is absorbed only from the upper segment of the GI tract, it is known to have a "window for absorption" [47]. For the drugs with window for absorption, adjusting drug release rate on different segments of the GI tract may be needed to compensate decreased absorption, in order to maintain relatively constant blood concentration. For example, to achieve a plateau-shaped profile of plasma concentrations at steady state throughout the 24 h dosing interval, nisoldipine coat core (CC) controlled release formulation releases drug slowly in the upper GI tract that has fast absorption and quickly in the colon that has decreased absorption rate [48]. Besides adjusting drug release rate, increasing the residence of drug formulations at or above the absorption window can also enhance the absorption for those drugs. Currently, two main approaches have been explored: bioadhesive microspheres that have a slow intestinal transit and the gastroretentive dosage system [49].

1.1.4.3 Presystemic Clearance For some drugs, presystemic clearance may occur at some sites of the GI tract and affect drug absorption. Degradation of orally administered drugs can occur by hydrolysis in the stomach, enzymatic digestion in the gastric and small intestinal fluids, metabolism in the brush border of the gut wall, metabolism by microorganisms in the colon, and/or metabolism in the liver prior to entering the systemic circulation (i.e., first pass effect). Such degradations may lead to highly variable or poor drug absorption into the systemic circulation. For example, digoxin undergoes microbial metabolism before absorption [50, 51]. For this type of drugs, for which presystemic clearance is determined by the site of absorption, drug bioavailability can be enhanced by restricting drug delivery to the upper segment of the gut, or to the stomach. For example, the same amount of metoprolol was administered at the same rate using a continuous 13.5 h intragastric infusion or a OROS® tablet; at 6–15 h after dosing, the intragastric infusion had higher plasma concentration than metoprolol OROS® tablet [52].

1.1.4.4 Poor Absorption of Peptide and Protein Drugs It is very difficult to develop oral formulations for peptide and protein drugs. First, peptide and protein drugs are unstable under the harsh conditions in the GI tract and can be degraded by enzymes. Second, even if the structures of peptide and protein drugs are maintained, absorption of high molecular weight drugs, for example, insulin, through the epithelial cells of the GI tract is very difficult at best. So far, much research has been done to develop new technologies for oral peptide and protein delivery [53–55]. For example, Emisphere's eligen® Technology has been used in the development of oral drug delivery for peptide and protein drugs such as salmon calcitonin, insulin, and recombinant human growth hormone (rhGH) (http://www.emisphere.com/pc_pp.asp).

1.1.4.5 Difficult In Vitro–In Vivo Correlation Establishing *in vitro–in vivo* correlation (*IVIVC*) will be very valuable in predicting drug *in vivo* performance based on *in vitro* dissolution tests. However, it is sometimes difficult to establish *IVIVC* for oral controlled release formulation due to many factors, such as variable transit time in different segments of the GI tract, nonuniform absorption abilities of different segments of the GI tract, and presystemic clearance.

1.2 PREFORMULATION AND BIOPHARMACEUTICAL CONSIDERATIONS FOR CONTROLLED RELEASE DRUGS

Many parameters of drug substances can affect the controlled release formulation design and drug absorption from the formulation. Many properties of drug substances, such as pH–solubility profile, pK_a, permeability, particle size distribution (PSD), thermal properties, hygroscopicity, compactibility, and excipient compatibilities, can affect oral CR formulation design, processing, storage, and drug absorption. However, in the *in vivo* process of drug release to drug absorption, the major factors are drug solubility, permeability, and stability. Based on drug aqueous solubility and gastrointestinal permeability, drugs have been classified into four biopharmaceutical drug classes [56].

1.2.1 Aqueous Solubility

For Biopharmaceutical Classification System (BCS) Class II drugs with poor water solubility but good permeability, the absorption of a drug is often limited by dissolution rate. If a drug's solubility is lower than 0.1 mg/mL, the drug is not a suitable candidate for diffusion-controlled formulation, but still feasible using other approaches such as dissolution-controlled or osmosis-based systems [57, 58]. Furthermore, drugs with solubility less than 0.01 mg/mL show dissolution-limited bioavailability, and thus have inherent controlled release property.

For drugs with high solubility, it is also pretty challenging to design controlled release formulations with high drug loading like more than 80% using the matrix dissolution system. Because the drug diffusion force from the high drug concentration in the matrix system will be strong, it will be difficult for the minimal amount of polymer to control the diffusion process. However, with new technology such as hot-melt extrusion, the drug loading can be significantly increased to even higher than 90%. Furthermore, BCS Class III drugs with high solubility but poor permeability are even more difficult to deliver in controlled release dosage forms. For those poorly permeable drugs, the drug absorption is controlled by the cellular absorption of the drug rather than the release from the dosage form, but localized high drug concentration can still benefit drug absorption.

Drugs with pH-dependent solubility can present difficulties in the drug delivery. Table 1.1 shows different pH values at different segments of the GI tract. One frequent challenge in developing poorly water-soluble free-base drug is that the free-base drug can dissolve to a greater extent in stomach due to acidic environment, but may precipitate out in the small intestine due to high pH; however, the maximum adsorption may occur in the small intestine. For some water-insoluble free-base drug compounds, when comparing the AUC (area under drug plasma concentration curve) of different CR formulations with different release profiles, sometimes the AUC of slow release can be higher than the AUC of fast release, which may be due to the precipitation of drug in the intestine, that is, neutral pH environment.

1.2.2   Permeability

For absorption to occur from the GI tract, drug molecules have to penetrate through the cellular membranes. The total drug absorption can be described as [56]

$$\text{Mass}(t) = \int_{0}^{t} \iint_{A} PwCw \, \mathrm{d}A \, \mathrm{d}t$$

where Mass is the total mass of drug absorbed, P and C are the permeability and drug concentration at certain time and location, respectively, and A is the adsorption surface area. Drug permeability can be affected by many factors, such as the location of the GI tract, drug concentration in the case of carrier-mediated transport, and so on. Considering the three major processes of drug absorption, transit flow, dissolution, and permeation, Lawrence Yu proposed an integrated absorption model to estimate the fraction of dose absorbed and to determine the causes of poor oral drug absorption [59].

To have desired bioavailability for the poorly permeable drugs, the equation shows the importance of localized high drug concentration, as well as the transit time of a drug in different segments of the GI tract. All these factors affect the absorption of poorly permeable drugs much more significantly than drugs with high permeability. The permeability and transit time of a drug can be affected by many factors, such as food, interpersonal variance, formulation design, and

so on, and all these variables make establishing *IVIVC* very difficult for poorly permeable drugs in oral controlled release formulations.

Uncharged form of a drug is preferentially absorbed through the cellular membrane. For charged form, the pH value of the environment, shown in Table 1.1, can affect the drug absorption significantly. If polymer membranes are used in oral controlled release formulations, the drug diffusion through polymer membranes can also be affected by environment pH based on the drug's pK_a.

1.2.3   Physicochemical Stability

The drugs unstable in acidic environment cannot be delivered in the stomach, and enteric coating has been widely used to release drugs in neutral small intestine environment. Drugs that are degraded in the GI tract may undergo more degradation when slowly released in stomach from the controlled release formulations.

Similar to immediate release formulations, both physical and chemical stability of drug substances are very important in formulation design, process development, and storage. The forced degradation studies of drug substances can check the drug stability under heating, acidic, basic, oxidative, and lighting (both UV and visible light) environment, which are very useful information in formulation design. Furthermore, excipient compatibility studies are commonly used to select suitable excipients. In process development, whether drug substances are moisture sensitive will be critical in wet granulation, and whether drug substances are heat sensitive will be critical in hot-melt extrusion granulation, and so on. For those drugs that have different crystal forms such as hydrous/anhydrous and polymorphisms (especially enantiotropic polymorphisms), the potential crystal form transformation during processing and storage has to be taken into careful consideration.

1.3   OPTIMAL FORMULATION AND PROCESS SELECTION FOR CONTROLLED RELEASE DRUGS

1.3.1   Controlled Release Formulation Mechanisms and Related Approaches

Although there are hundreds of commercial products based on controlled release technologies, there are only a few distinct mechanisms in controlled drug release. Oral controlled release formulations are designed mainly based on physical mechanisms, rather than chemical degradation, enzymatic degradation, and prodrug approach. Table 1.2 lists the types of controlled drug release mechanisms commonly used in oral controlled release formulations. All controlled release formulations are designed based on one or combination of a few mechanisms.

TABLE 1.2   Controlled Drug Release Mechanisms and Related Formulation

Mechanism	Related Formulation Approach
Dissolution	Encapsulated dissolution system (reservoir system)
	Matrix dissolution system
Diffusion	Reservoir system
	1. Nonporous membrane reservoir
	2. Microporous membrane reservoir
	Monolithic device
	1. Nonporous matrix
	a. Monolithic solution
	b. Monolithic dispersion
	2. Microporous matrix
	a. Monolithic solution
	b. Monolithic dispersion
Osmotic	
Ion exchange	

1.3.1.1 Dissolution-Controlled Formulations

In the encapsulated dissolution system (reservoir system), the drug release is determined by the thickness and the dissolution rate of the polymer membrane surrounding the drug core. Once the coated polymer membrane dissolves, all the drug will release like immediate release formulation. In general, small beads are designed based on this approach. Tablets are not preferred due to potential dose dumping if the tablet coating is broken. By adjusting membrane thickness on small beads, desired release profile can be achieved. The coated drug beads can be either compressed into tablets or filled into capsules.

In the matrix dissolution system, the most commonly used system in pharmaceutical industry, drug is homogeneously distributed throughout the polymer matrix. As the polymer matrix dissolves, drug molecules are released, also called "erosion controlled release." Actually, for both encapsulated dissolution system (reservoir system) and matrix dissolution system, drugs may release through diffusion mechanism as well based on the properties of drugs and polymers. In the matrix dissolution system, since the size of the matrix decreases as more drug is released, the amount of drug released is also decreased, that is, resulting in a nonzero-order release.

1.3.1.2 Diffusion-Controlled Formulations

In diffusion-controlled formulations, drug molecules have to diffuse through a polymer membrane [60] or a polymer matrix to be released. Diffusion-controlled formulations can be divided into reservoir and monolithic systems, depending on whether a drug is surrounded by a polymer membrane or distributed through the polymer matrix. Different diffusion-controlled reservoir systems have been shown in Figure 1.2. In nonporous reservoir systems, drug molecules have to diffuse

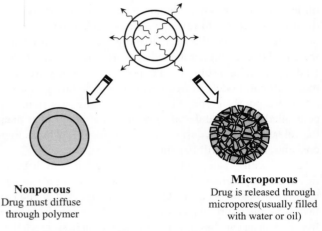

Nonporous
Drug must diffuse through polymer

Microporous
Drug is released through micropores(usually filled with water or oil)

FIGURE 1.2   Diffusion-controlled reservoir systems.

through the polymer membrane, but in microporous reservoir systems, drug molecules are released by diffusion through micropores that are usually filled with either water or oil.

In addition to nonporous and microporous systems, diffusion-controlled monolithic systems can be further classified based on the concentration of loaded drug. The monolithic system is called monolithic solution if a drug is loaded by soaking a polymer matrix in a drug solution, in which the drug concentration inside the matrix cannot be higher than the drug solubility, if the partition coefficient of a drug is 1. If the drug loading is higher than the drug solubility, shown as black dots in Figure 1.3, the monolithic system is called monolithic dispersion.

1.3.1.3 Osmosis-Based Formulations

Osmosis, the natural movement of water into a solution through a semipermeable membrane, has been used in the development of zero-order release drug delivery systems. Not solutes, only water can diffuse through the semipermeable membrane. For different polymer membranes, their water vapor transmission

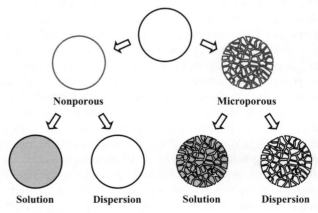

Nonporous   **Microporous**

Solution   **Dispersion**   **Solution**   **Dispersion**

FIGURE 1.3   Diffusion-controlled monolithic systems.

value can differ widely, and selection of a semipermeable membrane depends on the nature of the application. Overall, the release rate in osmosis-based systems depends on osmotic pressure of release medium.

In the development of osmosis-based controlled release formulations, cellulose acetate has been used most frequently [61–63]. Alza Corporation developed two different types of osmotic devices, known as OROS® osmotic therapeutic systems, as shown in Figure 1.4. Both OROS® osmotic systems can deliver drugs at continuously controlled rate for up to 24 h, independent of GI environment. However, the manufacturing processes for both OROS® osmotic systems are pretty complex. Compared to basic osmotic systems that can deliver only water-soluble drugs, "push–pull" osmotic systems can deliver water-insoluble drugs as well [64, 65].

Both the original Alza patents have expired, and many new approaches have been developed so far, such as modifying formulation compositions, alternate membrane coating, and so on [66]. For insoluble drugs such as nifedipine, EnSoTrol system of Shire Laboratories contains a nonswelling solubilizing agent that enhances the solubility of insoluble drugs and a nonswelling wicking agent dispersed throughout the composition that enhances the surface area

contact of the drug substances with the incoming aqueous liquid [67, 68]. The single composition osmotic tablet (SCOT®) of Andrx can deliver insoluble drugs as well [69, 70]. In the SCOT®, osmosis leads to swelling and disruption of coating; after membrane disruption, core matrix erodes and releases drug at controlled rate. In the swellable core technology of Pfizer, two model drugs, tenidap and sildenafil, have been released at similar rate despite significant differences in their physicochemical properties [71, 72]. For alternate membrane coatings, there are several unique approaches such as asymmetric membrane coating of Pfizer [73] and Merck osmotic delivery system [74].

1.3.1.4 Ion Exchange-Based Formulations

Ion exchange-controlled release systems use ion-exchange resins that are water-insoluble polymeric materials containing ionic groups [75]. Drug molecules can attach onto the ionic groups with opposite charge through electrostatic interaction. Thus, the drug molecules can be replaced with other ions with the same charge and released from the ion-exchange resin, as shown in Figure 1.5. The drug release from ion-exchange systems depends on replacement of the drug molecules by other electrolytes. To have a more predictable drug release, the ion-exchange resins can be coated with water-insoluble polymers such as ethylcellulose (EC) to provide diffusion-controlled drug release. Overall, the rate of drug release depends on the area of diffusion (i.e., surface area of resin particles), cross-linking density, ionic strength (i.e., concentration of replacing ions such as Na^+ or K^+ for cationic drugs and Cl^- for anionic drugs), and coating of the drug–resin complex.

There are a few advantages of the ion exchange-controlled systems. First, they are convenient to adjust individual dose especially for pediatrics and geriatrics. Second, the GI tract irritation is substantially reduced due to the slow release in

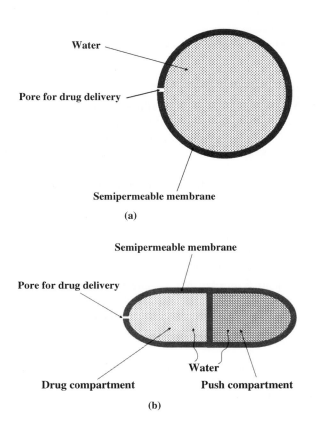

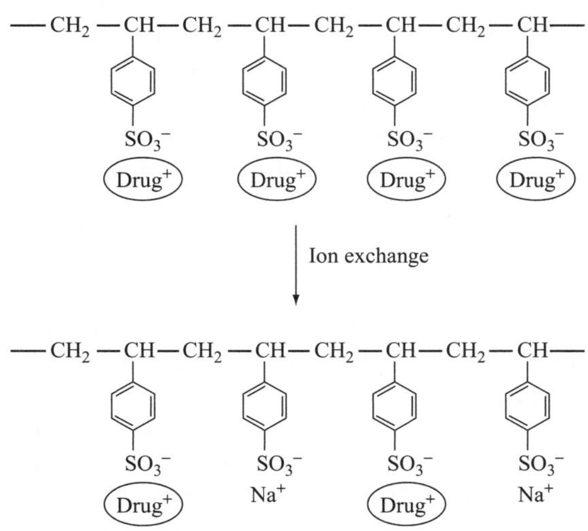

FIGURE 1.4 Brief description of Alza's two OROS® osmotic therapeutic systems: (a) basic osmotic pump; (b) push–pull OROS® system.

FIGURE 1.5 Ion exchange-controlled systems.

small quantities. They can effectively provide taste abatement because all drug molecules are initially bound to polymer chains. Suspension of ion-exchange resins was first developed by Pennwalt Pharmaceutical Company, and the system is called the Pennkinetic system [76, 77]. One of the Pennkinetic systems that is commercially available is Delsym® in which poly(styrene sulfonic acid) resins loaded with dextromethorphan are coated with ethylcellulose for delivery up to 12 h. Corsym® delivers codeine and chlorpheniramine. Nowadays, ion-exchange approaches have not only proved to be safe and effective, but also attracted more and more attention considering their uniqueness [78–80].

1.3.2 Various Process Approaches

Based on the dissolution-, diffusion-, and osmosis-based controlled release formulation mechanisms, oral CR formulations can be roughly divided into three approaches: matrix tablets, multiparticulates, and osmotic tablets [81]. Even though many processes can be used in different formulation approaches, the preferred processes for different formulation approaches are different.

1.3.2.1 Processes for Matrix Tablets

Matrix tablets contain both hydrophilic CR systems [82] and lipophilic CR systems [83, 84]. The drug release from hydrophilic systems involves both diffusion and dissolution (i.e., matrix erosion), and from lipophilic systems is only under diffusion control. Most traditional processes such as dry blend (direct compression), roller compaction, wet granulation, fluid bed granulation, foam granulation [85, 86], and melt extrusion granulation can be used to make both types of matrix tablets. The process selection for matrix tablets is similar to the process selection for immediate release tablets. The major factors involved in process selection are drug loading, flowability, and compactibility. Both wet granulation and fluid bed granulation may not be optimal for moisture-sensitive drugs, and melt extrusion granulation may not be suitable for thermally unstable drugs. For different processes, the maximal drug loading may follow approximately in the order of melt extrusion granulation > wet granulation > roller compaction ≈ fluid bed granulation > direct compression.

1.3.2.2 Multiparticulates

Multiparticulate CR systems contain both drug layered beads and microspheres. Fluid bed coating, very useful in preparing various multiparticulate CR systems, uses three different spraying methods: top spray, bottom spray (Wurster process), and tangential spray. The top spray method is commonly used for fluid bed granulation, sometimes for particle coating as well. The bottom spray (Wurster) coating is the usual method in particle/bead coating. For the multiparticulate CR systems, Wurster coating is very useful in drug layering on nonpareils as well as functional coating. The tangential spray (rotary) method can

achieve similar film quality as Wurster coating; however, it is more difficult to scale up.

In addition to fluid bed granulation, many other approaches have been used to prepare microspheres/beads, such as extrusion and spheronization, hot-melt extrusion granulation, spray congealing, and roller compaction. Extrusion–spheronization is a usual pelletization process for making pellets that are amenable for both immediate and controlled release formulation preparation [87]. Calcium can induce alginate to form beads, which have been widely used in controlled release formulation design. The beads can be collected by filtering and drying, or one-step spray drying [88, 89].

1.3.2.3 Osmotic Tablets

Preparation of osmotic tablets can be roughly divided into three parts: drug layer and/or sweller layer, membrane(s), and microscopic hole(s) for drug release. For drug layer and sweller layer, traditional processes can be used to make granules similarly. For elementary osmotic pump, that is, only drug layer, monolayer tablets can be compressed easily. For the "pull–push" osmotic pump, that is, with both drug layer and sweller layer, drug layer and sweller layer need to be compressed into bilayer tablets. After membrane(s) has been coated onto the core tablets, holes for releasing drug from membrane are normally created by laser drilling. However, in Merck osmotic delivery system [74], high concentrations of porosigens inside cellulose acetate will generate holes for drug release.

1.3.3 Computer-Aided Design

Computer-aided design (CAD) uses mathematical and numerical techniques to study drug release kinetics. With the help of CAD, it is possible to save both cost and time in drug development, as well as create better quality products. In order to use CAD in oral CR formulation development, the approximate workflow is listed below [90–93]:

- Understand delivery systems and related drug release mechanisms.
- Build models and determine related parameters.
- Execute numerical analysis.
- Identify discrepancies, adjust model, and analyze results.
- Design formulation-based computer-aided design to achieve desired release profiles.

Many formulation and process factors can affect drug release, and among them, five formulation factors may be most critical in influencing release kinetics. They include drug and excipient's properties, especially drug solubility, tablet shape and size (i.e., dimension), tablet surface area, drug loading, and coating (coating materials, coating thickness, etc.).

1.3.4 Scale-Up

After formulation and process has been determined, based on scientific understanding and collected process data, several DoEs (design of experiments) will be designed and executed to study the critical process parameters and their effective ranges. Nowadays, QbD (quality by design) becomes more and more important in developing robust manufacturing process to get qualified drug product. After DoEs, pilot size batches of plus and minus process set point conditions and/or set point conditions will be made to further understand process parameters to help scale up to commercial size manufacturing.

1.4 POLYMERS FOR CONTROLLED RELEASE FORMULATION DESIGN

1.4.1 Categories

Even though there are a lot of different synthetic polymers, not many have been used in pharmaceutical industry especially in oral CR formulation. Although biodegradable polymers such as poly(lactic acid) and poly(glycolic acid) are widely used in parenteral sustained release and implant drug release systems, they are not commonly used in oral CR formulations. Most common synthetic polymers used in oral CR formulation are poly(vinyl alcohol) (PVA), poly(acrylic acid), poly(ethylene oxide) (PEO), poloxamers, pluronics, and polymethacrylate. Poly(acrylic acid) and its derivatives are commonly used in enteric coating due to their insolubility at low pH. Carbopol is a high molecular weight cross-linked poly(acrylic acid) polymer [94]. Polymethacrylate and derivatives, main compositions of Eudragit®, are very commonly used for sustained release coating.

In pharmaceutical industry, much more natural polymers or their derivatives than synthetic polymers have been used in oral CR formulations. Among the three subclasses of natural polymers, proteins, polysaccharides, and nucleotides, only polysaccharides are widely used in oral CR formulations.

Cellulose derivatives such as hydroxypropylmethylcellulose (HPMC), hydroxypropylcellulose (HPC), hydroxyethylcellulose (HEC), ethylcellulose, and methylcellulose (MC) are the most commonly used polymers in oral CR formulations [95, 96]. For each cellulose derivative, different grades can also have significantly different properties in terms of molecular weight, viscosity, solubility, hydration, and so on; thus, different grades can be used for different purposes. Besides cellulose derivatives, many polysaccharides especially dietary fibers have been used in drug development. Table 1.3 lists the commonly used natural polymers or their derivatives in oral CR formulations.

Different from the polymers used in dissolution-controlled release systems, the polymers in diffusion-controlled systems are generally water insoluble. Some commonly used polymers for diffusion-controlled systems (reservoir and monolithic systems) are cellulose (e.g., ethylcellulose), collagen, nylon, poly(alkylcyanoacrylate), polyethylene, poly(ethylene-co-vinylacetate), poly(hydroxyethyl methacrylate), poly(hydroxypropylethyl methacrylate), poly(methyl methacrylate), polyurethane, and silicon rubber.

1.4.2 Polymer Properties

For polymers used in oral CR formulations, there are several important properties that can influence formulation design especially drug release rate, as shown in Table 1.4. Besides drug release kinetics, polymer properties can also affect process development. For example, in melt extrusion granulation, polymer glass transition temperature (T_g) is very important because the process temperature in melt extrusion granulation falls between polymer T_g and melting temperature T_m of drug substance. Other polymer properties such as flowability, compactibility, and so on are also very important in process development. The effects of polymer properties on

TABLE 1.3 Common Natural Polymers and Derivatives Used in Oral CR Formulations

Polymer	Comment
HPC	Used in matrix sustained release formulations
HPMC	Widely used in matrix sustained release formulations
EC	Insoluble in water. Widely used in coating for sustained release applications. Also used in matrix tablets for diffusion-controlled CR formulation, that is, lipophilic matrix
MC	Not as efficient as HPMC and HPC in slowing down drug release rate [97, 98]
Carboxymethylcellulose, Na	Sometimes used in matrix tablets together with HPMC [99, 100]
Alginate, Na	Besides thickening, gel-forming, and stabilizing properties, it can also easily gel in the presence of a divalent cation such as Ca^{2+} [101]
λ-Carrageenan	
Chitosan	pH-dependent hydrogelation of chitosan matrixes [102]
Heparin	
Xanthan gum	
Starch (thermally modified)	

TABLE 1.4 Polymer Properties Versus Drug Release Mechanisms

Mechanism	Polymer Property
Dissolution	Polymers such as HPMC, soluble in water: molecular weight, viscosity, hydration speed, solubility in water, and so on
Diffusion	Lipophilic polymers, such as ethylcellulose, poly(methyl methacrylate), poly(hydroxyethyl methacrylate), insoluble in water: molecular weight, viscosity, lipophilicity, and so on that can affect drug diffusion through them
Osmosis	Semipermeable membranes such as cellulose acetate: water permeability through them
Ion exchange	Cross-linked resins

oral CR formulation design and process development will be further discussed in a related chapter.

Plasticizer is a material that enhances the flexibility of the polymer with which it is mixed and reduces the T_g of the mixture. Examples of plasticizers are glycerin, glyceryl triacetate (triacetin), poly(ethylene glycol) (PEG), and propylene glycol. Plasticizers are commonly used in film coating to help polymer(s) achieve desired film quality. Besides, since plasticizers reduce the stiffness of polymer molecules, they can increase the diffusion rate of drug molecules through the polymer matrix or polymer membrane. Except preparing amorphous drug products, the process temperature for melt granulation should be below drug melting temperature and above polymer T_g. Therefore, by reducing polymer T_g through adding suitable plasticizer, the process temperature for melt granulation can be lowered.

1.5 PHARMCOKINETIC AND PHARMACODYNAMIC CONSIDERATIONS

1.5.1 PK/PD Principles

To design a controlled release drug product, besides disposition and absorption pharmacokinetic parameters, a complete understanding of the concentration–response relationship is also very important. Quantification of pharmacodynamics of a drug will enhance the development of a controlled release drug product and, importantly, should lead to faster regulatory approval.

The drug ADME (adsorption, distribution, metabolism, and elimination) process is a very complex process and can be affected by many factors such as physicochemical properties, physiologic constraints, and biochemical principles. To determine suitable dosage, the most useful PK parameter is total plasma clearance (CL). The CL is meaningful because of its relationship with dosing rate and plasma concentration at steady state (C_{ss}). For a zero-order input (R_0),

$$C_{ss} = R_0/CL$$

The above equation shows that the plasma concentration, which is important in achieving a desired effect and avoiding undesired toxicological effect, is mainly controlled by the drug input rate and the plasma clearance. Even though the equation is most routinely used for a constant i.v. infusion, it is still meaningful for oral controlled release formulations especially in the case of a zero-order release. For most drugs, the drug input rate is mainly controlled by the drug release rate.

The above equation is a very simplified scenario in which the drug is assumed to obey linear PK principles, that is, the clearance is constant and first-order elimination is observed at several doses. Considering the interpersonal variability, a range of input rates need be calculated to check whether there exist serious toxicities at upper range end of input rate. If the undesired toxicological effect does occur at relatively high input rate, the input rate should be adjusted toward lower end of the therapeutic range.

The drug plasma elimination half-life ($t_{1/2}$) is not directly needed for determining the input rate for a controlled release drug product. However, it will help evaluate the usefulness of a controlled release formulation. A compound with $t_{1/2}$ more than 12–24 h is considered a poor candidate for once-daily controlled release formulation. For a compound with $t_{1/2}$ relative short but not extremely short, its treatment can be improved by a controlled release formulation. For an immediate release formulation, the dosing interval is primarily determined by $t_{1/2}$. However, for a CR formulation, the dosing interval is more affected by the maximum drug input and the drug release rate.

Besides improving patient compliance, a controlled release formulation can help reduce undesired toxicological effects especially for drugs with narrow therapeutic window by reducing C_{max} [103, 104]. Oral controlled release formulation can help maintain drug blood concentration within its therapeutic window for longer time and decrease the plasma concentration fluctuation.

1.5.2 Dissolution Profiles and Testing Conditions for Target Dissolution Profiles

The *in vitro* release rate, that is, dissolution profile, for a dosage form, though not strictly a PK parameter, is clearly quite critical to the eventual regulatory approval and the effectiveness of the controlled release delivery system. From a PK perspective, this process is very important because in most scenarios it is the rate-determining step leading to drug in the plasma and at the active site. The drug release rate must be reproducible, and ideally be unaffected as much as possible by physiologic factors, and, as a final result, provide the desired therapeutic plasma concentration.

Dissolution tests not only are useful in formulation design and quality assurance, but can also help to get biowaiver for scale-up, postapproval changes. For the fixed dosage combinations, after high strength formulations have achieved bioequivalence with free combination formulations, dissolution tests can help low strength formulations to eliminate the need for bioavailability studies, that is, biowaiver. Note that the drug release dissolution method may not always be the same as the biorelevant dissolution method. The biorelevant dissolution method can predict plasma concentration curves based on changes of dissolution profiles. Ideally, dissolution specifications should be set so that all formulations whose dissolution profiles fall within specifications are bioequivalent [105].

Dissolution specifications are usually established based on drug substance properties and dissolution profile evaluation. For oral CR formulations, their dissolution tests meet regulatory requirements if the quantities of drug substance released conform to the harmonized USP⟨711⟩, or EuPharm 2.9.3 and JP 15 for EU and Japan, respectively. For oral CR formulations, dissolution apparatus can be USP Type I, II, IV, and VII, that is, rotating basket, rotating paddle, flow-through cell, and reciprocating holder. In dissolution medium selection, besides stability, sensitivity of assay, and so on, sink condition in which the final drug concentration is at least three times lower than a saturated concentration is needed.

Current USP dissolution tests exert minimal mechanical force on solid dosage forms, which is different from the reality in human gut that exerts about 1.5–1.9 N force on solid dosage forms [106–108]. *In vivo*, different from IR tablets, mechanical forces can obviously affect the dissolution process much more in sustained release matrix tablets [109]. A peristaltic dissolution test developed by Matthew Burke applied mechanical compression to simulate gastric contractions and peristalsis action in SR matrix formulations [110]. The test can be added to a USP Type II apparatus directly or utilized in established dissolution methods.

1.5.3 IVIVC

Like most drug substances, an adequate demonstration of the *in vitro* release profile should be established within appropriate limits for the controlled release formulations. A reasonable release profile must match with the claims of the particular product. Furthermore, adequate *in vivo* studies should be conducted to show the capability of the release system in controlling the drug release, that is, an *IVIVC* needs to be established. There are many models that may be used to establish the *IVIVC*. However, for drugs with complex adsorption scenarios such as drugs with low permeability or narrow GI adsorption window, it may be difficult and even not feasible to establish the *IVIVC* [111–113]. Besides, many

scenarios such as enter hepatic cycle, poor permeability, and so on make *IVIVC* very difficult. For drugs dissolving rapidly but with low permeability, no *IVIVC* may be expected [56].

However, there are still many reports of successfully establishing *IVIVC* for oral CR formulations [114, 115]. For example, Dutta et al. established an internally and externally validated level *IVIVC* model during the development of a once-daily extended release (ER) tablet of divalproex sodium, using multiple formulations with varying release rates. The *in vivo* absorption–time profile was inferred by deconvoluting of the PK profile against the unit disposition function (UDF). In the established *IVIVC* model, *in vivo* absorption was expressed as a function of *in vitro* drug dissolution profile. Therefore, plasma profiles of the extended release formulations could be established by convoluting *in vitro* drug release profiles with the UDF.

1.5.4 Others

Besides dissolution, visualization is also a good tool to evaluate oral CR formulations in human. There are several visualization tools currently available:

1. Wireless M2A Capsule Camera first designed by Given Imaging Inc. As the capsule travels through the GI tract, while light-emitting diodes flash the gastrointestinal tract, the microchip camera can capture thousands of images and send to a data recorder [116].

2. Gamma scintigraphy is a technique in which a short-lived radioactive isotope is incorporated into a dosage form, and the location of the dosage form in the GI tract can be noninvasively imaged with a gamma camera. Therefore, the observed transit of the dosage form can be correlated with the rate and extent of *in vivo* drug absorption [117] (http://www.scintipharma.com/html/what_is_gamma_scintigraphy_.htm).

3. Single photon emission computed tomography (SPECT) can be used together with any gamma imaging study, if a three-dimensional image is needed.

4. Magnetic resonance imaging (MRI), similar to gamma scintigraphy, also offers a powerful noninvasive method for picturing events inside controlled release dosage forms. It can keep track of drug release processes such as hydration, diffusion, and so on and help understand drug release mechanisms. Furthermore, the unique information collected from MRI studies may help in problem solving and formulation development [118].

1.6 NONCONVENTIONAL ORAL CONTROLLED DELIVERY SYSTEMS

Oral route is the most convenient and preferred route of administration. Besides sustained release formulations, there

TABLE 1.5 Devices Used a Platform for Gastric Retention [119, 120]

Device	Comment
Intragastric floating systems (low density)	To make the dosage forms with density less than 1, so that they can float on top of the gastric fluid [121, 122]
High-density systems	Increasing the density of the dosage forms from 1.0 to certain higher values can increase the average GI transit time [123, 124]
Mucoadhesive systems	Coating the dosage forms with mucoadhesive polymers such as poly(acrylic acid) that can adhere to the mucus layer of the gastric tissue [125–127]
Unfoldable, extendible, or swellable systems	Gastric retention by large dimension and rigidity can be used to improve bioavailability for drugs with narrow absorption window. By unfolding multilayer polymeric films, the extended absorption phase (>48 h) of riboflavin administered in a gastroretentive dosage form (GRDF) led to four fold increased bioavailability in dogs [128]. Cargill et al. reported that large single-unit dosage form with devices of different shapes could prolong gastric retention in dogs, and among them, tetrahedrons and rings are most efficient [129, 130]. InTec Pharma developed Accordion Pill™ platform using expanding geometries [131]
Superporous biodegradable hydrogel systems	The swelling of superporous biodegradable hydrogels can be very fast, with ratio over 1000. However, the traditional superporous hydrogels may not be strong enough to withstand gastric contraction. With IPN (interpenetrating network) with polyacrylonitrile, the mechanical properties of superporous hydrogels have been improved up to 50 times, and thus could be used to develop gastric retention devices for long-term oral drug delivery [132–135]

are several site-specific oral delivery systems, such as gastric retention devices, buccal controlled release formulations, and so on. For those drugs that may degrade in acidic environment, enteric coating can prevent drug release at low pH of the stomach and release drug at neutral environment of the small intestine. Overall, the drug absorption rates are best maintained by a delivery device that releases the drug at its optimum absorption sites.

1.6.1 Gastric Retention Devices

Because most drugs are absorbed in the upper small intestine, the absorption of the drugs will be improved if the oral controlled release dosage forms are maintained in the stomach. For drugs with a "window of absorption," gastric retention devices will be even more critical to achieve desired bioavailability (Table 1.5). As long as a drug is stable in acidic environment, using gastric retention devices, drugs can be delivered for longer period of time in the stomach than the conventional dosage forms. The approaches that have been used to achieve long-term gastric retention are listed in Table 1.6. The same topic is described in more depth in Chapter 12.

TABLE 1.6 Commercial Technologies for Gastric Retention Delivery Systems

System	Company
Superporous hydrogel systems	Kos Pharmaceuticals, Inc.
Gastric retention system	DepoMed
West gastroretentive system	West Pharmaceutical Services
OraSert™, OraSite®	KV Pharmaceutical

1.6.2 Colon-Specific Delivery Systems

To design a successful colon-specific drug delivery system for a drug, before the drug reaches the colon, it needs to be protected from degradation and release in the upper GI tract. Of course, the system needs to release the drug at the colon that has close to neutral pH [136]. Colon-specific drug delivery is critical for drugs intended for localized treatment, mainly inflammatory bowel diseases, irritable bowel syndrome, and colon cancer, or for drugs like proteins and peptides exhibiting maximal systemic absorption from the colon.

Several approaches have been generally used in designing colon-specific delivery systems [137–140], and they include pH-modulated/enteric systems [141], time-controlled (or time-dependent) systems, microbially controlled systems, and luminal pressure-controlled systems. Among these approaches, formulations that release drugs in response to colonic pH, enterobacteria, or time are most widespread formulation technologies. In microbially controlled systems, prodrugs and biodegradable polymers will be degraded by colon-specific enzymes. In time-controlled systems, drug release has to be delayed at least 5 h to reach the colon. Many factors make the development of colon-specific drug delivery systems very challenging, such as extreme pH conditions, gastric enzymes, diet, varying transit time in the GI tract, disease, and so on. For the pH-dependent systems, the drug may be prematurely released in lower small intestine (pH 7.5).

Based on the pH-dependent approach, Procter & Gamble developed Asacol® (mesalamine) delayed release tablets, in which the pH-sensitive polymer coating on tablets is designed to release drug at pH >7 in the terminal ileum and beyond, that is, to the colon (http://www.asacol.com/ulcerative-colitis-treatment/asacol.jsp) [142]. Based on programmed intervals, PORT® systems can deliver multiple doses and/or drugs and achieve colon-specific drug delivery based on timing [137].

CODES™ of Yamanouchi Pharmaceutical Co., Ltd. (now part of Astellas Pharma Inc.) is a colon-specific drug delivery system based on enzyme degradation. The drug core is prevented from degradation or drug release prior to the colon by first an enteric coating and then a cationic polymer coating for passage through the small intestine to the cecum. Note that a saccharide, lactulose, which will be degraded to organic acids when exposed to the enteric bacteria in a lower gastrointestinal tract, is included in the drug core formulation. pH changes in the large intestine trigger erosion of the cationic polymer, and lactulose diffuses through the cationic polymer and is degraded by cecal microflora. Organic acids produced by the microflora dissolve the cationic polymer and cause the drug to be released colon specifically [143]. In another oral controlled absorption system (OCAS™) of Yamanouchi, the coating gels rapidly and hydrates completely in the upper gastrointestinal tract, thus enabling the gradual drug release as the tablet system travels throughout the GI tract, including the colon where water is poorly available and the drug release is difficult to achieve. The hydrophilic gel-forming polymer matrix tablets of OCAS™ prevent degradation of drug prior to intestinal delivery and also allow the formulation of single daily doses and ensure stable drug release by minimizing the effects of individual differences and food consumption on drug absorption. Omnic-OCAS™ was developed by applying OCAS™ technology to tamsulosin for the treatment of functional symptoms of benign prostatic hyperplasia [144–146].

1.6.3 Buccal Controlled Release Dosage Forms

There are several main advantages of buccal delivery [147–149]. Barrier property of buccal membrane is much reduced compared to the skin, and extensive presystemic clearance occurring for some drugs after oral administration can be avoided. Also, it allows faster onset of drug action and is highly useful for a short duration ranging from a few minutes to several hours. These properties result in improved patient compliance, ease of dosage form removal in emergencies, robustness, and good accessibility.

The limitations and disadvantages in buccal drug delivery are limited surface area for absorption, concerns on taste and comfort in a highly innervated area, difficulties of adhesion to a mucosal surface for extended periods without the danger of swallowing or chocking of a device, potential bacterial growth, and blockage of salivary glands associated with prolonged occlusion.

TABLE 1.7 Commercial Fast-Dissolving Tablet Brands

Technology	Commercial Product
Freeze drying	Zydis® (Cardinal Health)
	Quicksolv® (Mediventure)
	Lyoc® (Lycos, Inc.)
	Pharmaburst™ (SPI Pharma)
Direct compression	DuraSolv®, OraSolv®, OraVescent® (Cima)
	Ziplets™, AdvaTab™ (Eurand)
	QDis™ (Phoqus)
Granulation and compression	WOWTAB® (Yamanouchi)
	Flashtab® (Ethypharm)
	Frosta® (Akina)
	Shear Form (Fuisz Tech)
	OraQuick™ (KV Pharmaceutical)
Molding/cotton-candy process	FlashDose® (Biovail)

Carbopol, a synthetic high molecular weight cross-linked water-soluble poly(acrylic acid), is commonly used as a bioadhesive in buccal, ophthalmic, intestinal, nasal, vaginal, and rectal applications [150–152]. Other polymers such as karaya gum, xanthan gum, and glycol chitosan also have strong adhesion to the mucosal membrane; however, concentrations greater than 50% (w/w) are needed for them to produce sustained drug release [153].

1.6.4 Fast-Dissolving Tablets

Fast-dissolving tablets are designed to take orally without water and without swallowing; therefore, they are very useful for patients with swallowing problem, and also during outside activities when there is no access to water [154–156]. In general, these tablets can disintegrate within 10–30 s. There are four technologies used in preparing fast-dissolving tablets, and each one has its advantages and disadvantages, which will be elaborated in a separate chapter in this book. Table 1.7 lists some commercial fast-dissolving products and their manufacturing technologies [154–156].

1.7 REGULATORY AND LEGAL ASPECTS OF CONTROLLED RELEASE

1.7.1 Applicable Guidelines and Acceptance Criteria

If a controlled release formulation is used for new chemical entity, the NDA filing will include the related information on oral controlled release formulations. However, if a controlled release formulation is used for an existing drug, a new drug application is required for approval. Of course, for an existing drug, data would be available to support any preclinical information as to what the drug would do and its safety.

However, if more dosage is used than previously approved dosage form, a safety level especially toxicity needs to be established. Overall, drug in a new controlled release formulation needs to demonstrate its safety and effectiveness before a new drug application can be filed. In general, time to review and approve the NDA will be shorter than the original NDA for the new chemical entity.

1.7.2 Patent Protection of Controlled Release Products

With regard to oral CR formulations, besides formulation compositions, some useful related information have been included to provide more patent protection [157]:

1. *Formulation*: type of preparation, excipients, *in vitro* drug release profile, chemical and physical stability, scale-up, and new processes for preparation of drug substance or formulation.
2. *In Vivo Performance*: PK/PD information, such as T_{max}, C_{max}, AUC (area under curve), efficacy, and side effects.
3. *Others*: controlled release profiles, enteric coating, fast-dissolving tablets, multiphase releases, and so on.

Considering the complex scenarios for patent protection for oral CR formulations, it is always valuable to plan early, discuss with patent attorney early, and file all necessary patents as early as possible. Besides orange book, it is also helpful to review all related patents and exclusivities.

1.7.3 Generic Versus Innovator Products

Even though brand drug companies face huge pressure from generic companies, controlled release formulation design is still a very useful tool in extending the life cycle for some drugs. For example, even though the drug substance of Ambien CR™ (Sanofi-Aventis), zolpidem tartrate, already lost patent protection, Ambien CR™ still garners sizeable revenue for the brand company. Developing a new oral CR formulation for an existing drug can extend drug life cycle through two aspects, exclusivity from regulatory agencies and formulation patent(s). Utilizing novel drug delivery technologies can substantially increase the overall efficacy of the drug and convenience and compliance by the patients. Such benefit should maintain the advantages of the innovator products even after their patent protections have expired.

REFERENCES

1. Abdul S, Poddar SS. A flexible technology for modified release of drugs: multilayered tablets. *J. Control. Release* 2004;97(3):393–405.

2. Bartholomaeus J, Ziegler I. Delayed-release formulation of 3-(3-dimethylamino-1-ethyl-2-methyl-propyl)phenol. WO Patent 2003035054, 2003.

3. Lerner EI, Flashner-Barak M, Achthovem EV, Keegstra H, Smit R. Delayed release formulations of 6-mercaptopurine. WO Patent 2005099666, 2005.

4. Prakash A, Markham A. Oral delayed-release mesalazine: a review of its use in ulcerative colitis and Crohn's disease. *Drugs* 1999;57(3):383–408.

5. Sharma SK, Ruggenenti P, Remuzzi G. Managing hypertension in diabetic patients—focus on trandolapril/verapamil combination. *Vasc. Health Risk Manage.* 2007;3(4):453–465.

6. Simon S. Opioids and treatment of chronic pain: understanding pain patterns and the role for rapid-onset opioids. *Med. Gen. Med.* 2005;7(4):54.

7. Alaux G, Lewis G, Andre F. Controlled-release dosage forms comprising a short acting hypnotic or a salt. EP Patent 1005863, 2000.

8. Hainer JW, Sugg J. Metoprolol succinate extended release/ hydrochlorothiazide combination tablets. *Vasc. Health Risk Manage.* 2007;3(3):279–288.

9. Ahlskog JE. Treatment of early Parkinson's disease: are complicated strategies justified? *Mayo Clin. Proc.* 1996;71(7): 659–670.

10. Divoll M, Greenblatt DJ, Ochs HR, Shader RI. Absolute bioavailability of oral and intramuscular diazepam: effects of age and sex. *Anesth. Analg.* 1983;62(1):1–8.

11. Salzman C, Shader RI, Greenblatt DJ, Harmatz JS. Long v short half-life benzodiazepines in the elderly. Kinetics and clinical effects of diazepam and oxazepam. *Arch. Gen. Psychiatry* 1983;40(3):293–297.

12. Swartz CM, Sherman A. The treatment of tricyclic antidepressant overdose with repeated charcoal. *J. Clin. Psychopharmacol.* 1984;4(6):336–340.

13. Hoffman A, Horwitz E, Hess S, Cohen-Poradosu R, Kleinberg L, Edelberg A, Shapiro M. Implications on emergence of antimicrobial resistance as a critical aspect in the design of oral sustained release delivery systems of antimicrobials. *Pharm. Res.* 2008;25(3):667–671.

14. Awan NA, Amsterdam EA, Vera Z, DeMaria AN, Miller RR, Mason DT. Reduction of ischemic injury by sublingual nitroglycerin in patients with acute myocardial infarction. *Circulation* 1976;54(5):761–765.

15. Anon. Oxymorphone—Endo/Penwest: EN 3202, EN 3203. *Drugs R&D* 2003;4(3):204–206.

16. Anon. Metformin extended release—DepoMed: metformin, metformin gastric retention, metformin GR. *Drugs R&D* 2004;5(4):231–233.

17. Arthur RMF, Mehmel H. A sustained release formulation of isosorbide-5-mononitrate with a rapid onset of action. *Int. J. Clin. Pract.* 1999;53(3):205–212.

18. Klein E. The role of extended-release benzodiazepines in the treatment of anxiety: a risk–benefit evaluation with a focus on extended-release alprazolam. *J. Clin. Psychiatry* 2002;63 (Suppl. 14):27–33.

19. Hutton JT, Morris JL. Long-acting carbidopa–levodopa in the management of moderate and advanced Parkinson's disease. *Neurology* 1992;42 (1 Suppl. 1):51–56; discussion 57–60.

20. Anonymous. Controlled-release budesonide in Crohn's disease. *Drug Ther. Bull.* 1997;35(4):30–31.

21. McCarberg B. Tramadol extended-release in the management of chronic pain. *Ther. Clin. Risk Manage.* 2007;3(3):401–410.

22. Michel MC. A benefit–risk assessment of extended-release oxybutynin. *Drug Saf.* 2002;25(12):867–876.

23. Pieper John A. Understanding niacin formulations. *Am. J. Manage. Care* 2002;8 (12 Suppl.):S308–S314.

24. Wagstaff AJ, Goa KL. Once-weekly fluoxetine. *Drugs* 2001;61(15):2221–2228.

25. Michelson EL. Calcium antagonists in cardiology: update on sustained-release drug delivery systems. *Clin. Cardiol.* 1991;14(12):947–950.

26. Khor S-P, Hsu A. The pharmacokinetics and pharmacodynamics of levodopa in the treatment of Parkinson's disease. *Curr. Clin. Pharmacol.* 2007;2(3):234–243.

27. Hoffman A, Stepensky D, Lavy E, Eyal S, Klausner E, Friedman M. Pharmacokinetic and pharmacodynamic aspects of gastroretentive dosage forms. *Int. J. Pharm.* 2004;277 (1–2):141–153.

28. Tuerck D, Wang Y, Maboudian M, Sedek G, Pommier F, Appel-Dingemanse S. Similar bioavailability of dexmethyl-phenidate extended (bimodal) release, dexmethyl-phenidate immediate release and racemic methylphenidate extended (bimodal) release formulations in man. *Int. J. Clin. Pharmacol. Ther.* 2007;45(12):662–668.

29. Shirwaikar AA, Srinatha A. Sustained release bi-layered tablets of diltiazem hydrochloride using insoluble matrix system. *Indian J. Pharm. Sci.* 2004;66(4):433–437.

30. Shah AC, Britten NJ, Olanoff LS, Badalamenti JN. Gel-matrix systems exhibiting bimodal controlled release for oral drug delivery. *J. Control. Release* 1989;9(2):169–175.

31. Maroni A, Zema L, Cerea M, Sangalli ME. Oral pulsatile drug delivery systems. *Expert Opin. Drug Deliv.* 2005;2 (5):855–871.

32. Ghimire M, McInnes FJ, Watson DG, Mullen AB, Stevens HNE. *In-vitro/in-vivo* correlation of pulsatile drug release from press-coated tablet formulations: a pharmacoscintigraphic study in the beagle dog. *Eur. J. Pharm. Biopharm.* 2007;67(2):515–523.

33. Mohamad A, Dashevsky A. *In vitro* and *in vivo* performance of a multiparticulate pulsatile drug delivery system. *Drug Dev. Ind. Pharm.* 2007;33(2):113–119.

34. Zou H, Jiang X, Kong L, Gao S. Design and evaluation of a dry coated drug delivery system with floating-pulsatile release. *J. Pharm. Sci.* 2008;97(1):263–273.

35. McCall TW, Baichwal AR, Staniforth JN. TIMERx oral controlled-release drug delivery system. *Drugs Pharm. Sci.* 2003;126:11–19.

36. Staniforth JN, Baichwal AR. TIMERx: novel polysaccharide composites for controlled/programmed release of drugs in the gastrointestinal tract. *Expert Opin. Drug Deliv.* 2005;2 (3):587–595.

37. Ritschel WA. Targeting in the gastrointestinal tract: new approaches. Methods Find. *Exp. Clin. Pharmacol.* 1991;13 (5):313–336.

38. Yu LX, Amidon GL. Characterization of small intestinal transit time distribution in humans. *Int. J. Pharm.* 1998;171 (2):157–163.

39. Yu LX, Crison JR, Amidon GL. Compartmental transit and dispersion model analysis of small intestinal transit flow in humans. *Int. J. Pharm.* 1996;140(1):111–118.

40. Youngberg CA, Berardi RR, Howatt WF, Hyneck ML, Amidon GL, Meyer JH, Dressman JB. Comparison of gastrointestinal pH in cystic fibrosis and healthy subjects. *Dig. Dis. Sci.* 1987;32(5):472–480.

41. Gruber P, Longer MA, Robinson JR. Some biological issues in oral, controlled drug delivery. *Adv. Drug Deliv. Rev.* 1987;1 (1):1–18.

42. Wagener S, Shankar KR, Turnock RR, Lamont GL, Baillie CT. Colonic transit time—what is normal? *J. Pediatr. Surg.* 2004;39(2):166–169; discussion 166–169.

43. Bouchoucha M, Devroede G, Faye A, Arsac M. Importance of colonic transit evaluation in the management of fecal incontinence. *Int. J. Colorectal Dis.* 2002;17(6):412–417; discussion 418–419.

44. Arhan P, Devroede G, Jehannin B, Lanza M, Faverdin C, Dornic C, Persoz B, Tetreault L, Perey B, Pellerin D. Segmental colonic transit time. *Dis. Colon Rectum* 1981;24 (8):625–629.

45. Grundy JS, Foster RT. The nifedipine gastrointestinal therapeutic system (GITS). Evaluation of pharmaceutical, pharmacokinetic and pharmacological properties. *Clin. Pharmacokinet.* 1996;30(1):28–51.

46. John VA, Shotton PA, Moppert J, Theobald W. Gastrointestinal transit of Oros drug delivery systems in healthy volunteers: a short report. *Br. J. Clin. Pharmacol.* 1985;19 (Suppl. 2):203S–206S.

47. Alvisi V, Gasparetto A, Dentale A, Heras H, Felletti-Spadazzi A, D'Ambrosi A. Bioavailability of a controlled release formulation of ursodeoxycholic acid in man. *Drugs Exp. Clin. Res.* 1996;22(1):29–33.

48. Heinig R, Ahr G, Hayauchi Y, Kuhlmann J. Pharmacokinetics of the controlled-release nisoldipine coat-core tablet formulation. *Int. J. Clin. Pharmacol. Ther.* 1997;35(8):341–351.

49. Davis SS. Formulation strategies for absorption windows. *Drug Discov. Today* 2005;10(4):249–257.

50. Lindenbaum J, Rund DG, Butler VP Jr, Tse-Eng D, Saha JR. Inactivation of digoxin by the gut flora: reversal by antibiotic therapy. *N. Engl. J. Med.* 1981;305(14):789–794.

51. Saha JR, Butler VP Jr, Neu HC, Lindenbaum J. Digoxin-inactivating bacteria: identification in human gut flora. *Science* 1983;220(4594):325–327.

52. Warrington SJ, Barclay SP, John VA, Shotton PA, Wardle HM, Good W. Influence of site of drug delivery on the systemic availability of metoprolol: comparison of intragastric infusion

and 14/190 Oros administration. *Br. J. Clin. Pharmacol.* 1985;19 (Suppl. 2):219S–224S.

53. Paul W, Sharma Chandra P. Tricalcium phosphate delayed release formulation for oral delivery of insulin: a proof-of-concept study. *J. Pharm. Sci.* 2008;97(2):863–870.

54. Senel S, Kremer M, Nagy K, Squier C. Delivery of bioactive peptides and proteins across oral (buccal) mucosa. *Curr. Pharm. Biotechnol.* 2001;2(2):175–186.

55. Soares AF, Carvalho RdA, Veiga F. Oral administration of peptides and proteins: nanoparticles and cyclodextrins as biocompatible delivery systems. *Nanomedicine* 2007;2 (2):183–202.

56. Amidon GL, Lennernaes H, Shah VP, Crison JR. A theoretical basis for a biopharmaceutic drug classification: the correlation of *in vitro* drug product dissolution and *in vivo* bioavailability. *Pharm. Res.* 1995;12(3):413–420.

57. Shokri J, Ahmadi P, Rashidi P, Shahsavari M, Rajabi-Siah-boomi A, Nokhodchi A. Swellable elementary osmotic pump (SEOP): an effective device for delivery of poorly water-soluble drugs. *Eur. J. Pharm. Biopharm.* 2008;68 (2):289–297.

58. Jacob JS, Bassett M, Schestopol MA, Mathiowitz E, Nangia A, Carter B, Moslemy P, Shaked Ze, Enscore D, Sikes C., Polymeric drug delivery system for hydrophobic drugs. WO Patent 2005084639, 2005.

59. Yu LX. An integrated model for determining causes of poor oral drug absorption. *Pharm. Res.* 1999;16 (12):1883–1887.

60. Backensfeld T. Pharmaceutical preparation with delayed release of an active substance. WO Patent 2000076484, 2000.

61. Liu L, Khang G, Rhee JM, Lee HB. Monolithic osmotic tablet system for nifedipine delivery. *J. Control. Release* 2000;67 (2–3):309–322.

62. Barzegar-Jalali M, Adibkia K, Mohammadi G, Zeraati M, Bolagh Behnaz Aghaee G, Nokhodchi A. Propranolol hydrochloride osmotic capsule with controlled onset of release. *Drug Deliv.* 2007;14(7):461–468.

63. Makhija Sapna N, Vavia Pradeep R. Controlled porosity osmotic pump-based controlled release systems of pseudo-ephedrine. I. Cellulose acetate as a semipermeable membrane. *J. Control. Release* 2003;89(1):5–18.

64. Wakode R, Bhanushali R, Bajaj A. Development and evaluation of push–pull based osmotic delivery system for prami-pexole. *PDA J. Pharm. Sci. Technol.* 2008;62(1):22–31.

65. Prabakaran D, Singh P, Kanaujia P, Jaganathan KS, Rawat A, Vyas SP. Modified push–pull osmotic system for simultaneous delivery of theophylline and salbutamol: development and *in vitro* characterization. *Int. J. Pharm.* 2004;284 (1–2):95–108.

66. Cardinal J. Formulation of controlled/modified release dosage forms. In *AAPS 41st Annual Pharmaceutical Technologies Arden Conference: Oral Controlled Release Development and Technology*, 2006.

67. Rudnic EM, Burnside BA, Flanner HH, Wassink SE, Couch RA, Pinkett JE.Soluble form osmotic dose delivery system for glipizide and other drugs. US Patent 6,361,796, 2002.

68. Rudnic EM, Burnside BA, Flanner HH, Wassink SE, Couch RA.Soluble osmotic drug delivery system. WO Patent 9818452, 1998.

69. Chen C-M, Chou JCH. Polymers for once daily pharmaceutical tablet having a unitary core. US Patent 6,485,748, 2002.

70. Sriwongjanya M, Weng T, Chou J, Chen C-M. Controlled release oral dosage form. WO Patent 9961005, 1999.

71. Thombre AG, Appel LE, Chidlaw MB, Daugherity PD, Dumont F, Evans LAF, Sutton SC. Osmotic drug delivery using swellable-core technology. *J. Control. Release* 2004;94 (1):75–89.

72. Thombre AG, Cardinal JR, Fournier LA. A delivery device containing a poorly water-soluble drug in a hydrophobic medium: ruminal delivery application. *J. Control. Release* 1992;18(3):221–233.

73. Johnson BA, Waterman KC.Asymmetric membrane polymeric coating for osmotic dosage form for varenicline, for treatment of nicotine dependency. US Patent 2007248671, 2007.

74. Haslam JL, Rork GS.Controlled porosity osmotic pump for controlled-release of diltiazem L-malate. EP Patent 309051, 1989.

75. Bajpai SK, Bajpai M, Saxena S. Ion exchange resins in drug delivery. *Ion Exchange Solvent Extr.* 2007;18:103–150.

76. Raghunathan Y. Controlled release pharmaceutical preparations. EP Patent 171528, 1986.

77. Raghunathan Y, Amsel L, Hinsvark O, Bryant W. Sustained-release drug delivery system I: coated ion-exchange resin system for phenylpropanolamine and other drugs. *J. Pharm. Sci.* 1981;70(4):379–384.

78. Heil MF, Wilson G. Multi-phase release methscopolamine compositions. US Patent 2008064694, 2008.

79. Jeong SH, Park K. Development of sustained release fast-dis-integrating tablets using various polymer-coated ion-exchange resin complexes. *Int. J. Pharm.* 2008;353(1–2):195–204.

80. Jadhav KR, Chandra M, Kurm SD, Kadam VJ. Ion exchange resins: a novel way to solve formulation problems. *Curr. Drug Ther.* 2007;2(3):205–209.

81. Thombre AG. Assessment of the feasibility of oral controlled release in an exploratory development setting. *Drug Discov. Today* 2005;10(17):1159–1166.

82. Vazquez MJ, Perez-Marcos B, Gomez-Amoza JL, Martinez-Pacheco R, Souto C, Concheiro A. Influence of technological variables on release of drugs from hydrophilic matrixes. *Drug Dev. Ind. Pharm.* 1992;18(11–12):1355–1375.

83. Cao Q-R, Kim T-W, Lee B-J. Photoimages and the release characteristics of lipophilic matrix tablets containing highly water-soluble potassium citrate with high drug loadings. *Int. J. Pharm.* 2007;339(1–2):19–24.

84. Huet de Barochez B, Lapeyre F, Cuine A. Oral sustained release dosage forms—comparison between matrixes and reservoir devices. *Drug Dev. Ind. Pharm.* 1989;15 (6–7):1001–1020.

85. Keary CM, Sheskey PJ. Preliminary report of the discovery of a new pharmaceutical granulation process using foamed aqueous binders. *Drug Dev. Ind. Pharm.* 2004;30(8):831–845.

86. Sheskey PJ, Keary CM. Process for coating solid particles. WO Patent 2003020247, 2003.

87. Iyer RM, Sandhu HK, Shah NH, Phuapradit W, Ahmed HM. Scale-up of extrusion and spheronization. *Drugs Pharm. Sci.* 2006;157:325–369.

88. Simonoska Crcarevska M, Glavas Dodov M, Goracinova K. Chitosan coated Ca-alginate microparticles loaded with budesonide for delivery to the inflamed colonic mucosa. *Eur. J. Pharm. Biopharm.* 2008;68(3):565–578.

89. Murata Y, Jinno D, Liu D, Isobe T, Kofuji K, Kawashima S. The drug release profile from calcium-induced alginate gel beads coated with an alginate hydrolysate. *Molecules* 2007;12 (11):2559–2566.

90. Huang J, Wong HL, Zhou Y, Wu XY, Grad H, Komorowski R, Friedman S. *In vitro* studies and modeling of a controlled-release device for root canal therapy. *J. Control. Release* 67 (2–3): 2000; 293–307.

91. Zhou Y, Wu XY. Theoretical analyses of dispersed-drug release from planar matrices with a boundary layer in a finite medium. *J. Control. Release* 2002;84(1–2):1–13.

92. Sune NM, Gani R, Bell G, Shirley I. Computer-aided and predictive models for design of controlled release of pesticides. *Comput. Aided Chem. Eng.* 2004;14:301–306.

93. Wu SXY. Computer-aided design for modified release dosage forms (CAD-MRDF). In *AAPS 41st Annual Pharmaceutical Technologies Arden Conference: Oral Controlled Release Development and Technology*, 2006.

94. Patel M, Patel B, Patel R, Patel J, Bharadia P, Patel M. Carbopol: a versatile polymer. *Drug Deliv. Technol.* 2006;6 (3):32–34, 36, 38, 40–43.

95. Alderman DA. A review of cellulose ethers in hydrophilic matrixes for oral controlled-release dosage forms. *Int. J. Pharm. Technol. Product Manuf.* 1984;5(3):1–9.

96. Salsa T, Veiga F, Pina ME. Oral controlled-release dosage forms. I. Cellulose ether polymers in hydrophilic matrixes. *Drug Dev. Ind. Pharm.* 1997;23(9):929–938.

97. Vueba ML, Batista de Carvalho LAE, Veiga F, Sousa JJ, Pina ME. Role of cellulose ether polymers on ibuprofen release from matrix tablets. *Drug Dev. Ind. Pharm.* 2005;31 (7):653–665.

98. Paavola A, Yliruusi J, Kajimoto Y, Kalso E, Wahlstroem T, Rosenberg P. Controlled release of lidocaine from injectable gels and efficacy in rat sciatic nerve block. *Pharm. Res.* 1995;12(12):1997–2002.

99. Obaidat AA, Rashdan LA, Najib NM. Release of dextromethorphan hydrobromide from matrix tablets containing sodium carboxymethyl cellulose and hydroxypropyl methyl cellulose. *Acta Pharm. Turc.* 2002;44(2):97–104.

100. Michailova V, Titeva S, Kotsilkova R. Rheological characteristics and diffusion processes in mixed cellulose hydrogel matrices. *J. Drug Deliv. Sci. Technol.* 2005;15(6):443–449.

101. Tonnesen Hanne H, Karlsen J. Alginate in drug delivery systems. *Drug Dev. Ind. Pharm.* 2002;28(6):621–630.

102. Miller DA, Fukuda M, McGinity JW. The properties of chitosan as a retardant binder in matrix tablets for sustained drug release. *Drug Deliv. Technol.* 2006;6(9):44, 46, 48–50, 52.

103. Bialer M. Pharmacokinetic evaluation of sustained release formulations of antiepileptic drugs. Clinical implications. *Clin. Pharmacokinet.* 1992;22(1):11–21.

104. Blondeau JM. Current issues in the management of urinary tract infections: extended-release ciprofloxacin as a novel treatment option. *Drugs* 2004;64(6):611–628.

105. Piscitelli DA, Young D. Setting dissolution specifications for modified-release dosage forms. *Adv. Exp. Med. Biol.* 1997;423:159–166.

106. Kamba M, Seta Y, Takeda N, Hamaura T, Kusai A, Nakane H, Nishimura K. Measurement of agitation force in dissolution test and mechanical destructive force in disintegration test. *Int. J. Pharm.* 2003;250(1):99–109.

107. Kamba M, Seta Y, Kusai A, Nishimura K. Comparison of the mechanical destructive force in the small intestine of dog and human. *Int. J. Pharm.* 2002;237(1–2):139–149.

108. Kamba M, Seta Y, Kusai A, Nishimura K. Evaluation of the mechanical destructive force in the stomach of dog. *Int. J. Pharm.* 2001;228(1–2):209–217.

109. Hayashi T, Kanbe H, Okada M, Suzuki M, Ikeda Y, Onuki Y, Kaneko T, Sonobe T. Formulation study and drug release mechanism of a new theophylline sustained-release preparation. *Int. J. Pharm.* 304(1–2): 2005; 91–101.

110. Burke MD, Maheshwari CR, Zimmerman BO. Pharmaceutical analysis apparatus and method. WO Patent 2006052742, 2006.

111. Dowell JA, Hussain A, Devane J, Young D. Artificial neural networks applied to the *in vitro–in vivo* correlation of an extended-release formulation: initial trials and experience. *J. Pharm. Sci.* 1999;88(1):154–160.

112. Young D. *In vitro–in vivo* correlation for modified release parenteral drug delivery systems. *Drugs Pharm. Sci.* 2007;165:141–151.

113. Piscitelli DA, Bigora S, Propst C, Goskonda S, Schwartz P, Lesko LJ, Augsburger L, Young D. The impact of formulation and process changes on *in vitro* dissolution and the bioequivalence of piroxicam capsules. *Pharm. Dev. Technol.* 1998; 3(4):443–452.

114. Mandal U, Ray KK, Gowda V, Ghosh A, Pal TK. *In-vitro* and *in-vivo* correlation for two gliclazide extended-release tablets. *J. Pharm. Pharmacol.* 2007;59(7):971–976.

115. Dutta S, Qiu Y, Samara E, Cao G, Granneman GR. Once-a-day extended-release dosage form of divalproex sodium III: development and validation of a level A *in vitro–in vivo* correlation (IVIVC). *J. Pharm. Sci.* 2005;94(9):1949–1956.

116. Meron GD. The development of the swallowable video capsule (M2A). *Gastrointest. Endosc.* 2000;52(6):817–819.

117. Podczeck F, Course N, Newton JM, Short MB. Gastrointestinal transit of model mini-tablet controlled release oral dosage forms in fasted human volunteers. *J. Pharm. Pharmacol.* 2007;59(7):941–945.

118. Melia CD, Rajabi-Siahboomi AR, Bowtell RW. Magnetic resonance imaging of controlled release pharmaceutical

dosage forms. *Pharm. Sci. Technol. Today* 1998;1(1): 32–39.

119. Talukder R, Fassihi R. Gastroretentive delivery systems: a mini review. *Drug Dev. Ind. Pharm.* 2004;30(10):1019–1028.

120. Hwang S-J, Park H, Park K. Gastric retentive drug delivery systems. *Crit. Rev. Ther. Drug Carrier Syst.* 1998;15 (3):243–284.

121. Nakagawa T, Kondo S-I, Sasai Y, Kuzuya M. Preparation of floating drug delivery system by plasma technique. *Chem. Pharm. Bull.* 2006;54(4):514–518.

122. Sriamornsak P, Thirawong N, Puttipipatkhachorn S. Emulsion gel beads of calcium pectinate capable of floating on the gastric fluid: effect of some additives, hardening agent or coating on release behavior of metronidazole. *Eur. J. Pharm. Sci.* 2005;24(4):363–373.

123. Clarke GM, Newton JM, Short MB. Comparative gastrointestinal transit of pellet systems of varying density. *Int. J. Pharm.* 1995;114(1):1–11.

124. Devereux JE, Newton JM, Short MB. The influence of density on the gastrointestinal transit of pellets. *J. Pharm. Pharmacol.* 1990;42(7):500–501.

125. Weon KY, Kim DW, Kim JS, Kim K.Gastric retention-type pellet and preparation. WO Patent 2008010690, 2008.

126. Jackson SJ, Bush D, Perkins AC. Comparative scintigraphic assessment of the intragastric distribution and residence of cholestyramine, Carbopol 934P and sucralfate. *Int. J. Pharm.* 2001;212(1):55–62.

127. Park H, Robinson JR. Mechanisms of mucoadhesion of poly(acrylic acid) hydrogels. *Pharm. Res.* 1987;4 (6):457–464.

128. Klausner EA, Lavy E, Stepensky D, Friedman M, Hoffman A. Novel gastroretentive dosage forms: evaluation of gastroretentivity and its effect on riboflavin absorption in dogs. *Pharm. Res.* 2002;19(10):1516–1523.

129. Cargill R, Engle K, Gardner CR, Porter P, Sparer RV, Fix JA. Controlled gastric emptying. II. *In vitro* erosion and gastric residence times of an erodible device in beagle dogs. *Pharm. Res.* 1989;6(6):506–509.

130. Cargill R, Caldwell LJ, Engle K, Fix JA, Porter PA, Gardner CR. Controlled gastric emptying. I. Effects of physical properties on gastric residence times of nondisintegrating geometric shapes in beagle dogs. *Pharm. Res.* 1988;5 (8):533–536.

131. Lapidot N, Afargan M, Kirmayer D, Kluev L, Cohen M, Moor E, Navon N.A gastro-retentive multi-layered system for the oral delivery of macromolecules. WO Patent 2007093999, 2007.

132. Omidian H, Park K, Rocca JG. Recent developments in superporous hydrogels. *J. Pharm. Pharmacol.* 2007;59 (3):317–327.

133. Yang S, Park K, Rocca JG. Semi-interpenetrating polymer network superporous hydrogels based on poly(3-sulfopropyl acrylate, potassium salt) and poly(vinyl alcohol): synthesis and characterization. *J. Bioactive Compat. Polym.* 2004;19 (2):81–100.

134. Qiu Y, Park K. Superporous IPN hydrogels having enhanced mechanical properties. *AAPS PharmSciTech* 2003;4 (4):406–412.

135. Gemeinhart RA, Park H, Park K. Effect of compression on fast swelling of poly(acrylamide-*co*-acrylic acid) superporous hydrogels. *J. Biomed. Mater. Res.* 2000;55(1):54–62.

136. Asghar LFA, Chandran S. Multiparticulate formulation approach to colon specific drug delivery: current perspectives. *J. Pharm. Pharm. Sci.* 2006;9(3):327–338.

137. Cardinal J. Non-conventional oral controlled drug delivery systems including bioadhesive and site specific systems. In *AAPS 41st Annual Pharmaceutical Technologies Arden Conference: Oral Controlled Release Development and Technology*, 2006.

138. Kaur G, Jain S, Tiwary AK. Recent approaches for colon drug delivery. *Recent Pat. Drug Deliv. Formul.* 2007;1(3):222–229.

139. Patel M, Shah T, Amin A. Therapeutic opportunities in colon-specific drug-delivery systems. *Crit. Rev. Ther. Drug Carrier Syst.* 2007;24(2):147–202.

140. Singh BN. Modified-release solid formulations for colonic delivery. *Recent Pat. Drug Deliv. Formul.* 2007;1(1):53–63.

141. Mukherji G, Wilson CG. Enteric coating for colonic delivery. *Drugs Pharm. Sci.* 2003;126:223–232.

142. Kelm GR, Kondo K, Nakajima A. Pharmaceutical dosage form with multiple enteric polymer coatings for colonic delivery. US Patent 5,914,132, 1999.

143. Watanabe S, Kawai H, Katsuma M, Fukui M.Colon-specific drug release system. WO Patent 9528963, 1995.

144. Korstanje C. The improved cardiovascular safety of omnic (tamsulosin) oral controlled absorption system (OCAS). *Eur. Urol. Suppl.* 2005;4(7):10–13.

145. Yang L, Watanabe S, Li J, Chu JS, Katsuma M, Yokohama S, Fix JA. Effect of colonic lactulose availability on the timing of drug release onset *in vivo* from a unique colon-specific drug delivery system (CODES). *Pharm. Res.* 2003;20(3):429–434.

146. Katsuma M, Watanabe S, Kawai H, Takemura S, Masuda Y, Fukui M. Studies on lactulose formulations for colon-specific drug delivery. *Int. J. Pharm.* 2002;249(1–2):33–43.

147. Birudaraj R, Mahalingam R, Li X, Jasti BR. Advances in buccal drug delivery. *Crit. Rev. Ther. Drug Carrier Syst.* 2005;22(3):295–330.

148. Yukimatsu K, Nozaki Y, Kakumoto M, Ohta M. Development of a trans-mucosal controlled-release device for systemic delivery of antianginal drugs pharmacokinetics and pharmacodynamics. *Drug Dev. Ind. Pharm.* 1994;20 (4):503–534.

149. Gandhi RB, Robinson JR. Oral cavity as a site for bioadhesive drug delivery. *Adv. Drug Deliv. Rev.* 1994;13 (1–2):43–74.

150. Patel M, Patel B, Patel R, Patel J, Bharadia P, Patel M. Carbopol: a versatile polymer. *Drug Deliv. Technol.* 2006;6 (3):32–34, 36, 38, 40–43.

151. Park H, Robinson JR. Mechanisms of mucoadhesion of poly (acrylic acid) hydrogels. *Pharm. Res.* 1987;4(6):457–464.

152. Park K, Robinson JR. Bioadhesive polymers as platforms for oral-controlled drug delivery: method to study bioadhesion. *Int. J. Pharm.* 1984;19(2):107–127.

153. Park CR, Munday DL. Evaluation of selected polysaccharide excipients in buccoadhesive tablets for sustained release of nicotine. *Drug Dev. Ind. Pharm.* 2004;30(6): 609–617.

154. Sandri G, Bonferoni MC, Ferrari F, Rossi S, Caramella C. Differentiating factors between oral fast-dissolving technologies. *Am. J. Drug Deliv.* 2006;4(4):249–262.

155. Sastry SV, Nyshadham J. Process development and scale-up of oral fast-dissolving tablets. *Drugs Pharm. Sci.* 2005;145:311–336.

156. Parakh SR, Gothoskar AV. A review of mouth dissolving tablet technologies. *Pharm. Technol.* 2003;27(11):92, 94, 96, 98, 100.

157. Davidson C. Oral controlled release development and technology—patent protection for controlled release products. In *AAPS 41st Annual Pharmaceutical Technologies Arden Conference: Oral Controlled Release Development and Technology*, 2006.

2

EVOLUTION OF ORAL CONTROLLED RELEASE DOSAGE FORMS

PING I. LEE

Department of Pharmaceutical Sciences, Faculty of Pharmacy, University of Toronto, Toronto, Ontario, Canada

JIAN-XIN LI

Evonik Degussa Corporation, Piscataway, NJ, USA

2.1 INTRODUCTION

Controlled release dosage forms provide additional temporal and spatial control of drug release not achievable by conventional or immediate release dosage forms. As a result, controlled release dosage forms are capable of achieving a variety of therapeutic benefits, including a more constant or prolonged therapeutic effect, an enhancement of activity duration of short half-life drugs, a reduction in dosing frequency and side effects, and an improved patient compliance. Over the past four decades, significant advances have been made in the area of oral controlled release dosage forms as evidenced by the proliferation of patents and publications, as well as a significant growth in marketed controlled release drug products. In addition to the above-mentioned therapeutic benefits, the potential to extend patent exclusivity and the opportunity to expand the market for existing drugs all contributed to the growing popularity of controlled release dosage forms. To a large extent, such growth has also benefited from the availability of a plurality of delivery technologies for the controlled release dosage forms through an interdisciplinary effort involving contributions from chemistry, materials science, engineering, and pharmaceutics. This chapter will provide an overview of the evolution of oral controlled release dosage forms, their current status, and future prospects.

2.2 HISTORICAL BACKGROUND

Oral drug preparations including pills, powders, and solutions have been in existence since before the medieval time. Coating of pills with gum mucilage and silvering and gilding of pills were practiced from the time of Rhazes and Avicenna (ninth to tenth century) well into the nineteenth century [1]. However, such pill coating or gilding for the most part was merely to prevent pills from sticking to each other or to the container box, and also to mask the offending taste of the medicine [2]. Other pill coating materials including magnesia, starch, licorice powder, lycopodium, gelatin, gum, and sugar became available before the mid-nineteenth century for the same intended utility [2]. It was not until 1884, did Dr. Paul Unna, a German physician, introduce keratin-coated pills that would stay intact in the stomach but dissolve or disintegrate in the intestine [3]. This represented the very initial attempt in affecting the release of a drug in the gastrointestinal (GI) tract. The idea of using granules or tablet matrices coated with wax, vegetable gum, or other enteric materials to regulate their hydration and drug release did not appear in the West until the 1930s [4–7]. Only after the emergence of modern-day slow release or timed release oral drug products in North America in the 1950s and 1960s did it set the stage ushering in the beginning of a new era of controlled release dosage forms in the early 1970s for a more precise control over the rate and location of drug delivery *in vivo* [8, 9].

Interestingly, much lesser known in the West, slow release medicine was practiced in traditional Chinese medicine much earlier in time as evidenced in excerpts from several surviving historical Chinese medical manuscripts. For example, in a book entitled "Medications Administered as Decoctions," the author Wang Hao-ku, a mid-thirteenth-century Chinese medical scholar, collected several essays from another book "Rules and Correspondences in the Use of Drugs" by his teacher, Li Kao (1180–1251 A.D.) [10]. In one of the essays, medical scholar Li Kao was quoted stating the following (English translation of passages from Wan Hao-ku's book; see original excerpt in Figure 2.1):

蜡圆者，取其难化，而旋旋取效也

Use wax pills for their resistance to dissolve thereby achieving (therapeutic) effect gradually and slowly

圆者缓也,..其用药之舒缓而治之意也

Such pills are slow acting . . . they provide the medicine gradually and slowly for treatment.

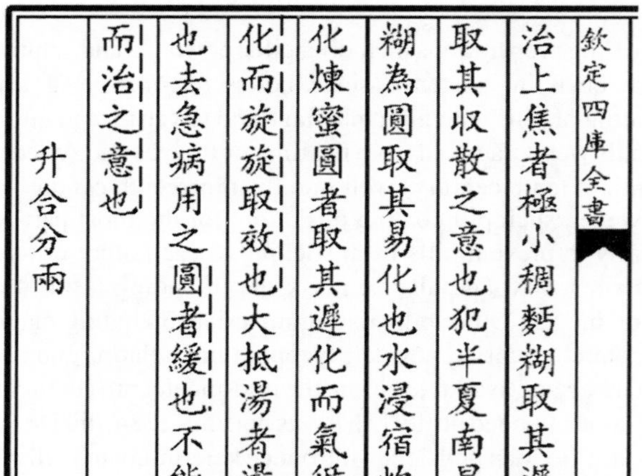
欽定四庫全書 湯液本草

治上焦者極小稠麪糊取其遲化直至下焦或酒或醋取其收散之意也犯半夏南星欲去濕者以生薑汁稀糊為圓取其易化也水浸宿炊餅又易化滴水圓又易化煉蜜圓者取其遲化而氣循經絡也蠟圓者取其難化而旋旋取效也大抵湯者盪也去大病用之散者散也去急病用之圓者緩也不能速去之其用藥之舒緩而治之意也

升合分兩

FIGURE 2.1 Highlighted quotes of Li Kao (1180–1251 A.D.) describing the use of wax pills for slow release therapeutic purpose in a book "Medications Administered as Decoctions" authored by the mid-thirteenth-century Chinese medical scholar, Wang Hao-ku.

To the best of our knowledge, this appears to be the earliest written record describing the use of wax pills for slow release therapeutic purpose. The realization of such utility may have occurred much earlier as the use of wax and fat pills in Chinese medicine dates back even further. For example, wax and fat pills were cited in an early medical treatise "Handbook of Prescriptions for Urgent Cases" authored by fourth-century physician, Ko Hung (281–341 A.D.) [10]. Furthermore, animal fat was used as a binder for pills as far back as second century B.C. as described in one of the earliest Chinese medical manuscripts unearthed from a Han dynasty tomb (ca. 168 B.C.) at Mawangdui in Hunan province during the early 1970s [11].

2.3 MORE RECENT HISTORY

Many of the present-day approaches for achieving controlled release from coated dosage forms are not new other than the involvement of newer coating materials and excipients. For example, American inventor Grover Miller patented in 1935 an enteric coating incorporating soluble, effervescent, or swellable agents for altering the structure of the coating, and upon activation by water, to release the medicine in the intestine at a desired time [5]. Using excipients known at the time (e.g., wax, shellac, nitrated cellulose lacquers, or gum benzoin as coating material; agar, cornmeal, or kaolin as hygroscopic swelling agent; mixture of tartaric acid and sodium bicarbonate as effervescent agent; sodium chloride or sugar as soluble additive), Miller found "when the coating is subject to the influence of water, the coating is either mechanically split away due to the expansion of the hygroscopic substance, the action of the effervescent substance, or is honeycombed or made porous by the dissolving away of the soluble substance incorporated . . . . the medicament may be liberated through the openings formed through the coating." Furthermore, Miller disclosed that the coating composition containing a mechanical breaking agent can be controlled so that "the time of its disruption or breaking down can be determined so that the liberation of medicine or medical compound may be timed for the point of digestive or colonic tract at which the application of the medicine is desired." Another American inventor Joseph Keller patented in 1937 a combination tablet based on an enteric coated tablet of one drug that is coated with a soluble coating containing a second drug [7]. The administration of such combination tablet *in vivo* "results in the action of the drugs of it two parts being separated by several hours, thereby increasing the efficiency or efficacy of the dosage." Amazingly, the ideas and approaches behind some of the present-day controlled release technologies such as controlled porosity tablet coatings and chronotherapeutic dosage forms for delayed release or colonic targeting seem not too different from those described by Miller and Keller over 70 years ago.

With the introduction of new semisynthetic polymers such as cellulose acetate phthalate (CAP) for more efficient enteric coatings in the 1940s, delayed action oral tablets that are impervious to gastric juice but can release the drug at a definite interval after ingestion became fashionable [1]. Thereafter, even before the fundamentals of biopharmaceutics were fully understood, the success of injectables in achieving prolonged absorption from the injection site with various medications such as penicillin and insulin further renewed interests in developing oral dosage forms that could provide similar prolonged drug absorption [9]. As a result, by late 1950s, the pharmaceutical market was more or less flooded with drug products that were claimed to provide delayed, repeat, prolonged, or sustained action. In 1959, it was said that there were 180 different products in oral sustained release dosage forms in the United States alone and the annual sales of such long-acting products exceeded $87 million [13]. Unfortunately, many of these early marketed prolonged action products suffered from lack of bioavailability and insufficient therapeutic rationale as criticized by Dragstedt [12–14]. This situation was best captured in Gilbaldi's statement that "The early history of the prolonged-release oral dosage form is probably best forgotten. Products were developed empirically, often with little rationale, and bioavailability problems were common. Many people viewed these dosage forms as little more than marketing inducements." [15]. It is not surprising today that many of these criticisms arose because the field of biopharmaceutics was still in its infancy during this early period.

Over the intervening years, biopharmaceutics matured into an important discipline and there was a better understanding of GI physiology as well as factors affecting oral absorption. With the invention of osmotic pump oral tablets in the early 1970s, various advanced oral controlled release products using this and other patented technologies were designed based on valid therapeutic rationale and appropriate biopharmaceutical considerations [16]. As a result, there are currently more than 50 oral controlled release prescription drug products in the United States as listed in the PDR (compared with 180 products in 1959 as mentioned above!) [17], and the annual sales of oral controlled release products grew to nearly $20 billion in 2000 (from the mere $87 million in 1959!) [18].

2.4 TERMINOLOGY AND POTENTIAL BENEFITS

The early pursuit of delayed or prolonged drug release in oral dosage forms resulted in a confusing array of terms in describing this class of products: extended release, sustained release, prolonged release, timed release, and controlled release are just few examples [12]. Since the successful introduction of controlled release drug products for the precise control of the rate and duration of drug release to achieve desired temporal patterns of drug concentration in the blood, these terms have been largely replaced by the more encompassing description of "controlled/modified release" [19]. In fact, "controlled release" and "modified release" have been used interchangeably for quite some time [19] and modified release dosage forms have been defined in FDA's guidance documents as those products that do not release the drug immediately including both delayed and extended release drug products [20]. For the rest of this chapter, the terms "controlled release" and "modified release" will be used synonymously.

By improving the way in which drugs are delivered to the body, that is, via additional temporal and spatial control of drug release, oral controlled release dosage forms are capable of achieving a variety of therapeutic benefits, including (a) maintenance of plasma concentrations within a desired therapeutic range with minimum fluctuations; (b) enhancement of activity duration of short half-life drugs; (c) reduction in side effects; (d) reduction in dosing frequency with improved patient compliance; (e) potential for site-specific drug delivery in the GI tract; and (f) potential for extended patent protection [21, 22]. A multitude of oral controlled release drug products are now available on the market, mostly based on proprietary controlled release technologies. In fact, more than 10% of the top 200 brand-name drugs in 2007 were oral controlled release products that had combined sales of over $12.6 billion, including the top 10 best selling brand-name controlled release products such as Effexor XR, Adderall XR, Oxycontin, Concerta, Wellbutrin XL, Torpol XL, Ambien CR, Detro LA, Depakote ER, and Allegra-D [23].

2.5 MAJOR HISTORICAL MILESTONES AFFECTING THE DIRECTION OF ORAL CONTROLLED RELEASE DOSAGE FORMS

Given the significant advances and market growth in oral controlled release technologies and products in recent decades, it is appropriate to examine in retrospect several major historical milestones that affected the direction, and provided real impetus to the further development, of oral controlled release dosage forms:

1. Availability of semi-synthetic and synthetic polymers for enteric coating (1940s through 1990s).
2. Introduction of new oral sustained release products by Smith Kline & French (SKF) using the Spansule® technology: Dexedrine (dextroamphetamine sulfate) (1952) and Contac, the cold remedy (1960).
3. Introduction of semisynthetic and synthetic hydrophilic gel-forming polymers for designing oral sustained release products (1950s and 1960s).
4. Invention and first commercialization of oral osmotic drug delivery system by Alza (1970s through 1980s)

These milestones will be discussed in more detail below.

2.5.1 Availability of Semisynthetic and Synthetic Polymers for Enteric Coating (1940s Through 1990s)

In addition to keratin-coated pills first introduced by Dr. Paul Unna [3], numerous substances and their combinations were also used for enteric coatings from the late 1880s to 1930s, including shellac, casein, zein, salol, tolu, stearic acid, gelatin–formadehyde product, tannic acid–gelatin product, cetyl alcohol, and so on [1]. It was not until after the introduction of semisynthetic pH-sensitive cellulose derivatives as tablet coating materials by Eastman Kodak in the early 1940s did cellulose acetate phthalate become the polymer of choice for enteric coatings because of its effectiveness over conventional materials [24–26]. Although cellulose acetate succinate (CAS) was also developed at the time, it did not attain widespread use in enteric coatings most likely due to limitations on its stability compared to CAP [27]. Subsequent introduction of polyvinyl acetate phthalate (PVAP), hydroxypropylmethylcellulose phthalate (HPMCP) and hydroxypropylmethylcellulose acetate succinate (HPMCAS) [28–30], methacrylate–methacrylic acid copolymers (e.g., Eudragit) [31], and various aqueous dispersions of the above enteric polymers [32, 33] made it possible to achieve more efficient and reproducible enteric coatings for tablets, capsules, beads, and pellets. The availability of different grades of synthetic enteric polymers such as Eudragit with different dissolution pH ranges made it possible to provide additional control in achieving delayed release and colonic targeting in the design of controlled release oral dosage forms [32, 33]. The more recent exploration of compatible blends of enteric and nonenteric polymers as controlled release coatings further increases the flexibility in achieving different degrees of release rate modulation from controlled release oral dosage forms [34].

2.5.2 Introduction of New Oral Sustained Release Products by Smith Kline & French Using the Spansule® Technology (1950s)

It is well recognized that the introduction of Dexedrine (dextroamphetamine sulfate) in 1952 and the Contact cold remedy in 1960 by Smith Kline & French Laboratories using their patented Spansule® technology was a historical milestone that motivated the further development of oral controlled release dosage forms [9, 35]. Based on a combination of wax–fat-coated pellets to provide a gradual drug release and uncoated pellets to provide an immediate release, the Spansule® technology was capable of achieving a variety of sustained release profiles by simply changing the wax–fat coating thickness or the proportion of the coated versus uncoated pellets [36]. Such Spansule® dosage forms containing multiple pellets were envisioned by its inventor to advantageously result in a "wide

dispersal throughout the intestines," thus overcoming "the problem occurring with the single large dosage form" [36]. This claimed advantage in GI transit for multiparticulate dosage forms over that of single-unit dosage forms was eventually confirmed many years later by the work of Davis et al. [37] through the use of gamma scintigraphy. Not surprisingly, the original wax–fat coatings have been mostly replaced by more reproducible and stable polymer coatings. Ironically, the availability of these and other types of sustained release dosage forms at SKF during the early years also stimulated the pioneering work of Swintosky [38], who conducted numerous human studies at SKF from the late 1950s to the early 1960s investigating the absorption, distribution, and fate of drug delivered by such dosage forms, thus culminating the start of the field of biopharmaceutics [12, 38].

2.5.3 Introduction of Semisynthetic and Synthetic Hydrophilic Gel-Forming Polymers for Designing Oral Sustained Release Products (1950s and 1960s)

A majority of current oral controlled release products are swellable matrix tablets based on certain hydrophilic rate-controlling polymers, usually identifiable from the listed inactive ingredients of a given product in the PDR [17]. This swellable tablet dosage form has been popular because of its low cost and ease of manufacture, requiring only the compression of a mixture of a gel-forming polymer and a drug. Typical hydrophilic gel-forming polymers found in these products include hydroxypropylmethylcellulose (HPMC), hydroxypropylcellulose (HPC), polyethylene oxide (PEO), cross-linked polyacrylic acid (Carbopol), sodium carboxymethylcellulose (NaCMC), and alginates. It is well accepted now that the drug release from such swellable matrix tablets is governed by both the swelling and erosion controlled mechanisms [39]. The advances in such swellable controlled release products would not have been possible without the introduction of semisynthetic and synthetic hydrophilic gel-forming polymers in the 1950s and 1960s by chemical companies such as Dow and Union Carbide [40]. The earliest examples of using swellable and erodible hydrophilic polymers such as HPMC were described in a patent by Christenson and Dale in 1962 [41], in which evidence was presented for the first time that a drug, whether water soluble or not, when mixed with a hydrophilic gel-forming polymer at a sufficiently large proportion and compressed into a tablet could be released at a relatively uniform rate over a prolonged period of time, often more than 8 h. Since the hydrophilic polymers in such tablets hydrate and swell after coming in contact with an aqueous medium, Christenson and Dale also envisaged that upon swelling, "a relatively water impermeable barrier is formed at the surface of the tablet [initial gel layer formation] which prevents further entry of water into the interior of the tablet. The soft mucilaginous gum gel barrier formed on the surface of the tablet is worn away [polymer erosion] by the motion of the tablet in the gastrointestinal tract, and some of the admixed medicinal agent is carried away with it and released. The fresh surface of the tablet that is exposed as the soft hydrated gum wears away becomes hydrated and swells thus renewing the protective coating [gel layer development]. As a result the tablet is slowly disintegrated [eroded] rather than dissolved and the medicament contained therein is released at a substantially uniform rate." ([41], notes and emphases added).

This description from almost half a century ago remains surprisingly consistent with the current understanding of swelling and erosion controlled mechanism of drug release from swellable matrix tablets. Nevertheless, it is now understood that the "relatively water-impermeable barrier" on the tablet surface is a result of the rapid formation of a gel layer on the tablet surface during the initial stage of the tablet swelling. This transforms the porous tablet surface region into a viscous gel layer, thus slowing the water penetration into, as well as the drug diffusion from, the tablet core. This has been regarded as the critical first step in achieving release rate control. It is well known now that, as the swelling continues, the polymer concentration near the gel surface decreases until it is below the so-called disentanglement concentration, at which time the erosion process takes place [42]. The resulting gel layer development is dictated by the relative movement of the penetrating swelling front near the tablet core and the erosion front at the gel surface. A synchronization of movement of the swelling and erosion fronts characterized by a constant gel layer thickness has been predicted to occur when the erosion rate dominates the diffusion rate and such front synchronization leads to a near constant rate of drug release [43]. Such dynamic gel layer development is depicted in Figure 2.2. Experimental verification of this predicted front synchronization behavior and the resulting constant rate drug release was reported for several hydrophilic swellable polymers such as NaCMC and PVA [44]. However, such front synchronization was not observed in HPMC matrices since their gel thicknesses appeared to increase continuously with time.

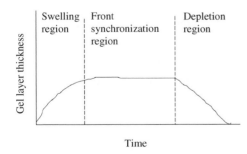

FIGURE 2.2 Dynamic gel layer thickness development in a swellable matrix tablet.

The interplay of front movement and the mechanism governing drug release from HPMC matrices was not conclusively elucidated until the work of Pham and Lee [45] who studied the transient dynamic swelling and dissolution behavior during drug release from HPMC matrices, utilizing a flow-through cell under well-defined hydrodynamic conditions. Their results demonstrate that, despite the continuous increase in gel layer thickness, polymer erosion does indeed occur irrespective of the HPMC viscosity grade; however, this erosion rate is much slower than the polymer swelling rate. Pham and Lee's results also suggest that for water-soluble drugs, the effect of HPMC erosion on drug release is insignificant and the release kinetics is mainly regulated by a swelling-controlled diffusion process. A corollary of this finding is that for water-insoluble drugs, the drug release will be predominantly controlled by HPMC erosion. These conclusions were later confirmed by Gao et al. [46], who also studied the effect of formulation variables on drug release from HPMC matrix tablets. Additional details of the front movement and drug release from HPMC matrix tablets have since been reported and predictive mathematical models established [47, 48]. Furthermore, there is a better understanding of the effect of polymer selection on the *in vitro* and clinical performance of HPMC-based controlled release tablets [49]. Representative examples of such swelling and erosion controlled matrix tablet products include Isoptin SR (verapamil; Abbott) and Trental (pentoxifyline; Aventis),

2.5.4 Invention and First Commercialization of Oral Osmotic Drug Delivery System by Alza (1970s through 1980s)

By all accounts, the invention and first commercialization of oral osmotic drug delivery system by Alza during mid-1970s through early 1980s was the real turning point that ushered in the new era of oral controlled release dosage forms. This was the first oral dosage form that could provide uniquely constant and extended drug release as well as plasma concentration profile that no other delivery technologies could offer at the time, and it required mostly conventional pharmaceutical manufacturing processes and equipment (except for the laser drilling of delivery orifice). It elevated the level of interest and enthusiasm in the science and technology of advanced drug delivery involving new chemistry, new materials, new delivery mechanisms, and new clinical applications that attracted scientists from a wide range of disciplines such as chemistry, materials science, engineering, and pharmaceutics. This eventually culminated the formation of the Controlled Release Society in the late 1970s.

Not widely known though, the very first oral osmotic tablet product was actually introduced by Ciba-Geigy (now Novartis) for the delivery of phenylpropanolamine HCl (Acutrim®) as an appetite suppressant in the early 1980s.

It was developed by Ciba-Geigy's internal R&D group based on the initial Alza patent for the so-called elementary osmotic pump design (OROS technology) [50]. The design consists of an osmotic drug core in a tablet surrounded by a cellulose acetate semipermeable membrane having a single laser-drilled delivery orifice. Upon activation in an aqueous environment, water is imbibed through the semipermeable membrane that dissolves the drug, generates additional osmotic influx of water, and builds up a hydrostatic pressure that in turn pumps the drug solution out of the tablet core through the orifice. In principle, this osmotic dosage form should be independent of environmental pH. The rate of drug release will be constant as long as the osmotic driving force or the influx of water is constant. For the elementary osmotic pump design, this is usually achieved with a saturated drug core having excess solid drug [51]. In this case, the drug release rate will decline once the reservoir concentration falls below saturation. The first OROS tablet product for phenylpropanolamine HCl was designed by taking into consideration appropriate pharmacokinetic parameters, and as a result, excellent *in vitro* and *in vivo* correlation was demonstrated [52]. This is a unique example for the rational design of controlled release dosage forms based on biopharmaceutical considerations.

For the delivery of relatively insoluble drugs, a variation of the elementary osmotic pump based on a push–pull OROS design was developed later, initially for the delivery of nifedipine for hypertension; the resulting product Procardia XL (Pfizer) gained FDA approval in 1989. In this design, the tablet core now contains a bilayer tablet with an active drug layer on top of a swellable polymer layer. Upon imbibing water, the drug layer converts to a suspension while the swellable layer expands as a result of the polymer swelling, thereby pushing the drug suspension through the orifice. Here, the constant rate of drug release is maintained by the constant rate of swelling of the hydrophilic polymer in the push layer. In addition to Procardia XL, most of the other OROS tablet products currently on the market in various therapeutic areas are also based on this bilayer push–pull design for the delivery of verapamil for hypertension (Covera-HS; Pfizer), glipizide for Type 2 diabetes mellitus (Glucotrol XL; Pfizer), and oxybutynin for overactive bladder (Lyrinel XL, previously Ditropan XL; Ortho-McNeil) [53]. More recently, a new multilayer core push–pull OROS tablet has been designed for methylphenidate (Concerta; Ortho-McNeil) that delivers the drug in an ascending profile for once-a-day administration in treating children with attention deficit hyperactivity disorder (ADHD) [53, 54]. Over the years, numerous variations of the osmotic delivery technology have also been developed, for example, the controlled porosity osmotic pump (CPOP), OROS for liquid formulations (L-OROS), and different swellable core configurations for the push–pull OROS, just to name a few [55–57].

2.6 OTHER MAJOR DEVELOPMENTS IMPACTING THE FIELD OF ORAL CONTROLLED RELEASE

Several major developments since the time of the Spansule® technology are worth noting that undoubtedly had helped shaping the field of oral controlled release. These include (1) the emergence of physical pharmacy and pharmacokinetics as new disciplines in the 1960s; (2) the establishment of controlled release as a field that has grown more interdisciplinary since the 1970s; (3) significant progress in the understanding of GI physiology and its impact on oral controlled release drug delivery; (4) integration of biopharmaceutics and pharmacokinetics into the rational design of oral controlled release dosage forms; (5) greater knowledge on material properties, drug release mechanisms, and physicochemical factors affecting the oral dosage form design and performance (e.g., establishment of Biopharmaceutic Classification System; BCS); (6) proliferation of novel drug delivery technologies; and (7) major advances in analytical chemistry, instrumentation, computer modeling, and process equipment and monitoring.

2.7 CLASSIFICATION AND MECHANISMS OF CURRENT ORAL CONTROLLED RELEASE DOSAGE FORMS

The proliferation of oral controlled release technologies in recent years is evident from the ever-growing list of acronyms derived from available drug delivery technologies [58, 59]. Since many such technologies will be covered in more detail in subsequent chapters of this book, only a general overview of their classification will be discussed here. Despite the seemingly endless combination of polymeric excipients and proprietary technologies in arriving at an ever-increasing number of oral controlled release products, physicochemical principles governing available release mechanisms from polymeric delivery systems are generally well established. To avoid confusion, a useful classification of current controlled release systems based on mechanistic considerations will be outlined below [39, 60]. This approach provides a systematic account of principles behind the design of various oral controlled release products:

1. *Membrane Systems*: Drug core surrounded by a rate-controlling membrane (e.g., microcapsules, coated pellets, beads, or coated tablets).
2. *Matrix Systems*: Drug dissolved or dispersed in a carrier matrix (e.g., beads, pellets, or tablets).
3. *Hybrid Systems*: A combination of membrane and matrix systems (e.g., coated pellets or beads embedded in a tablet matrix, core press coated tablets, or tablets in a capsule).

The membrane system, by virtue of its rate controlling membrane, is generally nondisintegrating. On the other hand, matrix systems and hybrid systems can be disintegrating or nondisintegrating. All of these systems can be in either single-unit or multiple-unit dosage forms.

2.7.1 Membrane Systems

Drug release from membrane systems is generally controlled by (1) *osmotic pumping* or (2) *solution–diffusion* mechanism. The osmotic pumping mechanism and product examples have already been discussed in Section 2.5. On the other hand, the *solution–diffusion* mechanism encompasses a majority of the coated dosage forms where the drug concentration gradient across the rate-controlling coating (or membrane) is the driving force for the diffusional drug release. If an excess of drug is present to maintain saturation on the upstream side, a constant rate of drug release will result. As the excess drug is depleted, the concentration gradient and therefore the drug release rate will decline. In addition to varying the membrane thickness, the drug release rate can also be regulated by modifying the membrane permeability. Similar to the early approaches described by Grove Miller in 1935 [5], this task can be accomplished by incorporating soluble additives or pore-forming agents in the polymer coating so as to increase the drug diffusion coefficient and thus the permeability [61]. Selected examples of coated beads or microcapsule products include Micro-K Extencaps (KCl; Robins) and Dilatrate SR (ISDN; Schwartz). An interesting outcome of incorporating pore-forming agent in a semipermeable polymer coating is that at the limit of a sufficiently small pore size, the osmotic pumping contribution will exceed that of the solution–diffusion mechanism. This limiting case becomes the controlled porosity osmotic pump discussed in Section 2.6 [55]. One major limitation of the membrane systems in general is the potential for dose dumping in the event of a breakage of the membrane coating.

2.7.2 Matrix Systems

Dosage forms based on matrix systems such as beads, pellets, and tablets are popular because of their low cost, ease of manufacture, and the lack of the potential for dose dumping. Drug release from matrix type of oral controlled release products is mostly regulated by mechanisms such as (1) *diffusion*, (2) *swelling/erosion*, (3) *geometry/area changes*, and (4) *nonuniform drug distribution*. In practice, a combination of these mechanisms often occurs. The *swelling/erosion* mechanism and product examples have already been discussed in Section 2.5. In *diffusion-controlled* systems, the drug release is mainly controlled by diffusion from nondisintegrating matrix tablets. The resulting release profiles are inherently first order in nature (square root of time

dependence in early times) with progressively diminishing release rates. This is due to the increasing diffusion distance and decreasing surface area at the penetrating diffusion front. Limited examples of pure diffusion-controlled tablet products include Fero-Gradumet (ferrous sulfate; Abbott) based on a methacrylate polymer matrix and Slow-K (KCl; Ciba) based on a wax matrix. In general, pure diffusion-controlled matrix tablets are rare, and more often diffusion occurs together with swelling and/or erosion mechanisms.

Another mechanism for regulating drug release from matrix type of oral controlled release products is via *geometry/area changes*. This is employed to compensate the continuously diminishing diffusional release rate inherent with matrix systems. An effective method, for example, is partially coating a swellable matrix tablet with a barrier coating to restrict the area initially available for swelling [62]. As time progresses, the tablet core swells and the exposed surface area increases rapidly to compensate the decrease in drug concentration gradient, thus resulting in a more uniform rate of drug release. Typical examples of marketed products based on this release mechanism include Dilacor XR (diltiazem HCl; Aventis) and Paxil CR (paroxetine, GSK). Other interesting approaches for controlling the geometry and area changes have also been proposed including the use of banded tablets or erosion-induced increases in tablet area [63, 64]; however, no commercial product has resulted most likely due to their technical complexity and manufacturing difficulties. Furthermore, *nonuniform drug distribution* has been pursued as an additional mechanism for regulating drug release from matrix systems. This is based on the concept of increasing drug loading from the surface toward the core of a matrix dosage form to compensate the inherent decreases in the diffusional release rate. The theoretical basis of this approach has been established and experimental demonstration has been accomplished in hydrogels and laminated dosage forms [65–67]. The latter lamination approach involves the increase of drug loading toward the core in a stepwise fashion, which can be readily accomplished via multilayered tablets, press-coated tablets, or multiple tablet/capsule coatings. When this concept is extended to erodible dosage forms, pulsatile release profiles can be easily achieved from alternating drug-loaded and drug-free layers [67, 68]. These approaches have been applied to the design of delayed or pulsatile oral dosage forms [69, 70]. A good example of marketed pulsatile product is Drixoral Cold & Allergy Sustained-Action Tablets (pseudoephedrine sulfate and dexbrompheniramine maleate; Schering-Plough) that produce two pulses of drug release separated by several hours.

2.7.3 Hybrid Systems

Hybrid type of dosage forms usually involves a combination of the above-mentioned drug release mechanisms. Representative examples of such tablet products involve Sular (nisoldipine; Zeneca), having a sustained release coat and an immediate release core, and Procanbid (procainamide HCl; Parke Davis), consisting of a wax core coated with a polymeric controlled release layer. Another type of hybrid dosage forms involves multi-unit systems in a single-unit presentation such as Theo-Dur (theophyline; Key) and Toprol XL (metoprolol succinate; AstraZenica), where coated controlled release pellets are imbedded in a tablet matrix. These hybrid dosage forms have been popular in developing chronotherapeutic dosage forms to provide drug release profiles that match the body's circadian rhythms. In practice, these dosage forms target drug release to the specific time of the day when there is maximal clinical manifestation of a disease (e.g., hypertension). Hybrid chronotherapeutic dosage forms have been designed based on OROS (controlled onset), membrane-coated beads (delayed release), press-coated tablets, or a combination of erodible polymer coating and a drug matrix (e.g., beads, pellets, and tablets) [71]. Typical chronotherapeutic product examples include Covera-HS (verapamil HCl; Pfizer), Verelan PM (verapamil HCl; Schwarz), and InnoPran XL (propranolol HCl; Reliant).

2.7.4   Comparison of Achievable Release Profiles

Given a target drug release profile satisfying the PK and PD requirements for a specific drug, there are generally more than a few release mechanisms available for the design of a suitable oral controlled release product. Table 2.1 illustrates achievable release profiles one can reasonably expect from the three types of dosage forms and drug release mechanisms discussed above. It is clear that hybrid systems based on

TABLE 2.1   Achievable Release Profiles

	Mechanism		
	First Order	Zero Order	Pulsatile/ Bimodal
Membrane systems			
Solution diffusion	×	×	
Osmotic pumping	×	×	
Matrix systems			
Passive diffusion	×		
Matrix swelling	×	×	
Matrix erosion	×	×	
Nonuniform concentration distribution	×	×	
Geometry/area changes	×	×	
Hybrid systems			
Erosion + nonuniform concentration	×	×	×
Erosion + geometry	×	×	×
Erosion + membrane	×	×	×
Osmotic pumping + nonuniform concentration	×	×	×

a combination of release mechanisms provide much greater flexibility in achieving the desired modulation of drug release profiles.

2.7.5   Biopharmaceutical Considerations

An important attribute of oral controlled release dosage forms is their ability to maintain a therapeutically effective plasma concentration over an extended period of time. Given the physiological constraints of GI transit, the desired dosage form should release the drug payload within 24 h, which is the total GI transit time in humans. Since during 80% of this time, the controlled release dosage form will be in the human colon rather than in the small intestine [37], one of the key considerations in determining the rationale for developing a controlled release product is whether there is sufficient absorption of the drug in the colon and whether there is any regional specific drug absorption taking place in the GI tract [72]. Similarly, a constant rate of drug release may provide little advantage over well-controlled first-order drug release under certain biopharmaceutic conditions such as a long biological half-life and a multiple dosing regimen. In some cases, a constant rate of drug delivery may not be optimum because of the potential for developing tolerance [73]. As mentioned earlier, for the rational design of controlled release oral dosage forms, one has to take into account relevant pharmacokinetic parameters for the absorption, distribution, and elimination of a drug in question as well as its *in vivo* release profile from the dosage form.

2.8   FUTURE PROSPECTS

Despite the significant advances made in various oral controlled release dosage forms, many hurdles remain. Among all the challenges, the following three areas will likely be at the forefront for the foreseeable future: (1) improving bioavailability of poorly soluble drugs; (2) prolongation of gastric residence time; and (3) oral delivery of peptide and protein drugs. First, as the number of poorly soluble new drug candidates from drug discovery increases, the enhancement of bioavailability of these compounds becomes a critical challenge in drug development. Various approaches including nanosizing, complexation, solubilization by surfactant or in lipid systems, and solid solutions or dispersions have been employed to tackle this challenge with varying degrees of success. These approaches will increasingly be explored and optimized for enhancing the rate of dissolution and the oral bioavailability of poorly soluble drugs [74–76]. Second, prolongation of the gastric residence time has been proposed as a means to achieve better bioavailability for drugs mainly absorbed in the upper GI tract (absorption window effect). Various approaches including floating dosage forms, expanding dosage forms, and mucoadhesive dosage forms have been investigated for this purpose. Drug products based on expanding dosage forms claiming to be "gastric retentive" have also been developed in recent years. However, as pointed out by Waterman [77], many of the reported clinical studies were not conclusive in terms of the significance of gastric retention because of the lack of a nondisintegrating, controlled release tablet as control to differentiate the contribution from inherent prolongation of gastric emptying in the fed state. Thus, more conclusive clinical studies will need to be carried out to provide evidence that the gastric retentive dosage form is indeed useful. Third, oral delivery of peptides and proteins drugs has been limited by their poor permeability, susceptibility to enzymatic degradation, rapid clearance, and physicochemical instability. Various technologies involving absorption enhancers, carriers, protease inhibitors, formulation approaches, colonic delivery, and surfactants (for minimizing aggregation) have been investigated with limited degree of success [78, 79]. At present, the achievable oral bioavailability is still rather low. Several formulations for small peptides such as heparin and calcitonin are at the stage of early clinical trials. Further success in the development of such oral products will hinge on whether the bioavailability of these oral dosage forms can be improved to achieve similar therapeutic efficacy to that of existing routes of administration and whether the higher oral dose required as a result of the low oral bioavailability translates to a much higher cost for the drug product.

More recently, there has been a surge of interest in the potential application of nanomaterials and/or nanoparticles for the oral delivery of drugs and vaccines [80, 81]. The carriers involved can range from biocompatible polymers with prior history of human use [82] to novel inactivated plant virus nanoparticles [83]. Since cellular uptake of intact nanoparticles has been observed after oral dosing, these nanoparticles may be promising for the oral delivery of antibodies, vaccines, and poorly soluble drugs. Nevertheless, more studies would be needed to further determine the extent of their applicability as well as their safety and efficacy in humans.

If the history of development of oral controlled release dosage forms is of any guide, one can be assured that the field will continue to thrive with better and more effective oral controlled release products that can provide better therapy. During this process, one can also expect that delivery technologies that are based more on hype than science and dosage forms that do not offer better *in vivo* performance and/or cost benefit over existing products will eventually disappear through the natural selection process in the evolution of oral controlled release dosage forms.

REFERENCES

1. Thompson HO, Lee CO. History, literature, and theory of enteric coatings. *J. Am. Pharm. Assoc.* 1945;34:135–138.

2. Mohr F, Redwood T, Procter W Jr. *Practical Pharmacy*, Lea & Blanchard, Philadelphia, PA, 1848, pp. 507–508.

3. Unna PG. Eine neue form medicamentoser einverleibung. *Fortschr. Med. (Berlin)* 1884;2:507–509.

4. Kirk WJ. Medical tablet. US Patent 1,881,197, 1932.

5. Miller GC. Coating for medical compound. US Patent 2,011,587, 1935.

6. Welin BP. Medicinal preparation and method of making the same. US Patent 2,146,867, 1939.

7. Keller JW. Pill or tablet. US Patent 2,099,402, 1937.

8. Shangraw RF. Timed-release pharmaceuticals. *Hosp. Pharm.* 1967;2:19–27.

9. Helfand WH, Cowen DL. Evolution of pharmaceutical dosage forms. *Pharm. Hist.* 1983;25:3–18.

10. Unschuld PU. *Medicine in China: A History of Pharmaceutics*, University of California Press, 1986, pp. 104–117.

11. Harper DJ. *Early Chinese Medical Literature: The Mawangdui Medical Manuscripts*, Kegan Paul International, 1998.

12. Lazarus J, Cooper J. Oral prolonged action medicaments: their pharmaceutical control and therapeutic aspects. *J. Pharm. Pharmacol.* 1959;11:257–290.

13. Lazarus J, Cooper J. Absorption, testing, and clinical evaluation of oral prolonged-action drugs. *J. Pharm. Sci.* 1961;50:715–732.

14. Dragstedt CA. Oral medication with preparations for prolonged action. *J. Am. Med. Assoc.* 1958;168:1652–1655.

15. Gibaldi M. *Biopharmaceutics and Clinical Pharmacokinetics*, Lea & Febiger, 1984, pp. 116.

16. Rosen H, Abribat T. The rise and rise of drug delivery. *Nat. Rev. Drug Discov.* 2005;4:381–385.

17. *Physicians' Desk Reference*, 62nd edition, Thomson PDR, 2008.

18. Hite M. Controlled release technologies for oral drug delivery: comparative advantages, market trends and available technologies. The Drug Delivery Companies Report Autumn/Winter 2003, PharmaVentures Ltd., pp. 30–32.

19. Skelly JP, Amidon GL, Barr WH, Benet LZ, Carter JE, Robinson JR, Shah VP, Yacobi A. *In vitro* and *in vivo* testing and correlation for oral controlled/modified-release dosage forms. *Pharm. Res.* 1990;7:975–982.

20. *Guidance for Industry SUPAC-MR: Modified Release Solid Oral Dosage Forms*, CDER, FDA, September 1997.

21. Cramer MP, Saks SR. Translating safety, efficacy and compliance into economic value for controlled release dosage forms. *Pharmacoeconomics* 1994;5:482–504.

22. Urquhart J. Can drug delivery systems deliver value in the new pharmaceutical marketplace? *Br J. Clin. Pharmacol.* 1997; 44:413–419.

23. http://drugtopics.modernmedicine.com/drugtopics/data/articlestandard//drugtopics/102008/500221/article.pdf.

24. Hiatt, GD. Enteric coating. US Patent 2,196,768, 1940.

25. Cook EF, Martin EW. *Remington's Practice of Pharmacy*, 10th edition, Mack Publishing, 1951, pp. 1400.

26. Luce GT. Cellulose acetate phthalate: a versatile enteric coating. *Pharm. Technol.* 1977;1:27–31.

27. Wilken LO, Kochhar MM, Bennett DP, Cosgrove FP. Cellulose acetate succinate as an enteric coating for some compressed tablets. *J. Pharm Sci.* 1962;51:484–490.

28. Nesbitt RU, Goodhart FW, Gordon RH. Evaluation of polyvinyl acetate phthalate as an enteric coating material. *Int. J. Pharm.* 1985;26:215–226.

29. Stafford JW. Enteric film coating using completely aqueous dissolved hydroxypropyl methyl cellulose phthalate spray solution. *Drug Dev. Ind. Pharm.* 1982;8:513–530.

30. Takahashi A, Kato T, Kamiya F. Dissolution mechanism for hydroxypropyl methylcellulose acetate succinate used in enteric coating of tablets. *Kobunshi Ronbunshu* 1985;42: 803–808.

31. Chang RK, Hsiao CH, Robinson JR. A review of aqueous coating techniques and preliminary data on release from theophylline product. *Pharm. Technol.* 1987;11:56–68.

32. Ibekwe VC, Fadda HM, Parsons GE, Basit AW. A comparative *in vitro* assessment of the drug release performance of pH-responsive polymers for ileo-colonic delivery. *Int. J. Pharm.* 2006;308:52–60.

33. Friend DR. New oral delivery systems for treatment of inflammatory bowel disease. *Adv. Drug Del. Rev.* 2005;57:247–265.

34. Siepmann F, Siepmann J, Walther M, MacRae RJ, Bodmeier R. Polymer blends for controlled release coatings. *J. Control. Release* 2008;125:1–15.

35. http://www.gsk.com/about/history.htm.

36. Blythe R.H. Sympathomimetic preparation. US Patent 2,738,303, 1956.

37. Davis SS, Hardy JG, Taylor MJ, Whalley DR, Wilson CG. A comparative study of the gastrointestinal transit of a pellet and a tablet formulation. *Int. J. Pharm.* 1984;21:167–177.

38. Swintosky JV. Personal adventures in biopharmaceutical research during the 1953–1984 years. *Drug Intell. Clin. Pharm.* 1985;19:265–276.

39. Lee PI. Oral ER technology: mechanism of release. In Amidon GL, Robinson JR, Williams RL,editors. *Scientific Foundations for Regulating Drug Product Quality*, AAPS, 1997, pp. 222–231.

40. Davison RL, editor. *Handbook of Water-Soluble Gums and Resins*, McGraw-Hill, 1980.

41. Christenson GL, Dale LB. Sustained release tablet. US Patent 3,065,143, 1962.

42. Lee PI, Peppas NA. Prediction of polymer dissolution in swellable controlled-release systems. *J. Control. Release* 1987;6:207–215.

43. Lee PI. Diffusional release of a solute from a polymeric matrix—approximate analytical solutions. *J. Membr. Sci.* 1980;7:255–275.

44. Conte U, Colombo P, Gazzaniga A, Sangalli ME, La Manna A. Swelling-activated drug delivery systems. *Biomaterials* 1988;9:489–493.

45. Pham AT, Lee PI. Probing the mechanisms of drug release from hydroxypropylmethyl cellulose matrices. *Pharm. Res.* 1994;11:1379–1384.

46. Gao P, Skoug JW, Nixon PR, Ju TR, Stemm NL, Sung K-C. Swelling of hydroxypropyl methylcellulose matrix tablets. 2. Mechanistic study of the influence of formulation variables on matrix performance and drug release. *J. Pharm. Sci.* 1996;85:732–740.

47. Kill S, Dam-Johansen K. Controlled drug delivery from swellable hydroxypropylmethylcellulose matrices: model-based analysis of observed front movements. *J. Control. Release* 2003;90:1–21.

48. Siepmann J, Peppas NA. Modeling of drug release from delivery systems based on hydroxypropyl methylcellulose (HPMC). *Adv. Drug Deliv. Rev.* 2001;48:139–157.

49. Mahaguna V, Talbert RL, Peters JI, Adams S, Reynolds TD, Lam FYW, Williams RO. Influence of hydroxypropyl methylcellulose polymer on *in vitro* and *in vivo* performance of controlled release tablets containing alprazolam. *Eur. J. Pharm. Biopharm.* 2003;56:461–468.

50. Theeuwes F, Higuchi T. Osmotic dispensing device for releasing beneficial agent. US Patent 3,845,770, 1974.

51. Theeuwes F. Elementary osmotic pump. *J. Pharm. Sci.* 1975;64:1987–1992.

52. Good WR, Lee PI. Membrane-controlled reservoir drug delivery systems, In Langer RS, Wise DL,editors. *Medical Applications of Sustained Release*, Vol 1, CRC Press, 1984, pp 1–39.

53. Conley R, Gupta SK, Sathyan G. Clinical spectrum of the osmotic-controlled release oral delivery system (OROS), an advanced oral delivery form. *Curr. Med. Res. Opin.* 2006;22:1879–1892.

54. Pelham WE, Gnagy EM, et al. Once-a-day Concerta methylphenidate versus three-times-daily methylphenidate in laboratory and natural settings. *Pediatrics* 2001;107:E105.

55. Zentner GM, Rork GS, Himmelstein KJ. The controlled porosity osmotic pump. *J. Control. Release* 1985;1:269–282.

56. Verma RK, Krishna DM, Garg S. Formulation aspects in the development of osmotically controlled oral drug delivery systems. *J. Control. Release* 2002;79:7–27.

57. Thombre AG, Appel LE, Chidlaw MB, et al. Osmotic drug delivery using swellable core technology. *J. Control. Release* 2004;94:75–89.

58. Verma RK, Garg S. Current status of drug delivery technologies and future directions. *Pharm. Technol.* 2001;25:1–14.

59. Rathbone MJ, Hadgraft J, Roberts MS,editors. *Modified-Release Drug Delivery Technology*, Marcel Dekker, 2003, Chapters 1–24.

60. Lee PI, Good WR. Overview of controlled-release drug delivery, In Lee PI, Good WR,editors. *Controlled-Release Technology*, ACS Symposium Ser. No. 348: American Chemical Society, 1987.

61. Lee EJ, Heimlich JM, Noack RM, Grant D. Method of preparing solid dosage forms coated in two layers comprising a water-insoluble polymer and a water-soluble pore former. PCT Application No. WO2004/010982, 2004.

62. Conte U, Maggi L, Colombo P, La Manna A. Multilayered hydrophilic matrices as constant release devices (Geomatrix® systems). *J. Control. Release* 1993;26:39–47.

63. Wong PSL, Edgren DE, Dong LD, Ferrari VJ. Banded prolonged release active agent dosage form. US Patent 5,667,804, 1997.

64. Cremer K. Device for the controlled release of active substances. US Patent 5,853,760, 1998.

65. Lee PI. A novel approach to zero-order drug delivery via immobilized non-uniform drug distribution in glassy hydrogels. *J. Pharm. Sci.* 1984;73:1344–1347.

66. Lee PI. Initial concentration distribution as a mechanism for regulating drug release from diffusion controlled and surface erosion controlled matrix systems. *J. Control. Release* 1986;4:1–7.

67. Lee PI.editor. Diffusion-controlled matrix systems. In Kydonieus A,editor. *Treatise of Controlled Drug Delivery: Physicochemical Basis, Optimization, Applications*, Marcel Dekker, 1992, pp. 155–197.

68. Xu X, Lee PI. Programmable drug delivery from an erodible association polymer system. *Pharm. Res.* 1993;10:1144–1152.

69. Conte U, Giunchedi P, Maggi L, et al. Ibuprofen delayed release dosage forms: a proposal for the preparation of an *in vitro/in vivo* pulsatile system. *Eur. J. Pharm. Biopharm.* 1992;38:209–212.

70. Ishino R, Yoshino H, Hirakawa Y, Noda K. Design and preparation of pulsatile release tablets as a new oral drug delivery system. *Chem. Pharm. Bull.* 1992;40:3036–3041.

71. Youan BC. Chronopharmaceutics; gimmick or clinically relevant approach to drug delivery? *J. Control. Release* 2004; 98:337–353.

72. Wilding IR. Site specific delivery in the gastrointestinal tract. *Crit. Rev. Ther. Drug Carrier Syst.* 2000;17:557–622.

73. Urquhart J. Controlled drug delivery: therapeutic and pharmacological aspects. *J. Intern. Med.* 2000;248:357–376.

74. Rabinow BE. Nanosuspensions in drug delivery. *Nat. Rev. Drug Discov.* 2004;3:785–796.

75. Pouton CW, Porter CJH. Formulation of lipid-based delivery systems for oral administration: materials, methods and strategies. *Adv. Drug Deliv. Rev.* 2008;60:625–637.

76. Leuner C, Dressman J. Improving drug solubility for oral delivery using solid dispersions. *Eur. J. Pharm. Biopharm.* 2000;50:47–60.

77. Waterman KC. A critical review of gastric retentive controlled drug delivery. *Pharm. Dev. Technol.* 2007;12:1–10.

78. Fix JA. Oral controlled release technology for peptides: status and future prospects. *Pharm. Res.* 1996;13:1760–1764.

79. Mahato RI, Narang AS, Thoma L, Miller DD. Emerging trends in oral delivery of peptide and protein drugs. *Crit. Rev. Ther. Drug Carrier. Syst.* 2003;20:153–214.

80. Galindo-Rodriguez SA, Allemann E, Fessi H, Doelker E. Polymeric nanoparticles for oral delivery of drug and vaccines: a critical evaluation of *in vivo* studies. *Crit. Rev. Ther. Drug Carrier Syst.* 2005;22:419–464.

81. Jain KK. *The Handbook of Nanomedicine*, Humana Press, 2008, Chapter 4, pp. 119–160.

82. Bowman K, Leong KW. Chitosan nanoparticles for oral drug and gene delivery. *Int. J. Nanomed.* 2006;1:117–128.

83. Rae CS, Khor IW, Wang Q, et al. Systemic trafficking of plant virus nanoparticles in mice via the oral route. *Virology* 2005;343:224–235.

3

BIOPHARMACEUTIC CONSIDERATION AND ASSESSMENT FOR ORAL CONTROLLED RELEASE FORMULATIONS

HUA ZHANG AND JEAN M. SURIAN

AstraZeneca Pharmaceuticals LP, Wilmington, DE, USA

3.1 INTRODUCTION

Assessing and understanding the biological performance of oral controlled release (CR) formulations are critical steps to achieve a successful product development. Over the past 10 years, significant advancements in the field of biopharmaceutics have been made to aid such an assessment. Significant efforts have been made to demonstrate and improve dose delivery understanding using *in silico* and *in vitro* assessments to supplement *in vivo* formulation assessment. Like immediate release (IR) formulation development, before a drug candidate is progressed into the CR dosage form development stage, its feasibility to be formulated in such a dosage form needs to be vigorously evaluated. Thus, besides the consideration of its efficacy, potency, and safety, relevant physicochemical and biopharmaceutic properties of a drug candidate, such as its gastrointestinal (GI) solubility, permeability, stability, and metabolism, need to be evaluated. Formulation assessment also needs to be conducted including product dissolution behavior in a testing paradigm that mimics the *in vivo* environment, simulation assessment of pharmacokinetics, and an assessment of the impact of patient population differences in utilizing the CR formulation.

This chapter presents an overview of the relevant biopharmaceutic factors that should be considered and assessed as part of the feasibility evaluation to ensure sufficient oral absorption of a drug candidate from a CR formulation and to enable a successful CR product.

3.2 BIOPHARMACEUTIC EVALUATION OF DRUG CANDIDATE FOR ORAL CR FORMULATION

For a drug to be absorbed from an oral delivery system, the drug must first release from the formulation, dissolve in the intestinal fluid, and then pass through the GI membrane (Figure 3.1). Unlike an IR formulation where the drug is released and essentially absorbed in the upper GI tract, the drug in the oral CR formulation is targeted to release and be absorbed continuously throughout the GI tract or targeted to a specific region. It is well known that the GI tract is physiologically complex and varies from stomach to colon in terms of composition, structure, and function, for example, pH, surface area, motility, and the variety of dosing conditions, for example, fed or fasting state [1–3]. These have a major impact on the rate and extent of absorption. Table 3.1 summarizes the major physiological differences from stomach to colon under fasting state. In evaluating CR, the drug properties must be considered as a function of position in the GI tract. These include solubility and dissolution over the entire GI pH range, regional permeability, and stability. These properties and the release characteristics of the delivery system in concert with GI transit rate should be evaluated using *in vitro* and *in vivo* methods [4].

3.2.1 Solubility

The aqueous solubility of a drug depends on its chemical structure, physicochemical properties of the functional

Oral Controlled Release Formulation Design and Drug Delivery: Theory to Practice, Edited by Hong Wen and Kinam Park
Copyright © 2010 John Wiley & Sons, Inc.

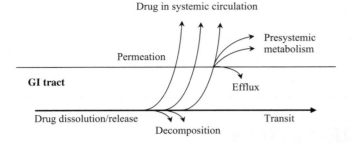

FIGURE 3.1   Oral absorption process along the GI tract.

TABLE 3.1   GI Physiology Relevant to Drug Absorption

GI Region	pH	Surface Area (m²)	Transit Time	Average Transit Time
Stomach	1–3	0.1	0–2 h	15–20 min
Duodenum	4–6.4	2	>1 min	~3–4 h
Jejunum	4–6.5	180	0.5–2 h	
Ileum	6.5–8	280	0.5–2.5 h	
Colon	5.5–7.5	1.5	1–72 h	~18 h

groups, any variation in its stereochemical configuration, and its solid-state properties [5–7]. It is one of the most important factors that can directly affect the oral bioavailability of the drug. As the drug has to dissolve in the GI tract to be absorbed into the blood stream for systemic pharmacological effect, absorption can be limited by its solubility and/or dissolution rate. The drug solubility term in the Noyes–Whitney expression (Equation 3.1) describes the impact of solubility on dissolution rate [8, 9]:

$$\frac{dC}{dt} = \frac{A \cdot D}{h}(C_s - C)$$

where dC/dt is the rate of drug dissolution at time t, expressed as concentration change in the dissolution fluid, D is the diffusion rate constant, A is the surface area of the particle, C_s is the concentration of drug (equal to solubility of drug) in the stagnant layer, C is the concentration of drug in the bulk solvent, and h is the thickness of the stagnant layer.

A drug with high solubility displays faster release, while drugs with poor aqueous solubility often demonstrate incomplete release because of the impact of solubility on dissolution rate from a formulation matrix. One challenge for a successful CR formulation is the need to match drug solubility and dissolution rate from the dosage form in line with the desired absorption profile. Since the GI pH varies between the regions of the intestine, drugs whose solubility is sensitive to pH change can be more challenging to formulate. This sensitivity is further complicated by the need to control dissolution rate in concert with formulation materials and design.

Depending on the drug's chemical structure and properties, it can demonstrate a pH-dependent solubility profile based on its ionizable groups. For example, a weakly acidic drug exhibits lower solubility at low gastric pH, whereas at a pH greater than its pK_a, it is much more soluble. This phenomenon is more critical for a weakly basic drug that might be soluble in low gastric pH but can precipitate at higher intestinal pH (major absorption site) to potentially limit its absorption. The drug exhibiting pH-dependent solubility, particularly in the GI pH range, is generally a poor candidate for a matrix-based oral CR delivery system [10]. Many efforts have been made in this area to adjust the microenvironmental pH to overcome nonuniform release [11, 12]. More detailed information on the approaches taken is discussed in Section 3.3.1.

In addition to pH differences in the GI tract, the impact of *in vivo* bile salts on drug solubilization in different GI regions should be considered. This can be important for a CR dosage form that releases a large fraction of the drug in the lower GI tract where bile salts are scarce. Hence, a comprehensive understanding and evaluation of solubility are necessary during the CR formulation feasibility assessment process.

The solubility of the drug can be estimated *in silico* based on its structure [13, 14]. It can also be determined *in vitro* using standard United States Pharmacopoeia (USP) buffers across the relevant *in vivo* pH range to create a pH solubility profile. Although it is difficult to draw a conclusion on which *in vitro* solubility data would truly reflect the *in vivo* solubility of a drug, research has progressed such that some advances to promote a better understanding of *in vivo* relevant *in vitro* solubility determination are now available. Specifically, simulated human intestinal fluids containing *in vivo* relevant bile salts and surfactants and real human intestinal fluids have been evaluated. Solubility tests in those media have been conducted [15–17]. Simulated and real human intestinal fluids are considered better media compared to simple buffers as they may offer more insightful information on solubility-associated absorption risk. As pointed out earlier, a weak base including its salt form could potentially precipitate upon entry in the small intestine causing incomplete absorption. Currently published information is limited on how to evaluate whether a drug can exist as a supersaturated solution [18] and how to assess the potential for drug precipitation in the GI tract and the impact of precipitation on absorption [19]. This is certainly a research area that needs more effort to improve the understanding.

There are a great number of CR technologies available for dosage form development [20–23]. One of the factors influencing the technology selection is to match the technology and biopharmaceutic properties of the drug candidate. A single solubility value is inadequate to make the judgment on feasibility and degree of difficulty of taking a particular CR formulation approach. Dose is critical. A sufficiently low

TABLE 3.2 Dose/Solubility Consideration in CR Approach Feasibility Assessment

Dose/Solubility Ratio (mL)[a]	Feasibility
<1	CR development straightforward, several technology options available
1–100	Average degree of difficulty
100–1000	Challenging but feasible
>1000	Need solubilization, difficulty increased
>10,000	Practically impossible

[a] Highest dose/lowest solubility in the pH range 1–7.5.

dose level is also a key requirement for a successful CR formulation [24]. In the feasibility assessment process, it is suggested not only to evaluate solubility but also to take proposed clinical dose into consideration, in other words, to use drug dose/solubility ratio to risk assess CR formulation potential within technology evaluation [24, 25]. Dose/solubility ratio defines a volume that is needed to dissolve the drug and create a drug solution for absorption [26]. In practice, this approach is more logical as it references the limitation of solubility in the process of absorption according to the drug dose range. According to Thombre [24] (Table 3.2), in general, the smaller the dose/solubility ratio, the more technologies are applicable, and CR formulation development is easier. For the commonly used CR formulation dosage forms, such as single unit and multiparticulate unit tablets/capsules, multiparticulate units offer more flexibility than single unit matrix tablets and are considered a better formulation platform for a drug that presents a dose/solubility challenge [24, 25]. Some attempts have been made to use a dose/solubility map as a simplified approach to evaluate technology selection. With growing experience, some key boundaries on the dose/solubility map linked to formulation technology selection can be further established, and confidence in using this approach should improve [24].

3.2.2 Permeability

In the oral absorption process, a drug must be permeable after dissolving in the GI lumen to be absorbed into the blood stream. Drugs administered into the intestinal lumen can cross the mucosal epithelium by the paracellular transport route (the main route of permeation for hydrophilic compounds) or the transcellular transport route (the main route of permeation for hydrophobic compounds). Generally, the transport pathways can be divided into (a) passive paracellular transport, (b) passive transcellular transport, (c) carrier-mediated transcellular transport, and (d) vesicular transport [27, 28]. The intestinal mucosa serves as a physiologically selective permeability barrier to control a drug's permeation. Properties such as molecular size and hydro-

phobicity can impact the drug's transport across the intestinal mucosa.

Due to the heterogeneous nature of the GI tract, it is now noticed that permeability can vary significantly from the jejunum to the colon. Site-dependent differences in drug permeability will not only lead to differences in absorption profiles from the upper small intestine to the colon in a single subject, but also cause a high degree of variability in the absorption data due to transit time variations between subjects. Given the broadest reasonable ranges of human GI residence time (Table 3.1), it is not uncommon for peak drug plasma concentrations from CR dosage forms to occur beyond 18 h after oral administration [29]. Therefore, for CR formulations, drugs not only need to release throughout the GI tract but also need to be absorbed throughout the GI tract. Their bioavailability and hence the biopharmaceutic feasibility may be dependent upon both the absorption mechanism and the colonic permeability of the drug candidate. The regional permeability and absorption of the drug as a function of position in the GI tract, especially colonic permeability and absorption, are critical that may determine whether a CR formulation approach is feasible. For many oral CR formulations, a required attribute is to achieve once-daily dosing. For drugs with a relatively short elimination half-life, poor colonic drug absorption may result in a CR formulation that cannot achieve once-daily dosing without a significant loss of bioavailability.

Although permeability limitation is a consideration for drugs administered in IR dosage forms, for CR formulations whose oral absorption rate is supposed to be controlled by dissolution or dosage release, the drug must have high permeability. Otherwise, the rate-limiting step would be the inherent drug absorption. CR development considerations usually require that the drug is well absorbed throughout the intestine, and preferably via passive diffusion rather than carrier-mediated transport. This is because differences in the level of expression of transporters along the GI tract could result in patient-to-patient variability. It should be noted that even though the Biopharmaceutic Classification System (BCS) is applicable only to IR dosage forms, some considerations and modifications have been proposed for CR dosage forms [29, 30]. Corrigan has proposed a biopharmaceutic drug classification for extended release formulations incorporating drug solubility and regional dependencies in drug permeability (Table 3.3) [29]. According to the BCS permeability classification and CR approach requirement, BCS Class III and IV drugs that carry low and probable region-dependent permeability may not be suitable candidates for CR formulation consideration. The general "rule of thumb" on relative colonic bioavailability ($Frel_{colon}$) versus that of small intestine is that if $Frel_{colon}$ is greater than 60%, then CR formulation development should be straightforward with a high confidence for success. When $Frel_{colon}$ is between 30% and 60%, CR formulation development may be

TABLE 3.3   A Proposed Biopharmaceutic Classification for CR Drugs

Classification	Criterion
Class I	Solubility high and independent of site
	(a) permeability: high and independent of site
	(b) permeability dependent on site: narrow absorption window
Class II	Solubility low and independent of site
	(a) permeability: high and independent of site
	(b) permeability dependent on site: narrow absorption window
Class III	Solubility high and independent of site
	(a) permeability: high and independent of site
	(b) permeability dependent on site: narrow absorption window

challenging but can be manageable. However, for $Frel_{colon}$ less than 30%, this formulation approach will become very difficult, and probably nontraditional approaches will be required to minimize low absorption. Information on relative colonic bioavailability can be obtained by performing the regional absorption studies as described in the following section.

Depending on the development phase of the drug, different techniques with different level of complexity and cost can be employed to risk assess permeability and GI regional absorption potential. Among them, *in vitro* and *in vivo* animal models are commonly utilized. Several review articles are available on this topic [28, 31, 32]. Among the *in vitro* methods, permeability measurement through Caco-2 monolayers has been used to predict absorption in humans [33, 34], and there is a good understanding of the circumstances under which the predictions are less accurate. It has been reported that human colonic permeability has been correlated to permeability in Caco-2 cells [35–37]. Since Caco-2 permeability of drug candidates is determined routinely, it would be worthwhile to collect the available data to build a database and correlate the data with the performance of CR dosage forms. Permeability data from marketed CR products can aid in establishing and assessing the correlation. Based on the database and the correlation, "rules of thumb" could be developed to aid in the selection of CR candidates. Besides Caco-2, permeability in Ussing chambers is also a popular *in vitro* tool [28]. The same approach proposed for Caco-2 systems to validate the model by testing the permeability of marketed CR drugs to draw a correlation is applicable to Ussing chambers.

To assess the permeability of a drug in different regions of the GI tract, perfused intestinal loops models have been used. The regional permeability of several drugs has been determined in the rat jejunum, ileum, and colon [38, 39]. This model can provide valuable insight into the absorption behavior of a drug in different regions of the GI tract and the existence of absorption windows for facilitated transport [40, 41].

As drug absorption in animals is generally believed to be a good predictor of absorption in humans due to the presence of relevant biological factors that may have impact on absorption, whole animal models have also been utilized to study permeability and absorption potential. In addition to traditional oral dosing regimens in animals, advanced models include colonic dosing in beagle dogs via an intestinal vascular access port [42] and dog colonoscopy-based methods [43]. The results from those studies indicate that the dog can be a good model as a surrogate for human regional absorption studies.

Direct and predictive *in vivo* human methods for assessing colonic absorption for CR consideration are also available [44]. Historically, the popular approaches to determine drug absorption from different intestinal GI regions in human are perfusion and intubation techniques [44, 45], which although invasive, provide a reliable measure of GI permeability. Over the past 20 years, several engineering-based capsule technologies have been developed and purposefully designed to deliver drugs noninvasively to the distal intestine to allow site-specific measurement of human absorption [46]. Among them, the Enterion™ capsule (Phaeton Research, Nottingham, UK) is a newer version that overcomes the limitations of the earlier devices and makes this technology more applicable and flexible to incorporate a wide variety of formulations [47–49].

Predicting human intestinal permeability is an active area of research. Different techniques are available to study permeation and absorption potential. In recent years, utilizing purely computational methods to assess absorption potential has become attractive and has been promoted to save money and reduce time and effort spent. While some positive outcomes have originated from the research effort [50–52], advanced progress is needed and reliable databases and diversified training sets are required to broaden their application.

3.2.3   Colonic Stability

Another important factor to colonic drug absorption and hence to CR formulation development is colonic stability. A unique feature of the human colon is that it contains a significantly higher amount of complex microorganisms than does the upper part of the GI tract, and the bacteria are essentially absent from the stomach and the small intestine [53]. Unlike the aerobic nature of the bacteria in the upper GI tract, colonic bacteria are predominantly anaerobic [53]. These anaerobes serve to control colonic pH, aid digestion, stimulate the immune system, and inhibit the growth of other food-poisoning or disease-causing bacteria. The colonic microflora are responsible for numerous metabolic reactions via a wide spectrum of enzymes and have the potential to metabolize drugs and other foreign compounds [54–56]. Since the colonic microbial environment is anaerobic, the metabolic activity of the microbial enzymes is different in the colon compared to metabolism in the liver or the small intestine brush border. Colonic metabolism occurs by reduction and hydrolysis to degrade xenobiotics, rather than by oxidative metabolism in the liver [57]. Several of these enzyme systems have been used to deliver drugs selectively to the colon, and the others have potential in the design of specific colon-targeted delivery systems [58]. More information on selective colonic delivery is described later in this book (Chapter 14).

This metabolic feature of the colon is important because some CR formulations release a significant fraction of the dose in this region, and bioavailability of the drug could be significantly reduced if it is not stable. Some drugs, such as ranitidine, are susceptible to degradation by colonic bacteria and tend to show reduced bioavailability from the colon [59]. Even though the colonic degradation phenomenon is well known, interest and advancing research in the area have been low. One reason may be that because the colonic microbial environment is anaerobic, establishing the *in vitro* experimental conditions to mimic the *in vivo* situation may be complex in order to maintain bacterial enzyme functionality to evaluate drug stability.

In vitro tests evaluating colonic drug degradation described in the literature include approaches such as fermentation studies with human feces or animal colonic contents [60, 61]. Unfortunately, little information is available in the literature with respect to *in vitro/in vivo* correlation of such test systems. Thus, results from these tests would provide only basic information in assessing drug colonic stability.

3.2.4 GI Metabolism and Elimination

The metabolism of a drug via presystemic extraction by intestinal and/or hepatic enzymes can affect bioavailability and hence the feasibility of the CR approach. In addition, data have suggested that drug absorption can be compromised by intestinal elimination mechanisms. Intestinal drug elimination by efflux transporters, such as P-glycoprotein (P-gp) [62] located in the intestinal enterocytes and possibly other mucosal membrane exporters or exchangers, is thought to limit the absorption of a number of drugs and their bioavailability. Intestinal metabolism and elimination may be of special implication when low drug concentrations do not saturate the metabolism and elimination pathways. This circumstance may exist often with the low luminal concentrations of drugs provided by CR dosage forms [63]. Therefore, intestinal metabolism and elimination impact the CR formulation approach.

As pointed out earlier, solubility and permeability are the key parameters in drug absorption. Oral absorption and

bioavailability of a drug from a CR formulation would be sufficient if the solubility and permeability criteria are met. However, this is only correct if the GI metabolism rate and elimination are constant over the region of intestine to which the drug is delivered. The role of gut wall metabolism enzymes cytochrome P450 (CYP450), especially the isoform CYP3A4, and the interplay between CYP3A4 and P-gp have recently become the focus of attention [64]. It has recently been suggested that intestinal CYP3A4 and P-gp provide a coupled pathway for intestinal elimination of orally administered drugs that affect their bioavailability [65].

The impact of CYP3A4 and P-gp on *in vivo* performance of CR formulations can be presented as concentration and/or position dependent [24]. First, intestinal metabolism and elimination processes are drug concentration dependent, which will vary depending on the drug release rate, for example, rapidly from an IR formulation where saturation of these processes is possible versus slowly from a CR formulation where a greater degree of GI metabolism/elimination may occur. Second, P-gp expression and CYP3A4 activity vary as a function of position in the GI tract. A study by Paine et al. has demonstrated that CYP3A4 activity in humans diminishes significantly from the jejunum to the ileum [66]. Therefore, if a drug is a CYP3A4 substrate, CR delivery to the ileum and colon regions can lead to an increase in absorption of the parent drug as seen in the case of oxybutynin chloride [67]. Similarly, CR delivery could reduce CYP3A4-related drug–drug interactions by delivering the drug mainly to the colon. P-gp synthesis, on the other hand, has been reported to increase from the proximal to the distal regions of the small intestine [68], resulting in a decrease in absorption for drugs that are P-gp substrates as they transit through the small intestine.

Thus, it is important to recognize the role of intestinal metabolism and elimination in the absorption process and its impact on CR formulation feasibility. A simple measure and assessment of permeability and solubility may not be adequate in such situations where a drug is CYP3A4 and/or P-gp substrate.

Overall, as presented in Table 3.4, prior to embarking on CR formulation development, the biopharmaceutic properties of the drug candidate with a focus on its GI solubility, permeability, and stability should be evaluated to guide the selection of the CR formulation approach to deliver the drug with the desired absorption profile. A traffic light system to capture the risk around its formulation feasibility can be applied.

3.3 BIOPHARMACEUTIC EVALUATION OF ORAL CR FORMULATION

The goal of CR delivery systems is to provide the desired delivery profile that can achieve predictable plasma levels.

TABLE 3.4   Feasibility and Risk Assessment of Colonic Drug Absorption Prior to Product Development

Level/Traffic Light	Risk Factor	Criteria	Implications
Green: no critical factor identified	Good absorption over entire GI tract	BCS Class I (high passive permeability/ high solubility) and stable in colon chyme (<10% degraded in 1 h)	Good candidate for CR development
Amber: acceptable risk	Risk for poor colon absorption	High permeability after saturation/ inhibition of efflux or intermediate solubility or rapid degradation in colon chyme (>10% degraded in 1 h)	Possible candidate for CR development but further evaluation recommended
Red: significant risk for development	Poor colon absorption expected	Low permeability according to BCS or volume needed to dissolve maximum dose at pH 5.5–7.5 >10 L	CR development not recommended

Utilizing carefully designed *in vitro* tools such as dissolution testing in assessing the CR formulation will help assess the robustness of consistent formulation delivery. Ideally for CR formulation, the release of the drug should occur in a controlled manner and not be affected by physiological conditions. But in reality, due to the changes in physiological conditions throughout the intestine, the *in vivo* release of the drug from the CR formulation may be affected along the GI tract. In this case, a change in dissolution pattern may result in a different pharmacological response. Dissolution testing is conducted not only to screen newly developed formulations but also to assess product quality, and most importantly, to ensure its clinical performance. For this reason, the dissolution testing to capture the release characteristics of the CR formulation should be conducted under realistic and *in vivo* relevant conditions to reflect its *in vivo* performance. With advancement in the field of simulation and modeling, it is also valuable to use the simulation tools available to simulate the pharmacokinetic (PK) profile based on the drug's properties and formulation dissolution behavior to guide formulation design.

3.3.1   Dissolution

CR formulations are designed such that the rate of drug absorption is controlled by the drug release rate from the formulation, resulting in dissolution as the rate-limiting step for absorption. Because dissolution is a critical attribute that can have significant impact on the clinical performance of CR formulations, dissolution testing becomes an important component in the evaluation package for CR formulation assessment. For this reason, a valid dissolution study design is needed. *In vitro* dissolution evaluation from the biopharmaceutic standpoint is addressed in the following section. A more detailed discussion on dissolution testing can be found in Chapter 15.

The most commonly used oral CR technologies are matrix systems, osmotic pumps, and multiparticulate systems. These technologies deliver drug release by diverse mechanisms. For instance, drug release from hydrophilic matrix tablets is controlled by diffusion through and/or erosion of the matrix; the release of drug from osmotic delivery systems relies on osmotic pressure to pump out a drug solution or suspension through delivery orifice; multiparticulate systems control drug release by diffusion through rate-controlling layers. Several *in vitro* dissolution test methods exist. No matter if it is USP paddle apparatus, basket apparatus, or flow-through cell, they offer useful means to capture release profiles to understand the effect of formulation composition and process procedures. Generally, flow-through systems offer a greater advantage over the paddle method because of the hydrodynamics in the continuous flow apparatus and the ease of pH change of the dissolution medium. However, if not applied appropriately, *in vitro* dissolution systems may not be indicative of the *in vivo* situation where pH and hydrodynamics can vary widely along the GI tract. Thus, when designing the *in vitro* dissolution test design for the purpose of biopharmaceutic evaluation of the CR formulation, any relevant parameters that can affect release and oral absorption should be considered, for example, physicochemical and biopharmaceutic properties of the drug and its product, the PK behavior of the parent compound, and, most importantly, the GI physiologic conditions/factors. Based on the understanding of the dissolution mechanism associated with each formulation technology, the more the *in vitro* dissolution conditions mimic those existing *in vivo*, such as pH, hydrodynamics, transit time, volume, and so on, the greater the chance that the results from *in vitro* dissolution will reflect the performance of the product *in vivo*. Biorelevant dissolution testing conditions, such as the use of simulated intestinal fluids that are used in the investigation of IR formulations [69, 70], may serve as a reasonable starting point in designing dissolution conditions for the initial evaluation of a CR formulation. To capture the release characteristics in colon, attempts have been made to design simulated colonic fluids based on published information on pH and composition of the contents of the proximal colon [71]. Since drug release from the CR formulation is expected to occur throughout the GI tract, utilizing one dissolution medium is most likely not sufficient to understand how dissolution occurs *in vivo*. Instead, multiple media or pH shift methods may offer more insightful information to understand the *in vivo* dissolution [72]. Besides defining the dissolution medium, hydrodynamics is another challenging aspect of *in vivo* relevant *in vitro* dissolution study design. At present, no dissolution apparatus is available that represents *in vivo* hydrodynamics in a meaningful way, although some effort and attempts have been made to understand the dynamic forces important for CR formulation dissolution [73, 74]. Due to the complexity of human GI physiology, it is no doubt a challenging area. Further research is needed to design *in vivo* predictive dissolution for better CR formulation evaluation.

Except in the case of enteric coated formulations, the drug release from the CR formulation should be unaffected by physiological conditions, for example, independent of the pH of the release medium. This principle can be applied to overcome the physiologic impact on dissolution and to achieve minimal biopharmaceutic variability [75]. As addressed earlier, solubility is a key parameter that can affect drug dissolution. For a drug demonstrating pH-dependent solubility, its release rate from CR formulations may vary as a function of dosage form movement in the various regions of the GI tract, leading to inefficient drug delivery and high intersubject variability. This has been shown to be very important for weakly basic and/or acidic drugs [76]. Efforts to overcome nonuniform release by adjusting the microenvironmental pH of weakly basic drugs in the formulations have been published [11, 12]. An example includes the utilization of blends of enteric and extended release polymers as film coatings [77]. Most approaches to achieve pH-independent delivery of weakly basic drugs are based on the inclusion of acidic excipients in the formulation. The acidic excipients maintain a constant and low-pH microenvironment in the formulation and thus make the solubility of the weakly basic drug high and irrespective of the dissolution medium [11, 12, 75]. For weakly acidic drugs, the same approach is used by including buffer salts in the formulation to maintain a high-pH microenvironment to prevent the nonuniform release [78].

3.3.2 *In Vivo* Study and PK Simulation

In addition to *in vitro* dissolution testing, the biopharmaceutic evaluation of CR dosage forms can be carried out by conducting *in vivo* release studies based on a correction between *in vitro* and *in vivo* release. This is also an alternative to guide *in vitro* dissolution study design. As CR products exhibit dissolution rate-controlled release, they present themselves as promising candidates for establishing *in vitro–in vivo* correlation (IVIVC) of their performance. Typically, CR formulations with two or more different *in vitro* release rates are evaluated in a crossover human PK study. The release of drug *in vivo* can be analyzed by deconvolution

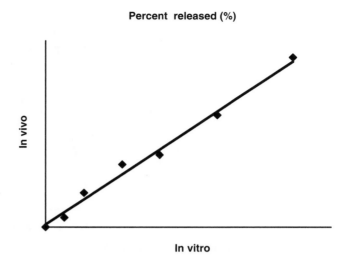

Percent released (%)

FIGURE 3.2   Illustration of Level A point-to-point comparison IVIVC.

of the plasma concentrations from the CR formulations with different release profiles. A common approach for deconvolution is to use an orally administered solution as a reference, which then allows one to calculate a unit impulse function from solution to obtain the input rate for the CR dosage form. Once the *in vivo* release profiles of the CR dosage forms are obtained, one can correlate *in vivo* release profiles to their corresponding *in vitro* release profiles to establish a point-to-point Level A correlation (Figure 3.2). The FDA guideline for controlled release oral dosage forms clearly defines acceptance criteria for successful IVIVC [79]. Once an IVIVC is successfully established, *in vitro* release characteristics can be modified to reflect the *in vivo* behavior of the dosage form [80–82].

In general, establishing an IVIVC is not a main design criterion for CR product development. IVIVC clinical trials involve the development and manufacturing of formulations with different release profiles as well as analysis of a large number of blood samples to obtain the PK data. It is not cost effective to evaluate formulation performance in the early phase of product development. Alternatively, an *in vivo* gamma scintigraphic study can be conducted in healthy human volunteers to evaluate the release behavior of the CR dosage form [4]. The formulation under investigation is labeled with a solute containing a gamma emitting radionuclide. The solute is used to mimic drug release by diffusion. In addition, an ion-exchange resin with a small particle size is utilized to represent the integrity of a delayed release of tablet or the erosion of a tablet matrix [4]. In the study, the computer system is employed to monitor the image of the delivery system and radioactivity is quantified. Then, the *in vivo* release profile created by monitoring the remaining radioactivity can be used to compare with the release profile

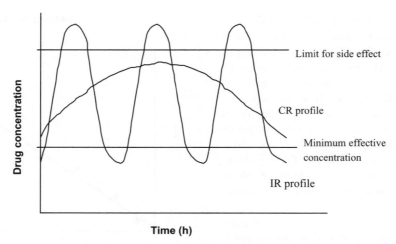

FIGURE 3.3 Simulation CR plasma concentration versus time profile based on IR formulation PK.

observed *in vitro*. The technique can serve as a complement to the PK investigation by providing information not only on the *in vivo* release characteristics, but also on the position of the formulation in the GI tract. From this type of study, the absorption window and sufficient or insufficient regional absorption, particularly in colon, can be obtained [83]. It has been reported that this technique can be used to evaluate CR matrix tablets [4] and osmotic pump [83] delivery systems.

In addition to the *in vitro* and *in vivo* assessment of the CR dosage form, PK simulation employing PK principles to guide design of a CR dosage form is a critical component of CR feasibility assessment. Simulation of CR oral dosage forms requires altering the dose and release rate in an attempt to maintain selected plasma concentrations within a desired dosing interval. During early CR formulation development, PK simulation is needed to assess whether a specific drug is a good candidate for CR formulation and whether it is possible to alter the release rate and dose to achieve that goal [84–86]. For this purpose, plasma concentrations following administration of various CR dosage forms can be simulated if the PK data from an IR dosage form are available (Figure 3.3). Generally, simulation is conducted by employing PK principles and concepts with custom-written computer programs [87]. Besides custom-written computer programs, several simulation software programs such as Kinetica™ (Innaphase), Berkeley Madonna™ (University of California at Berkeley), and GastroPlus™ (Simulations Plus) are commercially available [24]. A mathematical model using STELLA® has also been used to calculate drug absorption from different regions of the GI tract: duodenum, jejunum, ileum, and colon [88]. GastroPlus™ (Simulations Plus, Palmdale, CA), a compartmental absorption and transit time model, has the capacity of handling not only PK simulation but also regional absorption simulation [76, 89]. In the simulation, the program encompasses biopharmaceutic

properties of the drug, for example, solubility and permeability, and allows modeling of drug input rates, small and large intestinal transit time variations, and position-dependent absorption rates. The drug colonic absorption potential can be revealed from the simulation. A representative of the regional absorption simulation analysis is presented in Figure 3.4. By inputting the *in vitro* dissolution profile and drug PK parameters, projected performance of a CR formulation represented by a simulated plasma concentration profile can be illustrated. If information on colonic permeability or absorption is limited, then simulations can be carried out assuming low, medium, and high colonic absorption to explore the expected performance.

Overall, with commercially available or custom-written computer programs, the projected performance of a CR

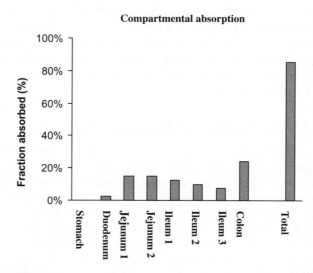

FIGURE 3.4 GastroPlus™ compartmental or regional absorption analysis. Fraction absorbed (%) at different regions along the GI tract.

formulation (single-dose or steady state PK) can be simulated. Simulation is conducted based on the physicochemical, biopharmaceutic, and/or PK properties of the drug. The results of simulations reveal the projected performance of a CR formulation and indicate the dose range and release rates that are reasonable to meet the clinical target.

3.4 PERFORMANCE OF API AND DOSAGE FORM: VARIABILITY CONSIDERATION

Consistent performance of the CR dosage form upon dosing is critical to a quality product. As stated earlier, a robust understanding of the drug substance properties with regard to solubility and permeability and a well-designed formulation are key to assessing the impact of biological factors on dose delivery.

Physiological differences in gastric emptying, pH, GI tract motility, fasting versus fed state, and particle size of the formulation can dramatically impact the dissolution profile and potential dosage form release mechanism, resulting in significant performance variability. Demonstrating the impact of food on CR dosage form behavior is a well-established practice. More recently, the impact of alcohol on potential dose dumping from CR formulations is being investigated with emphasis on release kinetics [90].

Gastric pH displays a circadian rhythm that can vary between individuals based on response to acid secretion on gastric emptying [91]. Low pH is prominent between midnight and early morning with the time depending on the last evening meal. Differences in gastric pH are apparent between gender and age. Basal acid secretion is reduced in women compared to men. When comparing age and population, acid secretion was found to increase to a greater extent in men compared to women (median age 51 years versus 33 years). The fasting gastric pH and duodenal pH in children are very similar to those in adults. This information is beneficial and more important for weakly basic or acidic drugs that possess pH-dependent solubility.

Gastric emptying also follows a circadian rhythm, with slow emptying after afternoon/evening meals compared to breakfast [92]. Posture relative to dose administration has been correlated with significant differences in rate and extent of absorption for some dosage forms such as pellets. Site-dependent differences in drug permeability, for example, will not only lead to absorption profiles that deviate from the *in vitro* release characteristics, but also exhibit a high degree of variability in absorption due to transit time differences among subjects.

Disease state can also play a role in absorption variability and affect formulation performance. For example, differences in oral absorption because of differences in intestinal blood flow have been shown for several cardiovascular drugs [93]. In patients with heart failure, several possible

mechanisms that impact drug absorption are speculated such as pathology to the intestinal wall, delayed gastric emptying, change in GI tract pH and motility, change in bacterial flora, and edema of the bowel [94]. Disease can often impact gastric motility and gastric emptying. A longer dwell time in the gastric media could have tremendous impact on dosage form dissolution. For CR products such as matrix systems, the impact of longer acid exposure on the polymers or other excipients controlling dissolution needs to be assessed. While testing for enteric coated systems typically verifies integrity of the coat for 1–2 h at low pH, a significant decrease in motility may result in great degree of dissolution variability. Damage to the intestinal mucosa, such as disease-related, drug-induced, or induced from surgical means, can impact the permeability of the drug substance released from the dosage form. Many nonsteroidal anti-inflammatory drugs are known to demonstrate increased permeability due to damage induced during their absorption [95]. Bowel diseases such as Crohn's disease or irritable bowel syndrome where diarrhea or constipation are variable can result in tremendous variability in CR delivery [96].

Variability in CR dosage form performance is also evident during lower GI tract delivery where colonic absorption is desired [97]. Diet significantly influences colonic mucosa, secretory content, and motility. Drug therapy that impacts colonic transit time will significantly impact the bioavailability of the CR dosage form, especially single unit erodible matrices. Administration time will also impact the transit time in the lower intestine, where dosages administered at bedtime will have a longer colonic transit than those administered in the morning. Dosage form size impacts the transit rate in that smaller particles are known to undergo colonic sieving. Larger particles tend to move faster, while smaller particles spread throughout the length of the colon.

As discussed above, the physiology of the GI tract and the manner in which it can be affected by disease conditions, age, gender, and coadministered drugs have a direct bearing on the performance of CR formulations. A good understanding of the impact of physiology, disease, and age on CR formulation significantly aids robust formulation design and development.

3.5 CONCLUSION

To assess *in vivo* feasibility of developing CR formulations, this chapter has focused on the assessment of biopharmaceutic properties of the drug candidate, evaluation of release behavior and *in vivo* performance of the drug product, and consideration of the impact of physiological and pathological factors in dosage form performance. Sufficient GI solubility, permeability, and stability, adequate colonic absorption with reasonable dose size, and appropriate dissolution behavior are key requirements for a successful CR formulation. Besides good understanding and significant advances made in the biopharmaceutic research area, gaps can be identified where improvement is needed for better CR dosage form development. These include the following: how to develop a robust CR formulation for a low solubility/high dose drug; how to design *in vivo* predictive *in vitro* dissolution tests; and how to optimize formulation performance to minimize the impact of physiological and disease factors.

ACKNOWLEDGMENTS

The authors thank Dr. Bertil Abrahamsson, Dr. Martin DeBerardinis, and Dr. Lisa Martin for their support, insights, and suggestions in preparation of this chapter.

REFERENCES

1. Macheras P, Argyrakis P. Gastrointestinal drug absorption: is it time to consider heterogeneity as well as homogeneity? *Pharm. Res.* 1997;14(7):842–847.

2. Davis SS. Assessment of gastrointestinal transit and drug absorption. In Prescott LF, Nimmo WS, editors. *Novel Drug Delivery and Its Therapeutic Application*, Wiley, Chichester, 1989.

3. Dressman JB, Berardi RR, Dermentzoglou LC, Russell TL, Schmaltz SP, Barnett JL, Jarvenpaa KM. Upper gastrointestinal (GI) pH in young, healthy men and women. *Pharm. Res.* 1990; 7(7):756–761.

4. Davis SS. The design and evaluation of controlled release systems for the gastrointestinal tract. *J. Control. Release* 1985;2:27–38.

5. Wassvik CM, Holmén AG, Draheim R, Artursson P, Bergström CAS. Molecular characteristics for solid-state limited solubility. *J. Med. Chem.* 2008;51(10):3035–3039.

6. Martin A, Swarbrick J, Cammarata A, editors. *Solubility and Distribution Phenomena. Physical Pharmacy*, 3rd edition, Lea & Febiger Philadelphia, PA, 1983, pp. 272–313.

7. Neau SH, Bhandarkar SV, Hellmuth EW. Differential molar heat capacities to test ideal solubility estimations. *Pharm. Res.* 1997;14(5):601–605.

8. Nernst W, Brunner E. Theorie der Reaktionsgeschwindigkeit in heterogenen systemen. *Z. Phys. Chem.* 1904;47:52–110.

9. Noyes AS, Whitney WR. The rate of solution of solid substances in their own solutions. *J. Am. Chem. Soc.* 1897; 19:930–934.

10. Varma MVS, Kaushal AM, Garg A, Garg S. Factors affecting mechanism and kinetics of drug release from matrix-based oral controlled drug delivery systems. *Am. J. Drug Deliv.* 2004; 2(1):43–57.

11. Streubel A, Siepmann J, Dashevsky A, Bodmeier R. pH-independent release of a weakly basic drug from water-insoluble and -soluble matrix tablets. *J. Control. Release* 2000; 67(1):101–110.

12. Martínez-González I, Villafuerte-Robles L. Influence of enteric citric acid on the release profile of 4-aminopyridine from HPMC matrix tablets. *Int. J. Pharm.* 2003;251(1–2):183–193.

13. Huuskonen J, Livingstone DJ, Manallack DT. Prediction of drug solubility from molecular structure using a drug-like training set. *SAR QSAR Environ. Res.* 2008;19(3–4):191–212.

14. Chen A, Mertz KM Jr. Prediction of aqueous solubility of a diverse set of compounds using quantitative structure–property relationships. *J. Med. Chem.* 2003;46(17):3572–3580.

15. Vertzoni M, Pastelli E, Psachoulias D, Kalantzi L, Reppas C. Estimation of intragastric solubility of drugs: in what medium? *Pharm. Res.* 2007;24(5):909–917.

16. Pedersen BL, Müllertz A, Brøndsted H, Kristensen HG. A comparison of the solubility of danazol in human and simulated gastrointestinal fluids. *Pharm. Res.* 2000;17(7):891–894.

17. Kalantzi L, Persson E, Polentarutti B, Abrahamsson B, Goumas K, Dressman JB, Reppas C. Canine intestinal contents vs. simulated media for the assessment of solubility of two weak bases in the human small intestinal contents. *Pharm. Res.* 2006;23(6):1373–1381.

18. Kostewicz ES, Wunderlich M, Brauns U, Becker R, Bock T, Dressman JB. Predicting the precipitation of poorly soluble weak bases upon entry in the small intestine. *J. Pharm. Pharmacol.* 2004;56:43–51.

19. Johnson KC. Dissolution and absorption modeling: model expansion to simulate the effects of precipitation, water absorption, longitudinally changing intestinal permeability, and controlled release on drug absorption. *Drug Dev. Ind. Pharm.* 2003;29(8):833–842.

20. Buri P, Doelker E. Formulation of prolonged release compressed tablets. II. Hydrophilic matrices. *Pharm. Acta Helv.* 1980;55:189–197.

21. Thombre AG, Appel LE, Chidlaw MB, Daugherity PD, Dumont F, Evans LAF, Sutton SC. Osmotic drug delivery using swellable-core technology. *J. Control. Release* 2004; 94(1):75–89.

22. Colombo P, Bettini R, Santi P, Peppas NA. Swellable matrices for controlled drug delivery: gel-layer behavior, mechanisms and optimal performance. *Pharm. Sci. Technol. Today* 2000; 3(6):198–204.

23. Liu J, Zhang F, McGinity JW. Properties of lipophilic matrix tablets containing phenylpropanolamine hydrochloride prepared by hot-melt extrusion. *Eur. J. Pharm. Biopharm.* 2001;52(2):181–190.

24. Thombre AG. Assessment of the feasibility of oral controlled release in an exploratory development setting. *Drug Discov. Today* 2005;17(10):1159–1166.

25. MacRae RJ. Candidate and technology selection for oral controlled release. In *SMI Controlled Release Annual Meeting*, London, UK, February 9–10, 2004.

26. Rinaki E, Valsami G, Macheras P. Quantitative Biopharmaceutics Classification System: the central role of dose/solubility ratio. *Pharm. Res.* 2003;20(12):1917–1925.

27. Smith PL, Wall DA, Gochoco CH, Wilson G. Routes of delivery: cases studies. (5) Oral absorption of peptides and proteins. *Adv. Drug Deliv. Rev.* 1992;8:253–290.

28. Hidalgo IJ. Assessing the absorption of new pharmaceuticals. *Curr. Top. Med. Chem.* 2001;1:385–401.

29. Corrigan OI. The biopharmaceutic drug classification and drugs administered in extended release (ER) formulations. *Adv. Exp. Med. Biol.* 1997;423:111–128.

30. Wilding IR. Evaluation of the Biopharmaceutics Classification System (BCS) to oral modified release (MR) formulations; what do we need to consider? *Eur. J. Pharm. Sci.* 1999;8:157–159.

31. Lennernas H. Human intestinal permeability. *J. Pharm. Sci.* 1998;87:403–410.

32. Lennernas H. Human jejunal effective permeability and its correlation with preclinical drug absorption models. *J. Pharm. Pharmacol.* 1997;49:627–638.

33. Walter E, Janich S, Roessler BJ, Hilfinger JM, Amidon GL. HT-29-MTX/Caco-2 cocultures as an *in vitro* model for the intestinal epithelium: *in vitro/in vivo* correlation with permeability data from rats and humans. *J. Pharm. Sci.* 1996;85 (10):1070–1076.

34. Yee S. *In vitro* permeability across Caco-2 cells (colonic) can predict *in vivo* (small intestinal) absorption in man—fact or myth. *Pharm. Res.* 1997;14(6):763–766.

35. Delie F, Rubas W. A human colonic cell line sharing similarities with enterocytes as a model to examine oral absorption: advantages and limitations of the Caco-2 model. *Crit. Rev. Ther. Drug Carrier Syst.* 1997;14:221–286.

36. Rubas W, Cromwell MEM, Shahrokh Z, Vilagran J, Nguyen T-N, Welton W, Nguyen T-H, Mrsny RJ. Flux measurements across Caco-2 monolayers may predict transport in human large intestinal tissue. *J. Pharm. Sci.* 1996;85(2):165–169.

37. Rubas W, Jezyk N, Grass GM. Comparisons of the permeability characteristics of a human colonic epithelial (Caco-2) cell line to colon of rabbit, monkey, and dog intestine and human drug absorption. *Pharm. Res.* 1993;10(1):113–118.

38. Ungell A-L, Nylander S, Bergstrand S, Sjöberg Å, Lennernas H. Membrane transport of drugs in different regions of the intestinal tract of the rat. *J. Pharm. Sci.* 1998;87(3):360–366.

39. Shah RB, Khan MA. Regional permeability of salmon calcitonin in isolated rat gastrointestinal tracts: transport mechanism using Caco-2 cell monolayer. *AAPS J.* 2004;6(4):1–5.

40. Stewart BH, Chan OH, Jezyk N, Fleisher D. Discrimination between drug candidates using models for evaluation of intestinal absorption. *Adv. Drug Deliv. Rev.* 1997;23:27–45.

41. Doluisio JT, Billups NF, Dittert LW, Sugita ET, Swintosky JV. Drug absorption I: an *in situ* rat gut technique yielding realistic absorption rates. *J. Pharm. Sci.* 1969;58(10):1196–1200.

42. Sinko PJ, Sutyak JP, Leesman GD, Hu P, Makhey VD, Yu H-S, Smith CL. Oral absorption of anti-AIDS nucleoside analogues: 3. Regional absorption and *in vivo* permeability of 2′,3′-dideoxyinosine in an intestinal-vascular access port (IVAP) dog model. *Biopharm. Drug Dispos.* 1997;18(8):697–710.

43. Sutton SC, Evans LA, Fortner JH, McCarthy JM, Sweeney K. Dog colonoscopy model for predicting human colon absorption. *Pharm. Res.* 2006;23(7):1554–1563.

44. Lennernas H, Ahrenstedt O, Hallgren R, Knutson L, Ryde M, Paalzow LK. Regional jejunal perfusion, a new *in vivo* approach to study oral drug absorption in man. *Pharm. Res.* 1992;9(10):1243–1251.

45. Knutson L, Odlind B, Hullgren R. A new technique for segmental jejunal perfusion in man. *Am. J. Gastroenterol.* 1989;84(10):1278–1284.

46. Gardner D, Casper R, Leith F, Wilding I. The InteliSite capsule: a new easy to use approach for assessing regional drug absorption from the gastrointestinal tract. *Pharm. Tech. Eur.* 1997;9:46–53.

47. Wilding I, Hirst P, Connor A. Development of a new engineering-based capsule for human drug absorption studies. *Pharm. Sci. Technol. Today* 2000;3(11):385–392.

48. Martin NE, Collison KR, Martin LL, Tardif S, Wilding I, Wray H, Barrett JS. Pharmacoscintigraphic assessment of the regional drug absorption of the dual angiotensin-converting enzyme/neutral endopeptidase inhibitor, M100240, in healthy volunteers. *J. Clin. Pharmacol.* 2003;43:520–538.

49. Hinderling PH, Karara AH, Tao B, Pawula M, Wilding I, Lu M. Systemic availability of the active metabolite hydroxy-fasudil after administration of fasudil to different sites of the human gastrointestinal tract. *J. Clin. Pharmacol.* 2007;47:19–25.

50. Wessel MD, Jurs PC, Tolan JW, Muskal SM. Prediction of human intestinal absorption of drug compounds from molecular structure. *J. Chem. Inform. Comput. Sci.* 1998;38(4):726–735.

51. Palm K, Stenberg P, Luthman K, Artursson P. Polar molecular surface properties predict the intestinal absorption of drugs in humans. *Pharm. Res.* 1997;14(5):568–571.

52. Oprea TI, Gottfries J. Toward minimalistic modeling of oral drug absorption. *J. Mol. Graph. Model.* 1999;17(5–6):261–274.

53. Gorbach SL. Intestinal microflora. *Gastroenterology* 1971;60:1110–1129.

54. Scheline RR. Metabolism of foreign compounds by gastrointestinal microorganisms. *Pharmacol. Rev.* 1973;25:451–523.

55. Ilett KF, Tee LBG, Reeves PT, Minchin RF. Metabolism of drugs and other xenobiotics in the gut lumen and wall. *Pharmacol. Ther.* 1990;46(1):67–93.

56. Shamat MA. The role of the gastrointestinal microflora in the metabolism of drugs. *Int. J. Pharm.* 1997;97(1–3):1–13.

57. Faigle JM. Drug metabolism in the colon wall and lumen. In Bieck PR, editor. *Colonic Drug Absorption and Metabolism*, Marcel Dekker, New York, 1993, p. 29.

58. Rubinstein A. Microbially controlled drug delivery to the colon. *Biopharm. Drug Dispos.* 1990;11:465–475.

59. Basit AW, Lacey LF. Colonic metabolism of ranitidine: implications for its delivery and absorption. *Int. J. Pharm.* 2001;227:157–165.

60. Silvester KR, Englyst HN, Cummings JH. Ileal recovery of starch form whole diets containing resistant starch measured

in vitro and fermentation of ileal effluent. *Am. J. Clin. Nutr.* 1995;62(2):403–411.

61. Siew LF, Basit AW, Newton JM. The properties of amylase-ethylcellulose films cast from organic-based solvents as potential coatings for colonic drug delivery. *Eur. J. Pharm. Sci.* 2000;11(2):133–139.

62. Wagner D, Spahn-Langguth H, Hanafy A, Koggel A, Langguth P. Intestinal drug efflux: formulation and food effects. *Adv. Drug Deliv. Rev.* 2001;50 (Suppl 1):S13–S31.

63. Zhou SY, Fleisher D, Pao LH, Li C, Winward B, Zimmermann EM. Intestinal metabolism and transport of 5-aminosalicylate. *Drug Metab. Dispos.* 1999;27(4):479–485.

64. Zhang Y, Benet LZ. The gut as a barrier to drug absorption: combined role of cytochrome P450 3A and P-glycoprotein. *Clin. Pharmacokinet.* 2001;40(3):159–168.

65. Zhang Y, Guo X, Lin ET, Benet LZ. Overlapping substrate specificities of cytochrome P450 3A and P-glycoprotein for a novel cysteine protease inhibitor. *Drug Metab. Dispos.* 1998; 26(4):360–366.

66. Paine MF, Khalighi M, Fisher JM, Shen DD, Kunze KL, Marsh CL, Perkins JD, Thummel KE. Characterization of interintestinal and intraintestinal variations in human CYP3A-dependent metabolism. *J. Pharmacol. Exp. Ther.* 1997;283(3):1552–1562.

67. Gupta SK, Sathyan G. Pharmacokinetics of an oral once-a-day controlled-release oxybutynin formulation compared with immediate-release oxybutynin. *J. Clin. Pharmacol.* 1999; 39(3):289–296.

68. Mouly S, Paine MF. P-glycoprotein increases from proximal to distal regions of human small intestine. *Pharm. Res.* 2003; 20(10):1595–1599.

69. Galia E, Nicolaides E, Hörter D, Löbenberg R, Reppas C, Dressman JB. Evaluation of various dissolution media for predicting *in vivo* performance of Class I and II drugs. *Pharm. Res.* 1998;15(5):698–705.

70. Kostewicz ES, Branus U, Becker R, Dressman JB. Forecasting the oral absorption behavior of poorly soluble weak bases using solubility and dissolution studies in biorelevant media. *Pharm. Res.* 2002;19(3):345–349.

71. Fotaki N, Symillides M, Reppas C. *In vitro* versus canine data for predicting input profiles of isosorbide-5-mononitrate from oral extended release products on a confidence interval basis. *Eur. J. Pharm. Sci.* 2005;24:115–122.

72. Klein S, Stein J, Dressman J. Site-specific delivery of anti-inflammatory drugs in the gastrointestinal tract: an *in-vitro* release model. *J. Pharm. Pharmacol.* 2005;57(6): 709–719.

73. Shameem M, Katori N, Aoyagi N, Kojima S. Oral solid controlled release dosage forms: role of GI-mechanical destructive forces and colonic release in drug absorption under fasted and fed conditions in humans. *Pharm. Res.* 1995; 12(7):1049–1054.

74. Kamba M, Seta Y, Takeda N, Hamaura T, Kusai A, Nakane H, Nishimura K. Measurement of agitation force in dissolution

test and mechanical destructive force in disintegration test. *Int. J. Pharm.* 2003;250(1):99–109.

75. Varma M, Kaushal AM, Garg S. Influence of micro-environmental pH on the get layer behavior and release of a basic drug from various hydrophilic matrices. *J. Control. Release* 2005;103:499–510.

76. Agoram B, Woltosz BWS, Bolger MB. Predicting the impact of physiological and biochemical processes on oral drug bioavailability. *Adv. Drug Deliv. Rev.* 2001;50:S41–S67.

77. Dashevsky A, Kolter K, Bodmeier R. pH-independent release of a basic drug from pellets coated with the extended release polymer dispersion Kollicoat® SR 30 D and the enteric polymer dispersion Kollicoat® MAE 30 DP. *Eur. J. Pharm. Biopharm.* 2004;58:45–49.

78. Riis T, Bauer-Brandl A, Wagner T, Kranz H. pH-independent drug release of an extremely poorly soluble weakly acidic drug from multiparticulate extended release formulations. *Eur. J. Pharm. Biopharm.* 2007;65:78–84.

79. Center for Drug Evaluation and Research (CDER). Guidance for Industry. Extended release oral dosage forms: development, evaluation, and application of *in-vitro/in-vivo* correlation, Food and Drug Administration, U.S. Department of Health and Human Services, Rockville, MD, 1997.

80. Corrigan OI, Devlin Y, Butler J. Influence of dissolution medium buffer composition on ketoprofen release from ER products and *in vitro–in vivo* correlation. *Int. J. Pharm.* 2003;254(2):147–154.

81. Huang Y-B, Tsai Y-H, Yang W-C, Chang J-S, Wu P-C, Takayama K. Once-daily propranolol extended-release tablet dosage form: formulation design and *in vitro/in vivo* investigation. *Eur. J. Pharm. Biopharm.* 2004;58(3): 607–614.

82. Qiu Y, Garren J, Samara E, Cao G, Abraham C, Cheskin HS, Engh KR. Once-a-day controlled-release dosage form of divalproex sodium II: development of a predictive *in vitro* drug release method. *J. Pharm. Sci.* 2003;92 (11):2317–2325.

83. Davis SS, Hardy JG, Taylor MJ, Stockwell A, Wilson CG. The *in vivo* evaluation of an osmotic device (Osmet) using gamma scintigraphy. *J. Pharm. Pharmacol.* 1984;36:740–742.

84. Zhou M, Notari RE. A nomogram to predict the best biological half-life values for candidates for oral prolonged-release formulations. *J. Pharm. Sci.* 1996;85(8):791–795.

85. Irvin JR, Notari RE. Computer-aided dosage form design. III. Feasibility assessment for an oral prolonged-release phenytoin product. *Pharm. Res.* 1991;8(2):232–237.

86. Chen Y, McCall TW, Baichwal AR, Meyer MC. The application of an artificial neural network and pharmacokinetic simulations in the design of controlled-release dosage forms. *J. Control. Release* 1999;59(1):33–41.

87. Cheung WK, Yacobi A, Silber BM. Pharmacokinetic approach to the rational design of controlled or sustained release formulations. *J. Control. Release* 1988;5(3):263–270.

88. Grass GM. Simulation models to predict oral drug absorption from *in vitro* data. *Adv. Drug Deliv. Rev.* 1997;23(1–3): 199–219.

89. Yu LX, Lipka E, Crison JR, Amidon GL. Transport approach to the biopharmaceutical design of oral drug delivery systems: prediction of intestinal absorption. *Adv. Drug Deliv. Rev.* 1996;19(3):359–376.

90. Khan MA. A quality by design perspective on dose dumping with alcohol. In *AAPS Workshop on the Role of Dissolution in QbD and Drug Product Life Cycle*, Arlington, VA, April 28–30, 2008.

91. Washington N, Washington C, Wilson CG. *Physiological Pharmaceutics: Barriers to Drug Absorption*, 2nd edition, Taylor & Francis, London, 2001, p. 84.

92. Washington N, Washington C, Wilson CG. *Physiological Pharmaceutics: Barriers to Drug Absorption*, 2nd edition, Taylor & Francis, London, 2001, pp. 92–93.

93. Wilkinson GR, Pharmacokinetics in disease states modifying body perfusion. In Benet LZ, editor. *The Effect of Disease States on Drug Pharmacokinetics*, American Pharmaceutical Association, Washington, DC, 1976, pp. 13–14.

94. Benet LZ, Greither A, Meister W. Gastrointestinal absorption of drugs in patients with cardiac failure. In Benet LZ, editor. *The Effect of Disease States on Drug Pharmacokinetics*, American Pharmaceutical Association, Washington, DC, 1976, pp. 33–48.

95. Washington N, Washington C, Wilson CG. *Physiological Pharmaceutics: Barriers to Drug Absorption*, 2nd edition, Taylor & Francis, London, 2001, p. 136.

96. Washington N, Washington C, Wilson CG. *Physiological Pharmaceutics: Barriers to Drug Absorption*, 2nd edition, Taylor & Francis, London, 2001, pp. 165–166.

97. Washington N, Washington C, Wilson CG. *Physiological Pharmaceutics: Barriers to Drug Absorption*, 2nd edition, Taylor & Francis, London, 2001, pp. 154–156.

4

PREFORMULATION CONSIDERATION FOR DRUGS IN ORAL CR FORMULATION

MANNCHING SHERRY KU

Pharmaceutical Development, Wyeth Research, Pearl River, NJ, USA

4.1 INTRODUCTION

Preformulation is nothing but the preparative work on the drug substance before formulating the drug product. A preformulation study tailored with the formulation in mind is especially helpful to achieve success in developing a stable and bioavailable formulation. Preformulation studies can be divided into four main categories: polymorphism/crystallinity, bulk characterization, solubility/dissolution/permeability, and stability/compatibility. This chapter is not intended to treat all areas of preformulation in an exhaustive manner; instead, it will concentrate on those properties impacting controlled release formulations, such as solubility, stability, excipient compatibility, and bulk powder mechanical properties. Polymorphism screening to select a thermodynamically stable form with a high degree of crystallinity and a low degree of hygroscopicity is well covered by other textbooks and will not be elaborated here. Neither covered is the deliberate selection of an insoluble salt form to effect a slow release and sustained blood levels. In addition, dissolution and permeability are covered in Chapter 3 Biopharmaceutics of Oral CR formulations and Chapter 15 Oral Targeted Drug Delivery Systems: Colon Delivery, respectively.

A successful formulation of a controlled release dosage form starts from the understanding of the physical, chemical, and biopharmaceutical properties of the drug substance. For instance, the selection of controlled release excipients, either polymer film or matrix former, to retard drug release may be based on the solubility of the drug substance. If drug solubility is highly pH dependent, a multiparticulate dosage form may be advantageous to maintain a constant release of

drug throughout the pH gradient of the gastrointestinal (GI) tract. Based on the chemical structure of the drug substance, excipients with possible incompatibility may be eliminated from candidacy at the early stage while only compatible excipients with nonreacting functional group prevail. From the viewpoint of processing, different methods of manufacturing can be prioritized based on the flow and compression properties of the drug substance. For high-dose drugs with poor flow and poor compressibility, wet granulation is a preferred process. On the other hand, dry granulation is more suitable for drug substances with poor flow but good compressibility. For water-sensitive drug substances, fluid bed granulation may be employed to limit the contact time with water.

In conclusion, sound preformulation studies lay a firm foundation for the success in formulation and prevent later formulation changes during clinical trials and the high cost for human pharmacokinetic studies to bridge from an old formulation to a new one. Furthermore, it is tightly knitted to what FDA and ICH advocate in quality by design where quality is built in by design and not tested into the product. It is an important element of risk management in the understanding and identification of critical raw material attributes.

4.2 PARTICULATE AND MECHANICAL PROPERTIES' CONSIDERATION FOR DRUG SUBSTANCES

The physical properties of drug substances have a large impact on the selection of controlled release formulation and manufacturing process. For low-dose drugs, content

Oral Controlled Release Formulation Design and Drug Delivery: Theory to Practice, Edited by Hong Wen and Kinam Park
Copyright © 2010 John Wiley & Sons, Inc.

uniformity is a critical quality attribute to both patient safety and efficacy. A tablet containing too much drug can elicit toxicity whereas that containing too little drug may not be efficacious. Content uniformity is particularly critical for narrow therapeutic drugs. To effect content uniformity, the drug particles have to be small enough to distribute evenly from one tablet to another tablet and mix well with the added excipients. A cohesive powder with higher surface energy tends to stick with itself and not to distribute among the excipients. Excipients such as silicon dioxide can effectively reduce the cohesiveness and help deaggregate and distribute the drug particles. Excipients such as microcrystalline cellulose can provide surfaces for attachment of the micronized drug particles and effect content uniformity. For high-dose drugs, a large percentage of the dosage form is the drug substance. Therefore, the flowability and compressibility of the drug substance are important. Increase in crystal size and a more rounded crystal shape can improve flowability. Particle size reduction often leads to poorer flow except when the crystal shape is modified from needle or plate to a more rounded geometry. Again, the addition of glidants such as silicon dioxide can help powder flow and consolidate better in encapsulation or compression processes. For insoluble drugs, the addition of pH modifiers or surfactants can improve wetting and dissolution. Finally, the determination of crystal habit during preformulation facilitates the selection and optimization of the final drug substance via crystal engineering or milling.

4.2.1 Particle Size Analysis

Oversized drug particles can cause a few overpotent dosage units with an overall low potency for the batch because these large particles are usually present in numbers too few to be found in every dose unit. This phenomenon has been observed experimentally and reported in the literature [1, 2]. Micronization to reduce particle size and, therefore, to improve content uniformity is routinely done. However, micronized material tends to agglomerate and a deagglomeration step would often be necessary to ensure that the drug is distributed evenly in the drug product. Prescreening can be an effective method of deagglomeration. However, for some drug substances with high surface energy, the particles can reagglomerate upon blending or after storage. Addition of excipients such as silicon dioxide can be beneficial to reduce the surface energy and prevent agglomeration during blending/conveying or upon storage. Setting a particle size specification that is small enough to effect content uniformity but large enough to avoid cohesive agglomeration would be most advantageous [3]. A computer program (INTELLIPHARM® CU software) is commercially available to simulate the content uniformity based on the assumption of ideal mixing and dispersion without segregation or agglomeration. The simulation reveals the minimum target dose or the maximum median size requirement under a given standard deviation to effect a uniform blend under ideal conditions. Modifications of the simulative algorithm were successfully made in the author's laboratory to accommodate particle size distribution deviating from log-normal to multimodal or in an arbitrary shape. In short, the simulation yields the best possible content uniformity results and the specified particle size distribution under ideal conditions.

Laser light scattering has become the industry norm for sizing the drug substance particle distribution in the past 20 years. However, the light scattering data can only be interpreted correctly when combined with microscopic crystal size and shape data. In addition, sieve analysis remains a useful method for rounded particles greater than $100\,\mu m$ in size, which are often too heavy to suspend sufficiently for light scattering analysis. These three methods for particle size distribution determination intended to be used in combination are described below.

4.2.1.1 *Microscopic Image Examination* Microscopic image examination may be the most direct way to determine particle size. However, it can be difficult for one to describe and communicate effectively the microscopic results because the 2D image of particles with varied sizes and shapes creates the conundrum of objectivity and quantification. Although a picture is worth a thousand words, establishment of a quantitative methodology in reporting microscopic particle size data is necessary. Computerized image analysis is a good way to translate microscopic data into quantifiable parameters. ImagePro® software is commercially available utilizing a standard algorithm to calculate the mean particle size and size distribution. Image analysis software usually measures the surface area of the 2D image. In addition, measurement of anisotropic particle lengths allows the estimation of aspect ratio.

Aspect ratio is a second parameter that can be obtained from microscopic examination. It is the ratio of the maximum length to the minimum width of a particle, which reflects the degree of symmetry (sphericity), an important determinant for flowability. The aspect ratio is 1 for a perfect sphere and 1.6 for a cube lying down flat. For needles, the aspect ratio could vary widely from 1 to several hundreds, when using milling to reduce the aspect ratio and improve the powder flowability is recommended.

Human attention must be paid when particles form agglomerates or overlap in the images. Computer software such as ImagePro® will count agglomerates as single particles since it cannot distinguish agglomerates from single crystals. Therefore, it is very important to make sure that the sample is well dispersed. A common practice shown in the example is to use 1% lecithin in mineral oil as the dispersant because most drugs have little solubility in mineral oil and 1% lecithin surfactant is usually sufficient to effect a fine dispersion.

A standard output from ImagePro® for an example compound with a needle crystal habit was milled using a comill as shown in Table 4.1. Out of the 2283 particles measured, 90% have aspect ratio below 6.11 with an average of 3.46 and standard deviation of 1.69. This outcome is typical when using grinding, hammer, or chopper mills. Use of a jet mill will be necessary to further reduce the aspect ratio. Jet mills, however, also micronize the material to a much finer size and result in poor flow. Balancing the views from both sides, a specification of aspect ratio not more than 3 has been successfully employed by the author for nearly 10 years.

4.2.1.2 *Laser Light Scattering Method*

Laser light scattering has become the industry norm for sizing the drug substance particle distribution in the past 20 years. It is preferred over the microscopic method because it measures the volume distribution over a large ensemble of particles directly. The microscopic method typically measures number distribution and tends to give smaller number mean size than the volume mean size because many small particles are needed to constitute the volume of a large particle. Volume distribution is equal to weight distribution if particle true density is constant, an assumption that is generally valid for drug substances. Weight distribution is important for content uniformity, which is nothing but the distribution of the API mass over the dosage units. The content uniformity software described at the beginning of this section uses the particle size distribution by weight.

The laser light scattering instrument consists of a laser source (typically He–Ne laser at $\lambda = 0.63 \, \mu m$), a set of photosensing detectors (16–32), and a sample chamber with a recirculating pump. Modern instruments use the Mie theory, which allows determination of a large particle size range of 0.1–1000 µm. It requires entry of refractive indices for the material and medium, which can be estimated or found in the literature. A typical data output from Malvern® Matersizer is shown in Table 4.2.

The light scattering method generates a volume distribution from the energy of diffraction, which is directly dependent on the particle size. These data are then converted into volume equivalent spherical diameters (volume of sphere = $(4/3)\pi r^3$). It differs from microscopy size because of the lack of data in particle shapes. For instance, for a rod-shaped crystal with a diameter of 2 µm and a length of 20 µm (aspect ratio of 10), the volume equivalent spherical diameter is only 5 µm. One must not take this derived particle size and surface area literally. The data can only be interpreted correctly when combined with complementary microscopy data for batch comparison.

Similar to the microscopic image method, it is advisable to prepare a suspension with a larger powder sample and then the suspension is pipetted into the Malvern reservoir in several replicate fashion. The pitfall of sample nonhomogeneity applies equally to both microscopic and light scattering methods. Because particles greater than 100 µm tend to settle in a suspension, particular care needs to be exercised to ensure representative sampling.

4.2.1.3 *Sieve Analysis*

Sieve analysis remains a useful method for rounded particles greater than 100 µm in size. For rod-shaped materials, sieve analysis data could be misleading since the rods may or may not pass through the sieves depending on their orientation at the moment hitting the sieve. Since small size particles tend to aggregate and agglomerate, the sieve analysis data appear larger than the actual due to the contribution of aggregates other than the primary particles. Although various methods including sonic and mechanical shaking are used, deagglomeration of small particles remains a problem for sieve analysis. A typical result is presented in Table 4.3.

4.2.2 Bulk Powder Properties

Bulk powder systems exhibit the most complex phenomenon encountered in pharmaceutical manufacturing. It is because no two particles are identical in a powder bed containing millions of particles that undergoes numerous unique arrangements during processing. The surface characteristic of a particle depends on not only the physiochemical properties of the molecule but also the spatial arrangement and orientation of the molecules of the particle. Pharmaceutical manufacturing processes such as blending, conveying, encapsulation, and tableting require movement of all ingredient particles against each other. Several commonly used tests are described here in the evaluation of bulk powder properties

TABLE 4.1 Microscopic Particle Size Analysis Results

Sample description	Compound X
Lot number	123
Date	11-MAR-2008
Analyst	A. Smith
Dispersing vehicle	1% Lecithin solution
Sample concentration (mg/mL)	5.62
Particle shape	Needle
Total fields acquired	6
Fields at 100×	2
Fields at 200×	2
Fields at 400×	2
Number of particles counted	2283

Results	Particle Size (µm)	Aspect Ratio
Average	23.18	3.46
Standard Deviation	29.94	2.11
10%	1.10	1.69
90%	58.67	6.11
Minimum	0.55	1.00
Maximum	337.11	19.47

TABLE 4.2 Malvern Particle Size Analysis Results

Result Analysis Report			
Sample name:	SOP name:	Measured:	
Sample source and type:	Measured by:	Analyzed:	
Sample bulk lot ref:	Result source: Averaged		
Particle name:	Accessory name: Hydro 2000S (A)	Analysis model: General purpose	Sensitivity: Normal
Particle RI: 1.450	Absorption: 0	Size range:: 0.020 to 2000.000 μm	Obscuration: 19.21%
Dispersant name: Water	Dispersant RI: 1.330	Weighted residual: 2.824 %	Result emulation: Off

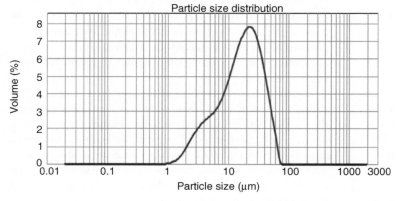

4.2.2.1 Powder Density

Carr's index is a simple test to evaluate flowability by comparing the pour density with tapped density using the following equation:

$$\text{Carr's index} = 100\% \times (\text{tapped density} - \text{pour density})/\text{tapped density}$$

Tapped density testers designed to meet or exceed current USP 31 <616> specifications are commercially available. These testers offer a standardized and repeatable way to measure the tapped volumes of powders. Some testers have simultaneous tapping and rotary motion to ensure that the material being tested is packed evenly throughout the graduated cylinder and that its surface is leveled for direct reading. A simplified equation may be used by canceling the weight of the powder sample as follows:

$$\text{Carr's index} = 100\% \times (\text{pour volume} - \text{tap volume})/\text{pour volume}$$

The Carr's index is referred to as compressibility index in the USP. The word "compressibility" is a misnomer since no compression force is applied and the powder bed is simply consolidated through repacking of particles under a tapping motion. An alternative index is the Hausner ratio, which is the tap density divided by the bulk density.

It may be helpful to consider some extreme cases in the interpretation of the Carr's index. For uniform smooth glass beads with a perfect flow, the Carr's index is nearly zero. On the other hand, the Carr's index may also be nearly zero for cotton balls with very poor flow that does not consolidate. Therefore, the Carr's index should be considered together with other tests such as flow meter. The pharmaceutical industrial empirical guide generally regards Carr's index greater than 33% as poor powder flow and less than 20% as good powder flow.

TABLE 4.3 Sieve Analysis Results

Sonic sifter: particle size distribution				Date: 5/1/2008	
Product name: ABC granulation				Sift: 5	
Batch no.: 123				Pulse: 5	
Sample weight: 30 g				Time: 5	

Sieve #	Microns	TARE (g)	Gross (g)	Net (g)	% On Screen	% Thru
16	1180	400.9	400.9	0	0.000	100.000
20	850	372.8	373	0.2	0.667	99.333
45	355	332.8	339	6.2	20.667	78.667
80	180	308.9	310.6	1.7	5.667	73.000
100	150	309	309.7	0.7	2.333	70.667
200	75	295.9	303.1	7.2	24.000	46.667
325	45	297.8	305.6	7.8	26.000	20.667
pan	<45	281.6	287.7	6.1	20.333	0.333
	Total:	2599.7	2629.6	29.9	99.667	

4.2.2.2 Angle of Repose

Stresses such as agitation or air force transmitted through a powder bed and the bed's response to the applied stress are reflected in various angles of friction and repose. Pouring a fixed quantity of powder onto a flat surface in a conical heap and measuring the inclined angle with the horizontal surface typically determines the angle of repose. Let us once again consider the extreme cases of glass beads and cotton balls. The glass beads with perfect flow may result in an angle of repose near 0° whereas the cotton balls will just pile up yielding values in excess of 90°. The pharmaceutical industrial empirical guide generally regards angle of repose greater than 40% as poor powder flow and less than 25% as good powder flow.

4.2.2.3 Flow Meter

Measurement of the rate of material flow seems a direct way to evaluate flowability. The European Pharmacopoeia (EP) has a monograph on flowability test. However, neither the USP nor the JP contains any method for measuring powder flow properties. It is probably because most drug substances yield no flow under EP testing conditions. The EP method involves pouring of 100 g of material into a vertically situated glass funnel. The bottom of the funnel is initially blocked and upon unblocking the time for the powder to flow out of the funnel is measured in seconds. Only highly flowable materials result in measurable flow rates. Excipients such as Starch 1500 and Avicel 102 result in no flow. As a result, commercial flow testers are equipped with various modes of tapping and vibration in an attempt to induce flow. An example tester Sotax Flow Tester FT-300 is shown in Figure 4.1. The powder samples can be preconditioned through the rotation of the upper pan to remove the memory from storage and transportation. Additional previbration of the upper pan can be used to deaggregate the sample. During the testing, the upper funnel has optional vibration modes to induce the powder flow. The gravimetric flow rate (time in seconds for 100 g of sample to flow out of the funnel) is measured. Schussele and Bauer-Brandl [4] have shown that the volumetric flow rate (using 100 mL instead of 100 g of the sample) is more reasonable in assessing samples of diverse powder densities. A change in the EP method from gravimetric to volumetric flow rate is recommended. The other output from the Sotax Tester is the ratio of the average angle α of the arctangent of the flow rate of the powder to that of the reference, which is sand quartz in most cases. The flowability is considered very good when $\alpha/\alpha_{ref} > 0.9$, good 0.8–0.9, satisfactory 0.7–0.8, medium 0.6–0.7, unsatisfactory 0.5–0.6, poor 0.4–0.5, and very poor <0.4.

4.2.2.4 Shear Cell Tester

Shear cell tester has gained popularity in recent years because experimentally it offers more control over the shear force and the shear rate as well as the degree of consolidation. The shear cell testers have three basic designs: translational, also known as a Jenike shear cell, in which a cell is pushed in a linear direction to shear the material against itself in a linear manner; a rotational "plate" cell, also known as Peschel shear cell; and the rotational annular cell, also known as the Schulze ring shear cell. A picture of a ring shear cell [5] is presented in Figure 4.2. A powder sample is loaded between the ring lid and

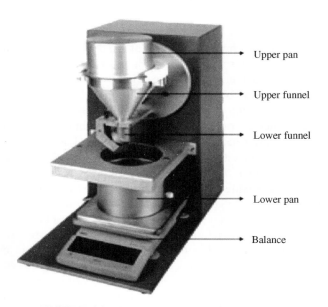

FIGURE 4.1 Sotax Powder Flow Tester FT-300.

Upper pan

Upper funnel

Lower funnel

Lower pan

Balance

FIGURE 4.2 Ring shear tester.

bottom. The lid is subjected to a vertically acting force (normal stress). The powder sample is subjected to a shear deformation by rotating the bottom ring of the shear cell whereby the lid is prevented from rotation. Both the bottom of the shear cell and the underside of the lid have vanes protruding into the powder. The unconfined yield stress f_c is a measure of stress or force necessary to cause a material unsupported in two directions to flow in the direction of the shear force. This is similar to the case when an arch collapses within the powder or at the hopper opening. The ratio of the consolidation stress σ_1 to the unconfined yield strength f_c is called the flowability ff_c:

$$ff_c = \sigma_1/f_c$$

This value usually increases with the consolidation stress σ_1; that is, the flowability depends on the stress level. Using flowability ff_c, the flow behavior of bulk solids can be compared. In this case, the bulk solid samples are tested at the same level of consolidation stress σ_1. In general, a higher flowability ff_c value is attributed to better flow. The individual powder can be further graded to a general description of flowability [6] as

Nonflowing	$ff_c < 1$
Nonflowing to bad flowing	$1 < ff_c < 2$
Cohesive	$2 < ff_c < 4$
Good flowing	$4 < ff_c < 10$
Free flowing	$10 < ff_c$

4.2.3 Compression Properties

A majority of the controlled release dosage forms are tablets although multiparticulate systems in capsules have gained some popularity in recent years due to more reproducible absorption profile since multiparticulates are dispersed in the GI tract. A typical controlled release tablet contains active, matrix former (polymers), and fillers. The choice of fillers may depend on the compression characteristic of the drug substance especially when the active exceeds 20% w/w. Characterization of the compression properties of the drug substance allows custom design of a tablet formulation by selecting excipients of complementary properties. For instance, if the drug substance is plastic, that is, capable of permanent deformation, excipients with brittle fracture such as lactose and calcium phosphate may be selected. Conversely, if the drug substance is brittle, plastic excipients such as microcrystalline cellulose and starch may be selected. Finally, if the drug substance is elastic, sustained release strategy other than polymer matrix tablet may be considered unless wet granulation with plastic material induces plasticity.

Instrumented press measuring force versus displacement is commonly used to differentiate plastic from elastic materials. However, it is more challenging to differentiate plastic

materials from those with brittle fractures. A simple method has been reported by J. I. Wells [7]. The drug substance is blended with 1% magnesium stearate for either 5 or 30 min. The blend is compressed in a Carver press at 1 ton for either 2 or 30 s. These compacts are held for 24 h before testing for tensile strength. Determination of plastic or brittle fracture is made according to the specific order of tensile strengths among the various compacts. The scheme is summarized below.

Procedure: Drug blended with 1% magnesium stearate. Compressed at 1 ton using Carver press. Hold compact for 24 h before measuring tensile strength

Sample ID	A	B	C
Blend time	5 min	5 min	30 min
Dwell time	2 s	30 s	2 s

Outcome: Plastic if $B > A > C$ Brittle fracture if $B = A = C$

When the material is plastic, it deforms by changing shape. Since there is no fracture, no new surfaces are generated during compression and a more intimate mix of magnesium stearate (sample C) leads to poorer bonding. Plastic material bonds by viscoelastic deformation, which is time dependent. An increase in the dwell time (sample B) increases bonding and therefore tensile strength of the compact. Therefore, the order of tensile strength is $B > A > C$ for a plastic material. Alternatively, if a material bonds predominantly by brittle fracture, neither the lubricant distribution nor the duration of force application affects the compact tensile strength, that is, $B = A = C$.

4.3 STABILITY AND COMPATIBILITY

A drug substance is usually more stable by itself than in a formulation with excipients, and as the drug concentration decreases, the stability deteriorates in a corresponding manner. Understanding chemical liability of the compound itself followed by consideration of interaction with the functional group of the excipient provides the basis for quality design of any formulation including controlled release dosage forms. To understand the liability of a compound, forced degradation is conducted in both solution and solid states. The two sets of data may reveal different degradation patterns providing further a way forward in formulation stabilization. The solid-state stability depends on crystal packing, and differences in crystal packing can result in certain functional groups exposed on the surface of the crystals. These functional groups are then more liable to react with environmental elements such as oxygen and moisture resulting in degradation. The surface defects created from milling or due to inclusion of impurities can become "hot spots" that have a high degree of reactivity

TABLE 4.4 Impurity Analysis Methodology and Detection Sensitivity

	Typical Detection Limit (%)
UV	1
HPLC	0.01–0.1
GC	0.01–0.1
HP-TLC	0.01–0.1
MS	0.1–1
DSC	5
XRD	5–10
IR/Raman	5–10
NMR	5–10

resulting in faster degradation especially in the presence of excipients.

The approach described in this section is in line with ICH Q1A that states "Stress testing of the drug substance can help identify the likely degradation products, which can in turn help establish the degradation pathways and the intrinsic stability of the molecule and validate the stability indicating power of the analytical procedures used. The nature of the stress testing will depend on the individual drug substance and the type of drug product involved." Therefore, the data generated are suitable for regulatory filing. In fact, the FDA Office of Generic Drugs has been asking for the stress testing and excipient compatibility data via question-based review (QbR) since a generic product is approved based on short-term (3 months) accelerated storage data. Demonstration of drug substance stability and excipient compatibility under stressed conditions lends additional credence to long-term storage stability of the drug product.

A large variety of analytical methodologies have been employed for preformulation stability testing including spectroscopy, chromatography, and thermal analysis. Table 4.4 summarizes the sensitivity of various analytical methods for impurity/degradant detection. These methodologies are described in detail in other textbooks [8]. In this chapter, emphasis is placed on the use of HPLC and LC–MS in the quantification and identification of degradants and degradation pathways.

4.3.1 Solution-State Forced Degradation Study

A forced degradation study is necessary to demonstrate that the HPLC method is stability indicating before any drug stability study can be conducted. Therefore, forced degradation is a part of HPLC method development that is usually the first step of preformulation studies. Forced degradation protocols encompass a comprehensive assessment of degradation liability under various stress conditions including acid, base, heat, light, and oxidative conditions. Once degradation occurs, the degradants are identified in order to elucidate degradation mechanism, which in turn facilitates selection of excipients and packaging materials suitable for long-term storage stability. Forced degradation in the solution and solid states will be discussed in this and the next sections, respectively.

The experimental conditions for solution forced degradation studies are summarized in Table 4.5. These conditions comply with ICH Q1A statement "Stress testing is likely to be carried out on a single batch of the drug substance. It should include the effect of temperatures (in 10°C increments (e.g., 50°C, 60°C, etc.) above that for accelerated testing), humidity (e.g., 75% RH or greater) where appropriate, oxidation, and photolysis on the drug substance. The testing should also evaluate the susceptibility of the drug substance to hydrolysis across a wide range of pH values when in solution or suspension." The two accelerated light conditions in Table 4.5 comply with ICH Q1B option 2 simulating a 2-year exposure under ambient light conditions.

The solution forced degradation study starts with dissolving the drug substance in a suitable solvent. For insoluble compounds, addition of an organic solvent may be necessary to dissolve the compound. Acetonitrile is preferred due to its inertness toward chemical reaction and its compatibility with HPLC systems. Methanol and ethanol may be used for lipophilic compounds if acetonitrile fails to solubilize

TABLE 4.5 Solution-State Forced Degradation Conditions

Study Type	Medium[a]	Temperature/Condition	Duration
Acid	0.1 N HCl	50, 60, 70, and up to 121°C	Daily up to 1 week
Base	0.1 N NaOH	50, 60, 70, and up to 121°C	Daily up to 1 week
Heat	Water	50, 60, 70, and up to 121°C	Daily up to 1 week
UV light	Water	1.2 million lux hours at 25°C	22 h and 15 min[b]
Fluorescence light	Water	1000 Wh/m^2 at 25°C	10 days and 8 h[b]
Oxidation 1	3–30% H_2O_2	25°C	Daily up to 1 week
Oxidation 2	Water	Pressurized O_2 with free radical initiator AIBN[c] at 25°C	Daily up to 1 week

[a] Acetonitrile is used up to 50% to solubilize drug if necessary. It may be replaced with alcohols or other suitable solvent to improve solubility.
[b] Minimum time required to meet ICH option 2 criteria.
[c] AIBN stands for 2,2'azobisisobutyrenitrile, a free radical initiator.

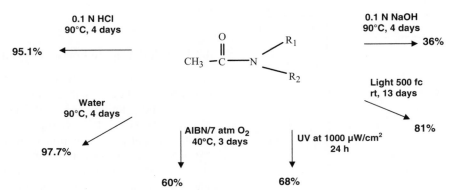

FIGURE 4.3 Solution-state forced degradation scheme.

the compound. Ester formation is possible when an alcohol is used as part of the solvent. The range of drug concentrations varies from 0.1 to 0.5 mg/mL depending on HPLC method sensitivity and the percentage of organic solvent necessary for drug dissolution. In general, the percentage of organic solvent is minimized to not more than 50% so that hydrolysis or water-mediated reactions do not slow down too much. The solution samples other than the oxygen chamber study are filled into ampules and sealed with air headspace.

There are two oxidative conditions: (1) hydrogen peroxide and (2) pressurized oxygen (100 psi). These two conditions sometimes give different degradation products and need to be employed for compounds with oxidation as the major degradation pathway. The oxygen pressure chamber studies are compared in the absence and presence of 2,2'-azobisisobutyronitrile (AIBN), a free radical initiator [9]. Samples are collected with time for the HPLC assay. For acid/base labile compounds, degradation is quenched by pH neutralization. A typical solution-state forced degradation scheme for a model acetamide compound is presented in Figure 4.3.

4.3.2 Solid-State Forced Degradation Study

Degradation of a solid drug particle is very different from that of the drug molecules in a solution. The degradation process of a solid drug particle can be described in four steps: (1) Surface molecules detach from the particles starting from the crystal defect or nucleation sites. (2) The loosened molecule reacts with neighboring molecules such as excipient or moisture/oxygen in the environment to form degradation products. (3) Or the loosened molecule dissolves in the surrounding water or solvent to form saturated solution. (4) The dissolved molecule reacts with the solvent, dissolved oxygen, or excipient to form the degradation product. The mechanism of degradation can be discerned by comparing the data between the solution and solid states.

Because degradation may occur as solid–gas, solid–solid, or in the saturated solution layer surrounding the solid particles, the solution-state degradation may or may not predict the solid-state stability. For water-soluble or hygroscopic compounds, the dissolved layer could be substantial leading to significant solution-state degradation, whereas dimer/adduct formation through nucleophilic attack seen at solid state may be prevented in solution since the nucleophile is stabilized by surrounding solvent molecules. Understanding of the solid-state degradation mechanism enables intelligent selection of excipients and packaging material, which in turn reduces unwanted experiments and expedites the development process.

The experimental conditions for solid-state forced degradation studies are summarized in Table 4.6. The effect of heat is discerned by comparing the closed bottle data at 40 and 56°C; the effect of moisture is discerned by comparing the closed and opened bottles at 40°C; and the effect of environmental oxygen or solvent vapor is discerned by comparing the closed and opened bottles at 56°C.

Oxidation is the most common degradation pathway for solid dosage forms. However, there is lack of standard forced

TABLE 4.6 Solid-State Forced Degradation Conditions

Study Type	Temperature/Condition	Duration
Dry heat	56°C open and closed bottles	Weekly up to 1 month
Moist heat	40°C/75% RH open and closed bottles	Weekly up to 3 month
UV light	1.2 million lux hours at 25°C, open dish	22 h and 15 min[a]
Fluorescence light	1000 Wh/m² at 25°C, open dish	10 days and 8 h[a]
Oxidation	Pressurized O₂ with free radical initiator AIBN at 25°C, open dish	Daily up to 1 week

[a] Minimum time required to meet ICH option 2 criteria.

TABLE 4.7   Comparison of Three Forced Oxidative Conditions

Model Compound	Solid API Under O₂ Pressure	API in Solution with H₂O₂	API in Solution with AIBN Under O₂ Pressure	API in Solution with KMnO₄
Cyclic secondary amine	No degradation	MW + 16, N-oxide	MW + 16, N-oxide	MW -2, epoxide
Linear secondary amine	MW + 28, carbamide	MW −220, −181, fragmentation	MW + 26, + 28, carbamide	MW + 28, + 14, −29, −59, carbamide, ketone, carboxylic acid
Aliphatic carboxylic acid	Not tested	MW + 16, lactone	MW + 16, lactone	MW + 16, + 28, lactone

oxidative conditions recommended by either regulatory authorities or industrial precedence. As a result, the author has embarked on a study to compare the oxidation products generated by hydrogen peroxide, molecular oxygen, and permanganate for three structurally diverse model compounds. The results are summarized in Table 4.7. The solution-state degradation in the oxygen pressure chamber study (molecular oxygen) best predicted the solid-state degradation. The author has since standardized early compound screening methodology using the oxygen pressure chamber.

4.3.3   pH–Stability Profile and Stabilization by Microenvironmental pH Modulation

Modulation of microenvironmental pH is an important strategy to stabilize drugs in formulations. It is especially useful for controlled release dosage forms since the drug is by design retained in a matrix or within coating membranes together with a pH modifier. There are three steps in the utilization of the microenvironmental pH strategy. The first step is to find the optimum pH from the solution pH–stability profile. The second step is to find a buffer solution compatible with the active. Finally, the selected buffer is mixed with the

active in the solid state to discern any stabilizing effect. Most buffer species can catalyze general acid–base reactions leading to accelerated drug degradation. Therefore, selection of a compatible buffer via a screening study should accompany the generation of pH–stability profile as illustrated in the following example. The detailed methodology for determining the pH–stability profile can be found in the literature [10] and will not be described here.

The pH–stability profile for a diacetate salt of a macrolide was generated using 50 mM phosphate buffer at a pH range of 4.5–8 at 40°C. Stability in water was also studied in parallel. In water, the natural pH of the diacetate salt is 5.5. Degradation followed pseudo-first-order kinetics as seen in the linearity shown in the semilogarithmic plot (Figure 4.4). The degradation rate constant (k) is plotted versus pH in Figure 4.5. It is estimated from the pH–rate profile that the rate constant is 0.178 at pH 5.5 in phosphate buffer, which is 20-fold higher than the rate constant observed in water. The results indicate that the presence of phosphate buffer species accelerates degradation whereas the presence of acetate counterions has no detrimental effect. The acetate salt has a self-buffering effect at its natural pH of 5.5. A formulation without a buffering agent is selected for further development.

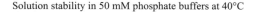

Solution stability in 50 mM phosphate buffers at 40°C

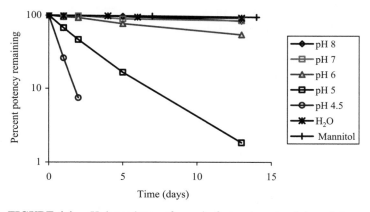

FIGURE 4.4   pH dependency of pseudo-first-order rate of degradation.

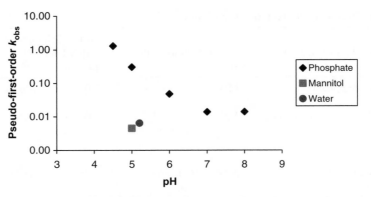

FIGURE 4.5 pH–rate profile at 40°C.

4.3.4 Degradant Identification

Forced degradation studies can not only provide the assurance that the analytical method is stability indicating but also generate degradants for identification of degradation pathways. Once the degradant is identified, the degradation mechanism can be discerned and the method of stabilization may be devised through the addition of stabilizer or the control of headspace and added moisture/oxygen protection from the barrier design of the packaging materials. LC–MS has gained popularity and became more affordable with lower prices in recent years. Any scientist working on forced degradation should have access to LC–MS to provide preliminary finding of degradant molecular weight. LC–NMR, LC–MS–MS, and accurate mass determinations can then follow in order to positively identify the chemical structure of the degradant. Isolation of the degradant using preparative LC followed by lyophilization is sometimes necessary to elucidate the structure while providing reference standard materials for future degradant analysis. These analytical techniques and instrumentation for the identification and isolation of degradants will not be described here but can be found in detail in pharmaceutical analysis textbooks.

4.3.5 Excipient Compatibility

Excipient compatibility gained importance when FDA started to ask for such data via QbR for generic drugs. It is because the ANDA only requires 1- and 3-month accelerated stability data upon filing and approval, respectively. A sound excipient compatibility study can ensure a longer term storage stability post ANDA approval.

Despite the importance of drug–excipient interaction, no standard method is generally accepted among pharmaceutical scientists and most methods reported in the literature have poor predictive values on final formulation stability [11]. There are a multitude of reasons for the poor predictability including poor sample uniformity, poor drug recovery, unrealistic active to excipient ratio, tertiary interaction with a second excipient, and so on. The results are often too varied to

be conclusive. It, at times, leads to unnecessary exclusion of key excipients such as magnesium stearate. Occasionally, a surprise is encountered during formal stability study where the excipient is no longer compatible because a second excipient changed the microenvironmental pH or plasticized the drug into a more amorphous nature with less stability.

A decision tree based on the inherent stability of the compound and in consideration of shortened development cycle is advocated herein and shown in Figure 4.6. It capitalizes on known incompatibility in the literature between functional groups of active and excipient and the inherent stability/reactivity of the drug molecule. The decision tree calls for upfront excipient compatibility study only for compounds with poor stability/high reactivity. For stable compounds, prototype formulation stability should start right away. In cases of "surprise" instability encountered, identification of the degradant should infer the "culprit" excipient that is then excluded from the next round of prototype stability. Alternatively, a stabilizer such as an antioxidant may be added to stabilize the formulation without removing essential excipients. This decision tree-based protocol is

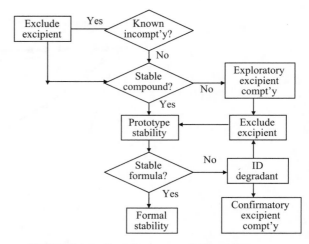

FIGURE 4.6 Excipient compatibility decision tree.

resource sparing and manages very well the risk without upfront excipient compatibility data.

An example protocol for excipient compatibility study composed of 16 excipients is shown in Table 4.8. The basic study design calls for a binary mixture weighed into an open bottle exposed to 40°C and 75% relative humidity for 2 and 4 weeks. A set of samples is assayed at time zero to obtain the baseline chromatography and to assess drug recovery. A second set of samples stored at 5°C in closed bottles is saved for troubleshooting. Two more sets of samples are stored for 2 and 4 weeks before HPLC analysis. This protocol comes to a total of 68 samples ($=17 \times 4$), each of which requires two weights. A larger study including a second storage condition may be desirable sometimes, which would incur greater labor and instrument costs.

Ideally, the control sample with active alone is not to incur any degradation, so appearance of a degradant in the binary mixture indicates incompatibility. A shorter term exposure to a less stressful condition may be considered for less stable

compounds, and vice versa. Various active to excipient ratios are employed to simulate the real-life use of excipients according to their functions. The sample composition is based on the basic product design, that is, 250 mg active in 500 mg formulation in a capsule. Each sample bottle contains a fixed amount of active (250 mg) while the quantity of excipients varies from 12.5 mg for lubricants to 250 mg for diluents per bottle. The mixture is gently mixed in a vortex before placing in the oven. The samples are removed periodically for HPLC analysis. Percent potency and impurity profile are reported by relative retention times.

It is a common practice to add 5–20% water into the excipient compatibility samples since moisture is involved in a lot of drug degradation pathways. However, it alters the nature of the solid sample to contain a saturated solution of the active and thus favors the solution-state degradation mechanism. A literature example [12] used 20% water at 50°C that led to excessive degradation and no excipients or excipient combinations were found compatible as acknowledged by the

TABLE 4.8   Protocol for Excipient Compatibility Study

Compound	Acidic, BCS Class 2, nonhygroscopic
Dosage form	250 mg capsule strength with 500 mg fill weight
Procedure	1. Weigh the active accurately into sample vials. Record the weight below
	2. Weigh the excipients with ±10% of target weight into the sample vials containing active
	3. Cap the sample vials and mix in a vortex with care not to dust the cap
	4. Uncap the vial and store the sample in 40°C/75% RH oven
	5. Store a set of control samples in a refrigerator
	6. Remove the sample from the oven at predetermined time and cap the vial
	7. Extract the active in a diluting solvent and assay by HPLC
	8. Enter the potency and impurity data in an excel sheet by relative retention time (RRT)

Sample Sheet

Excipient Name	Excipient/Active Ratio	Excipient (mg)	Active (mg)	Assay for Percent Potency and Impurity Profile			
				40°C/75% RH Open Bottle			5°C
				Initial	2 Weeks	4 Weeks	Control
1. Active	1 : 0	0	250				
2. Dicalcium phosphate, anhydrous	1 : 1	250	250				
3. Starch 1500 (pregelatinized starch)	1 : 1	250	250				
4. Lactose	1 : 1	250	250				
5. Mannitol	1 : 1	250	250				
6. Avicel PH101 (microcrystalline cellulose)	1 : 5	50	250				
7. Na starch glycolate	1 : 10	50	250				
8. Croscarmellose sodium	1 : 10	50	250				
9. Crospovidone	1 : 10	50	250				
10. Polyvinylpyrrolidone	1 : 20	12.5	250				
11. Magnesium stearate	1 : 20	12.5	250				
12. Stearic acid	1 : 20	12.5	250				
13. Lubritab (hydrogenated vegetable oil)	1 : 20	12.5	250				
14. Compritol (glyceryl behenate)	1 : 20	12.5	250				
15. Pruv (sodium stearyl fumarate)	1 : 20	12.5	250				
16. Talc	1 : 20	12.5	250				
17. Silicon dioxide	1 : 50	5	250				

authors "Since no suitable excipient was found, additional studies are needed." Overly stressful conditions taking no risk may work for stable compounds but can leave the formulator no usable excipients to formulate for a relatively unstable compound. A balanced approach based on inherent stability of the compound can benefit the formulation development while minimizing the stability risk. Our experience over 20 years shows the use of open bottle exposed to 75% relative humidity an effective and more plausible way to simulate the solid-state degradation in the final dosage form.

Excipient compatibility beyond binary mixture could lead to a very large number of samples and is not advisable at the exploratory phase when no information is available on individual excipient compatibility. If one excipient is found not compatible, all tertiary mixtures containing that excipient would not be stable and the resources in the preparation and analysis of those tertiary mixtures can be better utilized. Confirmatory studies with more than one excipient may be beneficial when excipient compatibility is demonstrated using a binary mixture but further tertiary interaction is suspected.

4.3.6 Solid Excipient Degradation Mechanism

It is important to understand the nature of solid–solid interaction and the mechanism of chemical degradation in the solid state in order to anticipate potential excipient incompatibility based on the chemical reactivity between the functional groups. Although chemical degradation in the solid state is less understood than that in solution, recent publication of the book [13] by Byrn, Pfeiffer, and Stowell provides a wealth of information and is a must-read for formulators working on solid dosage forms. The book categorizes the solid reactions into oxidation, addition, and those releasing gas, and those triggered by heat and light. It may be useful to categorize in a different way by functional groups of the active compounds. Table 4.9 categorizes this way the case studies of excipient incompatibility and the underlying solid-state degradation mechanisms experienced by the author of this chapter. A knowledge-based formulation strategy in anticipating and avoiding incompatible excipients is the key to success in any accelerated product development program.

4.4 API SOLUBILITY CONSIDERATION

Solubility is the most critical drug substance property for the development of controlled release dosage forms. The intent of controlled release dosage forms is to change the release mechanism from solubility limiting to formulation control. The design of the control mechanism requires understanding of drug solubility properties along the GI tract. For the formulation to be the rate-determining step, the solubility may need to be modified by addition of excipients that can boost or depress solubility. For drugs with high solubility, dissolution

can be slowed down by embedding the drug in a matrix or enclosing the drug within a film, whereas for drugs with low solubility it is more difficult to shift the controlling mechanism from solubility to formulation. Solubilizers may be necessary to promote dissolution of insoluble drugs so that drug release may be reproducibly controlled by the formulation and not the particle size or wetting properties of the drug particles.

The formulation challenge is higher when drug solubility is not only low but also pH dependent. For basic drugs, the solubility is high in acid and low in base, resulting in fast release in the upper GI and slow release in the lower GI. It is

TABLE 4.9 Solid Excipient Degradation Mechanism

Mechanism	Active Functional Group	Excipient
Hydrolysis	Ester, amide, peptide, β-lactam, hydroxamic acid	Buffer catalysis including phosphate, borate, citrate
Oxidation (N-oxide, S-oxide, epoxide, aldehyde/acid formation)	N and S containing, enol/keto, polyacene, steroid, conjugated compounds	Bleached excipient including PVP, crosPVP, MCC. Peroxide-containing excipients such as polysorbate. Free moisture-containing excipient
Decarboxylation/ dehydration	Diacid, β-lactam	Desiccants, desiccating excipient
Isomerization	Conjugated system and nucleophiles	Metal ion catalysis including magnesium stearate
Photolysis	Conjugated system and quinone, oxindone, pyrone	Metal ion catalysis including magnesium stearate
Transacylation	Acidic compound	Glycerides, VitE TPGS
Charge interaction, insoluble complex	Basic compounds, primary and secondary amines	Na starch glycolate, croscarmellose sodium, sodium lauryl sulfate
Maillard addition reaction, N-methylation, N-formylation	Basic compounds, primary and secondary amines	Reducing sugars, lactose starch, and MCC
Adduct addition, amide formation	Basic compounds, primary and secondary amines	Acids such as citric, ascorbic; conjugated acids such as fumaric, maleic; NaHCO$_3$

typical that when a slow release mechanism slows down the drug release in acid, the release in neutral pH of the lower GI becomes negligible resulting in partial release and lower bioavailability. More innovative methods are required to slow down the acid release without compromising the release at lower GI. The use of pH modifier and solubilizer has been employed and patented by the author for drugs of disparate solubility in gastric and intestinal fluids.

Multiparticulate dosage forms including beads, pellets, and minitabs are particularly useful for drugs with solubility that is highly pH dependent. Their growing popularity is borne out by the fact that FDA approvals of multiparticulate formulations have doubled over the past decade. This growth in the utilization of multiparticulate dosage forms is attributed to their superior clinical performance. The average GI transit time of multiparticulates is more uniform on a statistical basis than the single unit. In other words, the individual particle may have large variability in the transit time through each segment of the GI tract but when considered on average, the reproducibility between subjects and within subjects is much improved with a higher number of particles. Multiparticulate technologies provide a flexible dosing platform and a variety of release profiles, enabling the delivery of more than one API in a capsule, tablet, or sachet and a targeted release in a specific region of the gastrointestinal tract.

This section describes the fundamentals of solubility and its measurement. The factors influencing solubility and therefore dissolution rates are also described. The relationship between solubility and polymorphism will be examined. Finally, the prediction of food effect from solubility is discussed.

4.4.1 Equilibrium Solubility

When a solid surface is exposed to a liquid, molecules on the surface of the solid leave the surface and the molecules in the liquid return to the solid surface until equilibrium is established where the rate of molecules leaving is equal to the rate of molecules returning. In other words, equilibrium solubility is reached when the thermodynamic activity of the molecule in the solid surface equals that of the molecule dissolved in the solvent.

To further dissect the solution process, a cartoon in Figure 4.7 depicts the three steps of the dissolution process as follows:

1. Breaking of crystalline bonds holding the molecules to the solute.
2. Breaking of the solvent–solvent bonds creating a cavity in the solvent.
3. Occupying the solvent cavity by the solute molecules forming solute–solvent bonds.

This is why different crystalline forms give different equilibrium solubilities in a given solvent system reflecting

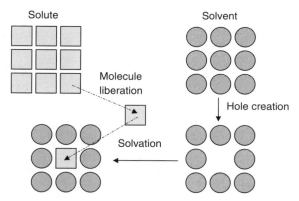

FIGURE 4.7   Solution process cartoon.

the differences in crystalline energies bonding the molecules together. It is also why solubility in a given solvent correlates to an extent with the heat of fusion or melting point, which is particularly true for analogues in a series since bond breaking from the crystal surface is the dominant component of solution energy. Solubility of the same compound is different in different solvents because the solution process depends on not only the heat of fusion but also the energy of solvation depicted in steps 2 and 3.

Equilibrium solubility varies among different polymorphs of a given compound whereas kinetic solubility measures the drug concentration where a polymorphic form precipitates from solution [14]. It is necessary to determine the absence of crystalline form changes when determining equilibrium solubility. If a form change begins to occur, solubility will not be equilibrated until the solid form transformation is complete. Therefore, the initial part of the solubility–time curve approximates the old polymorph whereas the later part of solubility–time curve approximates the new polymorph solubility. Equilibrium solubility should always be reported referencing a particular crystalline form with specific sample/batch number. Equilibrium solubility in simulated gastric and intestinal fluids at 37°C is required by FDA for BCS classification.

4.4.1.1 Solubility Unit and Terms   Solubility data may be expressed in several different ways. The most common way utilized by pharmaceutical scientists nowadays is mg/mL, μg/mL, or even ng/mL for very insoluble compounds. However, biological testing is usually expressed in molarity (number of moles per liter) as mM, μM, or nM. An easy conversion is illustrated as follows: for a compound with a molecular weight of 500, 1 mg/mL is equal to 2 mM with an easy conversion factor of 1000/molecular weight = 2.

Moreover, for organic chemists, solubility expressed as approximate volume of solvent necessary to dissolve 1 g of solute remains a popular way of reporting. The descriptive terms for approximate solubilities are defined in NF 31/USP 26 Reference Tables/Description and Solubility, page 840. The corresponding units in mg/mL are tabulated in Table 4.10.

TABLE 4.10 USP Descriptive Solubility Terms

USP31/NF26 Descriptive Solubility Term	Parts of Solvent Required for 1 Part of Solute	Solubility Cutoff (mg/mL)
Very soluble	Less than 1	>1000
Freely soluble	From 1 to 10	>100
Soluble	From 10 to 30	>33.3
Sparingly soluble	From 30 to 100	>10
Slightly soluble	From 100 to 1000	>1
Very slightly soluble	From 1000 to 10,000	>0.1
Practically insoluble, or insoluble	Greater than or equal to 10,000	>0.01

4.4.1.2 Method of Determining Equilibrium Solubility

The shake flask method, which remains largely manual, is still the standard way to measure equilibrium solubility. The procedure used in the author's laboratory is described here. An excess amount of solid is added to the solvent and the suspension is equilibrated in a rotator at a specified temperature in an oven or a laboratory with controlled room temperature for at least 24 h to reach equilibrium. The suspension is centrifuged and the supernatum is withdrawn using a syringe fitted with a 0.45 μm filter tip. The first milliliter of the filtrate is discarded in case of drug loss to the filter membrane and filter housing. The second milliliter is collected and further diluted with a diluting solvent for the HPLC assay. The pellet remaining in the centrifuge tube is resuspended with as little solvent as possible. The resulting slurry is scanned using a powder X-ray diffractometer (pXRD) to confirm the absence of polymorphic conversion.

For early compound development, the amount of sample is frequently insufficient to discard the first milliliter of the solubility sample, so a compatible filter has to be found before solubility can be determined. A standard set of filter compatibility results is shown in Table 4.11. Seven filters made of various polymers were screened. The results are tabulated in an ascending order of loss per unit surface area. (The most compatible filter is listed first.) Only two out of the seven filters did not show significant drug loss. The percentage remaining data of the two compatible filters (99.91% and 99.23%) are within the HPLC analytical variability. They are cellulose acetate and mixed cellulose ester. The other filter membrane polymers such as Teflon, PTFE, and Durapore that are generally considered as low binding showed significant binding to the drug. In fact, Durapore was almost as bad as Nylon. In terms of filter housing, both polypropylene and polyvinyl chloride were deemed compatible resulting in no drug loss from the two compatible filters. Polyethylene may also be fine but it is not conclusive since we cannot differentiate whether the drug was lost from the filter or the housing. We may conclude from this example that drug loss to filter cartridges can significantly alter the solubility data. Not only the filter membrane but also the filter housing can adsorb drug and result in drug loss. Use of compatible filter cartridges needs to be checked before conducting solubility experiments.

For compounds with low solubility and high adsorption to filters, centrifugation at high gravity force may be used instead. The solubility values determined may be falsely high due to the very fine particles not being removed by centrifugation. In this case, triplicate samples are used and the lower solubility values with demonstrated reproducibility may be reported.

A second method is to create a supersaturated solution by dissolving the solute at elevated temperatures and then cooling with agitation to induce precipitation. Seeding with trace amount of solute with the right crystalline form may also be used to induce precipitation. This method has the further complication of degradation at elevated temperatures and polymorphic conversion in the precipitate. This method is of value when very long equilibration time is required for viscous solvent or for compounds with very low solubility.

4.4.1.3 Effect of Temperature

It is important to understand the effect of temperature on solubility because manufacturing processes usually involve elevated temperatures during milling, drying, and compression. It is also because the life cycle of a drug product that is stored at room temperature will end with exposure to body temperature upon ingestion.

TABLE 4.11 Filter Compatibility Study Results

Brand Supplier	Filter Polymer	Housing	Pore Size (μm)	Diameter (mm)	Filter Area (cm²)	Percent Remaining (HPLC)	Percent Loss (HPLC)	Loss/Area (μg/cm²)
Corning	Cellulose acetate	PP	0.45	25	3.9	99.91	0.09	0.04
Millipore	Mix cellulose ester	PVC	0.22	25	3.9	99.23	0.77	0.38
Whatman	Polypropylene	PP	0.45	25	3.9	80.02	19.98	8.61
National Scientific	Teflon	PP	0.45	25	3.9	75.88	24.12	10.40
National Scientific	PTFE	PP	0.45	25	3.9	36.53	63.47	27.37
Millipore	Durapore PVDF	PE	0.45	25	3.9	17.21	82.79	35.70
National Scientific	Nylon	PP	0.45	25	3.9	11.41	88.59	38.20

PVDF: polyvinylidene fluoride; PTFE: polytetrafluoroethylene.

The Gibbs free energy of solution may be expressed in two parts as the heat-related enthalpy, ΔH, and the order-related entropy, ΔS. The free energy is related to equilibrium constant that is solubility (K_s) in this case. The overall equation may be expressed as

$$\Delta G = \Delta H - T\Delta S = -2.3RT \log K_s$$

where T is absolute temperature and R is the gas constant. The equation may be rearranged into the van't Hoff equation that describes a linear relationship of $\log K_s$ versus $1/T$ with a slope equal to $-\Delta H/2.3RT$ and an intercept equal to ΔS:

$$\log K_s = -1/2.3R(\Delta H/T - \Delta S)$$

When ΔH is positive, heat is absorbed upon dissolution and solubility increases with increasing temperature and the plot of $\log K_s$ versus $1/T$ is negative as shown in Figure 4.8. Once solubilities are determined at several temperatures, the heat of solution is estimated from the slope. Solution calorimetry can measure the heat of solution directly. However, the heat of solution determined from calorimetry differs to some extent from that determined from the van't Hoff equation. It is related to the volume changes in the solution process ($P\Delta V$). Calorimetry is a more direct method whereas the van't Hoff method provides additional information on transition temperature for polymorphic conversion as described in Section 4.4.3.

4.4.2 Kinetic Solubility

With the advent of high-throughput screening, kinetic solubility is routinely generated for discovery compounds nowadays [15]. The compound is first dissolved in an organic solvent, typically DMSO, and diluted gradually in an aqueous

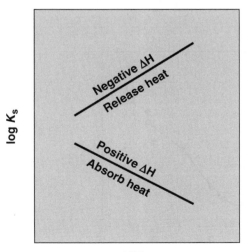

FIGURE 4.8 Effect of temperature on equilibrium solubility.

buffer, typically a pH 7.4 phosphate buffer, until precipitation is registered by in-line instruments such as UV detectors or laser nephelometers [16]. The concentration where the turbidity occurs is reported as kinetic solubility.

Kinetic solubility is generally greater than equilibrium solubility because of the lack of crystalline surfaces since compound is first dissolved in DMSO. It often leads to supersaturation until nucleation occurs. Therefore, kinetic solubility is less reproducible depending on factors affecting nucleation, such as cleanliness, the contact surface texture, and the impurities present.

Kinetic solubility can be a useful approximation of the solubility for an amorphous material that requires less energy input because of the lack of crystalline bonding energies. Kinetic solubility may also reflect the in-gut concentration when in-gut precipitation occurs. Basic compounds, with high solubility in the stomach acid, may crash out when traveling down to the lower GI where pH returns to neutral. This phenomenon is particularly important for controlled release dosage forms since a long absorption duration throughout the small and large intestines is usually required to effect a once-a-day dosage form.

4.4.3 pH–Solubility Profile

The most studied drug property influencing oral bioavailability is probably the pH dependence of aqueous solubility. For instance, salt forms of many basic and acidic drugs, based on the improved solubility in gastrointestinal pH, have optimized bioavailability and thus pharmacological or therapeutic activity. For controlled release formulation, the pH–solubility properties are even more important because absorption by design is to take place throughout the entire gastrointestinal tract with a pH gradient from 1 to 8. Therefore, dissolution may continuously change with the change of GI pH environment. To achieve pH-independent dissolution for a drug with pH-dependent solubility is the major challenge for the development of a controlled release dosage form.

A review of the literature reveals no standard way to determine pH–solubility profile [17]. Many attempted to use buffer to control the pH. It often failed to control the sample pH when solubility is high above the buffer capacity. Even when the pH is successfully controlled by the buffer for low-solubility compound, the resulting pH–solubility profile fails to follow the Henderson–Hasselbalch equation due to the depression of solubility by the buffer. The buffer effect is discussed in detail in Section 4.4.6. A second method utilizing pH STAT with HCl/NaOH as titrants has limited success because each pH sample requires a minimum of 24 h instrument time and the establishment of pH– solubility profile requires up to 15–25 samples. A third method utilizing HCl/NaOH buffers at various concentrations has successfully been employed by the author and pH–solubility profiles have been generated for hundreds of research compounds.

4.4.4 Determination of pK_a Value from pH–Solubility Profile

Determination of pK_a values from pH–solubility profile is routinely done in the author's laboratory. These pK_a values showed good agreement with either calculated pK_a values from chemical structures based on the Hammet–Taft equation or experimental values by pH titration. The pH–solubility profile is fitted using nonlinear regression to the Henderson–Hasselbalch equation:

$$pH = pK_a + \log_{10}\left(\frac{[\text{base}]}{[\text{acid}]}\right)$$

where pK_a is $-\log_{10}(K_a)$ and K_a is the acid dissociation constant. For an acid, its conjugated base is the deprotonated form, and for a base, its conjugated acid is the protonated form. A more useful form of the equation is

$$\text{For base,} \quad S_T = S_0(1 + 10^{(pK_a - pH)})$$

$$\text{For acid,} \quad S_T = S_0(1 + 10^{(pH - pK_a)})$$

The situation is more complicated when more than one pK_a value exists. This is illustrated with a model zwitterionic compound with both acidic and basic functional groups. The two pK_a values are 6.6 and 7.4 for hydroxamic acid and piperidine groups, respectively. The pH–solubility profile was determined in HCl/NaOH buffers in the pH range of 1.2–11.2. The pH–solubility profile is presented in Figure 4.9. The lowest solubility was found at the isoelectric pH of 7 (0.004 mg/mL). In the basic region, the molecule has a negative charge with solubility increasing logarithmically with increasing pH. In the acidic region, the molecule has a positive charge showing an initial logarithmical increase with decreasing pH. It reaches a plateau at pH

4.25, achieving a maximum solubility of 0.558 mg/mL. When more HCl was added to lower the pH further, solubility decreased indicating an interaction between the HCl salt and the chloride ions, known as the common ion effect. It was finally reduced to 0.012 mg/mL at the stomach pH of 1.2. Based on this information, two crystalline salt forms were made to increase the solubility over the free form. Simple capsule formulation was manufactured and dosed to the dogs. The acetate salt was found 40% more bioavailable than the HCl salt that has depressed solubility in the stomach acid due to common ion effect.

4.4.5 Solubility in Simulated Biological Fluid

The use of biorelevant solubility and dissolution tests has gained popularity in recent years. Clinically relevant food effect has been linked with increased drug solubility in the intestinal fluid under fed condition. Several GI surfactants including bile salts and lecithin are known to increase drug solubility. Upon food intake, more surfactants are secreted to alter the intestinal fluid composition in addition to acid secretion that alone could have profound impact on solubility for ionizable drugs. In addition, high-fat meal itself can modify the intestinal fluid composition.

In a 1997 workshop, Dressman et al. [18] were the first to propose to add sodium lauryl sulfate to the USP/NF simulated gastric fluid (SGF) as fasting-state SGF (FaSSGF) and sodium taurocholate and lecithin to the simulated intestinal fluid as fasting-state SIF (FaSSIF). The fed-state SIF is characterized by a higher amount of sodium taurocholate and lecithin. In the meantime, Charman et al. modified FaSSGF to fed-state SGF (FeSSGF) by increasing the pH from 2 to 5 [19]. For once-a-day formulation, the majority of the GI transit is in colon, so the understanding of colonic fluid composition is also important. Simulated fluids for upper and lower colons have also been devised [20].

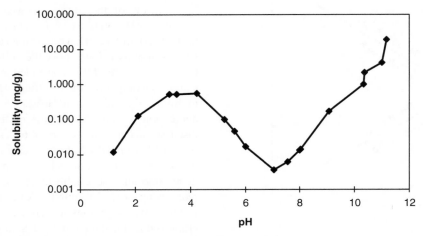

FIGURE 4.9   pH–solubility profile of a model zwitterion with overlapping pK_a values at 6.6 and 7.4.

TABLE 4.12   Physiological Buffer System

Substance (mM)	SGF	SIF	FaSSGF	FeSSGF	FaSSIF	FeSSIF	FeSSIF (HF)	SUCF	SLCF
NaCl	3.4		50	80	100	200	102	17	17
HCl	867		20						
KH$_2$PO$_4$, monobasic		3.9					50		
NaOH		38			9	100			
Sodium lauryl sulfate									
Acetic acid				10		144		70	25
NaH$_2$PO$_4$·H$_2$O					1.7				
Sodium taurocholate					8	12			
Lecithin					0.75	2	3.7		
Glycocholic acid							150		
Oleic acid							43		
Propionic acid								30	10
Final pH	1.2	6.8	1.7	5.0	6.5	5.0	6.5	6.2	7.0

SGF: simulated gastric fluid without 0.32% w/v pepsin (Sigma, P-7000); SIF: simulated intestinal fluid without 1% w/v pancreatin (Sigma, P-1625); FaSSGF: fasting-state simulated gastric fluid; FeSSGF: fed-state simulated gastric fluid; FeSSIF (HF): fed-state simulated intestinal fluid (high fat); SUCF: simulated upper colon fluid; SLCF: simulated lower colon fluid.

The various simulated gastric and intestinal fluid compositions are summarized in Table 4.12. For low-solubility compounds, comparison of solubility in these simulated biological fluids in addition to the pH–solubility profile will help to predict the release of drug at different segments of GI. Such information will aid in the design of controlled release dosage form with dissolution profiles tailored to the target GI segments. Moreover, a high solubility ratio between fed- and fasting-state simulated GI fluids may suggest the presence of food effect.

These biorelevant solubility tests have been used in a physiologically based absorption model (GastroPlus) to predict clinical food effect with limited success [21]. A high solubility ratio (>7) between the fed- and fasting-state simulated GI fluids is correlated with high food effect when the dose volume is high (8–18 L). However, the rank order of food effect does not necessarily correlate with the rank order of solubility ratio and dose volume.

4.4.6   Effect of Salt/Buffer and Common Chloride Ion Effect in the Stomach

Salts and buffers have a great effect on solubility especially of ionic compounds via the following four possible mechanisms:

1. Reducing the amount of solvent available to dissolve the compound due to salt–solvent interaction (salting out).
2. Association of salt with the compound leading to increased solubility (salting in).
3. Solubility product (K_{sp}) between the compound and the buffer species is low (K_{sp} control).

4. Buffer works as hydrogen donor or acceptor to effect ionization of the compound (pH control).

This salt/buffer phenomenon has led to a large discrepancy in solubility values reported in the literature. Wu and Benet [22] found conflicting reports of solubility data during the process of establishing BCS classification for common medicines such as furosemide, hydrochlorothiazide, methotrexate, ciprofloxacin, and erythromycin. They are reported as high-solubility drugs by one author but low-solubility drugs by another author. Therefore, it is very important to report the exact composition of the solvent and its salt content when reporting solubility values.

The common ion effect with chloride ion that is present in the stomach acid is a demonstration of the K_{sp} controlled solubility phenomenon (mechanism 3). It has physiological significance in reducing drug dissolution and solubility in the stomach. It can further reduce bioavailability as found for the example zwitterion compound reported in Section 4.4.4. The acetate salt was found 40% more bioavailable than the HCl salt that has depressed solubility in the stomach acid.

Common ion effect can be easily detected by measuring solubility in various percentages of saline solutions as illustrated in Figure 4.10. To clarify, the common chloride ion effect historically refers to the interaction between a HCl salt and chloride ions, thus with common chloride ions in both species. However, the slow down of dissolution is seen for free base and other acid salts not limiting to the HCl salt. More recent studies showed that fast interfacial counterion exchange and the presence of chloride ions quickly inhibit dissolution of free base and other acid salts not limiting to the HCl salt. For the lack of a better term, common chloride ion effect is used indiscriminately for all solid forms not limiting to the HCl salt.

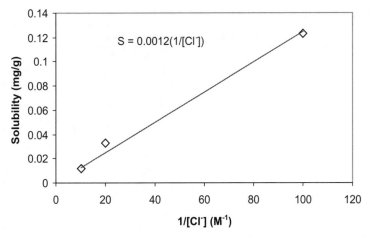

$$S = 0.0012(1/[Cl^-])$$

FIGURE 4.10   Common Ion Effect: Solubility as a function of chloride ion concentration

4.4.7   Solubility and Food Effect

FDA and most other Boards of Health (BOHs) require or recommend evaluation of the effect of food intake on bioavailability of controlled release dosage forms. In the 2002 Guidance for Industry on Food-Effect Bioavailability and Fed Bioequivalence Studies, FDA states "We recommend that food-effect BA and fed BE studies be performed for all modified-release dosage forms." In the same Guidance for Industry, FDA further states "We recommend that a food-effect BA study be conducted for all new chemical entities (NCEs) during the IND period." These statements highlighted the importance of delineating food effect in drug development. It would be advantageous if one can predict the possible food effect based on the physiochemical and biopharmaceutical properties of the drug substance. A formulation approach with BCS based decision trees to minimize food effect is described by M. S. Ku [23]. Armed with this understanding, the controlled release formulation may be designed to circumvent the potential food effect at the outset of formulation development.

The literature on food effect of common medicines was extensively reviewed by Goetz Leopold, who divided the drugs into four categories with reduced, delayed, unaffected, and increased drug absorption [24]. Comparing the solubility of these drugs, a trend can be discerned in that water-soluble compounds tend to have reduced bioavailability with food whereas water-insoluble compounds tend to have increased bioavailability with food. As nature's way to promote absorption of nutrients, food intake tends to slow down gastrointestinal motility, increase blood flow, and promote secretion of gastric acid and enzymes. These changes benefit the poorly absorbed compounds with low solubility and/or low permeability. On the other hand, food dilutes the concentration of those compounds that are well absorbed and therefore slows the rate of absorption. Similar trends have been observed by the author in that food tends to have little effect on BCS Class 1 compounds, delay or reduce bioavailability for BCS Class 3 compounds, but improve bioavailability for BCS Class 2 and 4 compounds.

Moreover, drugs with low and pH-dependent solubility tend to have a greater food effect due to increased gastric secretion and prolonged gastric emptying. Food initially brings the stomach pH from 1–2 to 5–7, then with increased gastric secretion, the stomach pH is lowered and emptying of the acidic food meal brings down the duodenum pH. Understanding of this complex interplay between gastric emptying and pH changes is critical to the prediction of food effect.

4.4.8   Rate of Solution (Dissolution Rate)

It is important to understand the physical model of solid particle dissolution process that serves as the basis for microenvironmental pH or surfactant modulation, a technique frequently used in controlled release formulations [25]. The model has long been depicted as a solid particle surrounded by a stagnant layer with a concentration gradient of the solute in regard to the surrounding bulk solvent (Figure 4.11).

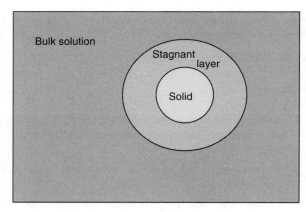

FIGURE 4.11   Fick's law diffusion model.

According to Fick's law, the rate of solution (dm/dt) or flux (J) is proportional to the particle surface area (A) and the concentration gradient ($C_s - C_b$) and is inversely proportional to the stagnant layer thickness (h), which is expressed in the equation

$$dm/dt = J = DA(C_s - C_b)/h$$

where D is the diffusion coefficient. It is clear that smaller particle size with a larger surface area increases dissolution rate and that increase of stir rate reduces the stagnant layer thickness and thus increases the dissolution rate. It is less clear what happens when a pH modifier or a surfactant is added to the dissolution medium. An acid–base reaction may occur at the interface between the stagnant and bulk solvent layers. Similarly, micellar formation may also take place at the interface and carry the drug away from the remaining solid. For powder dissolution, the addition of surfactants can help to wet hydrophobic particles and increase the effective surface area enhancing the dissolution rate.

4.4.9 Intrinsic Dissolution

The selection of pH modulators and surfactants can be made via intrinsic dissolution testing that is independent of the size and the surface area of drug particles. A variety of pH modulators or surfactants can be tested at several concentrations with little drug substance. The most effective pH modulator and surfactant can then be selected for formulation development. The intrinsic dissolution apparatus is depicted in Figure 4.12. A powder sample is loaded into a die and compressed using a Carver press to a solid compact at a relatively high pressure (~500 lb for 4–5 min). The compact surface needs to be smooth and free of visible pores. The die

is loaded into a flat bottom dissolution vessel with 1 in. distance from the paddle. Dissolution is performed and sample assayed per normal procedure. After the run, it is advisable to examine the die to discern that no significant erosion occurs to impede or promote dissolution due to surface change. It is also advisable to run pXRD on the sample recovered from the die if polymorphic conversion or salt exchange is suspected during dissolution. The intrinsic dissolution profile for an insoluble compound at three different concentrations of sodium lauryl sulfate is illustrated in Figure 4.13.

4.4.10 Solubility and Thermodynamically Stable Polymorph

Determination of the thermodynamically stable polymorph can easily be done by comparing equilibrium solubility among the different polymorphs. The polymorph with the lowest solubility is generally the most stable form. When the pure neat forms are not available for solubility comparison or solubility is too low to be determined accurately, a slurry test using a mixture with the two polymorphs can instead be used. The rate of solution for the more soluble form is faster than that for the less soluble form leading to disappearance of the more soluble form and the appearance of the less soluble form. Upon equilibration, the solid is isolated by filtration followed by washing and drying. pXRD pattern of the isolated solid is determined to identify the thermodynamically stable form. Sometimes, polymorphic conversion is so fast that the solubilities determined among different polymorphs are identical since all converted to the thermodynamically stable form. In this case, solubility comparison is of no use and the slurry test should be employed.

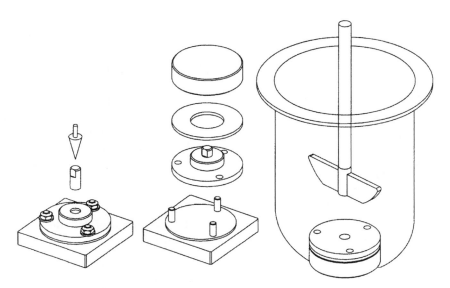

FIGURE 4.12 Intrinsic dissolution apparatus.

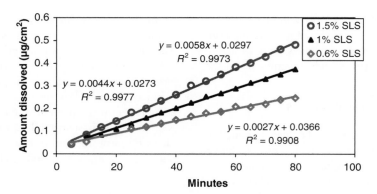

FIGURE 4.13 Intrinsic dissolution profile of an insoluble compound at various concentrations of sodium lauryl sulfate.

Polymorphic conversion in the solid state is different from that in slurry. The polymorph pairs that convert in the solid state are referred to as enantiotropic and the polymorph pairs that do not convert are referred to as monotropic. Thermal analysis with a mixture of the two polymorphs is the best way to discern if there is conversion below the melting points. In the case when polymorphic conversion is seen around the melting point, it is difficult to discern whether the material melts first or converts first. The construction of van't Hoff equations for the polymorph pairs is another good method of determination. As illustrated in Figure 4.14, lines A and B intersect at the transition temperature where A converts to B. Form A is the thermodynamically stable form above the transition temperature whereas form B is the thermodynamically stable form below the transition temperature. Forms A and B are enantiotropic. Line C is parallel to line A. Forms A and C are monotropic with no solid-state polymorphic conversion but in slurry form C can convert to the more stable form A that is at a lower energy state.

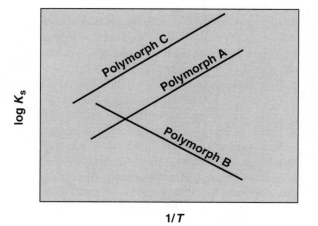

FIGURE 4.14 van't Hoff plot of enantiotropic and monotropic polymorphs.

4.5 PARTITION COEFFICIENT

Biological membranes are predominantly lipophilic, and unionized drugs can penetrate these barrier membranes through passive diffusion. This process is described as pH partition theory since 1950s [26]. Other processes such as transport of ionic drugs via sodium channels, transport of small molecules via aqueous pores in the membrane, facilitated lipid absorption via chylomicron, active transport via transporters, and the interplay between transporter and enterocyte microsomal enzymes have since been discovered one by one. It renders the pH partition theory less important. Nevertheless, pH partition still plays a role and correlates with GI permeability for some drugs.

The partition coefficient is defined as the partition of a neutral noncharged species into a lipid phase, a process that is not pH dependent [27]. Whereas, for ionic compounds, the distribution constant, which accounts for partitioning of both charged and noncharged species, is pH dependent and can be measured at various pH values. The distribution constant in pH 7.4 is most relevant to biological systems. For oral absorption, pH partition profile over a pH range of 1–8 is pertinent to the GI absorption window. For controlled release, pH partition over pH 1–8 is even more pertinent since a wide absorption window utilizing the entire GI tract is needed. This section will discuss the estimation of partition coefficient from chemical structure as well as the experimental measurement of distribution constants by the shake flask method. Other methods including the use of reversed-phase HPLC are detailed in a book edited by Dunn, Block, and Pearlman [28].

4.5.1 Estimation of Partition Coefficient from Chemical Structure

Similar to pK_a, estimation of partition coefficient from chemical structures is well established nowadays and inherent to most chemistry computational softwares. It is based on the assumption that $\log P$ is an additive constitutive property. The chemical substituent effect is determined over small increments of structural changes similar to the Hammet equation. A large number of constants for a variety of substituents were derived in the early 1970s [29]. Some correction factors were also derived to account for interactions between the substituents and the presence of charges. Finally, databanks were established and computer programs were written such as the one from Pomona College group, that is, CLogP. For ionic compounds, once $\log P$ is estimated for the free base or free acid, the distribution constant at pH 7.4 is calculated based on the degree of dissociation at pH 7.4 assuming partition is dominated by the neutral species.

4.5.2 Determination of Distribution Coefficient

Distribution coefficient are typically determined using n-octanol as the oil phase and a buffer as the aqueous phase via the shake flask method. The buffer concentration is selected to maintain a constant pH. Both aqueous and n-octanol phases are presaturated with each other by mixing for a period of time followed by separation so that volume corrections would not be necessary after equilibration. A sufficiently low initial drug concentration in n-octanol needs to be utilized so that drug will not precipitate upon transfer to the aqueous phase. When precipitation occurs, the aqueous phase concentration is no longer controlled by partition equilibrium but supersaturation kinetics that induces precipitation. The volume ratio of aqueous medium to n-octanol is selected to effect significant mass transfer. For instance with a $\log D$ value of 2, only 1% drug will transfer from the n-octanol phase into the aqueous phase when the volume ratio is 1. A fivefold increase of aqueous volume will effect an increase in mass transfer to approximately 5%. The mixture is vigorously agitated and then equilibrated in a rotator before phase separation by centrifugation. Drug concentrations in each phase are determined by HPLC. The total drug recovery should be near 100% to ensure no systematic error (no loss due to adsorption). The $\log D$ value is calculated as the ratio of drug concentration in n-octanol phase to that in the aqueous phase. The $\log P$ value is then calculated from pK_a and pH using the following equation:

For acid, $\text{Log } P = \log D + \log(1 + 10^{(\text{pH}-pK_a)})$

For base, $\text{Log } P = \log D + \text{Log}(1 + 10^{(pK_a-\text{pH})})$.

4.5.3 Pitfalls in Partition Coefficient Measurement

For compounds with high $\log P$, there are two reasons that prevent accurate determination. First, the analytical method may not be sensitive enough to detect the low drug concentration in the aqueous phase. For a compound soluble in n-octanol at 1 mg/mL, the aqueous phase concentration is 1 ng/mL at $\log P = 6$. Most HPLC methods will not be able to quantify accurately 1 ng/mL. Second, it is difficult to separate the two phases of disparate concentration drop without contamination. The presence of trace amount of n-octanol phase in the aqueous phase can result in an artificially higher aqueous concentration and lower $\log P$ than the true value.

As an alternative method, the equilibrium solubility can be measured separately in the water and n-octanol phases and the ratio of the two is the partition coefficient since the thermodynamic activities at saturation in each phase are equivalent to the solid and therefore to each other. Should the solubility be below the HPLC detection limit, estimation of partition coefficient from chemical structures is the only option, which is not uncommon nowadays with drug pipelines dominated by insoluble compounds.

The second pitfall is the use of buffer species that forms ion pair leading to artificially high partition coefficients. The author has seen that $\log D$ changed by an order of magnitude with the addition of chloride ions. Difference in $\log D$ is also seen when sodium ion is replaced with potassium ion. It is advisable to select buffer species that do not depress solubility ahead of the experiment.

In all, the experimental procedure for the determination of partition coefficient needs to be tailored to individual compound. The automation of $\log P$ measurement in high-throughput screening often leads to erroneous results due to the lack of fine tuning of experimental conditions.

4.5.4 Determination of pK_a from pH Partition Profile

A pH partition profile is presented in Figure 4.15 for a model zwitterion for which pH–solubility profile is presented in Section 4.4.3. An isoelectric point with maximum partition and minimum solubility was detected at pH 7. The pH

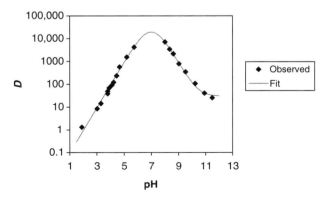

FIGURE 4.15 pH partition profile of a model zwitterion with overlapping pK_a values at 6.6 and 7.4 corresponding to hydroxamic acid and piperidine functional groups, respectively.

partition profile was determined in phosphate buffer/*n*-octanol over a pH range of 1.9–11.5. A biased bell-shaped curve representing varying partition of the three species was observed. The data were fitted using a nonlinear regression model and the pK_a values were determined to be 6.6 and 7.4 for the hydroxamic acid and piperidine functional groups, respectively. The cation was found to have no significant partition into the *n*-octanol phase. The log partition coefficients (log *P*) for the anion and the neutral species were 1.5 and 4.5, respectively. The nonlinear fitted pH–distribution constant (*D*) profile is presented as the solid line.

This model compound exhibited a high Caco-2 cell permeability that is consistent with the high partition coefficient. The Caco-2 permeability is 49.8 nm/s indicating that the compound is readily absorbed once dissolved. Since dissolution in the GI tract is the rate-limiting step, the more soluble acetate salt resulted in an increase of oral bioavailability in dogs whereas the HCl salt with chloride ion common effect is less soluble in gastric fluid and less bioavailable in dogs.

This example illustrates that pH partition profile remains a useful tool to predict poor oral absorption for passively diffused drugs. Drugs with low partition at colonic pH 7–8 are likely to exhibit low absorption in colon, which is essential for once-a-day formulation. Early detection of this issue can efficiently direct the development efforts toward gastric retentive formulations to overcome the narrow GI absorption window.

REFERENCES

1. Zhang Y, Johnson KC. Effect of drug particle size on content uniformity of low-dose solid dosage forms. *Int. J. Pharm.* 1997; 154:179–183.

2. Rohrs BR, Amidon GE, Meury RH, Secreast PJ, King HM, Skoug CJ. Particle size limits to meet USP content uniformity criteria for tablets and capsules. *J. Pharm. Sci.* 2006;95: 1049–1059.

3. Huang CY, Ku SM. Prediction of drug particle size and content uniformity in low-dose solid dosage forms. *Int. J. Pharm.* 2010;383(1–2):70–80.

4. Schussele A, Bauer-Brandl A. Note on the measurement of flowability according to the European Pharmacopoeia, *Int. J. Pharm.* 2003;257:301.

5. Standard Shear Test Method for Bulk Solids Using the Schulze Ring Shear Tester. ASTM D6773-02.

6. Amidon GE, Houghton ME. Powder flow testing in preformulation and formulation development. *Pharm. Manuf.* 1985 (July); 21–31.

7. Wells JI. *Pharmaceutical Preformulation*, Ellis Harwood Limited, 1988.

8. Murphy D, Rabel S. Thermal analysis and calorimetric methods for the characterization of new crystal forms. In Adeyeye MC,

Brittain HG, editors. *Preformulation in Solid Dosage Form Development*, Informa Healthcare USA Inc., 2008.

9. Waterman KC, et al. Stabilization of pharmaceuticals to oxidative degradation. *Pharm. Dev. Technol.* 2002;7(1):1–32.

10. Connors KA, Amidon GL, Stella VJ. Chemical stability of pharmaceuticals: a handbook for pharmacists. Wiley, Second edition,1986.

11. Monkhouse DC, Maderich A. Whither compatibility testing? *Drug Dev. Ind. Pharm.* 1989;15:2115–2130.

12. Serajuddin A, Thakur A, Ghoshal R, Fakes M, Ranadive S, Morris K, Varia S. Selection of solid dosage form composition through drug–excipient compatibility testing. *J. Pharm. Sci.* 1999;88:696–704.

13. Byrn SR, Pfeiffer RR, Stowell JG. *Solid State Chemistry of Drugs*, 2nd edition, SSCI Inc., 1999, pp. 305–441.

14. Ku, MS. Use of the Biopharmaceutical Classification System in Early Drug Development. *The AAPS Journal* 2008;10 (1):208–212.

15. Avdeef A. High throughput measurements of solubility profiles. In Testa B, van de Waterbeemd H, Folkers G, Guy R, editors. *Pharmacokinetic Optimization in Drug Research: Biological, Physiological, and Computational Strategies*, Verlag Helvetica Chimica Acta, Zurich, 2001, pp. 305–326.

16. Pan L, Ho Q, Tsutsui K, Takahashi L. Comparison of chromatographic and spectroscopic methods used to rank compounds for aqueous solubility. *J Pharm. Sci.* 2001;90:521–529.

17. Fiese EF, Hagen TA. Preformulation. In Lachman L, Lieberman HA, Kanig JL, editors. *The Theory and Practice of Industrial Pharmacy*, Lea & Febiger, 1986, pp. 96–171.

18. Dressman JB, Amidon GL, Reppas C, et al. Dissolution testing as a prognostic tool for oral drug absorption: immediate release dosage forms. *Pharm. Res.* 1998;15(1):11–22.

19. Charman WN, Porter CJH, Mithani S, et al. Physicochemical and physiological mechanisms for the effects of food on drug absorption: the role of lipids and pH. *J. Pharm. Sci.* 1997;86(3): 269–282.

20. Galia E, Nicolaides E, Horter D, et al. Evaluation of various dissolution media for predicting *in vivo* performance of class I and II drugs. *Pharm. Res.* 1998;15(5): 698–705.

21. Jones HM, Parrott N, Ohlenbusch G, Lave T. Predicting pharmacokinetic food effects using biorelevant solubility media and physiologically based modelling. *Clin. Pharmacokinet.* 2006;45:1–15.

22. Wu CY, Benet LZ. Predicting drug disposition via application of BCS: transport/absorption/elimination interplay and development of a biopharmaceutics drug disposition classification system. *Pharm. Res.* 2005;22:11–23.

23. Ku, MS. An oral formulation decision tree based on the biopharmaceutical classification system for first in human clinical trials. *Bull. Tech.* Gattefosse. 2006;99:89.

24. Leopold G. Experimental factors influencing the results of drug product bioavailability/bioequivalency studies in humans. In Smolen VF, Ball LA, editors. *Controlled Drug Bioavailability:*

Bioavailability Methodology and Regulation, Vol. 2, Wiley, 1984, pp. 7–9.

25. Farag SI, Munir B, Hussain A. Minireview: microenvironmental pH modulation in solid dosage forms. *J. Pharm. Sci.*, 2007;96:948–959.

26. Martin A. *Physical Pharmacy*, 4th edition, Lea & Febiger, 1993, pp. 342–351.

27. Ku, MS, Berge, S., Bansal, P. Overlapping pKa Determination by Non-Linear Regression Analysis of pH-Dependent Partition Measurements. *APhA* 1985;15(2):85.

28. Dunn WJ, Block JH, Pearlman RS. *Partition Coefficient: Determination and Estimation*, Pergamon Press, Oxford, 1986.

29. Rekker RE. *The Hydrophobic Fragment Constant*, Elsevier, Amsterdam, 1976.

5

POLYMERS IN ORAL MODIFIED RELEASE SYSTEMS

JIASHENG TU AND YAN SHEN

Department of Pharmaceutics, School of Pharmacy, China Pharmaceutical University, Nanjing, China

RAVICHANDRAN MAHALINGAM

Formurex, Inc., Stockton, CA, USA

BHASKARA JASTI AND XIAOLING LI

Department of Pharmaceutics and Medicinal Chemistry, Thomas J. Long School of Pharmacy and Health Sciences, University of the Pacific, Stockton, CA, USA

5.1 INTRODUCTION

Peroral administration of drugs is the preferred route for drug delivery. Drug products administered via oral route comprise 40% of all the drug products in USP 27. Due to the physiological characteristics of the gastrointestinal tract (GIT), oral drug delivery systems/dosage forms encounter a harsh environment and the residence time is dictated by the stomach emptying rate and intestinal movement. To achieve optimal therapeutic outcomes after oral administration, the delivery system should be programmed to release a predetermined amount of drug and reside sufficient time at a desired gastrointestinal location for complete absorption. It is, therefore, essential to control both the release rate of drug and the residence time of system within the gastrointestinal tract. In designing such modified release oral drug delivery systems, pharmaceutical excipients play an important role. Among the pharmaceutical excipients, polymers are extensively used for developing various modified release pharmaceutical products.

The release rate of a drug from an oral delivery system can be controlled by designing either matrix systems or membrane-controlled release systems. Both systems are fabricated with polymers (Table 5.1). For matrix systems, hydrophilic polymers and erodible polymers are used to form swellable and erodible matrices. In membrane-controlled systems, water-insoluble or enteric polymers are widely used as coating materials to modulate the rate or time of drug release.

The residence time of modified release systems can be prolonged by retaining the system in stomach, increasing the floating time in GIT fluid, and/or increasing the bioadhesion to the GIT mucus wall. Various polymers are used to design gastric floating and bioadhesive systems. Gastric floating systems are usually matrix-type systems incorporated with effervescent chemicals. Bioadhesive systems are composed of bioadhesive polymers that can adhere to the surface of GIT mucus.

In general, polymers from natural and synthetic sources are used for developing oral modified release pharmaceutical products. Synthetic polymers are more useful than natural polymers due to their consistency in quality and other properties. An overview of the polymers used in oral drug products is shown in Table 5.2. In this chapter, polymers widely used for formulating oral modified release drug delivery systems are discussed with an emphasis on the mechanism of drug release.

5.2 POLYMERS FOR ORAL MATRIX SYSTEMS

Hydrophilic matrix systems have been the most popular and widely used systems to achieve controlled drug release since

TABLE 5.1   Types of Oral Modified Release Delivery Systems and Related Polymers

System	Dosage Form	Polymers
Matrix systems	Hydrophilic systems	Methylcellulose, hydroxypropyl methylcellulose (hypromellose), hydroxypropyl cellulose, carboxymethyl cellulose sodium, alginate, chitosan, gelatin
	Erodible systems	Stearic acid, beeswax, stearyl alcohol, microcrystalline wax, white wax, yellow wax, carnauba wax
	Insoluble systems	Polyethylene, polypropylene, polyvinyl chloride
	Bioadhesive systems	Hydroxypropyl cellulose, Carbopol, alginate, chitosan, pectin, polyethylene oxide
Membrane-controlled systems	Osmotic pumps, coated tablets, coated pellets/spheroids in capsules	Cellulose acetate, poly(meth)acrylates, ethylcellulose, ethylene vinyl acetate (EVA) copolymer, carboxymethyl cellulose sodium, cellulose acetate phthalate, chitosan, hydroxyethyl cellulose, hydroxyethyl methylcellulose, hydroxypropyl cellulose, hypromellose, hypromellose phthalate, methylcellulose, polyvinyl acetate phthalate, hypromellose acetate succinate
Bioadhesive systems	Tablets, pellets, capsules	Carbopol, polyethylene oxide, hydroxypropyl cellulose, hypromellose, carboxymethyl cellulose, alginate, pectin, carrageenan

1950s. Hydrophilic matrix systems are robust, flexible, and easy to manufacture. A typical hydrophilic matrix system is composed of one or more active pharmacological ingredients (APIs), hydrophilic polymers, and other excipients that modulate API release rate and/or tablet characteristics.

Polymers for oral hydrophilic matrix systems are water-soluble, gel-forming, and/or swellable. During the past two decades, a variety of synthetic, semisynthetic, and natural hydrophilic swellable polymers have been used in matrix tablet formulations. Natural hydrophilic swellable polymers include chitosan, pectin, and alginate. The synthetic or semisynthetic hydrophilic polymers include cellulose derivatives and carbomers. Drug release from hydrophilic matrix tablets is controlled by the formation of a hydrated viscous layer around the dosage form, which acts as a barrier to drug release, by opposing penetration of water into tablet. This subsequently controls the release of dissolved solutes out of the matrix. A key functional characteristic of hydrophilic polymer matrix is its gel-forming capability. Water-soluble drugs are released primarily by diffusion of dissolved drug molecules across the gel layer, whereas poorly water-soluble drugs are released predominantly by erosion mechanisms. The contribution of each release mechanism to the overall drug release process is influenced not only by drug solubility but also by the physical and mechanical properties of the gel barrier. The hydration characteristics of the polymer and the physical properties of the hydrated gel layer may critically affect the drug release. Therefore, matching the hydrophilic polymers and other additives with drug is critical for the successful control of drug release in oral modified release formulations.

5.2.1 Chitosan

Chitosan, a poly(amino saccharide), has been used as a hydrophilic matrix material due to its swellability and erodibility. Chitosan is produced by partial alkaline deacetylation

of chitin. The term chitosan refers to a family of aminopolysaccharides, which differ according to their degree of deacetylation (DA, 2–60%) and molecular weight (50–2000 kDa) [1].

5.2.1.1 *Structure and Properties*   Chemically, chitosan is a linear, binary heteropolysaccharide consisting of randomly distributed (1–4)-linked 2-acetamido-2-deoxy-β-D-glucose and 2-amino-2-deoxy-β-D-glucose molecules. Chitosan can be the fully or partially deacetylated form of chitin. In addition, a chitosan molecule has amino and hydroxyl groups that can be chemically modified to provide a high versatility (Figure 5.1).

Chitosan is a weak base due to the large quantities of amino groups on its chain with a pK_a value of approximately 6.2–7.0. Chitosan exhibits a pH-dependent solubility: it dissolves easily at low pH and is insoluble at higher pH ranges. Therefore, chitosan is soluble in dilute HCl, HBr, HI, HNO_3, and $HClO_4$. Chitosan is also slightly soluble in dilute H_3PO_4, but sparingly soluble in H_2SO_4 at room temperature. Chitosan sulfate dissolves in water on heating and becomes insoluble on cooling. Before complete dissolution, chitosan undergoes pH-sensitive swelling due to the protonation of amine groups at low-pH conditions. Upon dissolution in acidic media, the free amino groups are fully protonated, resulting in a highly charged and a fully extended flexible conformation of the chitosan backbone. The solubility behavior of chitosan can be attributed to the existence of intramolecular hydroxy and amino groups. The poor solubility of chitosan in water and common organic solvents limits its application. Therefore, structural modifications of chitosan, such as modification by carboxylation, acylation, alkylation, and etherification, have been investigated to improve its solubility. *N,O*-Carboxymethyl chitosan is soluble in water over a wide pH range, but is almost insoluble at its isoelectric point (pH 5.5). *N*-Alkylated chitosans such as *N*-trimethyl chitosan chloride, *N*-*N*-propyl-*N,N*-dimethyl

TABLE 5.2 Polymers Used in Oral Modified Release Formulations

Name	Molecular Weight	Official Monograph	Commercial Name
Albumin	About 66,500	BP, USP, PhEur	Plasbumin, Pro-Bumin, Albumisol, Albuspan, Albutein
Alginic acid, sodium alginate	20,000 to 240,000	BP, USPNF, PhEur	Kelcosol, Keltone, Algin, Protanal, Sodium Polymannuronate
Carbomers	7×10^5 to 4×10^9	BP, USPNF, PhEur	Polyacrylic Acid, Carboxyvinyl Polymer, Pemulen, Carbopol, Ultrez
Carboxymethyl cellulose calcium	90,000 to 700,000	BP, JP, USPNF, PhEur	Carboxymethyl Cellulose, Calcium CMC, Calcium Nymcel ZSC
Carboxymethyl cellulose sodium	90,000 to 700,000	BP, JP, USPNF, PhEur	Aquasorb, CMC Sodium, Sodium CMC, Akucell, Tylose CB
Carrageenan	>100,000	USPNF	Genu, Hygum TP-1, Irish Moss Extract, Seaspen PF, Gelcarin, Viscarin
Chitosan	100,00 to 1,000,000	BP, PhEur	Deacetylated Chitin, Deacetylchitin, Poly-D-Glucosamine
Dextrin	4500 to 85,000	BP, JP, USPNF, PhEur	Canary Dextrin, Crystal Gum, British Gum, Starch Gum
Ethylcellulose	Wide range	BP, USPNF, PhEur	Aqualon, Ethocel, Aquacoat ECD, Surelease
Gelatin	15,000 to 250,000	BP, JP, USPNF, PhEur	Instagel, Cryogel, Solugel
Guar gum	Approx. 220,000	BP, USPNF, PhEur	Jaguar Gum, Meyprogat, Galactosol, Meyprofin
Hydroxyethyl cellulose	Wide range	BP, USPNF, PhEur	HEC, Ethylose, Hyetellose, Natrosol, Oxycellulose, Tylose PHA
Hydroxypropyl cellulose	50,000 to 1,250,000	BP, JP, USPNF, PhEur	Klucel, Methocel, Cellulose, Hyprolose, Nisso HPC
Hypromellose	10,000 to 1,500,000	BP, JP, USPNF, PhEur	HPMC, Methocel, Hydroxypropyl Methylcellulose, Metolose, Tylopur
Maltodextrin	900 to 9000	BP, USPNF, PhEur	Glucodry, Maldex, Maltrin, C*Pharmdry, Glucidex, Star-Dri
Methylcellulose	10,000 to 220,000	BP, JP, USPNF, PhEur	Methocel, Benecel, Culminal MC, Metolose
Microcrystalline cellulose	Approx. 36,000	BP, JP, USPNF, PhEur	Avicel PH, Emcocel, Celphere, Fibrocel, Pharmacel
Poly(methyl vinyl ether/maleic anhydride)	200,000 to 2,000,000	None	Gantrez
Polycarbophil	700,000 to 4 billion	USP	Noveon
Polydextrose	1200–2000	None	Polydextrose A, Polydextrose K
Polyethylene glycol	190 to 9000	BP, JP, USPNF, PhEur	Carbowax Sentry, Lipoxol, Carbowax, Lutrol E, PEG
Polyethylene oxide	100,000 to 7,000,000	USPNF	Polyox, Polyoxyethylene, Polyoxirane
Polyoxyethylene stearates	549 to 4690	JP, USPNF	PEG Fatty Acid Esters, Marlosol, PEG Stearates
Polyvinyl alcohol	20,000 to 200,000	USP, PhEur	Elvanol, Lemol, Polyvinol, Alcotex, PVA
Povidone	2500 to 3,000,000	BP, JP, USP, PhEur	PVP, Polyvinylpyrrolidone, Kollidon, Plasdone, Polyvidone
Powdered cellulose	Approx. 243,000	BP, JP, USPNF, PhEur	Elcema, Arbocel, Sanacel
Tragacanth	840,000	BP, JP, USPNF, PhEur	Gum Dragon, Gum Benjamin
Xanthan gum	Approx. 2,000,000	BP, USPNF, PhEur	Keltrol, Rhodigel, Vanzan NF, Corn Sugar Gum, Xantural

chitosan, and *N*-furfuryl-*N*,*N*-dimethyl chitosan have higher solubility than chitosan.

Chitosan exists in one of the three forms, hydrated, anhydrous crystal, or noncrystal, each with distinct powder diffraction pattern. The hydrated crystal shows a strong reflection at an angle (2θ) of 10.4° and other weak peaks at 20° and 220°. The anhydrous crystal exhibits strong peaks at 15° and 20°. Amorphous form does not show any significant reflection, and only a broad halo around 20° can be seen. The dissolution of chitosan involves a progressive disappearance of the peak at 220° suggesting that swelling reduces the crystallinity of the excipient.

5.2.1.2 Applications of Chitosan in Hydrophilic Matrix Systems Chitosan has been used as a pharmaceutical excipient in conventional oral dosage forms as a binding,

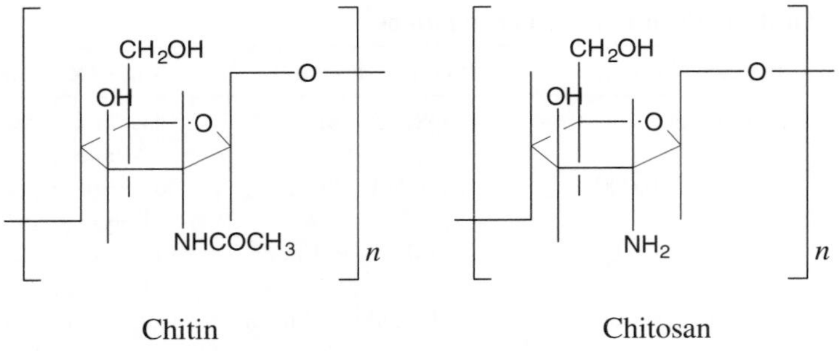

FIGURE 5.1 Chemical structures of chitin and chitosan.

disintegrating, and coating agent. Because of its multitude of functional characteristics, various chitosan-based oral controlled systems have been developed as summarized in Table 5.3. Chitosan has been studied as a swelling hydrophilic material and a bioadhesive polymer in many oral controlled release formulations. Since chitosan compacts poorly and is soluble only in a few dilute acid solutions, it is formulated in combination with other polymers such as alginates.

Protonated chitosan can interact with negatively charged mucin layers on mucosal tissues or cell membrane proteoglycans. This bioadhesion process leads to increased contact between drugs and mucosal tissues with minimal and reversible membrane damages compared with other agents such as bile acids and surfactants. A reversible opening of the tight junctions is expected as the probable mechanism of action of chitosan as a transmucosal absorption enhancer via paracellular route. Because of these properties, chitosan is believed to enhance oral bioavailability of drugs.

There has been a growing interest in chemical modification of chitosan to improve its compaction property and solubility and widen its applications [1]. Chemically modified chitosans possess controlled release and targeted delivery properties of almost all classes of bioactive molecules. Tablets composed of unmodified chitosan, which exhibit a low degree of order, will have a weak tablet mechanical strength. To increase the hardness of chitosan tablets, several structural modifications on chitosan were made. N-Acylation of chitosan modified with various fatty acids (C_6–C_{16}) showed increased hydrophobicity, which can be used to sustain the release of hydrophobic drugs and induce changes in tablet characteristics. The aqueous swelling of acylated chitosan decreases as the chain length of alkyl group increases from C_6 to C_{16}. Tablets prepared with palmitoyl chitosan (substitution degree 40–50%) tablets showed ideal mechanical characteristics and drug release properties for acetaminophen. Palmitoyl chitosan with higher substitution (e.g., 69%) sustains acetaminophen release for 90 h [30].

Phosphorylated chitosan, chitosan succinate, and hydroxamated chitosan succinate also form polyelectrolyte complex gel beads and are studied for oral controlled release of ibuprofen [43]. The release rates of ibuprofen from phosphorylated chitosan gel beads increased with the pH of dissolution medium due to the ionization of phosphate groups and high solubility of ibuprofen at higher pH. It is advantageous for bypassing acidic gastric fluid region of the

TABLE 5.3 Application of Chitosan in Oral Controlled Release Formulations

Dosage Form	Origin of Chitosan	Application	References
Polymeric beads	Chitosan succinate, hydroxamated chitosan succinate, methyl chitosan, N,O-carboxymethyl chitosan	Controlled release of theophylline, ASA, acetohydroxamic acid, insulin	[2–8]
Various experimental dosages	Chitosan, N-trimethyl chitosan chloride, thiolated chitosan, palmitoyl glycol chitosan, pegylated chitosan, N-sulfonato-N,O-carboxymethyl chitosan	Absorption enhancement of heparin, insulin, calcitonin, gene, saquinavir, acyclovir, ganciclovir, naproxen, denbufylline, clodronate	[9–23]
Tablets	Chitosan, N-carboxymethyl chitosan, thiolated chitosan, N-acylated chitosan	Sustained release oral formulations of alpha-lipoic acid, prednisolone, calcitonin, insulin, tobramycin sulfate, naproxen	[24–31]
Floating microspheres	Chitosan/dioctyl sulfosuccinate	Gastroretentive controlled release delivery system of acetohydroxamic acid	[8, 32]
Colon specific capsules	chitosan	Delivery of calcitonin, n-dodecyl-beta-D-maltopyranoside, ASA, prednisolone, budesonide, calcitonin	[3, 13, 33–42]

stomach and for controlled drug delivery after oral administration.

Protonated chitosan forms complex with protein drugs and is recognized as a potential carrier for oral delivery of protein drugs such as salmon calcitonin and insulin. It sustained the release and protected the protein from degradation and also enhanced its paracellular absorption. Mono-*N*-carboxymethylated chitosan is soluble in water and is used as a carrier of low molecular weight heparin by complexation, which increases the permeation of heparin across the Caco-2 intestinal cell monolayer. Similar properties are also observed with *N,O*-carboxymethyl chitosan.

5.2.2 Alginate

Alginate is one of the most studied and applied natural polysaccharide polymers in oral controlled delivery systems [1, 44]. Alginate is abundant in nature and is found as a structural component of marine brown algae and as capsular polysaccharides in some soil bacteria. Pharmaceutical grade alginates are extracted from brown seaweed.

5.2.2.1 *Structure and Properties* Alginate is composed of sequential β-D-mannuronic acid monomers (M-blocks), α-L-guluronic acid (G-blocks), and interspersed M and G units [1] (Figure 5.2). The length of M- and G-blocks and sequential distribution along the polymer chain vary depending on the source of alginate.

Alginate is a hydrophilic polymer and forms hydrogel in water. Due to its polyacidic structure, alginates undergo reversible gelation in aqueous solution. Divalent cations such as Ca^{2+} and Zn^{2+} can cooperatively bind between the G-blocks of adjacent alginate chains and create ionic interchain bridges. This results in a reduction of the permeability of different solutes, delaying the release of entrapped API in alginate hydrogel.

Alginate-based materials are pH-sensitive [44]. Release of drugs from alginate-based matrix in low-pH solutions is significantly reduced, which is beneficial in the development of a delivery system. Theoretically, alginate shrinks at low pH and the encapsulated or embedded drugs would not be released. Upon contact with intestinal fluid, alginate matrix

will undergo an ion exchange between divalent cation and sodium resulting in disintegration of the matrix and release of drugs. This pH-dependent behavior of alginate is exploited to tailor release profiles and in the development of intelligent systems. However, the limited control on gel dissolution and significant burst release effect have prevented alginate from being used as a sole component in oral controlled release delivery systems.

5.2.2.2 *Applications in Oral Controlled Release Systems* The ability of sodium alginate to rapidly form viscous solutions in contact with aqueous media and to form gels in contact with acid or di- or trivalent cationic ions is ideal for its use as a hydrophilic matrix in oral controlled release dosage forms. Matrices in the form of microparticles or tablets incorporating alginate salts or a combination of alginate with other polymers have been successfully employed to prolong release of many drugs as shown in Table 5.4.

Alginate microparticles have been investigated for loading both small molecules and macromolecules. Because of insolubility under acidic conditions, alginate hydrogel is an excellent hydrophilic matrix system for highly acid-soluble drugs. Since the swelling and dissolution of alginate matrix is pH dependent, significant burst release (up to 30–40%) from the matrix is generally observed when the system is in contact with intestinal fluid. Composite matrix systems of alginate and chitosan [1] or polyacrylate [45, 46] were used to optimize the release rate. Chitosan-coated alginate microparticles were shown to have potential as carriers for poorly absorbable hydrophilic drugs and possibility to increase their oral bioavailability [47].

Several methods have been reported for the preparation of alginate microparticles. The spray congealing method has been used to encapsulate peptides, but it produces microparticles of almost 1 mm size. Alginate microparticles of micron size are produced by emulsification/internal gelation and investigated as a promising carrier for insulin delivery [48]. Alginate microparticles with size less than 3 μm were found to be absorbed via Peyer's patches [49]. Enteric release pattern is reported for alginate microparticles [44, 47]. Alginate can be used for microencapsulating cells

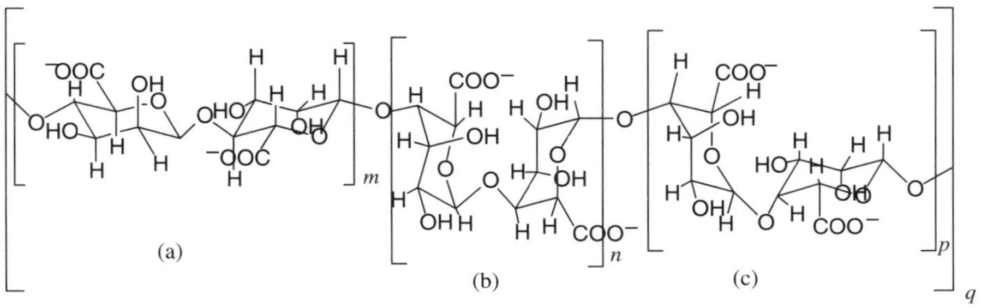

FIGURE 5.2 Chemical structure of alginate (a: M-blocks; b: G-blocks; c: interspersed M and G units).

TABLE 5.4   Application of Alginate in Oral Controlled Release Formulations

Classification	Origin of Alginate	Application	References
Microcapsules for cells	Alginate deposition on cells	Bifidobacteria, *Lactococcus lactis* cells, vaccine	[50, 53–55]
Microparticles of peptides	Thiolated alginate	Subtilisin, insulin, heparin	[45, 46, 48, 51]
Microparticles of small molecules	Alginate	Gliclazide, sulindac, isoniazid, polymyxin B	[47, 49, 56–58]
Matrix tablets	Alginate	Baclofen	[52, 56, 59]

to protect cellular intact. Such cells can be used as carriers for peptide drugs [50]. Alginate microspheres are also shown to be an effective delivery vehicle for oral immunization of ruminants [51].

Alginate shows good compressibility and tablets composed of alginate microparticles have been widely investigated for oral controlled delivery. Modified release alginate tablet formulations of baclofen were prepared by wet granulation. These tablets showed good mechanical properties (hardness and friability) and sustained the rate of baclofen release [52].

5.2.3 Pectins

Pectin is widely studied as a hydrophilic matrix material in recent years. It shows excellent swelling and erosion properties and is useful for colon target delivery [38, 60–62].

5.2.3.1 *Structure and Properties*   Pectin is a complex polysaccharide consisting mainly of esterified D-galacturonic acid residues linked as an α-(1–4) chain. The acid groups

along the chain are largely esterified with methoxy groups in the natural product (Figure 5.3). There can also be acetyl groups present on the free hydroxy groups. The galacturonic acid main chain also has an occasional rhamnose group that disrupts the chain helix formation. Depending on the origin, pectin also contains other neutral sugars such as xylose, galactose, and arabinose. The side chains, known as hairy regions, may appear in groups.

Pectins form gel in acidic media or by cross-linking with calcium ion, via oxidization of the feruloylester substituents, or by reacting with alginate. During the acid-induced gelling, the solution pH is lowered, repressing the ionization of carboxylate groups on pectin. The acid-induced calcium cross-linking has been studied for the development of pectin-based drug delivery systems.

The gelation characteristics of pectins depend on their structure and are divided into two main types: high-ester gelation and low-ester gelation (Table 5.5). Gelation of high-ester pectin usually occurs at a pH of less than 3.5 and a total solid content of more than 55%. Low-ester pectin is gelled with calcium ions and hence is not dependent on the presence

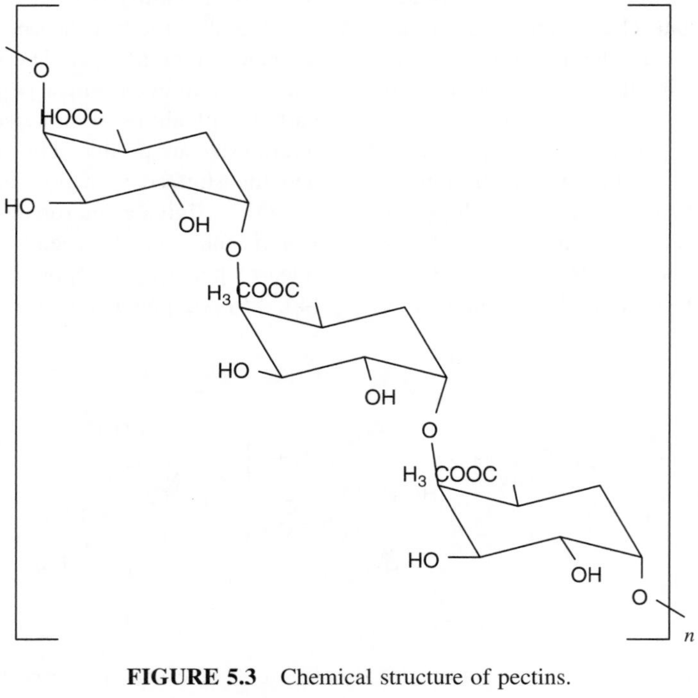

FIGURE 5.3   Chemical structure of pectins.

TABLE 5.5   Type of Pectin and Gelation Characteristics

Type	Methylation Level	Amidation Level	Gelation Characteristics
High methoxy	74–77	0	Ultrarapid set
High methoxy	71–74	0	Rapid set
High methoxy	66–69	0	Medium rapid set
High methoxy	58–65	0	Slow set
Low methoxy	40	0	Slow set
Low methoxy	30	0	Rapid set
Amidated	35	15	Slow set
Amidated	30	20	Rapid set

of acid or a high solid content. The low-ester pectins are more sensitive for acid gelation. Amidation can interfere and slow down the gelation process. Amidated pectin gels have the property of rehealing after shearing. For coating, such as mechanical strength and permeation, amidated calcium pectin would be best in all kinds of pectins.

Pectin gels can be degraded by pectinase enzyme that is abundant in colon, and therefore, it is useful in colon delivery systems.

5.2.3.2 Applications in Oral Controlled Delivery Systems

As an excipient, pectin is used as an adsorbent, emulsion stabilizer, and bulk-forming agent. Recently, pectin has been used in hydrophilic matrix formulations for the oral sustained drug delivery (Table 5.6). Pectins have been explored in colon-biodegradable systems for retarding the onset of drug release [63]. Gamma scintigraphy studies on the gastrointestinal transit and *in vivo* drug release behavior of pectin-coated tablets in humans showed that the amount of radioactive tracer released from the labeled tablets was minimal when the tablets were in the stomach and the small intestine. The radioactivity increased when the tablets reached the colon due to increased degradation of the film coatings by pectinase.

Pectin gel beads formed by ionotropic gelation were studied for drug delivery where drug loading was achieved under mild conditions. Hardening agents such as glutaraldehyde and calcium chloride were used to prepare rigid cross-linked gel matrices and extend the drug release. Cross-linking of pectin inhibited the release of incorporated drug from pectin tablets. Pectin beads cross-linked with calcium ions disintegrated after oral administration due to replacement with other ions in the gastrointestinal tract.

The drug release rate decreased with exposure to increasing concentrations of calcium chloride during preparation. The beads disintegrated in the phosphate buffer (pH 6.8) and the rate of disintegration depended on the calcium chloride concentration used in the preparation. In addition, the drug release rate in buffer solution decreased as the rate of gel erosion decreased.

Zinc pectin microparticles are reported to be better than calcium pectin microparticles in colon delivery systems [64, 65]. Ketoprofen-loaded microparticles exhibited drug release profiles that have 5–38 times slower release of ketoprofen in simulated gastric fluid (pH 7.4) compared to the conventional calcium pectinate beads [66]. Pectin-based matrices with varying degrees of esterification have also been evaluated in oral controlled release tablets.

5.2.4 Hydroxypropyl Methylcellulose

Hydroxypropyl methylcellulose (HPMC) or hypromellose refers to soluble methylcellulose ethers. The structure of HPMC is depicted in Figure 5.4. HPMC is used as a thickening agent, binder, film former, and hydrophilic matrix material. Methocel and Metolose are trademarks of the Dow Chemical Company and Shin-Etsu Chemical Company, respectively, for a line of HPMC products. HPMC polymers for hydrophilic matrix systems are available in various viscosity grades ranging from 4000–100,000 mPa s. Grades may be distinguished by their apparent viscosity. Methocel K4M indicates that the apparent viscosity of a 2% aqueous dispersion at 20°C is 4000 mPa s. In the USP, HPMC is defined according to the substitution type and designated by a four-digit number to the nonproprietary name (e.g., hypromellose 2206). The first two digits refer to the approximate percentage content of the methoxy group ($-OCH_3$). The last two digits refer to the approximate percentage content of the hydroxypropoxy group ($-OCH_2CH(OH)CH_3$), calculated on a dry weight basis (Table 5.7).

5.2.4.1 Properties   HPMC is a white or light gray pellet or fibril power, which is odorless and tasteless. It sparingly dissolves in anhydrous ethanol, ether, or acetone and swells as a clear or slightly muddy colloid solution in cold water.

HPMC has good compressibility and is particularly suitable for direct compression. HPMC matrix systems are considered as simple, low-cost, and easy-to-make sustained release systems. However, the static charge on the surface of

TABLE 5.6   Application of Pectins in Oral Controlled Release Formulations

Classification	Origin of Pectins	Application	References
Coating for tablets or pellets	Pectin, calcium pectin, amidated pectin	Colon delivery of 5-FU, 5-ASA, theophylline	[60, 63, 67–70]
Microparticles or beads	Calcium pectinate, zinc pectinate	Sustained release of catechin; colon delivery of ketoprofen	[64–66, 71]

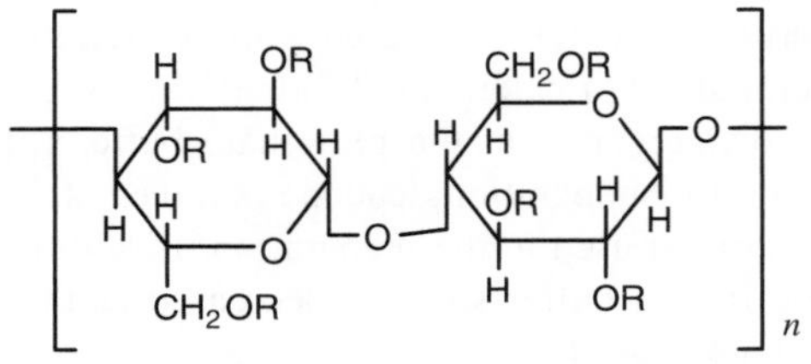

R: -H,-CH$_3$,-(CH$_2$CH(CH$_3$)O)$_x$-H, or-(CH$_2$CH(CH$_3$)O)$_x$-CH$_3$

FIGURE 5.4 Chemical structure of HPMC.

TABLE 5.7 Commercial HPMC Products and USP Nomenclature

Hypromellose Type	Methocel	Metolose	Methoxy Content (%)	Hydroxypropyl Content (%)
1828	None		16.5–20.0	23.0–32.0
2208	K	90SH	19.0–24.0	7.0–12.0
2906	F	65SH	27.0–30.0	4.0–7.5
2910	E	60SH	28.0–30.0	7.0–12.0

HPMC sometimes results in nonuniformity among the tablets during scale-up. When the drugs are highly charged, the release kinetics of the scale-up products may change significantly from the laboratory-made products. In such situations, wet granulation is used to improve the uniformity.

HPMC matrix systems allow greater control and reproducible drug release profiles by manipulation of its chemical and physical properties. HPMC matrix systems absorb water and swell and drug release takes place by diffusion through and/or by erosion of the gel layer. The hydration and gel formation depend on the substitution and molecular weight of polymer, and follow the order HPMC 2208 > HPMC 2910 > HPMC 2906 or Methocel K > Methocel E > Methocel F. HPMC 2208 and HPMC 2910 are used more frequently than HPMC 2906 due to faster hydration and gel formation.

5.2.4.2 Applications in Oral Controlled Release Systems

HPMC is a popular matrix material in oral controlled delivery systems [72]. Its applications include buccal delivery, gastroretentive systems, and colon delivery systems (Table 5.8). HPMC matrices show sustained release pattern by two release mechanisms, that is, diffusion and erosion of the gel

layer. The viscosity of the polymer affects the diffusion pathway. HPMC can be used as a matrix for controlling the release of both hydrophilic and hydrophobic drugs. When hydrophobic drugs are incorporated into HPMC matrix systems, solubilization of the drug should be considered first. For example, controlled release of nifedipine was achieved using a combination of hydroxypropyl methylcellulose (HPMC 2208) and PEG 6000. The crushing strength of the tablet can be markedly increased by increasing compression force in tableting. The release of nifedipine from the PEG–HPMC tablets is significantly delayed with an increase in the amount or viscosity of HPMC.

Buccal delivery systems made of HPMC have been shown to be suitable for drugs that are subjected to first-pass metabolism or unstable within the rest of the gastrointestinal tract. Buccal adhesive tablets and films composed of HPMC show good buccal retention and controlled release characteristics. Propranolol hydrochloride buccal adhesive tablets composed of HPMC K4M, polycarbophil, and lactose showed sustained release kinetics by both diffusion and erosion mechanisms. Mixtures of different polymers with HPMC, such as Carbopol 934, have been used successfully to produce gastroretentive floating delivery systems. An effervescent-based gastric floating delivery system of diltiazem hydrochloride showed 24 h floating and drug release in accordance with the USP dissolution criteria for extended release capsules.

HPMC has been employed as a crystallization inhibitor to prepare supersaturated self-emulsifying drug delivery systems (SEDDS) and increase bioavailability of drugs [73]. HPMC containing SEDDS formulation of paclitaxel showed minimal precipitation and improved pharmacokinetics after oral dosing in rats. They showed approximately 10-fold higher maximum concentration (C_{max}) and 5-fold higher oral bioavailability (F approximately 9.5%) than the orally dosed Taxol formulation (F approximately 2.0%) and a SEDDS formulation without HPMC (F approximately 1%). HPMC is also used for developing time-, pH-, and enzyme-controlled colonic drug delivery systems [74]. Compression coats containing a combination of HPMC and spray-dried chitosan acetate have shown to control the release of 5-aminosalicylic acid until the dosage form reaches the colon. Erosion rates and drug release profiles from these capsules depend on the concentration of HPMC.

HPMC capsules are being used to replace gelatin capsules. HPMC capsules are stable and do not cross-link like

TABLE 5.8 Application of HPMC in Oral Controlled Release Systems

Classification	Polymer	Application	References
Buccal adhesive tablets	HPMC K4M	Pentazocine, propranolol HCl, ketorolac, chlorhexidine, omeprazole, insulin	[76–83]
Gastroretentive systems	HPMC, HPMC K100M	Ofloxacin, ketorolac	[59, 73, 84–86]

the gelatin capsules and are suitable to encapsulate oxidative and hygroscopic drugs. Although HPMC capsules have a tendency to stick to esophagus like gelatin capsules, the force needed to detach the HPMC capsules from esophagus is significantly lower than that needed for the gelatin capsules [75].

5.2.5 Other Cellulose Ethers

Cellulose derivatives such as hydroxypropyl cellulose (HPC) and hydroxyethyl cellulose (HEC) are also used to obtain desirable drug release rates in oral controlled delivery systems (Figure 5.5). They are nonionic water-soluble cellulose ethers with remarkable combinations of properties, for example, organic solvent solubility, thermoplasticity, and surface activity. They dissolve at ordinary temperature in water or in any organic solvents, such as methanol, ethanol, isopropanol, propylene glycol, methylene chloride, acetone, chloroform, and toluene. The molecular weights depend on the degrees of polymerization of the cellulose backbone, which, in turn, controls the viscosity of hydroxypropyl cellulose. The order of mesh size of matrices can be used for predicting drug diffusion and release from cellulose matrices. The order proposed by Peppas and coworkers is HPC < HEC < HPMC K100M < HPMC K4M [87]. Desired released rate can be obtained by combining different cellulose derivatives.

Higher viscosity grades of HPC and HEC have been used as hydrophilic matrix materials for several purposes such as bioadhesive sustained release systems and gastric buoyant sustained release tablets.

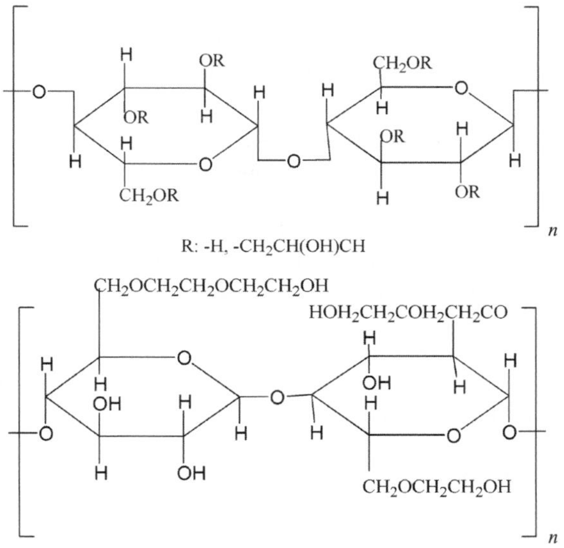

FIGURE 5.5 Chemical structures of HPC (upper) and HEC (lower).

5.3 POLYMERS FOR ORAL MEMBRANE-CONTROLLED SYSTEMS

Osmotic pump delivery systems have popularized membrane-controlled release systems. An osmotic pump is composed of a core and a semipermeable coating, where the core contains drug and osmotic agents. The semipermeable membrane controls the release rate of drug. Polymers used for semipermeable coating are generally insoluble and have adequate film-forming characteristics. Numerous natural, semisynthetic, and synthetic polymers are being used to fabricate osmotic pumps or microosmotic delivery systems.

5.3.1 Cellulose Acetate

Cellulose acetate is an insoluble cellulose derivative and is widely used in oral pharmaceutical products (Figure 5.6). It is regarded as a nontoxic, nonirritant, and biodegradable material. It is heat-resistant and less hygroscopic. Cellulose acetate is partially acetylated cellulose, in which the acetyl content ranges from 29.0% to 44.8%, corresponding to mono-, di-, and triacetate. The content of free hydroxyl is less than that in cellulose. The solubility behavior of cellulose acetate and cellulose diacetate is significantly different from that of cellulose triacetate (Table 5.9).

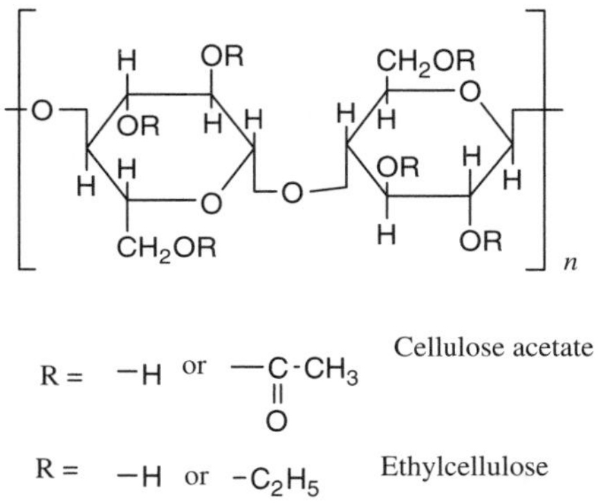

FIGURE 5.6 Chemical structures of cellulose acetate and ethyl cellulose.

TABLE 5.9 Solubility of Cellulose Acetate

Solvents	Cellulose Triacetate	Cellulose Mono- or Diacetate
CH_2Cl_2	Soluble	Soluble
CH_2Cl_2/CH_3OH (9:1)	Soluble	Soluble
CH_2Cl_2/isopropanol (9:1)	Soluble	Soluble
Acetone/CH_3OH (9:1)	Insoluble	Soluble
Acetone/C_2H_5OH (9:1)	Insoluble	Soluble
Acetone	Insoluble	Soluble
Cyclohexanone	Insoluble	Soluble

Cellulose acetate is used for forming semipermeable coating on tablets, especially in osmotic pump-type tablets and microparticles for controlled release of drugs [88, 89]. Cellulose acetate films, in conjunction with other materials, offer controlled release without any necessity of drilling a hole in the coating as with typical osmotic pumps. A controlled porosity osmotic pump delivery system for sodium ferulate containing cellulose acetate as a semipermeable membrane, polyethylene glycol 400 as a pore-forming agent, and dibutylphthalate as a plasticizer and release controller was reported [90].

Cellulose acetate shows excellent compatibility with thermotropic liquid crystal (LC) substances such as *N*-heptyl cyanobiphenyl. A thermosensitive osmotic pump prepared by coating tablets with cellulose acetate incorporated with *n*-heptyl cyanobiphenyl (K21) showed a phase transition temperature of 41.5°C from nematic to isotropic [91]. K21 is a thermosensitive material and used to modulate drug permeation of methimazole and paracetamol (acetaminophen), which were used as hydrophilic and hydrophobic drug models, respectively. The release data indicated that upon changing the temperature of the system around the transition point, cellulose acetate membranes without LC showed no temperature sensitivity to drug permeation, whereas the LC entrapped membranes exhibited a distinct increase in permeability when temperature was raised to above the transition point of the liquid crystal for both drug models. The thermoresponsive drug permeation through the membranes is reversible, reproducible, and follows zero-order kinetics. Compared to cellulose nitrate membrane, LC embedded cellulose acetate membranes showed higher temperature sensitivity, apparently due to higher LC loading in cellulose acetate membrane. The pattern of on–off permeation through LC embedded membranes is more distinguished for methimazole compared to paracetamol due to its lower molecular weight.

5.3.2 Ethylcellulose

Ethylcellulose is a hydrophobic cellulose ether used extensively as a coating material, tablet binder, and matrix former in microcapsules, microspheres, and controlled release dosage forms [92–95] (Figure 5.6). Because of its excellent film-forming characteristics, ethylcellulose became one of the most important coating polymers in controlled release dosage forms. Ethylcellulose membrane is not swellable and is compatible with most plasticizers and pore-forming substances, and therefore, it is possible to adjust the release rate of drugs by optimizing the coating formulation without laser drilling for osmotic pump delivery systems.

Ethylcellulose is dissolvable in organic solvents such as acetone, isopropanol, ethanol, methanol, dichloromethane, and chloroform. The degree of substitution of commercial ethylcellulose grades ranges from 2.25 to 2.81, and ethoxy

TABLE 5.10 Degree of Substitution and Solubility of Ethylcellulose

Degree of Substitution	Solubility
0.5	Soluble in 4–8% NaOH solution
0.8–1.3	Dispersible in water
1.4–1.8	Swellable in water
1.8–2.2	Soluble in polar solvents
2.2–2.4	Soluble in nonpolar solvents
2.4–2.5	Very soluble in nonpolar solvents
2.5–3.0	Only soluble in nonpolar solvents

content ranges from 44% to 52.5%. Ethylcellulose with higher degree of substitution is more hydrophobic in nature (Table 5.10).

The glass transition temperature of ethylcellulose ranges from 106 to 133°C, and its softening point is 156°C. The softening point can be manipulated using plasticizers. The great flexibility of ethylcellulose film over a wide range of temperatures, from −70°C to its softening point, is one of the most marked characteristics. Ethylcellulose films without plasticizers are too tough to be used for coating. Therefore, plasticizers or softening agents are added to ethylcellulose to obtain proper degree of suppleness, lower the softening point, and improve its thermoplasticity. Ethylcellulose is somewhat softer than cellulose acetate, and therefore requires proportionately less plasticizer to attain the desired degree of softness. When ethylcellulose is used as a coating material for pellets, pore-forming agents such as polyethylene glycol are usually included.

Organic solution of ethylcellulose and plasticizers is used in the coating process. Explosion and solvent residue are common problems when the organic solvents are used. Aqueous dispersions of ethylcellulose (e.g., Surelease) are being used in oral CR formulations. They are latex systems of fully plasticized ethylcellulose dispersions with about 30% (w/w) solid content. They also contain dibutyl sebacate and oleic acid as plasticizers and fumed silica as an antiadherent in ammoniated water. The advantages of aqueous dispersions include avoidance of organic solvents and high solid content.

Ethylcellulose aqueous dispersions are sometimes used as granulating agents for formulating nondisintegrating-type prolonged release matrix tablets [96]. The drugs release from tablets either via water-filled pores or by diffusion through membrane, depending on the levels of dispersion. Ethylcellulose is also used in the matrix-controlled delivery systems by the solid dispersion technique. The application of ethylcellulose solid dispersion has proved to be a valuable method in controlling the drug release rate. As the amount of ethylcellulose increases, the drug release rate decreases and the drug release kinetics follows zero-order kinetics.

5.3.3 Polymethacrylates

Acrylic resins such as polymethacrylates have excellent film-forming characteristics and are used for coating of tablets, pellets, capsules, and granules. Acrylic resins are pharmacologically inactive and have good compatibility with skin and mucosal membranes. Eudragit is a trade name of copolymers derived from esters of acrylic and methacrylic acids, whose properties are determined by functional groups (Figure 5.7). Eudragit grades differ in their proportion of neutral, alkaline, or acid groups resulting in different physicochemical properties and offer excellent choice for the formulating scientists.

The glass transition temperature of gastrosoluble polymethacrylate grade (e.g., Eudragit E or NE) is $-8°C$, enteric grade (Eudragit L) is over $160°C$, and insoluble one (Eudragit RL/RS) is about $55°C$. Plasticizers lower the glass transition temperature by 15–25°C resulting in optimal film-forming properties of polymethacrylate polymers. For obtaining a proper degree of suppleness, 10% plasticizer is needed for insoluble grade and 40% plasticizer for enteric grade, but no plasticizer is needed for gastrosoluble grade methacrylate polymers.

Enteric-type methacrylate polymers such as Eudragit L/S coatings provide a better barrier against drug release in the stomach and enable controlled release in the intestine. Enteric-type Eudragit FS is employed for controlled release of drugs in the colon. Efforts are being made to target therapeutic peptides to colon using such polymers [97]. Eudragit RS and RL types are widely used in controlled release coating. The amount of quaternary ammonium groups in RS type is between 4.5% and 6.8% and in RL type is between 8.8% and 12%. Both RS and RL types are insoluble at physiological pH values and capable of swelling. Eudragit RL100 and RS100 are pH-independent cationic polymers with low to chemical reactivity [98]. Eudragit RL films are more permeable than the Eudragit RS films. The skillful use of combination of different Eudragit polymers offers ideal solutions for controlled drug release in various pharmaceutical and technical applications. Traditionally, these meth-

acrylate polymers are dissolved in organic solvents such as acetone and isopropanol for coating applications. Organic coating dispersions of methacrylates are generally turbid and disturb the coating process. Addition of a small amount of water will improve the appearance of organic coating dispersions.

Aqueous dispersions of polymethacrylates (e.g., Eudragit RL 30D and Eudragit RS 30D) are being used for coating applications in recent years. Use of aqueous dispersions as coating materials always results in stickiness, and therefore, glidants such as talc and plasticizers such as triethyl citrate are included prior to coating application. The dispersions are also diluted with required amount of water to adjust polymethacrylate content.

5.4 POLYMERS FOR BIOADHESIVE CONTROLLED RELEASE SYSTEMS

A bioadhesive polymer adheres to a biological surface such as mucus and provides an extended time for drug absorption. Bioadhesive controlled delivery systems are emerging as potential drug delivery systems due to their applicability for prolonging gastrointestinal residence time of dosage forms, enhancing contact between the drug and its absorbing surface, and improving bioavailability [99]. Many hydrophilic polymers such as Carbopol, alginate, polyethylene oxide, and pectin adhere to mucosal surfaces upon hydration at the mucus gel layer of the epithelial surface. This mechanism of adhesion is known as "adhesion by hydration."

Bioadhesion of a polymer depends on the porosity, flexibility, and hydration rate. The force of attachment is related to the depth of interpenetration of polymers and mucous. Nonhydrogel-forming polymers that contain polar/charged groups or active interaction sites can also show intense bioadhesive properties by formation of chemical bonds such as ionic bonds, van der Waals interactions, or hydrogen bonds with the mucus. Many synthetic polymers such as cellulose phthalate, carboxyethyl cellulose, hydroxypropyl cellulose, hydroxyethyl cellulose, hydroxypropyl methylcellulose, hydroxyethyl methacrylate, polyvinylpyrrolidone, polyvinyl alcohol, pectin, polycarbophil, polyacrylic acid, and polymethacrylic acid and natural polymers such as chitosan and sodium alginate have shown to exhibit bioadhesive properties [100]. Recently, thiolated polymers have been investigated as mucoadhesive materials. Polyacrylic acid polymers (e.g., carbomers) and polyethylene oxides (e.g., Polyox) have been successfully used in such bioadhesive drug delivery systems and commercialized.

5.4.1 Carbomers

Carbomers are high molecular weight, cross-linked acrylic acid-based polymers. They show good mucoadhesion, espe-

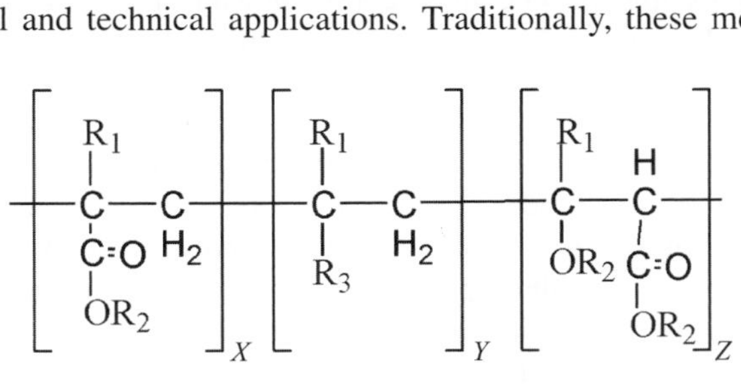

$R_1=CH_3; H$

$R_2=CH_3; CH_2CH_3$

$R_3=COOH;$ Eudraigit L&S

$R_3=CH_2CH_2N^+(CH_3)Cl^-;$ Eudragit RL &RS

FIGURE 5.7 Chemical structure of Eudragit.

cially at low pH where they are in the protonated state. The USP, European Pharmacopoeia, British Pharmacopoeia, United States Adopted Names (USAN) Council, and International Nomenclature for Cosmetic Ingredients (INCI) have adopted the generic (nonproprietary) name *carbomer* for various Carbopol homopolymers. The Japanese Pharmacopoeia designates Carbopol homopolymers as *carboxyvinyl polymer* and *carboxy polymethylene*. Depending upon the degree of cross-linking and manufacturing conditions, various grades of carbomers are available. Each grade has its significance and usefulness in pharmaceutical dosage forms. Carbopol 934 is cross-linked with allyl sucrose and is polymerized in benzene. Carbopol 71, 971, and 974 are cross-linked with allyl pentaerythritol and polymerized in ethyl acetate. Carbomer polymers contain 56–68% of carboxylic acid (–COOH) groups and 0.75–2% of cross-linking agents. Though Carbomer 971 and Carbomer 974 are manufactured by same process under similar conditions, Carbomer 971 has slightly lower level of cross-linking agent than Carbomer 974. Carbomer 71G has the granular form and is used in tablet manufacturing. The general structure of carbomers is shown in Figure 5.8.

Carbomers absorb water readily upon contact with water, get hydrated, and swell. In aqueous vehicles at a pH range of 4–6, they swell up to 1000 times of their original volume and 10 times their original diameter to form a gel. Because the pK_a value of these polymers is 6.0–6.5, the carboxylate moiety on the polymer ionizes, resulting in repulsion between the negative charges, which adds to the swelling of the polymer. The glass transition temperature of carbomers is 105°C in powder form. It decreases significantly as the polymer comes into contact with water. The polymer chains start gyrating and the radius of gyration becomes increasingly larger. Macroscopically, this phenomenon appears as swelling.

In a pH 7.2 phosphate buffer, the carboxylic groups of carbomers are highly dissociated. The repulsion between the negatively charged carboxyl groups causes uncoiling and expansion of molecules and results in gel formation. The gel consists of closely packed swollen particles, which delay drug release from the matrices. As the concentration of drug

increases, its release from the gel increases, showing concentration dependency.

In tablets, carbomers are employed as a binder and control release excipient. It is suitable for both wet and dry granulation processes. Aqueous, hydroalcoholic, anhydrous organic solvent dispersions of carbomers are used in wet granulation processes. Certain cationic inert components are included to reduce the tackiness of wet mass and improve mixing. Talc is added to reduce tackiness of aqueous dispersions of carbomer. Carbomer matrix is significantly different from HPMC matrix in that it undergoes pH-dependent gelling and swelling when it comes in contact with the GI fluid and the gel remains intact during the swelling process. Zero-order release kinetics may be achieved by manipulation of carbomer concentration [96, 101–103].

Carbomers are also used in the preparation of controlled release matrix beads and microspheres [104]. Studies have shown that bioadhesive oral formulations containing carbomers increase the gastrointestinal residence time and improve bioavailability of drugs. Albumin microspheres with 30% polycarbopol particles adhered to stomach mucosa and improved the oral bioavailability of chlorothiazide, a drug that shows site-specific absorption only in proximal parts [83]. Scintigraphic studies have shown threefold enhancement in gastroretention of ethylcellulose–carbomer granules compared to ethylcellulose granules. Magnetic granules containing carbomers have shown potential for site-specific and controlled drug delivery to the esophagus [105, 106].

5.4.2 Polyethylene Oxide

Polyethylene oxide is a nonionic homopolymer of ethylene oxide and represented by the formula $(CH_2CH_2O)_n$. The letter n denotes the average number of oxyethylene groups. It is prepared by the polymerization of ethylene oxide with a suitable catalyst. Up to 3% of silicon dioxide may also be present in polyethylene oxide. Polyethylene oxides from Dow excipients are commercially known as Polyox and are available in different molecular weight grades. The molecular weights of Polyox 301, Polyox coagulant, and Polyox 303 are 4,000,000, 5,000,000, and 7,000,000, respectively.

Polyethylene oxide is an excellent bioadhesive polymer and used as a binder and controlled release component in tablet formulations [107–110]. These polymers show pH-independent swelling, a property that is considered superior than some other bioadhesive and swelling type polymers. Polymeric compacts made of hydroxypropyl methylcellulose, Carbopol 974, and polyethylene oxide show regional variations in their adhesion with porcine buccal, gastric, and intestinal mucosa. The swelling behavior of Polyox hydrogels is not influenced by the pH or ionic strength of medium and shows similar bioadhesive trend in acidic and neutral pH environments. In contrast, Carbopols show pH-dependent

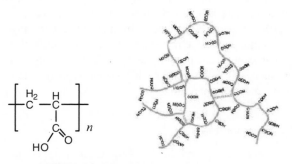

FIGURE 5.8 General structure of carbomers.

swelling and variations in bioadhesion at different locations of the gastrointestinal tract. Studies have shown that the swelling ratio of PEO is approximately twice that of HPMC matrices in acidic and neutral pH environments.

The relationship between the molecular weight and swelling capacity is evaluated when selecting specific grades for immediate release or sustained release matrix formulations. High molecular weight grades are generally used to provide delayed drug release in hydrophilic matrix systems and higher bioadhesion and gastroretention. An increase in the molecular weight of Polyox from 6×10^5 to 7×10^6 decreases the rate of drug release due to increased degree of entanglement of the macromolecules, decreased water imbibition, and reduced polymer chain mobility. Like molecular weight, the amount of Polyox in formulation also affects the rate of drug release. In general, increasing the polymer content reduces the rate of drug release from Polyox matrices due to a decrease in the fluid-filled channels that are required for the mobility of dissolved drug molecules in the polymer matrix. The swelling, bioadhesion, and rate of drug release from Polyox matrices can be manipulated by mixing with other hydrophilic polymers such as polyvinylpyrrolidone [111].

5.5 SUMMARY

Advances in synthetic organic chemistry and biotechnology are enabling the development of a wide range of novel polymeric materials for developing controlled drug delivery systems vehicles. Novel polymers provide new dimensions for the advancement and development of oral controlled release delivery systems. A thorough understanding of the polymer molecular properties as well as the underlying drug release mechanisms is highly desirable for the successful development of oral controlled release delivery systems.

REFERENCES

1. George M, Abraham TE. Polyionic hydrocolloids for the intestinal delivery of protein drugs: alginate and chitosan—a review. *J. Control. Release* 2006;114(1):1–14.

2. Aiedeh K, Taha MO. Synthesis of iron-crosslinked chitosan succinate and iron-crosslinked hydroxamated chitosan succinate and their *in vitro* evaluation as potential matrix materials for oral theophylline sustained-release beads. *Eur. J. Pharm. Sci.* 2001;13(2):159–168.

3. Atyabi F, Majzoob S, Dorkoosh F, Sayyah M, Ponchel G. The impact of trimethyl chitosan on *in vitro* mucoadhesive properties of pectinate beads along different sections of gastrointestinal tract. *Drug Dev. Ind. Pharm.* 2007;33(3):291–300.

4. Chandy T, Rao GH, Wilson RF, Das GS. Delivery of LMW heparin via surface coated chitosan/peg-alginate microspheres prevents thrombosis. *Drug Deliv.* 2002;9(2):87–96.

5. Iruin A, Fernandez-Arevalo M, Alvarez-Fuentes J, Fini A, Holgado MA. Elaboration and *"in vitro"* characterization of 5-ASA beads. *Drug Dev. Ind. Pharm.* 2005;31(2):231–239.

6. Lin YH, Liang HF, Chung CK, Chen MC, Sung HW. Physically crosslinked alginate/N,O-carboxymethyl chitosan hydrogels with calcium for oral delivery of protein drugs. *Biomaterials* 2005;26(14):2105–2113.

7. Onal S, Zihnioglu F. Encapsulation of insulin in chitosan-coated alginate beads: oral therapeutic peptide delivery. *Artif. Cells blood Substit. Immobil. Biotechnol.* 2002;30(3):229–237.

8. Rajinikanth PS, Mishra B. Preparation and *in vitro* characterization of gellan based floating beads of acetohydroxamic acid for eradication of *H. pylori. Acta Pharm. (Zagreb, Croatia)* 2007;57(4):413–427.

9. Andersson M, Lofroth JE. Small particles of a heparin/chitosan complex prepared from a pharmaceutically acceptable microemulsion. *Int. J. Pharm.* 2003;257(1–2):305–309.

10. Cano-Cebrian MJ, Zornoza T, Granero L, Polache A. Intestinal absorption enhancement via the paracellular route by fatty acids, chitosans and others: a target for drug delivery. *Curr. Drug Deliv.* 2005;2(1):9–22.

11. Foger F, Kafedjiiski K, Hoyer H, Loretz B, Bernkop-Schnurch A. Enhanced transport of P-glycoprotein substrate saquinavir in presence of thiolated chitosan. *J. Drug Target.* 2007;15(2):132–139.

12. Garcia-Fuentes M, Prego C, Torres D, Alonso MJ. A comparative study of the potential of solid triglyceride nanostructures coated with chitosan or poly(ethylene glycol) as carriers for oral calcitonin delivery. *Eur. J. Pharm. Sci.* 2005;25(1):133–143.

13. Lamprecht A, Yamamoto H, Takeuchi H, Kawashima Y. pH-sensitive microsphere delivery increases oral bioavailability of calcitonin. *J. Control. Release* 2004;98(1):1–9.

14. Maestrelli F, Zerrouk N, Chemtob C, Mura P. Influence of chitosan and its glutamate and hydrochloride salts on naproxen dissolution rate and permeation across Caco-2 cells. *Int. J. Pharm.* 2004;271(1–2):257–267.

15. Martien R, Loretz B, Schnurch AB. Oral gene delivery: design of polymeric carrier systems shielding toward intestinal enzymatic attack. *Biopolymers* 2006;83(4):327–336.

16. Martin L, Wilson CG, Koosha F, Uchegbu IF. Sustained buccal delivery of the hydrophobic drug denbufylline using physically cross-linked palmitoyl glycol chitosan hydrogels. *Eur. J. Pharm. Biopharm.* 2003;55(1):35–45.

17. Palmberger TF, Hombach J, Bernkop-Schnurch A. Thiolated chitosan: development and *in vitro* evaluation of an oral delivery system for acyclovir. *Int. Pharm.* 2008;348(1–2):54–60.

18. Prego C, Torres D, Fernandez-Megia E, Novoa-Carballal R, Quinoa E, Alonso MJ. Chitosan-PEG nanocapsules as new carriers for oral peptide delivery. Effect of chitosan pegylation degree. *J. Control. Release* 2006;111(3):299–308.

19. Raiman J, Tormalehto S, Yritys K, Junginger HE, Monkkonen J. Effects of various absorption enhancers on transport of clodro-

nate through Caco-2 cells. *Int. J. Pharm.* 2003;261(1–2): 129–136.

20. Ross BP, Toth I. Gastrointestinal absorption of heparin by lipidization or coadministration with penetration enhancers. *Curr. Drug Deliv.* 2005;2(3):277–287.

21. Shah P, Jogani V, Mishra P, Mishra AK, Bagchi T, Misra A. Modulation of ganciclovir intestinal absorption in presence of absorption enhancers. *J. Pharm. Sci.* 2007;96(10): 2710–2722.

22. Thanou M, Henderson S, Kydonieus A, Elson C. *N*-Sulfonato-*N,O*-carboxymethylchitosan: a novel polymeric absorption enhancer for the oral delivery of macromolecules. *J. Control. Release* 2007;117(2):171–178.

23. van der Merwe SM, Verhoef JC, Kotze AF, Junginger HE. *N*-Trimethyl chitosan chloride as absorption enhancer in oral peptide drug delivery. Development and characterization of minitablet and granule formulations. *Eur. J. Pharm. Biopharm.* 2004;57(1):85–91.

24. Bernkop-Schnurch A, Pinter Y, Guggi D, Kahlbacher H, Schoffmann G, Schuh M, et al. The use of thiolated polymers as carrier matrix in oral peptide delivery—proof of concept. *J. Control. Release* 2005;106(1–2):26–33.

25. Bernkop-Schnurch A, Schuhbauer H, Clausen AE, Hanel R. Development of a sustained release dosage form for alpha-lipoic acid. I. Design and *in vitro* evaluation. *Drug Dev. Ind. Pharm.* 2004;30(1):27–34.

26. Di Colo G, Baggiani A, Zambito Y, Mollica G, Geppi M, Serafini MF. A new hydrogel for the extended and complete prednisolone release in the GI tract. *Int. J. Pharm.* 310(1–2): 2006 154–161.

27. Guggi D, Kast CE, Bernkop-Schnurch A. *In vivo* evaluation of an oral salmon calcitonin-delivery system based on a thiolated chitosan carrier matrix. *Pharm. Res.* 2003;20(12): 1989–1994.

28. Hombach J, Hoyer H, Bernkop-Schnurch A. Thiolated chitosans: development and *in vitro* evaluation of an oral tobramycin sulphate delivery system. *Eur. J. Pharm. Sci.* 2008;33 (1):1–8.

29. Krauland AH, Guggi D, Bernkop-Schnurch A. Oral insulin delivery: the potential of thiolated chitosan-insulin tablets on non-diabetic rats. *J. Control. Release* 2004;95(3): 547–555.

30. Le Tien C, Lacroix M, Ispas-Szabo P, Mateescu MA. *N*-Acylated chitosan: hydrophobic matrices for controlled drug release. *J. Control. Release* 2003;93(1):1–13.

31. Mura P, Zerrouk N, Mennini N, Maestrelli F, Chemtob C. Development and characterization of naproxen–chitosan solid systems with improved drug dissolution properties. *Eur. J. Pharm. Sci.* 2003;19(1):67–75.

32. El-Gibaly I. Development and *in vitro* evaluation of novel floating chitosan microcapsules for oral use: comparison with non-floating chitosan microspheres. *Int. J. Pharm.* 2002;249 (1–2):7–21.

33. Fetih G, Fausia H, Okada N, Fujita T, Attia M, Yamamoto A. Colon-specific delivery and enhanced colonic absorption of [Asu(1,7)]-eel calcitonin using chitosan capsules containing various additives in rats. *J. Drug Target.* 2006;14(3): 165–172.

34. Fetih G, Lindberg S, Itoh K, Okada N, Fujita T, Habib F, et al. Improvement of absorption enhancing effects of *n*-dodecyl-beta-D-maltopyranoside by its colon-specific delivery using chitosan capsules. *Int. J. Pharm.* 2005;293(1–2): 127–135.

35. Margulis EB, Moustafine RI. Swellability testing of chitosan/Eudragit L100-55 interpolyelectrolyte complexes for colonic drug delivery. *J. Control. Release* 2006; 28; 116(2):e36–e37.

36. McConnell EL, Murdan S, Basit AW. An investigation into the digestion of chitosan (noncrosslinked and crosslinked) by human colonic bacteria. *J. Pharm. Sci.* 2008;97(9): 3820–3829.

37. Park HS, Lee JY, Cho SH, Baek HJ, Lee SJ. Colon delivery of prednisolone based on chitosan coated polysaccharide tablets. *Arch. Pharm. Res.* 2002;25(6):964–968.

38. Patel M, Shah T, Amin A. Therapeutic opportunities in colon-specific drug-delivery systems. *Crit. Rev. Ther. Drug Carrier Syst.* 2007;24(2):147–202.

39. Shimono N, Takatori T, Ueda M, Mori M, Higashi Y, Nakamura Y. Chitosan dispersed system for colon-specific drug delivery. *Int. J. Pharm.* 2002;245(1–2):45–54.

40. Simonoska Crcarevska M, Glavas Dodov M, Goracinova K. Chitosan coated Ca-alginate microparticles loaded with budesonide for delivery to the inflamed colonic mucosa. *Eur. J. Pharm. Biopharm.* 2008;68(3):565–578.

41. Smoum R, Rubinstein A, Srebnik M. Chitosan–pentaglycine–phenylboronic acid conjugate: a potential colon-specific platform for calcitonin. *Bioconjug. Chem.* 2006;17(4): 1000–1007.

42. Tozaki H, Fujita T, Odoriba T, Terabe A, Suzuki T, Tanaka C, et al. Colon-specific delivery of R68070, a new thromboxane synthase inhibitor, using chitosan capsules: therapeutic effects against 2,4,6-trinitrobenzene sulfonic acid-induced ulcerative colitis in rats. *Life Sci.* 1999; 64(13):1155–1162.

43. Win PP, Shin-ya Y, Hong K-J, Kajiuchi T. Formulation and characterization of pH sensitive drug carrier based on phosphorylated chitosan (PCS). *Carbohydr. Polym.* 2003;53: 305–310.

44. Tu J, Bolla S, Barr J, Miedema J, Li X, Jasti B. Alginate microparticles prepared by spray-coagulation method: preparation, drug loading and release characterization. *Int. J. Pharm.* 2005;303(1–2):171–181.

45. Scocca S, Faustini M, Villani S, Munari E, Conte U, Russo V, et al. Alginate/polymethacrylate copolymer microparticles for the intestinal delivery of enzymes. *Curr. Drug Deliv.* 2007;4 (2):103–108.

46. Greimel A, Werle M, Bernkop-Schnurch A. Oral peptide delivery: *in-vitro* evaluation of thiolated alginate/poly(acrylic acid) microparticles. *J. Pharm. Pharmacol.* 2007;59(9): 1191–1198.

47. Rastogi R, Sultana Y, Aqil M, Ali A, Kumar S, Chuttani K, et al. Alginate microspheres of isoniazid for oral sustained drug delivery. *Int. J. Pharm.* 2007;334(1–2):71–77.

48. Reis CP, Ribeiro AJ, Neufeld RJ, Veiga F. Alginate microparticles as novel carrier for oral insulin delivery. *Biotechnol. Bioeng.* 2007;96(5):977–989.

49. Coppi G, Iannuccelli V, Sala N, Bondi M. Alginate microparticles for Polymyxin B Peyer's patches uptake: microparticles for antibiotic oral administration. *J. Microencapsul.* 2004;21(8):829–839.

50. Lawuyi B, Chen H, Afkhami F, Kulamarva A, Prakash S. Microencapsulated engineered *Lactococcus lactis* cells for heterologous protein delivery: preparation and *in vitro* analysis. *Appl. Biochem. Biotechnol.* 2007;142(1):71–80.

51. Simi CK, Emilia Abraham T. Encapsulation of crosslinked subtilisin microcrystals in hydrogel beads for controlled release applications. *Eur. J. Pharm. Sci.* 2007;32(1):17–23.

52. Abdelkader H, Abdalla OY, Salem H. Formulation of controlled-release baclofen matrix tablets: influence of some hydrophilic polymers on the release rate and *in vitro* evaluation. *AAPS PharmSciTech* 2007;8(4):E100.

53. Ajdary S, Dobakhti F, Taghikhani M, Riazi-Rad F, Rafiei S, Rafiee-Tehrani M. Oral administration of BCG encapsulated in alginate microspheres induces strong Th1 response in BALB/c mice. *Vaccine* 2007;25(23):4595–4601.

54. Cui JH, Cao QR, Lee BJ. Enhanced delivery of bifidobacteria and fecal changes after multiple oral administrations of bifidobacteria-loaded alginate poly-L-lysine microparticles in human volunteers. *Drug Deliv.* 2007;14(5):265–271.

55. Kim B, Bowersock T, Griebel P, Kidane A, Babiuk LA, Sanchez M, et al. Mucosal immune responses following oral immunization with rotavirus antigens encapsulated in alginate microspheres. *J. Control. Release* 2002;85(1–3):191–202.

56. Al-Kassas RS, Al-Gohary OM, Al-Faadhel MM. Controlling of systemic absorption of gliclazide through incorporation into alginate beads. *Int. J. Pharm.* 2007;341(1–2):230–237.

57. Silva CM, Veiga F, Ribeiro AJ, Zerrouk N, Arnaud P. Effect of chitosan-coated alginate microspheres on the permeability of Caco-2 cell monolayers. *Drug Dev. Ind. Pharm.* 2006;32(9):1079–1088.

58. Yegin BA, Moulari B, Durlu-Kandilci NT, Korkusuz P, Pellequer Y, Lamprecht A. Sulindac loaded alginate beads for a mucoprotective and controlled drug release. *J. Microencapsul.* 2007;24(4):371–382.

59. Coviello T, Matricardi P, Marianecci C, Alhaique F. Polysaccharide hydrogels for modified release formulations. *J. Control. Release* 2007;119(1):5–24.

60. Chourasia MK, Jain SK. Pharmaceutical approaches to colon targeted drug delivery systems. *J. Pharm. Pharm. Sci.* 2003;6(1):33–66.

61. Liu L, Fishman ML, Kost J, Hicks KB. Pectin-based systems for colon-specific drug delivery via oral route. *Biomaterials* 2003;24(19):3333–3343.

62. Sande SA. Pectin-based oral drug delivery to the colon. *Expert Opin. Drug Deliv.* 2005;2(3):441–450.

63. Ofori-Kwakye K, Fell JT, Sharma HL, Smith AM. Gamma scintigraphic evaluation of film-coated tablets intended for colonic or biphasic release. *Int. J. Pharm.* 2004;270(1–2):307–313.

64. Chambin O, Dupuis G, Champion D, Voilley A, Pourcelot Y. Colon-specific drug delivery: influence of solution reticulation properties upon pectin beads performance. *Int. J. Pharm.* 2006;321(1–2):86–93.

65. Dupuis G, Chambin O, Genelot C, Champion D, Pourcelot Y. Colonic drug delivery: influence of cross-linking agent on pectin beads properties and role of the shell capsule type. *Drug Dev. Ind. Pharm.* 2006;32(7):847–855.

66. El-Gibaly I. Oral delayed-release system based on Zn-pectinate gel (ZPG) microparticles as an alternative carrier to calcium pectinate beads for colonic drug delivery. *Int. J. Pharm.* 2002;232(1–2):199–211.

67. Wei H, Qing D, De-Ying C, Bai X, Li-Fang F. *In-vitro* and *in-vivo* studies of pectin/ethylcellulose film-coated pellets of 5-fluorouracil for colonic targeting. *J. Pharm. Pharmacol.* 2008;60(1):35–44.

68. Sriamornsak P, Kennedy RA. Development of polysaccharide gel-coated pellets for oral administration: swelling and release behavior of calcium pectinate gel. *AAPS PharmSciTech* 2007;8(3):E79.

69. Sriamornsak P, Burton MA, Kennedy RA. Development of polysaccharide gel coated pellets for oral administration. 1. Physico-mechanical properties. *Int. J. Pharm.* 2006;326 (1–2):80–88.

70. Xu C, Zhang JS, Mo Y, Tan RX. Calcium pectinate capsules for colon-specific drug delivery. *Drug Dev. Ind. Pharm.* 2005;31(2):127–134.

71. Lee JS, Chung D, Lee HG. Preparation and characterization of calcium pectinate gel beads entrapping catechin-loaded liposomes. *Int. J. Biol. Macromol.* 2008;42(2):178–184.

72. Li CL, Martini LG, Ford JL, Roberts M. The use of hypromellose in oral drug delivery. *J. Pharm. Pharmacol.* 2005;57 (5):533–546.

73. Gao P, Rush BD, Pfund WP, Huang T, Bauer JM, Morozowich W, et al. Development of a supersaturable SEDDS (S-SEDDS) formulation of paclitaxel with improved oral bioavailability. *J. Pharm. Sci.* 2003;92(12):2386–2398.

74. Nunthanid J, Huanbutta K, Luangtana-Anan M, Sriamornsak P, Limmatvapirat S, Puttipipatkhachorn S. Development of time-, pH-, and enzyme-controlled colonic drug delivery using spray-dried chitosan acetate and hydroxypropyl methylcellulose. *Eur. J. Pharm. Biopharm.* 2008;68(2):253–259.

75. Honkanen O, Laaksonen P, Marvola J, Eerikainen S, Tuominen R, Marvola M. Bioavailability and *in vitro* oesophageal sticking tendency of hydroxypropyl methylcellulose capsule formulations and corresponding gelatine capsule formulations. *Eur. J. Pharm. Sci.* 2002;15(5):479–488.

76. Akbari J, Nokhodchi A, Farid D, Adrangui M, Siahi-Shadbad MR, Saeedi M. Development and evaluation of buccoadhesive propranolol hydrochloride tablet formulations: effect of fillers. *Farmaco* 2004;59(2):155–161.

77. Choi H, Jung J, Yong CS, Rhee C, Lee M, Han J, et al. Formulation and *in vivo* evaluation of omeprazole buccal adhesive tablet. *J. Control. Release* 2000;68(3):405–412.

78. Alsarra IA, Alanazi FK, Mahrous GM, Abdel Rahman AA, Al Hezaimi KA. Clinical evaluation of novel buccoadhesive

film containing ketorolac in dental and post-oral surgery pain management. *Pharmazie* 2007;62(10):773–778.

79. Carlo Ceschel G, Bergamante V, Calabrese V, Biserni S, Ronchi C, Fini A. Design and evaluation *in vitro* of controlled release mucoadhesive tablets containing chlorhexidine. *Drug Dev. Ind. Pharm.* 2006;32(1):53–61.

80. Hosny EA, Elkheshen SA, Saleh SI. Buccoadhesive tablets for insulin delivery: *in-vitro* and *in-vivo* studies. *Boll. Chim. Farm.* 2002;141(3):210–217.

81. Agarwal V, Mishra B. Design, development, and biopharmaceutical properties of buccoadhesive compacts of pentazocine. *Drug Dev. Ind. Pharm.* 1999;25(6):701–709.

82. Ali J, Hasan S, Ali M. Formulation and development of gastroretentive drug delivery system for ofloxacin. *Methods Find. Exp. Clin. Pharmacol.* 2006;28(7):433–439.

83. Harris D, Fell JT, Taylor DC, Lynch J, Sharma HL. Oral availability of a poorly absorbed drug, hydrochlorothiazide, from a bioadhesive formulation in the rat. *Int. J. Pharm.* 1989;56(2):97–102.

84. McConville JT, Ross AC, Florence AJ, Stevens HN. Erosion characteristics of an erodible tablet incorporated in a time-delayed capsule device. *Drug Dev. Ind. Pharm.* 2005;31(1):79–89.

85. Chavanpatil M, Jain P, Chaudhari S, Shear R, Vavia P. Development of sustained release gastroretentive drug delivery system for ofloxacin: *in vitro* and *in vivo* evaluation. *Int. J. Pharm.* 2005;304(1–2):178–184.

86. Chavanpatil MD, Jain P, Chaudhari S, Shear R, Vavia PR. Novel sustained release, swellable and bioadhesive gastroretentive drug delivery system for ofloxacin. *Int. J. Pharm.* 2006;316(1–2):86–92.

87. Baumgartner S, Kristl J, Peppas NA. Network structure of cellulose ethers used in pharmaceutical applications during swelling and at equilibrium. *Pharm. Res.* 2002;19(8):1084–1090.

88. Bindschaedler C, Gurny R, Doelker E. Osmotically controlled drug delivery systems produced from organic solutions and aqueous dispersions of cellulose acetate. *J. Control. Release* 1986;4(3):203–212.

89. Khan MA, Sastry SV, Vaithiyalingam SR, Agarwal V, Nazzal S, Reddy IK. Captopril gastrointestinal therapeutic system coated with cellulose acetate pseudolatex: evaluation of main effects of several formulation variables. *Int. J. Pharm.* 2000;193(2):147–156.

90. He L, Gong T, Zhao D, Zhang ZR, Li L. A novel controlled porosity osmotic pump system for sodium ferulate. *Pharmazie* 2006;61(12):1022–1027.

91. Atyabi F, Khodaverdi E, Dinarvand R. Temperature modulated drug permeation through liquid crystal embedded cellulose membranes. *Int. J. Pharm.* 2007;339(1–2):213–221.

92. Dabbagh MA, Ford JL, Rubinstein MH, Hogan JE. Effects of polymer particle size, compaction pressure and hydrophilic polymers on drug release from matrices containing ethylcellulose. *Int. J. Pharm.* 1996;140(1):85–95.

93. Katikaneni PR, Upadrashta SM, Neau SH, Mitra AK. Ethylcellulose matrix controlled release tablets of a water-soluble drug. *Int. J. Pharm.* 1995;123(1):119–125.

94. Lindstedt B, Sjöberg M, Hjärtstam J. Osmotic pumping release from KCl tablets coated with porous and non-porous ethylcellulose. *Int. J. Pharm.* 1991;67(1):21–27.

95. Majid Khan G, Bi Zhu J. Ibuprofen release kinetics from controlled-release tablets granulated with aqueous polymeric dispersion of ethylcellulose II: influence of several parameters and coexcipients. *J. Control. Release* 56(1–3): 1998 127–134.

96. George HK, Evone SG, Stuart CP, Joseph BS. Formulation of Controlled Release Matrices by Granulation with a Polymer Dispersion. *Drug Dev. Ind. Pharm.* 1990;16(9):1473–1490.

97. Brunner M, Lackner E, Exler PS, Fluiter HC, Kletter K, Tschurlovits M, et al. 5-Aminosalicylic acid release from a new controlled-release mesalazine formulation during gastrointestinal transit in healthy volunteers. *Aliment. Pharmacol. Ther.* 2006;23(1):137–144.

98. Dillen K, Vandervoort J, Van den Mooter G, Ludwig A. Evaluation of ciprofloxacin-loaded Eudragit RS100 or RL100/PLGA nanoparticles. *Int. J. Pharm.* 2006;314(1):72–82.

99. Helliwell M. The use of bioadhesives in targeted delivery within the gastrointestinal tract. *Adv. Drug Deliv. Rev.* 1993;11(3):221–251.

100. Salamat-Miller N, Chittchang M, Johnston TP. The use of mucoadhesive polymers in buccal drug delivery. *Adv. Drug Deliv. Rev.* 2005;57(11):1666–1691.

101. Owens TS, Dansereau RJ, Sakr A. Development and evaluation of extended release bioadhesive sodium fluoride tablets. *Int. J. Pharm.* 2005;288(1):109–122.

102. Petrovic A, Cvetkoviæ N, Trajkovic S, Ibric S, Popadic D, Djuric Z. Mixture design evaluation of drug release from matrix tablets containing carbomer and HPMC. *J. Control. Release* 2006;116(2):e104–e106.

103. Tapia-Albarran M, Villafuerte-Robles L. Assay of amoxicillin sustained release from matrix tablets containing different proportions of Carbopol 971P NF. *Int. J. Pharm.* 2004;273(1–2):121–127.

104. Bommareddy GS, Paker-Leggs S, Saripella KK, Neau SH. Extruded and spheronized beads containing Carbopol® 974P to deliver nonelectrolytes and salts of weakly basic drugs. *Int. J. Pharm.* 2006;321(1–2):62–71.

105. Ito R, Machida Y, Sannan T, Nagai T. Magnetic granules: a novel system for specific drug delivery to esophageal mucosa in oral administration. *Int. J. Pharm.* 1990;61(1–2):109–117.

106. Nagano H, Machida Y, Iwata M, Imada T, Noguchi Y, Matsumoto A, et al. Preparation of magnetic granules containing bleomycin and its evaluation using model esophageal cancer. *Int. J. Pharm.* 1997;147(1):119–125.

107. Kojima H, Yoshihara K, Sawada T, Kondo H, Sako K. Extended release of a large amount of highly water-soluble diltiazem hydrochloride by utilizing counter polymer in polyethylene oxides (PEO)/polyethylene glycol (PEG) matrix tablets. *Eur. J. Pharm. Biopharm.* 2008;70(2):556–562.

108. Maggi L, Bruni R, Conte U. High molecular weight polyethylene oxides (PEOs) as an alternative to HPMC in controlled release dosage forms. *Int. J. Pharm.* 2000;195(1–2): 229–238.

109. Maggi L, Segale L, Torre ML, Ochoa Machiste E, Conte U. Dissolution behaviour of hydrophilic matrix tablets containing two different polyethylene oxides (PEOs) for the controlled release of a water-soluble drug. *Dimensionality study. Biomaterials* 2002;23(4):1113–1119.

110. Wu N, Wang L-S, Tan DC-W, Moochhala SM, Yang Y-Y. Mathematical modeling and *in vitro* study of controlled drug release via a highly swellable and dissoluble polymer matrix: polyethylene oxide with high molecular weights. *J. Control. Release* 2005;102(3):569–581.

111. Mahalingam R, Jasti B, Birudaraj R, Stefanidis D, Killion R, Alfredson T, et al. Evaluation of Polyethylene oxide compacts as gastroretentive delivery systems. *AAPS PharmSciTech*, published online: January 16 2009.

6

ORAL EXTENDED RELEASE HYDROPHILIC MATRICES: FORMULATION AND DESIGN

XIAOGUANG WEN

WuXi AppTec Inc., Shanghai, China

ALI NOKHODCHI

Medway School of Pharmacy, Universities of Kent and Greenwich, Kent, UK

ALI RAJABI-SIAHBOOMI

Colorcon Ltd., West Point, PA, USA

6.1 EXTENDED RELEASE MATRIX TECHNOLOGY

6.1.1 Background to Matrix Systems

In recent years, extended release (ER) dosage forms have been extensively used because of their significantly improved clinical efficacy and patient compliance [1]. Hydrophilic matrix tablets are the most frequently manufactured and used ER dosage forms for oral administration [2]. Hydrophilic ER matrices do not disintegrate and are formulated in such a way that the drug is released over a defined period of time following exposure to water or after oral administration. An oral extended release matrix allows a reduction in dosing frequency compared to a conventional dosage form [3].

Preparation of matrix tablets that may involve direct compression of a blend of a drug with release retardant polymers and other process-aid additives is the simplest approach for extended release delivery of drugs by the oral route [4]. Various polymeric materials have been explored as release retarding agents in hydrophilic matrix systems. Although different oral matrix technologies such as inert matrices and wax or hydrophobic matrices have been investigated by scientists, some of which have made their way to the marketplace, in this chapter the main focus will be on the formulation and design of hydrophilic matrices.

In brief, wax or lipid matrices are prepared by adding the drug and excipients to the molten fat or wax, congealing, granulating, and compressing into matrix cores. Substances that produce these matrices include carnauba wax, fatty alcohol, glycerol palmitostearate, stearyl alcohol, beeswax, aluminum monostearate, and glycerol monostearate [5]. The mechanism of drug release from these matrices may be diffusion of drug through liquid-filled pores. On the other hand, there are erodible (through digestion) lipid-based matrix systems that control the release of drug through combination of diffusion and erosion. Inert matrices using polymers such as ethylcellulose, methylacrylate, methylmethacrylate, polyvinyl chloride, and polyvinyl acetate are prepared through wet granulation and compression into matrices. Drug release from these matrices is by simple diffusion through water-filled pores.

6.1.2 Hydrophilic Matrices

Hydrophilic polymers were first introduced for matrix (HM) extended release applications in 1962 [6]. The ingredients of a hydrophilic matrix can be either directly compressed or granulated to aid flow and compression or improve content uniformity. Upon contact with water, the hydrophilic polymer on or near the surface of the matrix hydrates to form a gel

Oral Controlled Release Formulation Design and Drug Delivery: Theory to Practice, Edited by Hong Wen and Kinam Park
Copyright © 2010 John Wiley & Sons, Inc.

layer. The gel layer controls water ingress into the matrix and influences the mechanism of drug release. Therefore, the mechanism of drug release from HM systems is a combination of hydration and swelling of the tablet, drug dissolution, diffusion, and outer matrix surface erosion [7, 8]. Polymers used in the manufacture of hydrophilic matrices singularly or in combinations may include

- *Cellulose Derivatives*: methylcellulose (MC), hydroxypropylmethylcellulose (hypromellose, HPMC), sodium carboxymethylcellulose (Na CMC), and hydroxypropylcellulose (HPC).
- *Noncellulose Derivatives*: sodium alginate, xanthan gum, guar gum, carrageenan, and carbomers.

There are other hydrophilic polymers that are used in matrices, but they may not form a gel structure, and therefore, they may fall under a different classification. The choice of the polymer used in the matrix formulation depends on the chemistry of the drug, desired release profile, additional functionalities, availability, and global regulatory acceptability.

Hypromellose is available commercially from Dow Chemical Company under the trade name METHOCEL™ (Dow excipients, Methocel™ products, http://www.dow.com/dowexcipients/products/methocel.htm). METHOCEL is available in different chemistries depending on the degree of hydroxypropoxyl and methoxyl group substitutions (Figure 6.1). METHOCEL E (hypromellose 2910, USP) and K (hypromellose 2208, USP) are most widely used in extended release formulations and are distributed worldwide by Colorcon Inc. The USP classification code is based on the substitution type with first two digits represent-

ing the mean percent methoxyl substitution and the last two digits representing the mean percent hydroxypropyl substitution (USP Monographs: Hypromellose, http://www.uspnf.com/uspnf/login). Hypromellose is the most commonly used polymer for the preparation of hydrophilic matrix systems [9] mainly due to its fast and uniform gel formation to protect the matrix from disintegration, formation of a strong, viscous gel layer to control drug release [4, 10]. In addition, HPMC has a long history of application in marketed products with wide global regulatory acceptance.

As described above, HPMC polymers may differ in their degree of methoxyl or hydroxypropoxyl substitution and/or degree of polymerization (Figure 6.1). Varying the ratios of methoxyl and hydroxypropoxyl substitution and molecular weight influences properties such as organic solubility, thermal gelation temperature of their aqueous solutions [15], swelling [16], powder flow [17], compressibility and compactability [18–20], diffusion [17], and drug release [16].

6.1.3 Mechanism of Drug Release from Hydrophilic Matrices

In order to describe the mechanism of drug release from hydrophilic matrices, the basic structure formation and consequent drug release have been widely investigated and reported in the literature. Controlled drug release from HM systems is achieved through rapid hydration of the polymer (e.g., HPMC) on the outer tablet surface to form a gelatinous layer. This rapid formation of a gelatinous layer is critical to retain structural integrity, prevent water ingress to the interior of the matrix, and inhibit disintegration of the tablet. Once the protective gel layer is formed, it controls the water movement in the gel layer and further ingress into the tablet [11]. For this

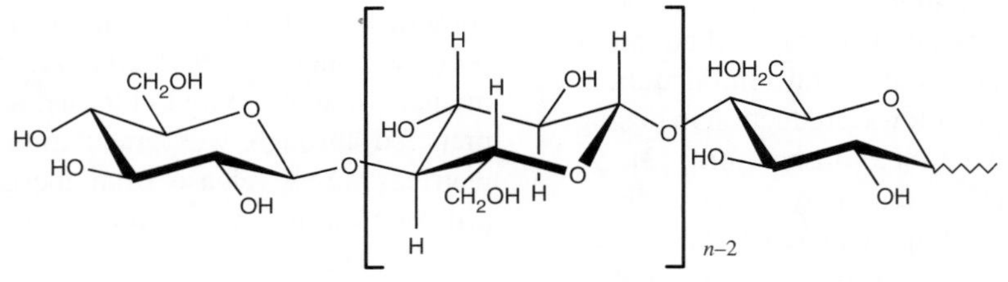

Dow Chemical Co	USP	Methoxyl	Hydroxypropoxyl
METHOCEL™ E	2910	28–30	7–12

Viscosity grades (2% aqueous solution at 20°C): 3, 5, 6, 15, 50, 4000, 10,000, 15,000 (cP)

METHOCEL™ K	2208	19–24	7–12

Viscosity grades (2% aqueous solution at 20°C): 3, 100, 4000, 15,000, 100,000 (cP)

FIGURE 6.1 General structure of cellulose with three possible substitution sites (hydroxyl groups) on each D-anhydroglucose monomer. HPMC contains methoxyl (CH_3-O) and hydroxypropoxyl ($CH_3CHOHCH_2$-O) substituents.

reason, hydrophilic polymers are usually supplied in small particle size range to ensure rapid and consistent hydration of the polymer. As the outer gel layer fully hydrates, the polymer disentangles from the surface [12, 13], which is continuously replaced with the hydrated polymer from within the core to control drug release. HPMC and polyethylene oxide (POLYOX™) are well known for their rapid hydration and gel formation [14].

The release of drug from a hydrophilic matrix system relies on swelling of the matrix, dissolution of the drug, and diffusion and erosion properties of the gel layer. The solubility and dose of the drug, type and quantity of fillers, and the polymer influence the mechanism of drug release. The release behavior from an insoluble inert matrix (illustrated in Figure 6.2) can be mathematically expressed by the following equation:

$$dM/dh = C_0\, dh - (C_s/2) \qquad (6.1)$$

where dM is the change in the amount of drug release per unit area and dh denotes the change in the thickness of the zone of matrix of the drug that has depleted. C_0 and C_s are the total amount of drug in a unit volume of matrix and saturated concentration of the drug within the matrix, respectively.

According to diffusion theory, dM is proportional to the diffusion coefficient (D_m) and C_s; therefore,

$$dM = (D_m C_s/h)dt \qquad (6.2)$$

Combination and integration of Equations 6.1 and 6.2 lead to

$$M = [C_s D_m(2C_0 - C_s)t]^{0.5} \qquad (6.3)$$

When the amount of drug is in excess of the saturated concentration, Equation 6.3 can be refined to

$$M = (2C_s D_m t)^{0.5} \qquad (6.4)$$

This equation is known as the Higuchi equation and initially was valid only for planar matrix systems, and later it was

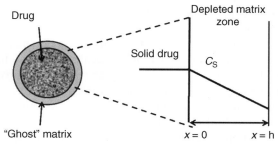

FIGURE 6.2 Schematic representation of an inert monolithic matrix release system.

modified to consider different geometrical shapes and matrix characteristics including porous structures [21–25]. It is important to keep in mind that the classical Higuchi equation was derived under pseudo-steady-state assumptions and cannot be applied to real controlled release systems [26]. The final equation shows that if a system is predominantly diffusion controlled, then it is expected that a plot of the drug release against square root of time will result in a straight line.

For the purpose of data treatment, the Higuchi equation can be simply written as

$$M = kt^{0.5} \qquad (6.5)$$

where M is the portion of drug released at time t and k is a constant. If the predominant mechanism of drug release from this type of matrix is diffusion controlled, then the drug release can be controlled by varying the factors such as porosity, tortuosity, initial concentration of drug in the matrix, solubility of the drug, and polymer system forming the matrix.

Since a hydrophilic matrix structure in an aqueous medium undergoes dynamic alterations due to the polymer hydration and swelling, drug dissolution, diffusion, and erosion, mathematical models describing drug release from these systems are complex. But for simplicity a series of transport phenomena involved in drug release have been reported. Figure 6.3 summarizes a schematic representation of three different fronts as have been described in the literature [27–29]: the swelling front, the erosion front, and the diffusion front. Each front indicates a change in physical condition from the adjoining front. The swelling front separates the rubbery region (gel layer) from the glassy region (dry core) and has absorbed enough water to allow macromolecular mobility and swelling of polymer. The diffusion front that separates the areas of nondissolved drug from the area of dissolved drug is located between the swelling and erosion fronts. Finally, the erosion front is the border between the unstirred gel layer and the well-stirred medium separating the matrix from the bulk solution [27–29].

A large number of mathematical models have been developed to describe drug release profiles from matrices [29, 30–34]. The simple and more widely used model is the one derived by Korsmeyer et al. [35]:

$$M_t/M_\alpha = kt^n \qquad (6.6)$$

where M_t/M_α is the fraction of drug released, k is the diffusion rate constant, t is the release time, and n is the release exponent indicative of the mechanism of drug release. The equation was modified by Ford et al. [36] to account for any lag time (l) or initial burst release of the drug:

$$M_t/M_a = k(t-l)^n \qquad (6.7)$$

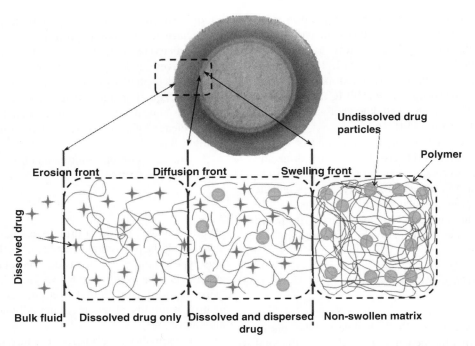

FIGURE 6.3   Schematic representation of different front positions in a hydrating swellable hydrophilic matrix tablet.

It is clear from Equations 6.6 and 6.7 that when the exponent n takes a value of 1.0, the drug release rate is independent of time. This case corresponds to zero-order release kinetics (also known as case II transport). Here, the polymer relaxation and erosion [37] are the rate-controlling steps. When $n = 0.5$, Fickian diffusion is the rate-controlling step (case I transport). Values of n between 0.5 and 1 indicate the contribution of both the diffusion process and polymer relaxation in controlling the release kinetics (non-Fickian, anomalous, or first-order release). It should be noted that the two extreme values of $n = 0.5$ or 1 are valid only for slab geometry. For cylindrical tablets, these values range from $0.45 < n < 0.89$ for Fickian, anomalous, or case II transport [31].

In order to describe relaxation transport, Peppas and Sahlin [38] modified Equation 6.7 to account for relaxation transport:

$$Q = k_1 t^n + k_2 t^{2n} \tag{6.8}$$

where k_1 and k_2 were Fickian diffusion constant and relaxation mechanism constant, respectively. If the surface area of the system is fixed, which is unlikely, the value of n should be 0.5 and the above equation is transformed to the following equation:

$$Q = k_1 t^{0.5} + k_2 t \tag{6.9}$$

The first term of the equation represents the diffusional phenomenon while the second term accounts for polymer erosion.

6.1.4 Prediction of Drug Release Mechanism from Hydrophilic Matrices

Mechanism and rate of drug release from hydrophilic matrices depend not only on the type and level of the polymer, choice of filler, and size of the matrix, but also on the properties of the drug [39–41]. For example, mechanisms of release for propranolol HCl and indomethacin from HPMC matrices are significantly different, due to their different solubilities [39, 40]. In a study of drug release from hydrophilic matrices, it was shown that the release rates of potassium chloride, phenylpropanolamine hydrochloride, and bovine serum albumin decreased as the molecular weight of the drug increased [42]. Baveja et al. investigated release characteristics of six water-soluble bronchodilators from HPMC K4M matrices in order to examine correlation between release rate and their molecular geometry for drug release prediction [43]. They showed that despite almost identical aqueous solubilities, different drug molecules showed different release rates from the matrices, which were related to the accessible surface areas of the drugs.

Shah et al. made an attempt to predict the fraction of drug release as a function of HPMC concentration and release time [44]. Gao et al. utilized Higuchi theory to derive a mathematical model with a capability to predict the relative change in drug release rate as a function of formulation composition [45]. Siepmann et al. have shown that water transportation into HPMC is important and on the basis of the water transportation into matrix tablets, mathematical models have been suggested to calculate the required shape and size of the HPMC tablet to achieve a desired drug release profile [27, 35, 46]. Recently, researchers have used quantitative structure–property relationship (QSPR) methods for prediction of mechanism and rate of drug release from HPMC matrices [47, 48]. QSPR is a valuable tool that employs specialized statistical techniques to relate the property under investigation to the molecular structure of the chemicals represented by physicochemical properties or structural descriptors. The resulting QSPR models facilitate understanding of the property in terms of chemical structure, ultimately enabling the investigators to estimate the property for other similar compounds. Ghafourian et al. [48] used structural descriptors consisting of molecular mechanical, quantum mechanical, and graph theoretical parameters, as well as the partition coefficient and the aqueous solubility of the drugs, to develop various models for the prediction of drug release. They showed that the most important factors determining the release profile from HPMC matrices were the aqueous solubility of drugs (which could be substituted efficiently by dipole moment) and the size of the drug molecules. Comparison of drug release from matrices prepared using HPMC showed very distinct differences for some drugs, as evaluated by the similarity factor. The results indicated that the source of the difference could be sought in the drug properties (as exemplified by the aqueous solubility and surface area) and the rate of erosion (that depends mainly on the polymer type). The most recent research on the prediction of drug release led to introduction of two unified mathematical models suitable for predicting a broad array of release behaviors on the basis of bulk and surface erosions and kinetics of hydration [49, 50]. These unified models can be used to acquire the matrix specifications that will yield a desired release profile.

Extrapolation of all the findings shown in the literature to broad range of drugs with different chemical structures, solubilities, and doses is generally a challenge. However, attempts have been made to develop models to assist the pharmaceutical scientists with a starting formulation for hydrophilic matrix tablets. For example, Colorcon Inc. has developed a predictive formulation service, called HyperStart® (HyperStart formulation service, http://www.colorcon.com/pharma/mod_rel/methocel/hyperstart_text.html), to simplify the development process and reduce the time to market.

6.2 APPLICATION OF HYDROPHILIC MATRICES

In this section, a few examples of approaches used to modulate drug release from hydrophilic matrices are provided. These technologies include coating of hydrophilic matrices with an insoluble barrier membrane to reduce burst effect of highly soluble drugs, multilayered tablets to achieve constant release (zero order), and polymer blends to modulate release and provide control of microenvironmental pH within the gel layer of the matrix. The microenvironmental pH control may enhance solubility or stability of the formulated drug in the matrix and within the gel layer of a hydrated system.

Compression coating (tablet in tablet) technology containing hydrophilic polymers has been applied to generate lag time followed by either fast release or slow release. Mini-matrices (mini-tabs) are examples of application of hydrophilic matrices in multiparticulate formulations. Table 6.1 shows a summary of studies published in the literature, where various technologies have been investigated to modulate drug release from hydrophilic matrices.

6.2.1 Barrier Membrane Coating to Modulate Release

For soluble and highly soluble drugs, hydrophilic matrices exhibit an initial fast or burst release, which may be due to release of drug from surface or near the surface of the matrix. Dias et al. [51] have studied the application of an insoluble but permeable ethylcellulose barrier membrane coat onto the matrix tablet in order to reduce the burst effect. In this study, a highly water-soluble drug, venlafaxine (solubility of 572 mg/mL), was used as a model drug and HPMC K15M was used as rate-controlling polymer. The burst effect was significantly reduced by the aqueous ethylcellulose (Surelease®, an aqueous ethylcellulose dispersion) barrier membrane coating (weight gain of 4% w/w) of the matrix. The reduced burst effect may be described as follows. During dissolution, the water uptake through the barrier membrane allows dissolution of the drug as well as hydration of the hydrophilic polymer. Drug release is initially retarded by the barrier membrane as rapid water ingress into the matrix is prevented and thus drug from surface and adjacent to the surface is only released via diffusion through the membrane. As the hydration and swelling of the matrix occurs, the axial swelling of tablet causes the barrier membrane to rupture on the edges and around the belly of the matrix. Ayres [52] has shown that by careful formulation composition, configuration of the matrix, and coating composition and level, it is possible to obtain consistent and reproducible rupture of the barrier membrane after a defined dissolution time. Drug release modulation can be achieved by choice of insoluble polymer, coating formulation (plasticizer type and level, presence and level of pore former – permeability enhancer), and coating level. It is possible to design the coated hydrophilic matrix in order to achieve zero-order release for highly soluble drugs.

In a study, Colombo et al. applied a coating of cellulose acetate propionate as an impermeable membrane onto HPMC matrix tablets to physically restrict the swelling of the dosage form [53]. The coating was performed on one side or both sides of the tablet surface to modify the release mechanism by controlling the swelling ratio between the

TABLE 6.1 A Summary of Studies Published in the Literature, Where Various Technologies Have Been Investigated to Modulate Drug Release from Hydrophilic Matrices

Release Modulation	Drug Candidate	Polymers and Functional Additives	Literature References
Bimodal	Salbutamol sulfate	Different viscosities of HPMC	51
	Aspirin, ibuprofen	Different viscosities of HPMC	57
Zero order	Alfuzosin hydrochloride	HPMC, HPC, PEO, and NaHCO$_3$	69
	Metoprolol tartrate	PEO, sodium carbonate pentasodium tripolyphosphate	90
	Diltiazem HCl	HPMC, PEO, HPC, EC	91
	Pseudoephedrine HCl, aminophylline	HPMC, carnauba wax	68, 71
	4-Aminopyridine	HPMC, Na CMC	92
	Caffeine, ibuprofen	HPMC, Na CMC, EC	93
	Trapidil, ketoprofen, nicardipine HCl	HPMC	66
Lag time	Ibuprofen	HPMC, potassium carbonate	75
	Salbutamol sulfate	HPMC, carnauba wax, white beeswax	73
	Diltiazem HCl	HPMC, EC, CAP	61
Micro-pH control	Pelanserin HCl	HPMC, citric acid	85
	Verapamil HCl	HPMC, PVAP, Carbopol	79
	Oxybutynin HCl	HPMC, Carbopol, xanthan gum, fumaric acid	83
	Dipyridamole	HPMC, fumaric acid, succinic acid, citric acid	81, 82
	Felbinac	HPMC, sodium citrate	84
Minimatrices	Ibuprofen	HPMC, xanthan gum, karagy gum	94
	Pseudoephedrine HCl	HPMC, HPC, EC	87
	Metoprolol tartrate	PEO, EC, xanthan gum	88
	Ibuprofen	HPMC, EC	86
Multilayer tablets	Levodopa methyl ester, carbidopa	HPMC	95
	Metoprolol tartrate	HPMC, guar gum	96
	Pseudoephedrine	HPMC, carnauba wax	71
	Diltiazem HCl	HPMC, EC	67
	Caffeine, ibuprofen	HPMC, Na CMC, EC	93
	Nifedipine	HPMC, Carbopol	70
Polymer blends	Theophylline	PEO, Carbopol, sodium chloride	76
	Diltiazem HCl	PEO, Carbopol, cross-linked carboxyvinyl polymer	97
	Diclofenac Na	HPMC, Carbopol	78
	Ibuprofen	HPMC, carrageenans	97
	Propranolol HCl	HPMC, Eudragit L 100-55, Eudragit HPMC, Na CMC	98
Compression coating	Salbutamol sulfate	Different viscosities of HPMC	51
	Ibuprofen	HPMC, EC	86
	Salbutamol sulfate	HPMC, carnauba wax, white beeswax	73

axial and radial directions. By physically restricting the matrix from swelling in all directions, the mechanism of drug release was modified and a close to constant drug release profile was obtained [53]. Sangalli et al. [54] studied matrix tablet consisting of low viscosity grade HPMC compressed with a specially designed press tool to generate a circular central hole (perforated tablet). These perforated tablets were then coated on the outer surface (excluding the central parts) with ethylcellulose as a permeable insoluble barrier membrane. As a result, drug was released only through diffusion and erosion from the central hole. In cases where the coating

of ethylcellulose reached the inner surface of the hole on the perforated tablets, a significant lag time was generated.

6.2.2 Multilayer Matrix Technology to Modulate Drug Release

There have been different approaches to achieve zero-order drug release from dosage forms for sustained plasma concentration. Osmotic technology has been the major technology for constant drug release rate (zero-order release) and its application has led to a number of commercial products [55].

Achieving zero-order release for highly soluble drugs from matrices may be challenging (as described above).

Among different approaches to achieve zero-order drug release from hydrophilic matrix technologies, multilayer matrices have been widely evaluated and developed for commercial products under the trade name of Geomatrix. The technology makes use of bilayer or trilayer tablets to modulate the release and to achieve constant release [56–62]. It consists of a hydrophilic matrix middle layer or core layer, containing the active ingredient, and one or two impermeable or semipermeable polymeric coatings applied on one or both bases of the core [58]. The external layers reduce the surface area of the middle core layer available to media and change the properties of hydration and swelling and thus drug release from the middle core tablet [63]. Papadokostaki et al. [64] developed mathematical models to simulate the operation of polymer-based symmetrical three-layer (ABA) or equivalently two-layer (AB with the B surface blocker) ER matrix devices.

Conte et al. conducted a number of studies to evaluate multilayered hydrophilic matrices as constant release devices [63, 65, 67]. Diltiazem hydrochloride, a water-soluble drug, was formulated with the high viscosity grade HPMC, K100M, and the release profile manifested a first-order release. A single layer compression coating (resulting into a bilayer tablet) with ethylcellulose reduced the release rate and a two-layer compression coating (resulting into a trilayer tablet) resulted in a zero-order release profile [65]. Multilayer matrix technology is a flexible approach to modulate drug release through choice and level of polymer used on each layer. Polymers used for compression coating to obtain multilayer matrices may include ethylcellulose, carnauba wax [65, 68], gel-forming polymers such as high viscosity grade HPMC, POLYOX™, and Carbopol [66, 69, 70], and erodible polymers such as low viscosity grade HPMC and POLYOX™ [66, 70]. The polymer selection to be used in outer layers of multilayer matrices depends on drug solubility and desired release profile. For example, a three-layer device using HPMC K100M in the outer layers changed the first-order trapidil release to a zero-order release. Use of erodible layers using HPMC E5 changed the first-order release of nicardipine to a zero-order release [66]. The difference is due to the fact that the trapidil is freely soluble while nicardipine is sparingly soluble and therefore outer layers contain high and low viscosity grade HPMC, respectively. The gel-forming polymers more efficiently modulate release of water-soluble drugs than the insoluble polymers. In addition to the use of outer layer as a way to modulate drug release, the formulation of the middle core layer has also been evaluated. The hydrophobic middle layer instead of the hydrophilic middle layer was studied in order to achieve zero-order release [68, 71]. In these studies, carnauba wax was used as the hydrophobic polymer and hydrophobic barrier while HPMC K4M, K15M, and K100M were used as hydrophilic layers. It was shown that by changing the material used and sequence of each layer, drug release modulation was obtained through different configurations of tablet design, HMH, HML, and LML, where M stands for the hydrophobic middle layer, H stands for the hydrophilic layer, and L stands for the hydrophobic layer [71].

6.2.3 Press Coating of Matrices to Modulate Release

Press coating technology of matrices has been investigated to achieve time-dependent delayed release; that is, the drug is released after a predetermined lag time [72, 73].

Halsas et al. studied the press-coated tablet for time-controlled release using HPMC and found that the lag time was controlled by coating thickness and viscosity grades of the polymer used in the compression coating [74, 75]. The core tablet containing 80 mg ibuprofen with a tablet weight of 140 mg was press coated with a combination of HPMC K100LV and K4M. The coating layers contained 200 and 20 mg of ibuprofen, respectively. This tablet was tested on healthy volunteers and showed *in vivo* lag time [74].

6.2.4 Use of Polymer Blends to Modulate Drug Release

Use of polymer blends may provide an alternative approach to modulate drug release compared to conventional single polymer matrices. Matrices containing POLYOX™ and Carbopol were studied by El-Malah and Nazzal [76]. In this comprehensive study, the effects of molecular weight of POLYOX™ (from 600,000 to 7,000,000), polymer concentrations (POLYOX™ and Carbopol), addition of sodium chloride and citric acid, and compression force on the release of theophylline were evaluated. Two key factors, polymer concentrations (POLYOX™ and Carbopol) and POLYOX™ molecular weight, had a significant impact on drug release rate and profile [76, 77].

Samani et al. studied matrices containing polymer blends of HPMC, Carbopol, and diclofenac sodium [78]. They found that the polymer blends extended the duration of drug release. It has been postulated that there may be strong hydrogen bonding between the carboxyl groups of Carbopol and hydroxyl groups of HPMC leading to stronger interactions between the two polymers and therefore slower drug release than a matrix with a single polymer.

Blends of HPMC and polyvinyl pyrrolidone (PVP) in a hydrophilic matrix showed a biphasic release of caffeine as a water-soluble model drug [56]. The authors concluded that the initial release was controlled by HPMC, but because of the faster release of drug from the matrix, the PVP becomes rich progressively. The breakup of the HPMC gel by enriched PVP resulted in a bimodal release profile (faster release at the terminal phase). The breakup occurred at different time points depending on the PVP level in the matrix.

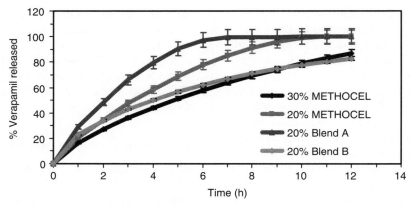

FIGURE 6.4   Verapamil HCl release from hydrophilic matrices [48% drug, 20% or 30% HPMC METHOCEL™ K4M or blends (10% K4M + 2–8% PVAP + 8–2% Carbopol), q.s.% filler and 0.5% each of lubricant and glidant]. USP apparatus II at 100 rpm and 900 mL of water.

Blending of polyvinyl acetate phthalate (PVAP), an enteric polymer, HPMC, and Carbopol resulted in a significant reduction in verapamil HCl release rate [79]. Figure 6.4 shows that by changing the ratio of the polymers in the blend, it was possible to modulate the drug release profile. This approach may be utilized for high-dose drugs where the overall large size of the tablet may not be acceptable by patents. Use of polymer blends at lower concentrations, providing the desired release profiles and maintaining formulation robustness, would be a suitable approach for high-dose drugs (e.g., drug doses of greater than 400 mg). The retardation of drug release has been attributed to the synergistic interactions between PVAP, Carbopol, and HPMC leading to formation of a stronger gel layer and slower diffusion and erosion rates.

6.2.5   Microenvironmental pH Control

Formulation of drugs with pH-dependent solubility in hydrophilic matrices may lead to release profiles that vary depending on the media pH. In most situations, a desirable release profile should be pH independent to withstand the physiological pH changes in the gastrointestinal tract. Buffers and polymers may be added to the formulation to maintain the pH of the microenvironment within the gel structure of the hydrophilic matrix [80–82].

Varma et al. [83] studied the influence of microenvironmental pH on the release of oxybutynin hydrochloride in a matrix system with two nonionic polymers (HPMC K4M and K15M) and two anionic polymers (Carbopol and xanthan gum). The oxybutynin HCl has a high water solubility at low pH values and low solubility at higher pH values. Fumaric acid was added to the formulation to maintain a low-pH environment during the release and 10% was found to be effective to achieve pH-independent release profiles in media with pH 1.2 and 6.8.

The majority of the frequently used low molecular weight pH modifiers tend to diffuse out of the hydrated matrix faster than the drug and maintaining the desired pH over the entire duration of release can be a challenge. Solubility of weak acid dipyridamole is highly pH dependent. The effectiveness of different types of acids, fumaric acid, citric acid, succinic acid, and ascorbic acid, to maintain a microenvironmental pH for the duration of release was evaluated [82, 84]. The quantity of pH modifier in a HPMC matrix system was evaluated by Espinoza [85]. Because of the relatively lower solubility and lower pK_a, fumaric acid was the most effective acid in improving the dipyridamole solubility at pH 6.8. The release profiles of different acids show that the fumaric acid is the slowest in release. For situations where a low molecular weight pH modifier leaches out of the matrix system too fast, ionic polymers that contribute to microenvironmental pH are used and due to their high molecular weight they stay within the gel structure until eroded from the surface of the matrix (see Section 6.2.4).

6.2.6   Minimatrices as an Alterative Modulation Approach

Minitablets (mini-tabs), with a diameter around 2–3 mm [86], can be encapsulated or compressed into larger tablets as final dosage forms [87, 88]. Mini-tabs are becoming popular amongst formulation scientists due to advantages offered such as reduced inter- and intrasubject variability [89], formulation flexibility, and appropriate dosage forms for drug delivery for children.

Minimatrices (2.5 mm in diameter and 12 mg in weight) have been evaluated for biphasic drug release (fast/slow) where both HPMC K100M and ethylcellulose (ETHOCEL™) were used as release retardant matrix formers [86].

Minimatrices (3 mm in diameter and 2 mm in height) using polymer blends of POLYOX™ and polyethylene glycol

(PEG) and ethylcellulose containing metoprolol tartrate have been investigated [88]. The minimatrices were manufactured by hot melt extrusion and 24 h drug release profiles were achieved.

6.3 SUMMARY AND CONCLUSIONS

Hydrophilic matrix systems have been widely studied and many successful products in the market utilize this simple, robust, and versatile extended release technology. Different chemistries and viscosity grades of hydrophilic polymers, and more specifically different grades of HPMC, allow the formulators to reach for this technology for ER formulations of drugs with wide range of solubilities and dose strengths. Various approaches have been used in order to modulate drug release to obtain pharmacologically desired profiles. HPMC matrices offer a platform for blending other polymers to provide flexibility to the formulator in achieving their desired formulation characteristics. Ionic, nonionic, and insoluble polymers have been used in HPMC matrices as blends or as film or compression coating to successfully modulate the release profile of various drugs. The addition of ionic polymers may not only modify the drug release profile but also allow microenvironmental pH control of the gel layer, which may enhance solubility or stability of drugs.

REFERENCES

1. Merkus FWHM. Controlled and rate-controlled drug delivery: principal characteristics, possibilities and limitations. In Struyker-Boudier HAJ,editor. *Rate-Controlled Drug Administration and Action*, CRC Press, Boca Raton, FL, 1986, pp. 15–47.

2. Cardinal JR. Matrix systems. In Langer RS, Wise DL, editors. *Medical Applications of Controlled Release: Classes of Systems*, Vol. 1, CRC Press, Boca Raton, FL, 1984, pp. 41–67.

3. Qiu Y, Zhang G. Research and development aspects of oral controlled-release dosage forms. In Wise DL,editor. *Handbook of Pharmaceutical Controlled Release Technology*(), Marcel Dekker Inc., 2000.

4. Tiwari SB, Rajabi-Siahboomi AR. Extended release oral drug delivery technologies: monolithic matrix systems. In Jain K, editor. *Drug Delivery Systems*, Humana Press, 2008.

5. Chiao CSL, Robinson JR. Sustained release drug delivery systems. In *Remington: The Science and Practice of Pharmacy*, 19th edition, Mack Publishing Company, 1995.

6. Huber HE, Dale LB, Christensen GL. Utilization of hydrophilic gums for the control of drug release from tablet formulations. I. Disintegration and dissolution behaviour. *J. Pharm. Sci.* 1966;55:974.

7. Rajabi-Siahboomi RR, Jordan MP. Slow release HPMC matrix systems. *Eur. Pharm. Rev.* 2000;5(4):21–23.

8. Alderman DA. A review of cellulose ethers in hydrophilic matrices for oral controlled release dosage forms. *Int. J. Pharm. Technol. Prod. Manuf.* 1984;5:1–9.

9. Levina M, Rajabi-Siahboomi AR. The influence of excipients on drug release from hydroxypropylmethylcellulose matrices. *J. Pharm. Sci.* 2004;93(11):2746–2754.

10. The Dow Chemical Company. Formulation for controlled release with METHOCEL cellulose ethers, 1987.

11. Rajabi-Siahboomi AR, Bowtell RW, Mansfield P, Davies MC, Melia CD. Structure and behavior in hydrophilic matrix sustained release dosage forms. 4. Studies of water mobility and diffusion coefficients in the gel layer of HPMC tablets using NMR imaging. *Pharm. Res.* 1996;13:376–380.

12. Lee PI, Peppas NA. Prediction of polymer dissolution in swellable controlled-release systems. *J. Control. Release* 1987;6:207–215.

13. Narasimhan B, Peppas NA. Molecular analysis of drug delivery systems controlled by dissolution of the polymer carrier. *J. Pharm. Sci.* 1997;86:297–304.

14. Yang L, Johnson B, Fassihi R. Determination of continuous changes in the gel layer of poly(ethylene oxide) and HPMC tablets undergoing hydration: a texture analysis study. *Pharm. Res.* 1998;15:1902–1906.

15. Sarker N. Thermal gelation properties of methyl and hydroxypropyl methylcellulose. *J. Appl. Polym. Sci.* 1979;24:1073–1087.

16. Rajabi-Siahboomi AR, Bowtell RW, Mansfield P, Henderson A, Davies MC, Melia CD. Structure and behavior in hydrophilic matrix sustained-release dosage forms. 2. NMR imaging studies of dimensional changes in the gel layer and core of HPMC tablets undergoing hydration. *J. Control. Release* 1994;31: 121–128.

17. Nokhodchi A, Rubinstein MH. An overview of the effects of material and process variables on the compaction and compression properties of hydroxypropyl methylcellulose and ethylcellulose. *STP Pharm. Sci.* 2001;11:195–202.

18. Nokhodchi A, Ford JL, Rowe PH, Rubinstein MH. The effects of compression rate and force on the compaction properties of different viscosity grades of hydroxypropyl methylcellulose 2208. *Int. J. Pharm.* 1996;129:21–31.

19. Rajabi-Siahboomi AR, Nokhodchi A. Compression properties of methylcellulose and hydroxylpropyl methylcellulose polymers. *Pharm. Pharmacol. Commun.* 1999;5:67–71.

20. Malamataris S, Karidas T. Effect of particle size and sorbed moisture on the tensile strength of some tableted hydroxypropyl methylcellulose (HPMC) polymers. *Int. J. Pharm.* 1994;104: 115–123.

21. Lapidus H, Lordi NG. Some factors affecting the release of a water-soluble drug from compressed hydrophilic matrices. *J. Pharm. Sci.* 1966;55:840–843.

22. Higuchi T. Mechanism of sustained-action medication: theoretical analysis of rate of release of solid drugs dispersed in solid matrices. *J. Pharm. Sci.* 1963;52:1145–1149.

23. Desai SJ, Simonelli AP, Higuchi WI. Investigation of factors influencing release of solid drug dispersed in inert matrices. *J. Pharm. Sci.* 1965;54:1459–1464.

24. Desai SJ, Singh P, Simonelli AP, Higuchi WI. Investigation of factors influencing release of solid drug dispersed in inert matrices II. *J. Pharm. Sci.* 1966;55:1224–1229.

25. Lapidus H, Lordi NG. Drug release from compressed hydrophilic matrices. *J. Pharm. Sci.* 1968;57:1292–1301.

26. Peppas NA. Mathematical modelling of diffusion processes in drug delivery polymeric systems. In Smolen VF, Ball L, editors. *Controlled Drug Bioavailability*, Vol. 1, Wiley, New York, 1984, pp. 203–237.

27. Siepmann J, Kranz H, Bodmeier R, Peppas NA. HPMC-matrices for controlled drug delivery: a new model combining diffusion, swelling, and dissolution mechanisms and predicting the release kinetics. *Pharm. Res.* 1999;16:1748–1756.

28. Colombo P, Bettini R, Santi P, Peppas NA. Swellable matrices for controlled drug delivery: gel-layer behaviour, mechanisms and optimal performance. *Pharm. Sci. Technol. Today* 2000;3:198–204.

29. Siepmann J, Siepmann F. Mathematical modelling of drug delivery. *Int. J. Pharm.* 2008;364:328–343.

30. Siepmann J, Peppas NA. Hydrophilic matrices for controlled drug delivery: an improved mathematical model to predict the resulting drug release kinetics (the "sequential layer" model). *Pharm. Res.* 2000;17:1290–1298.

31. Siepmann J, Peppas NA. Modeling of drug release from delivery systems based on hydroxypropyl methylcellulose (HPMC). *Adv. Drug Deliv. Rev.* 2001;48:139–157.

32. Siepmann J, Peppas NA. Mathematical modeling of controlled drug delivery. *Adv. Drug Deliv. Rev.* 2001;48:137–138.

33. Siepmann J, Podual K, Sriwongjanya M, Peppas NA, Bodmeier R. A new model describing the swelling and drug release kinetics from hydroxypropyl methylcellulose tablets. *J. Pharm. Sci.* 1999;88:65–72.

34. Siepmann J, Streubel A, Peppas NA. Understanding and predicting drug delivery from hydrophilic matrix tablets using the "sequential layer" model. *Pharm. Res.* 2002;19:306–314.

35. Korsmeyer RW, Gurny R, Doelker E, Buri P, Peppas NA. Mechanisms of potassium chloride release from compressed, hydrophilic, polymeric matrices: effect of entrapped air. *J. Pharm. Sci.* 1983;72:1189–1191.

36. Ford JL, Mitchell K, Rowe P, Armstrong DJ, Elliott PNC, Rostron C, Hogan JE. Mathematical modelling of drug release from hydroxypropylmethylcellulose matrices: effect of temperature. *Int. J. Pharm.* 1991;71:95–104.

37. Bajwa GS, Hoebler K, Sammon C, Timmins P, Melia CD. Microstructural imaging of early gel layer formation in HPMC matrices. *J. Pharm. Sci.* 2006;95:2145–2157.

38. Peppas NA, Sahlin JJ. A simple equation for the description of solute release. III. Coupling of diffusion and relaxation. *Int. J. Pharm.* 1989;57:169–172.

39. Ford JL, Rubinstein MH, Hogan JE. Propranolol hydrochloride and aminophylline release from matrix tablets containing hydroxypropylmethylcellulose. *Int. J. Pharm.* 1985;24:339–350.

40. Ford JL, Rubinstein MH, Hogan JE. Dissolution of a poorly water soluble drug, indomethacin, from hydroxypropylmethyl-

41. Tahara K, Yamamoto K, Nishihata T. Application of model-independent and model analysis for the investigation of effect of drug solubility on its release rate from hydroxypropyl methylcellulose sustained release tablets. *Int. J. Pharm.* 1996;133:17–27.

42. Korsmeyer RW, Gurny R, Doelker E, Buri P, Peppas NA. Mechanism of solute release from porous hydrophilic polymers. *Int. J. Pharm.* 1983;15:25–35.

43. Baveja SK. Ranga Rao KV, Singh A, Gombar VK. Release characteristics of some bronchodilators from compressed hydrophilic polymeric matrices and their correlation with molecular geometry. *Int. J. Pharm.* 1988;41:55–62.

44. Shah N, Zhang G, Apelian V, Zeng F, Infeld MH, Malick AW. Prediction of drug release from hydroxypropylmethylcellulose (HPMC) matrices: effect of polymer concentration. *Pharm. Res.* 1993;10:1693–1695.

45. Gao P, Nixon PR, Skoug JW. Diffusion in HPMC gels. Prediction of drug release rates from hydrophilic matrix extended release dosage forms. *Pharm. Res.* 1995;12:965–971.

46. Siepmann J, Kranz H, Peppas NA, Bodmeier R. Calculation of the required size and shape of hydroxypropyl methylcellulose matrices to achieve desired drug release profiles. *Int. J. Pharm.* 2000;201:151–164.

47. Fu XC, Wang GP, Liang WQ, Chow MSS. Prediction of drug release from HPMC matrices: effect of physicochemical properties of drug and polymer concentration. *J. Control. Release* 2004;95:251–259.

48. Ghafourian T, Safari A, Adibkia K, Parviz F, Nokhodchi A. A drug release study from hydroxypropylmethylcellulose (HPMC) matrices using QSPR modelling. *J. Pharm. Sci.* 2007;96(12):3334–3351.

49. Rothstein SN, Federspiel WJ, Little SR. A simple model framework for the prediction of controlled release from bulk eroding polymer matrices. *J. Mater. Chem.* 2008;18:1873–1880.

50. Rothstein SN, Federspiel WJ, Littlee SR. A unified mathematical model for the prediction of controlled release from surface and bulk eroding polymer matrices biomaterials. *Biomaterials* 2009;30:1657–1664.

51. Dias VD, Fegely K, Rajabi-Siahboomi AR. Modulation of drug release from Hypromellose (HPMC) matrices: Suppression of the initial burst effect, AAPS 2006, San Antonio, TX.

52. Ayres JW. Coated, platform-generating tablet. US Patent 6,733,784 B1, 2004.

53. Colombo P, Conte U, Gazzaniga A, Maggi L, Sangalli ME, Peppas NA, La Manna A. Drug release modulation by physical restrictions of matrix swelling. *Int. J. Pharm.* 1990;63(1):43–48.

54. Sangalli ME, Giunchedi P, Gazzaniga A, Conte U. Erodible perforated coated matrix for extended release of drugs. *Int. J. Pharm.* 1993;91(2–3):151–156.

55. Wong PSL, Gupta SK, Stewart BE. Osmotically controlled tablets. In Rathbone MJ, Hadgraft J, Roberts MS, editors.

Modified Release Drug Delivery Technology, Vol. 126, Marcel Dekker, New York, 2003, p. 102.

56. Hardy IJ, Windberg-Baarup A, Neri C, Byway PV, Booth SW, Fitzpatrick S. Modulation of drug release kinetics from hydroxypropyl methyl cellulose matrix tablets using polyvinyl pyrrolidone. *Int. J. Pharm.* 2007;337(1–2):246–253.

57. Shah AC, Britten NJ, Olanoff LS, Badalamenti JN. Gel-matrix systems exhibiting bimodal controlled release for oral drug delivery. *J. Control. Release* 1989;9(2):169–175.

58. Streubel A, Siepmann J, Peppas NA, Bodmeier R. Bimodal drug release achieved with multi-layer matrix tablets: transport mechanisms and device design. *J. Control. Release* 2000;69(3):455–468.

59. González-Rodriguez ML, Fernández-Hervás MJ, Caraballo I, Rabasco AM. Design and evaluation of a new central core matrix tablet. *Int. J. Pharm.* 1997;146(2):175–180.

60. Fukui E, Uemura K, Kobayashi M. Studies on applicability of press-coated tablets using hydroxypropylcellulose (HPC) in the outer shell for timed-release preparations. *J. Control. Release* 2000;68(2):215–223.

61. Colombo P, Conte U, Gazzaniga A, Maggi L, Sangalli ME, Peppas NA, La Manna A. Drug release modulation by physical restrictions of matrix swelling. *Int. J. Pharm.* 1990;63(1):43–48.

62. Colombo P, Conte U, Caramella C, Gazzaniga A, La Manna A. Compressed polymeric mini-matrices for drug release control. *J. Control. Release* 1985;1(4):283–289.

63. Conte U, Maggi L. A flexible technology for the linear, pulsatile and delayed release of drugs, allowing for easy accommodation of difficult *in vitro* targets. *J. Control. Release*, 2000;64(1–3):263–268.

64. Papadokostaki KG, Stavropoulou A, Sanopoulou M, Petropoulos JH. An advanced model for composite planar three-layer matrix-controlled release devices. Part I. Devices of uniform material properties and non-uniform solute load. *J. Membr. Sci.* 2008;312(1–2):193–206.

65. Conte U, Maggi L, Colombo P, La Manna A. Multi-layered hydrophilic matrices as constant release devices (Geomatrix™ Systems). *J. Control. Release* 1993;26(1):39–47.

66. Conte U, Maggi L. Modulation of the dissolution profiles from Geomatrix® multi-layer matrix tablets containing drugs of different solubility. *Biomaterials* 1996;17(9):889–896.

67. Conte U, Maggi L, Torre ML, Giunchedi P, Manna AL. Press-coated tablets for time-programmed release of drugs. *Biomaterials* 1993;14(13):1017–1023.

68. Qiu Y, Chidambaram N, Flood K. Design and evaluation of layered diffusional matrices for zero-order sustained-release. *J. Control. Release*, 1998;51(2–3):123–130.

69. Liu Q, Fassihi R. Zero-order delivery of a highly soluble, low dose drug alfuzosin hydrochloride via gastro-retentive system. *Int. J. Pharm.* 2008;348(1–2):27–34.

70. Hong SI, Oh SY. Dissolution kinetics and physical characterization of three-layered tablet with poly(ethylene oxide) core matrix capped by Carbopol. *Int. J. Pharm.* 2008;356(1–2):121–129.

71. Chidambaram N, Porter W, Flood K, Qiu Y. Formulation and characterization of new layered diffusional matrices for zero-order sustained release. *J. Control. Release* 1998;52(1–2):149–158.

72. Fukui E, Miyamura N, Kobayashi M. An *in vitro* investigation of the suitability of press-coated tablets with hydroxypropyl-methylcellulose acetate succinate (HPMCAS) and hydrophobic additives in the outer shell for colon targeting. *J. Control. Release* 2001;70(1–2):97–107.

73. Pozzi F, Furlani P, Gazzaniga A, Davis SS, Wilding IR. The time clock system: a new oral dosage form for fast and complete release of drug after a predetermined lag time. *J. Control. Release* 1994;31(1):99–108.

74. Halsas M, Hietala J, Veski P, Jürjenson H, Marvola M. Morning versus evening dosing of ibuprofen using conventional and time-controlled release formulations. *Int. J. Pharm.* 1999;189(2):179–185.

75. Halsas M, Ervasti P, Veski P, Jurjenson H, Marvola M. Biopharmaceutical evaluation of time-controlled press-coated tablets containing polymers to adjust drug release. *Eur. J. Drug Metab. Pharmacokinet.* 1998;23:190–196.

76. El-Malah Y, Nazzal S. Hydrophilic matrices: application of Placket–Burman screening design to model the effect of POLYOX–carbopol blends on drug release. *Int. J. Pharm.* 2006;309(1–2):163–170.

77. Kojima H, Yoshihara K, Sawada T, Kondo H, Sako K. Extended release of a large amount of highly water-soluble diltiazem hydrochloride by utilizing counter polymer in polyethylene oxides (PEO)/polyethylene glycol (PEG) matrix tablets. *Eur. J. Pharm. Biopharm.* 2008;70(2):556–562.

78. Samani SM, Montaseri H, Kazemi A. The effect of polymer blends on release profiles of diclofenac sodium from matrices. *Eur. J. Pharm. Biopharm.* 2003;55(3):351–355.

79. Tiwari SB, Rajabi-Siahboomi A. Applications of complementary polymers in HPMC hydrophilic extended release matrices. *Drug Deliv. Technol.* 2009;9(7):20–27.

80. Sriamornsak P, Thirawong N, Korkerd K. Swelling, erosion and release behavior of alginate-based matrix tablets. *Eur. J. Pharm. Biopharm.* 2007;66(3):435–450.

81. Siepe S, Herrmann W, Borchert H-H, Lueckel B, Kramer A, Ries A, Gurny R. Microenvironmental pH and microviscosity inside pH-controlled matrix tablets: an EPR imaging study. *J. Control. Release* 2006;112(1):72–78.

82. Siepe S, Lueckel B, Kramer A, Ries A, Gurny R. Strategies for the design of hydrophilic matrix tablets with controlled microenvironmental pH. *Int. J. Pharm.* 2006;316(1–2):14–20.

83. Varma MVS, Kaushal AM, Garg S. Influence of micro-environmental pH on the gel layer behavior and release of a basic drug from various hydrophilic matrices. *J. Control. Release* 2005;103(2):499–510.

84. Pygall SR, Kujawinski S, Timmins P, Melia CD. Mechanisms of drug release in citrate buffered HPMC matrices. *Int. J. Pharm.* 2009;370(1–2):110–120.

85. Espinoza R, Hong E, Villafuerte L. Influence of admixed citric acid on the release profile of pelanserin hydrochloride from HPMC matrix tablets. *Int. J. Pharm.* 2000;201(2):165–173.

86. Lopes CM, Lobo JMS, Pinto JF, Costa P. Compressed mini-tablets as a biphasic delivery system. *Int. J. Pharm.* 2006;323 (1–2):93–100.

87. Ishida M, Abe K, Hashizume M, Kawamura M. A novel approach to sustained pseudoephedrine release: differentially coated mini-tablets in HPMC capsules. *Int. J. Pharm.* 2008;359 (1–2):46–52.

88. Verhoeven E, De Beer TRM, Schacht E, Van den Mooter G, Remon JP, Vervaet C. Influence of polyethylene glycol/polyethylene oxide on the release characteristics of sustained-release ethylcellulose mini-matrices produced by hot-melt extrusion: *in vitro* and *in vivo* evaluations. *Eur. J. Pharm. Biopharm.* 2009;72(2):463–470.

89. De Brabander C, Vervaet C, Fiermans L, Remon JP. Matrix mini-tablets based on starch/microcrystalline wax mixtures. *Int. J. Pharm.* 2000;199(2):195–203.

90. Pillay V, Fassihi R. A novel approach for constant rate delivery of highly soluble bioactives from a simple monolithic system. *J. Control. Release* 2000;67(1):67–78.

91. Kim C-J. Release kinetics of coated, donut-shaped tablets for water soluble drugs. *Eur. J. Pharm. Sci.* 1999;7(3):237–242.

92. Juárez H, Rico G, Villafuerte L. Influence of admixed carboxymethylcellulose on release of 4-aminopyridine from hydroxypropyl methylcellulose matrix tablets. *Int. J. Pharm.* 2001;216 (1–2):115–125.

93. Sundy E, Paul Danckwerts M. A novel compression-coated doughnut-shaped tablet design for zero-order sustained release. *Eur. J. Pharm. Sci.* 2004;22(5):477–485.

94. Cox PJ, Khan KA, Munday DL, Sujja-areevath J. Development and evaluation of a multiple-unit oral sustained release dosage form for $S(+)$-ibuprofen: preparation and release kinetics. *Int. J. Pharm.* 1999;193(1):73–84.

95. Bettini R, Acerbi D, Caponetti G, Musa R, Magi N, Colombo P, Cocconi D, Santi P, Catellani PL, Ventura P. Influence of layer position on *in vitro* and *in vivo* release of levodopa methyl ester and carbidopa from three-layer matrix tablets. *Eur. J. Pharm. Biopharm.* 2002;53(2): 227–232.

96. Krishnaiah YSR, Karthikeyan RS, Bhaskar P, Satyanarayana V. Bioavailability studies on guar gum-based three-layer matrix tablets of trimetazidine dihydrochloride in human volunteers. *J. Control. Release* 2002;83(2):231–239.

97. Nerurkar J, Jun HW, Price JC, Park MO. Controlled-release matrix tablets of ibuprofen using cellulose ethers and carrageenans: effect of formulation factors on dissolution rates. *Eur. J. Pharm. Biopharm.* 2005;61(1–2):56–68.

98. Takka S, Rajbhandari S, Sakr A. Effect of anionic polymers on the release of propranolol hydrochloride from matrix tablets. *Eur. J. Pharm. Biopharm.* 2001;52(1):75–82.

7

COATING SYSTEMS FOR ORAL CONTROLLED RELEASE FORMULATIONS

LINDA A. FELTON

College of Pharmacy, University of New Mexico, Albuquerque, NM, USA

Pharmaceutical solids can be coated with polymeric materials for decorative, protective, and functional purposes. These coatings can improve the aesthetic appearance of a drug product. Unique appearances can be created for easy identification and to reduce the potential for counterfeiting. Coatings can be used to enhance the chemical stability of a drug by providing a physical barrier to environmental storage conditions such as light, oxygen, or water vapor [1–4]. The physical strength and fracture resistance of the dosage form can be increased through the application of such coatings [5]. One drug can be coated to prevent its reaction with a secondary chemical within a single multicomponent product. The taste and odor of drugs may be masked by the application of a coating [6]. In addition, coatings can be used to alter the release characteristics of the drug from the dosage form [7–10].

With coatings, a number of modified release mechanisms can be used to achieve the desired drug release patterns. For example, a sustained release effect can be achieved by a simple diffusional reservoir system, where drug inside the tablet core diffuses out through a semipermeable polymer membrane. In contrast, the osmotic pump system relies on an increase in osmotic pressure within the core as the driving force for drug release [11]. Enteric coatings release the drug at elevated pH due to dissolution of the coating [9, 12]. Rupture-type systems for pulsatile delivery [13, 14], coated floating products for gastroretention [15], and other novel release approaches [16–18] can be realized with the selection of appropriate coating and core materials. This chapter will discuss the film formation process, materials used for coating applications, formulation variables, equipment considerations, and techniques to assess coating performance. Issues of substrate–coating interactions and common coating defects will also be addressed.

7.1 FUNDAMENTALS OF FILM COATING

In the pharmaceutical industry, coatings are generally applied to solid substrates using a spray atomization technique. The coating material is either dissolved or dispersed in an aqueous or an organic solvent prior to spraying. The substrates are often preheated in the coating equipment prior to the initiation of the coating process. The coating solution or dispersion is atomized with air into small droplets, which are then delivered to the substrate surface. Upon impingement, the droplets spread across the surface and the coating forms as the solvent evaporates. For aqueous-based dispersions, the polymer particles densely pack on the surface of the solid and, upon further evaporation, the particles flow together due to cohesive forces between the polymer spheres and ultimately coalesce to form the film [19]. A schematic of the coalescence process is shown in Figure 7.1. Heat is generally applied to the coating equipment to facilitate solvent evaporation and film formation.

Curing or postcoating storage at elevated temperature is often employed to promote coalescence of the film [20, 21]. During this curing stage, the microstructure of the polymer is altered [20] and without sufficient curing, the polymer chains will continue to intertwine and densify over time, resulting in changes in drug release [19]. The time required to form a fully coalesced film has been shown to be dependent on

---

Oral Controlled Release Formulation Design and Drug Delivery: Theory to Practice, Edited by Hong Wen and Kinam Park
Copyright © 2010 John Wiley & Sons, Inc.

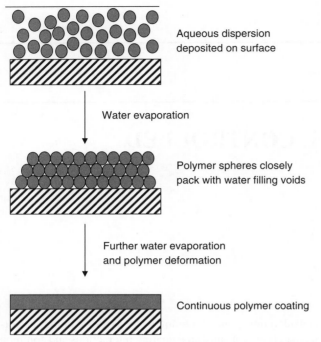

FIGURE 7.1 Schematic of the coalescence process for aqueous polymeric dispersions.

a number of factors, including coating formulation variables and processing conditions [22–27].

7.2 MATERIALS USED IN COATING PROCESSES

Historically, a number of materials have been used in coating processes and these are typically characterized by aqueous

solubility. The development of coating processes in the pharmaceutical industry likely began with sugar coating, a process that is still used today. Since sugar is a water-soluble material, these types of coatings cannot be used to alter drug release characteristics. Other commonly used water-soluble materials include hydroxypropyl methylcellulose, polyvinyl alcohol, hydroxypropyl cellulose, polyvinyl pyrrolidone–vinyl acetate copolymer, and polyvinyl alcohol–polyethylene glycol copolymer. More relevant for this text on oral controlled release systems are the water-insoluble and the pH-dependent soluble polymers, since these materials can be used to modify drug release.

Water-insoluble polymers are typically used to achieve a sustained release of drug over an extended time period. These systems reduce dosing frequency, potentially improving patient compliance to the therapeutic regimen. Such sustained release products reduce blood level fluctuations of the active ingredient, thus eliminating high peak concentrations often associated with side effects and minimizing subtherapeutic plasma levels. While these advantages may lead to better disease state management, the potential for dose dumping, where the entire payload of drug is rapidly released, is a major concern. An example of an *in vitro* dissolution profile from such sustained release systems compared to that of an immediate release product is shown in Figure 7.2.

The polymeric film of these sustained release systems controls the rate at which fluid from the gastrointestinal tract diffuses into the core substrate. This gastrointestinal fluid dissolves the drug and the drug then diffuses out of the dosage form due to the concentration gradient, as shown in Figure 7.3. Films are generally between 50 and 150 μm thick, with thicker films creating a longer, tortuous diffusional pathway [7, 19, 28, 29]. Examples of polymers used to achieve such sustained drug release include ethylcellulose,

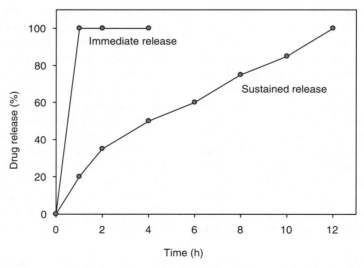

FIGURE 7.2 Comparison of *in vitro* drug release from a model immediate release and a sustained release product.

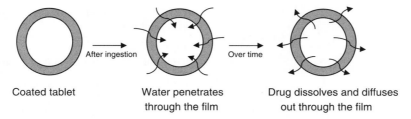

FIGURE 7.3 Schematic showing the mechanism by which a reservoir diffusional system functions to release a drug.

polymethacrylates with quaternary ammonia groups, and polyvinyl acetate. Polymer may also be blended to adjust film permeability or mechanical properties [3, 22, 30]. Hydrophilic materials may be included in the coating to create pores and facilitate drug release, an approach that is especially useful for drugs with limited water solubility [31].

One specific application of coating with water-insoluble polymers to achieve a sustained drug release is the osmotic pump system. These delivery systems contain osmotic agents such as sodium chloride within the core. The cores are coated with water-insoluble polymers and a laser is used to drill a delivery orifice through the film. Water diffuses into the core through the film coating and dissolves the osmotic agents,

which increases the osmotic pressure. The drug is then released through the orifice. The coating used for these systems must be sufficiently strong to withstand the hydrostatic forces reached during gastrointestinal transit [32]. One of the most common polymeric materials used for osmotic pump delivery systems is cellulose acetate [33, 34]; ethylcellulose has also been used [18]. There have been several variations of the osmotic pump reported in the literature. For example, a bilayer tablet design, referred to as a push–pull osmotic pump, utilizes a swellable layer to maintain the hydrostatic pressure within the device for a more complete release of the drug payload [34]. A schematic of this system is shown in Figure 7.4. In another type of system, additives

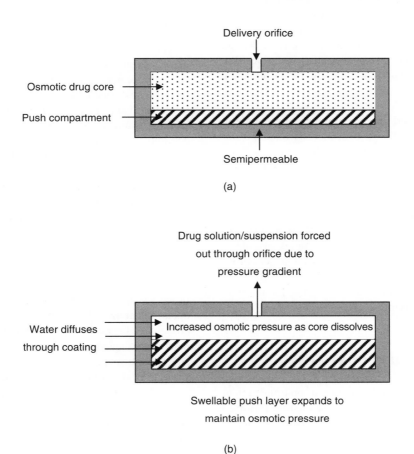

FIGURE 7.4 Schematic of (a) the components of a bilayer osmotic pump system and (b) the mechanism of drug release from such a system.

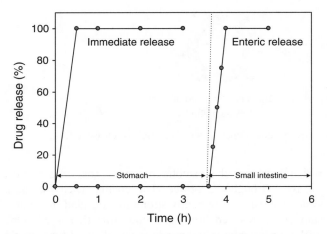

FIGURE 7.5 Comparison of *in vitro* drug release from a model immediate release and an enteric product.

in the coating dissolve after ingestion to create pores in the film and this circumvents the need for a delivery orifice [35, 36].

In contrast to the water-insoluble agents, polymers with pH-dependent solubility are typically employed to target drug release to the small intestines. These polymers have ionizable functional groups that are unionized and thus insoluble in the low pH of the stomach. When the dosage form passes into the higher pH of the small intestines, the functional groups ionize and the polymer dissolves, releasing the substrate core. An example of an *in vitro* dissolution profile from a coated enteric system is shown in Figure 7.5. Since no drug is released in the stomach, these enteric coatings are used to protect the gastric mucosa from exposure to an irritating drug or protect an acid-labile compound from the harsh environment of the stomach. These systems have also been used to target drug release to a particular region of the intestinal tract [37]. With such enteric systems, there is a minimum film thickness necessary to prevent drug release in the stomach and higher levels of coating do not significantly affect drug release [38]. Examples of enteric polymers include phthalate derivatives such as cellulose acetate phthalate and polyvinyl acetate phthalate, hydroxypropyl methylcellulose acetate succinate, and methacrylic acid–ethyl acrylate copolymers.

7.3   EXCIPIENTS USED IN COATINGS

Irrespective of the reason for coating a pharmaceutical product, excipients may be required in the coating formulation to facilitate film formation, modify the permeability or mechanical properties of the polymer, or aid in processing. A list of common excipients is provided in Table 7.1 and each is briefly discussed below.

TABLE 7.1   List of Common Excipients in Film Coating Formulations

Type of Excipient	Function(s) of Excipient	Example
Plasticizers	Increase flexibility of the film	Triethyl citrate
	Reduce brittleness	Tributyl citrate, diethyl phthalate, dibutyl sebacate
Antiadherents	Reduce tackiness of the film	Talc
	Prevent substrate agglomeration	Glyceryl monostearate
Pigments	Add color	Iron oxides
	Enhance attractiveness of product	Aluminum lakes
Surfactants	Emulsify water-insoluble plasticizers	Polysorbate 80
	Improve substrate wettability	Sorbitan monooleate
	Stabilize suspensions	Sodium dodecyl sulfate

7.3.1   Solvents

As mentioned previously, most polymeric films are formed upon solvent evaporation and thus a few words regarding solvents are necessary. While there are a number of solvents that have historically been used in film coating processes with several still currently employed, film coating technology has shifted toward aqueous-based systems due to environmental, economic, and toxicologic reasons. With water-based systems, the risk of explosion is diminished, the costs associated with the purchase and disposal of organic solvents are reduced, and potential toxicity issues due to residual solvents are eliminated. Moreover, many commercial polymeric systems are available as aqueous-based dispersions.

7.3.2   Plasticizers

Many of the polymers used in pharmaceutical film coating processes are brittle and require the addition of a plasticizing agent. Plasticizers function by weakening the intermolecular attractions between the polymer chains. The addition of a plasticizing agent to a coating formulation generally causes an increase in the flexibility and a decrease in the tensile strength of the film. Plasticizer type and concentration have been shown to significantly influence drug release from coated solids [39, 40] as well as the mechanical [41, 42] and adhesive properties of the film [43] and are thus key parameters in coating performance.

Plasticizers are generally nonvolatile, miscible substances and many materials can function to plasticize polymers, including water. Phthalate esters, sebacate esters, and citrate esters are commonly used as plasticizing agents. Various glycol derivatives, such as propylene glycol and polyethylene glycol, have also been employed as plasticizers. In addition, surfactants, preservatives, and other compounds have been shown to plasticize the cellulosic and acrylic polymers [44, 45].

A plasticizer should exhibit little or no tendency for evaporation or volatilization, since the film will become more brittle if the plasticizer vacates the film. The plasticizer must be miscible with the polymer and solubility parameters can be used to predict such compatibility. In order to plasticize the polymer, the agent must partition from the solvent phase into the polymer phase and subsequently diffuse throughout the polymer to disrupt the intermolecular interactions. The rate and extent of this partitioning for an aqueous dispersion is dependent on the solubility of the plasticizer in water and its affinity for the polymer phase [46]. The partitioning of water-soluble plasticizers in aqueous dispersions occurs quite rapidly, whereas longer mixing times are required for plasticizer uptake of water-insoluble agents [47, 48]. Water-insoluble plasticizers are typically emulsified with surfactants before being added to the aqueous polymeric dispersion. If the time allotted for plasticizer uptake is insufficient, unincorporated plasticizer droplets will be sprayed onto the substrate, resulting in uneven plasticizer distribution within the film. Siepmann and Bodmeier developed a mathematical model to predict the minimum mixing time necessary for plasticizer partitioning [49]. Curing or a postcoating thermal treatment may reduce or eliminate uneven plasticizer distribution effects [26].

The effectiveness of a plasticizing agent is dependent to a large extent on the amount of plasticizer added to the coating formulation and the extent of interaction with the polymer. Forces involved in polymer–plasticizer interactions include hydrogen bonding, dipole–dipole and dipole-induced dipole interactions, and dispersion forces. Although many experimental methods have been reported in the pharmaceutical literature, the change in glass transition temperature of the polymer is one of the most common techniques to evaluate the effectiveness of a plasticizer [50]. Most polymers used in the pharmaceutical industry are amorphous and thus exhibit a glass transition temperature (T_g) or a temperature where the behavior of the film changes from hard and brittle to soft and elastic. Table 7.2 shows the glass transition temperature of an acrylic polymer as a function of both plasticizer type and plasticizer concentration [51]. Triethyl citrate was found to be the most effective plasticizing agent of those evaluated, as demonstrated by the lower T_g values. Furthermore, higher plasticizer concentrations were shown to further lower the T_g.

Determination of the mechanical properties of the polymeric film is another technique used to evaluate plasticizer efficacy. While there are a number of testing methods

TABLE 7.2 Glass Transition Temperature (T_g) of Eudragit®️ RS 30 D as a Function of Plasticizer

Plasticizer	10% Plasticizer	20% Plasticizer
Triethyl citrate	34.3°C	12.8°C
Acetyl triethyl citrate	37.0°C	17.5°C
Tributyl citrate	38.2°C	20.5°C
Acetyl tributyl citrate	38.2°C	22.2°C
Triacetin	42.2°C	27.4°C

From Ref. 51.

available to the pharmaceutical scientist, tensile testing of free films is one of the most common [50]. The tensile test is a relatively simple, straightforward analysis that provides information regarding the strength, elasticity, and toughness of the film. For the experimental setup, a free film is placed between two grips and stretched at a specific rate. Load and displacement data are collected and can be mathematically converted to stress and strain. Data are typically presented as a stress versus strain graph, as shown in Figure 7.6. The tensile strength is the maximum stress at which the film specimen breaks. Strain, typically expressed as a percent elongation, is the deformation the film undergoes. Work of failure is the work required to break the specimen and represents the film toughness. Young's modulus, sometimes referred to as elastic modulus, is a measure of the stiffness of the film and is calculated as the slope of the linear region of the stress versus strain graph. A more detailed description of the experimental parameters for this and other mechanical tests can be found elsewhere [50]. The increased elasticity associated with plasticization is attributed to an increase in the free volume of the polymer [52].

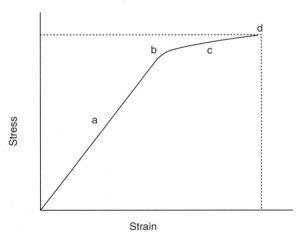

FIGURE 7.6 Example of a stress–strain diagram. (a) Elastic deformation, where stress is proportional to strain; (b) yield point; (c) plastic deformation, where polymer chains orient themselves with the applied stress; (d) breakage of the film.

7.3.3 Antiadherents

Antiadherents are added to film coating formulations to prevent agglomeration of the substrate during both the coating process and subsequent storage. The degree of tackiness correlates with the minimum film-forming temperature of the polymer or the minimum temperature necessary for coalescence to occur and has been shown to increase with increasing plasticizer concentration [53]. One of the most common antiadherents for pharmaceutical coating formulations is talc. The high levels of talc (generally 25–100%, based on dry polymer weight) necessary to reduce film tackiness may result in clogging of the spray nozzle during coating and particle sedimentation. In addition, talc has been shown to decrease water vapor permeability and the dissolution rate of drugs [54, 55], presumably due to its hydrophobic nature. Other studies have shown that the addition of talc to a coating formulation affects the mechanical and adhesive properties of the film [56–58]. Glyceryl monostearate at a concentration of 2–10% (based on dry polymer weight) has been suggested as an alternative to talc [59].

7.3.4 Pigments

Pigments are added to film coating formulations to produce a more pharmaceutically elegant dosage form and to provide easy product identification for both healthcare workers and patients. Opacifying agents such as titanium dioxide have been incorporated into coatings to improve the stability of photolabile or light-sensitive drugs [1, 60]. The most common pigments used in the pharmaceutical industry are the water-insoluble lakes and the iron oxides. Color migration and stability issues have diminished the use of water-soluble dyes in film coating formulations [61, 62].

The pharmaceutical scientist should be aware that the addition of pigments to a coating formulation can significantly affect the mechanical properties and permeability of the film as well as the drug release characteristics [56, 62–65]. Pigments differ significantly in their physical properties, such as particle size, morphology, and density, and these differences contribute to the complex relationship with aqueous film coatings. An inverse relationship between the particle size of the pigment and film–tablet adhesion has been reported [62]. The researchers theorized that larger particles disrupt the interfacial bonding between the polymer and the surface of the tablet to a greater extent than the smaller pigment particles. In another study, the increased elastic modulus of polymeric films was found to be dependent on the shape of the pigment particle and the extent of polymer–particle interaction [66]. Surface polarity of pigments has been shown to influence drug release [64]. Chemical incompatibilities between the pigment and the polymer may also arise, predominantly related to the size and surface charge of the components and the pH of the medium [65].

TABLE 7.3 CPVC of Hydroxypropyl Methylcellulose Films as Determined by Gloss Measurements

Pigment	CPVC (% v/v)
Black iron oxide	7.0–8.5
Red iron oxide	8.5–10.0
Yellow iron oxide	10.0–12.0
Titanium dioxide	13.5–15.0
Aluminum lake yellow No. 6	12.0–13.5
Talc (surface area = 2.99 m^2/g)	12.0–15.0
Talc (surface area = 14.33 m^2/g)	25.0–35.0

From Ref. 61.

As with antiadherents, the water-insoluble coloring agents may clog the spray gun during coating and may settle in the mixing vessel over time. Sodium carboxymethylcellulose has been used to stabilize pigmented dispersions by a mechanism of steric hindrance and increased viscosity [63].

One important concept related to the addition of pigments or other insoluble agents in coating formulations is the critical pigment volume concentration (CPVC) or the maximum concentration (based on volume) of pigment that can be incorporated into a film. If the CPVC is exceeded, there is insufficient polymer present to surround all the insoluble particles and marked changes occur in the mechanical properties, the appearance, and the permeability of the film [61]. Table 7.3 clearly shows the wide variations in CPVC of various pigments as well as talc.

7.3.5 Surfactants

Surfactants may be added to film coating formulations to emulsify water-insoluble plasticizers, to improve substrate wettability and facilitate spreading of the polymer-containing droplets on the substrate surface, or to stabilize suspensions [7, 67, 68]. The concentration of surfactants in such pharmaceutical coatings is typically very low, generally 0.25–1%. As with all other additives discussed, these agents can also affect various polymer properties and influence drug release from the coated substrate [40]. For example, Lindholm and coworkers reported that drug release was dependent on the concentration of polysorbate 20 in ethylcellulose films [45]. These researchers suggested that the hydrophilic surfactant leached during the dissolution test and created pores in the film, thus accelerating drug release.

7.4 PROCESSING EQUIPMENT

There are three major types of equipment currently used to apply coatings to pharmaceutical solids: conventional pans, perforated pans, and fluidized beds. More detailed descriptions of the various coating equipment can be found elsewhere [69, 70]. All equipment require a method to deliver the polymeric material to a spray gun, the atomization of the coating liquid, and movement of the substrate through the spray zone. Heat is generally applied to facilitate solvent evaporation and some type of exhaust system removes the solvent-laden air. One of the most common pumping systems is a peristaltic pump that is considered ideal for delivering latex and pseudolatex dispersions that may coagulate at high pressures.

Polymeric solutions and dispersions are generally atomized with air to create a fine mist for deposition on the substrate surface. Pneumatic nozzles pass high-pressure air across the exiting fluid stream to atomize the liquid whereas hydraulic nozzles rely on the fluid being pumped at relatively high pressures through a small opening. Droplet size of the coating liquid is of critical importance in the coating process. If the droplets are too large, the surface will become overwet, which could potentially result in surface dissolution of the substrate [71]. In contrast, very fine droplets may dry before spreading on the substrate surface and the resulting film will have a rough appearance similar to that of an orange peel. In addition, spray drying, where the liquid evaporates completely before the droplet impinges on the surface, can result. The atomized droplet size can be controlled independently of the polymer flow rate with the pneumatic nozzle.

The conventional coating pan system consists of a round drum that rotates on an inclined axis, as shown in Figure 7.7a. The rotational movement of the drum causes the substrates to tumble and make multiple passes through the spray application zone. Heat is blown across the surface of the tumbling tablets and exhaust air is withdrawn. Originally used in sugar coating operations, syrups were ladled onto the substrates. Aqueous-based polymeric coating, however, requires more rapid solvent evaporation and perforations within the pan significantly improve drying efficiency by forcing the drying air through the tablet bed. A schematic of the perforated coating pan is shown in Figure 7.7b. Baffles are often placed in the coating pans to facilitate tablet tumbling within the pan for a more uniform film distribution.

In contrast to the coating pans, small particles such as beads, pellets, and even powders are generally coated in a fluidized bed apparatus. These multiparticulate systems offer a number of advantages over larger, single unit dosage forms such as more uniform gastrointestinal transit and reduced potential for dose dumping [72]. Coated multiparticulates can be packaged into capsules or compressed into tablets. If the coated particles are to be compressed, obviously the compressional force used during tableting and the mechanical strength of the coating are critical [73]. A faster drug release may occur if the coating cracks or fractures during compression [74], while slower drug release may result from the coating fusing together to form a matrix [75]. Excipients such as microcrystalline cellulose may be employed to

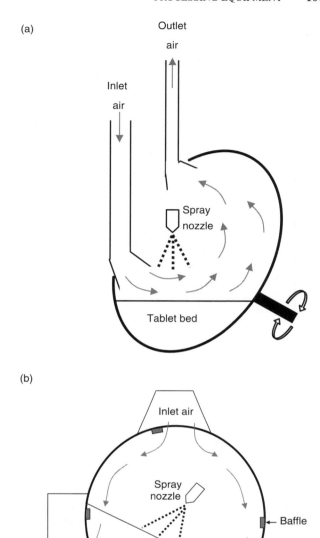

FIGURE 7.7 Schematics of (a) conventional coating pan and (b) perforated coating pan. Gray arrows represent air flow in the apparatus. Adapted from Refs 69 and 70.

prevent direct contact of the coated pellets and to dissipate the compressional force and prevent film fracture [73]. Bodmeier provides an excellent review on the tableting of coated pellets [72].

Fluidized bed coating processes use a carrier gas, usually air, to keep the particles in motion. The high air flow makes this coating technique more efficient at water removal in comparison to the coating pans. As with the pans, the coating material is applied using a spray atomization technique. With fluidized bed coaters, the nozzle can be positioned

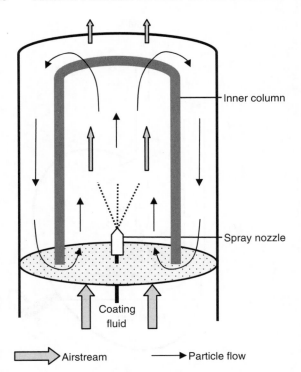

Inner column

Spray nozzle

Coating fluid

⇨ Airstream → Particle flow

FIGURE 7.8 Schematic of a fluidized bed coating apparatus. Adapted from Ref. 69.

above the particle bed such that the polymeric material is sprayed concurrent to the fluidized particles or the nozzle can be located at the base of the apparatus. The top spray process, also known as the granulator mode, is used predominately for taste and odor masking purposes, where drug release rates are not critical, since films produced by this method are not uniform in thickness [19].

In contrast to the top spray mode, the bottom spray technique is one of the most common methods for application of rate-controlling polymers and a schematic of the apparatus is shown in Figure 7.8. A perforated base plate allows air to pass into the vertical, cylindrical chamber. The size and shape of the base plate as well as the fluidization air pressure and the size and shape of the substrates influence particle movement within the apparatus. Particles move up the chamber with the atomized polymer droplets through a slightly conical insert, know as a Wurster column. After passing through the column, the particles enter an expansion chamber, where gravitational forces cause the particles to fall to the bottom of the apparatus and the process is repeated until the desired amount of polymer is applied. Continued movement in the heated apparatus after application of the polymer can be used for postcoating curing. In addition to the top spray and bottom spray methods, the rotary or tangential spray technique adds centrifugation to the forces of fluidization and gravity, allowing for even more efficient solvent evaporation and higher spray rates [19].

7.5 DRUG AND SUBSTRATE CONSIDERATIONS

The chemical properties of the active pharmaceutical ingredient (API) must be considered before initiating any coating project. For example, the API must be stable at the processing temperatures employed. The stability of a hydrolytic API may be adversely affected by the application of an aqueous-based coating system. The shelf life of an acid-labile drug could also be significantly compromised by coating with an enteric polymer due to the acidic nature of the film. Subcoats may be useful to prevent such interactions [7, 76]. In addition, acidic drugs or excipients can extend the dissolution time of an enteric product by creating an acidic microenvironment, whereas alkaline drugs could cause premature failure in gastric fluid.

Not every substrate is suitable for application of a polymeric coating and some thought should be given to the design of the core. First, and maybe most obvious, the substrate must be physically strong to withstand the movement in the coating apparatus. The dosage form should also be stable at the processing temperatures to be employed. For example, gelatin may soften at typical processing temperatures and, combined with the spraying of an aqueous-based polymer, could dissolve [77]. There are other issues to consider when coating capsules. For hard shell capsules, the bodies and caps should be band-sealed prior to coating to ensure adequate film formation at this joint [78, 79]. For soft gelatin capsules, a prewarming step to raise the temperature of the fill liquid to that of the coating bed is necessary to achieve uniform drying of the film [9].

The size and shape of the solid substrate can influence processing, especially movement within the coating equipment. Generally, a biconvex tablet is preferred, as flat-faced tablets tend to agglomerate during coating. Tablet shape can influence coating performance as well. Cunningham and coworkers showed that the sharp edge of a shallow concave tablet abraded during processing and higher levels of coating and a subcoat were required to achieve enteric protection of the dosage form [80]. In contrast, standard concave tablets were enterically coated at lower theoretical weight gains. Surface properties should also be considered since wettability and ultimately film adhesion can be affected. Selection of excipients with thermal expansion coefficients similar to that of the polymer can reduce internal stresses and improve polymer adhesion [81].

During the coating process, the atomized polymer-containing droplets may cause dissolution of the outermost surface of the substrate, which may in fact be necessary for film adhesion. However, the API could potentially interact with the polymer or an excipient in the coating. Substrate components could migrate into the coating and influence the mechanical, adhesive, and drug release properties of the product. For example, Bodmeier and Paeratakul found migration and subsequent recrystallization of propranolol HCl in Eudragit NE 30 D films [82]. Felton and coworkers demonstrated that the fill liquid in a soft gelatin capsule influenced both adhesion and the performance of an enteric polymeric film, presumably due to migration from the capsule [83]. A subcoat may prevent such migration and may be particularly necessary for highly water-soluble drugs when aqueous-based coatings are applied [7, 84, 85].

7.6 POLYMER ADHESION AND THE FILM–TABLET INTERFACE

Adhesion of a polymeric film to the substrate surface is a major prerequisite in the film coating of pharmaceutical dosage forms. The two major forces that influence adhesion are the strength of the interfacial bonds and the internal stresses within the film [86]. Hydrogen bond formation is the primary type of interfacial interaction, although dipole–dipole and dipole-induced dipole interactions also occur. Factors that affect the type or number of bonds formed between the polymer film and the solid surface will influence adhesion. Internal stresses within a film tend to weaken adhesion. These stresses include stress due to shrinkage of the film upon solvent evaporation, thermal stress due to the differences in thermal expansion of the film and the substrate, and volumetric stress due to swelling during storage [87–89]. Additives in the coating formulation [43, 62, 67, 90, 91], substrate variables [92–95], and processing conditions [96] have been shown to influence film–substrate adhesion.

The small size of a dosage form and the nonuniform surface roughness have presented significant challenges in the assessment of polymer adhesion. In the 1970s, the peel test was used to quantify adhesion properties, where a modified tensile tester peeled the film from the tablet surface at a 90° angle [77]. More recently, several variations of the butt adhesion technique, where the entire film is removed normal to the surface of the tablet, have been reported. Data from a butt adhesion test can be plotted as force versus deflection, similar to the stress–strain diagram generated from tensile testing of free films [43, 93]. This graph permits the visualization of the development of the force within the sample during the testing process. In addition to the force of adhesion, the elongation at adhesive failure, the modulus of adhesion, and the adhesive toughness of the film can be determined. The elongation at adhesive failure is the distance the upper platen traveled at film separation and reflects the ductility of the coating. The modulus of adhesion is the slope of the linear region of the force–deflection graph and is analogous to the Young's modulus obtained in tensile testing of free films. The area under the force–deflection curve represents adhesive toughness or the work required to remove the film from the tablet surface.

As mentioned earlier, the atomized polymer-containing droplets may cause dissolution of the outermost surface of

the substrate, allowing for physical mixing at the interface, which may facilitate polymer adhesion. Our understanding of the film–substrate interfacial region, however, is still quite limited and a number of techniques have been proposed to study film–substrate interactions. Confocal laser scanning microscopy provides a nondestructive method to investigate migration of fluorescent compounds [85, 97]. X-ray photoelectron spectroscopy has also been used to quantify the thickness of the film–tablet interfacial region [71, 98] and could potentially be used to follow subsequent migration of components over time.

7.7 NOVEL DRY COATING TECHNIQUES

To circumvent some of the problems associated with aqueous- or solvent-based coating, several techniques involving the deposition of powdered particles onto a substrate have been developed. With these methods, the potential for surface dissolution and drug migration are significantly reduced. Since no solvents are employed in the process, there are no issues related to residual solvents. Moreover, these methods are suitable for coating hydrolytic drugs since no water is employed during processing.

Electrostatic coating, based on principles of photocopy production, allows a resistive powder coating to be deposited onto a tablet containing a conductive material by applying a known energy field [99]. The tablet is then exposed to radiant heat to fuse the coating to the core. Obviously, the selection of substrate and coating material is critical and the particle size of the coating polymer must be uniform. This method coats individual tablets in a continuous process and very precise quantities of polymer can be applied to every tablet [100].

Another dry powder coating process uses a modified coating apparatus or spheronizer to apply the polymer [6, 101, 102]. Initial attempts at dry powder coating relied on the powdered polymer and a liquid plasticizer being applied simultaneously through separate nozzles [103]. However, significantly higher weight gains were reportedly required with such systems. More recently, scientists have coprocessed the polymer powder and the plasticizer by hot melt extrusion and used the finely ground extrudate to deposit onto the substrate surface. Additional excipients, such as polyethylene glycol, may be necessary to promote adhesion [104].

7.8 DEFECTS IN THE COATING

Following the coating process, substrates are generally evaluated visually to detect defects in the film. Table 7.4 provides a list of common defects found in coatings. Some defects may be related solely to the aesthetic appearance of the product while others are more serious and may compromise

TABLE 7.4 Common Defects in Film Coating

Defect	Description
Blistering	Film becomes locally detached, forming a blister
Bridging of the intagliation	Film pulls out of intagliation or monogram, forming a bridge across the mark
Chipping	Film becomes chipped
Color variation	Intertablet variation in color
Cracking	Film cracks across the crown of the tablet
Cratering	Volcanic-like craters in the film
Flaking	Film flakes off, exposing tablet surface
Infilling	Intagliation filled by solidified foam
Mottling	Uneven distribution of color (intratablet)
Orange peel	Film surface rough and nonglossy, like skin of orange
Peeling	Film peels back from edge, exposing tablet surface
Picking	Isolated areas of film pulled away from surface
Pitting	Pits in surface of tablet core without disruption of film
Roughness	Film rough and nonglossy
Wrinkling	Film with wrinkled appearance

From Ref. 106.

the functionality of the film. Manipulating processing parameters, adjusting the coating formulation, or redesigning the substrate may be necessary in order to resolve these problems. For example, picking, where tablets stick together briefly during the coating process, is often associated with an overwet tablet bed and may be eliminated by reducing the spray rate, increasing the bed temperature, or adding an antiadherent to the coating formulation. More serious for rate-controlling films, cracking or peeling of the coating may occur when the internal stresses exceed the tensile strength of the film and reformulation of the coating is often required [105, 106]. The incidence of chipping may be related to the mechanical strength of the substrate and can be reduced by slowing the rotational pan speed of the coating apparatus [106].

Two particularly important defects that are not evident by visual inspection yet result in changes in the drug release properties of the product over time include incomplete coalescence and physical aging. Plasticizer concentration and storage conditions significantly influence both processes. As described earlier in this chapter, the polymer spheres must interdiffuse to form the film [19]. This process occurs very slowly at room temperature but is accelerated at elevated temperatures. Curing or postcoating drying is thus undertaken after the coating process to fully coalesce the film. If the film has not completely coalesced, viscous flow of the polymer spheres will continue and drug release will change over time. In contrast, physical aging results when the free volume of the amorphous polymer slowly relaxes toward a lower free energy state, resulting in a densification of the

film [107]. This process occurs slowly at temperatures below the glass transition temperature of the coating. Several techniques have been proposed to prevent physical aging, including the use of immiscible polymers and very high concentrations of talc, both of which prevent the polymer chains from moving within the film matrix [22, 55, 108].

7.9 CONCLUSIONS

Polymer coating systems have been widely used in the pharmaceutical industry for decorative, protective, or functional purposes. The selection of the polymer is dependent on the purpose of the coating, with water-insoluble materials used for sustained release applications while polymers soluble at elevated pH employed for enteric protection. The water-soluble polymers tend to be used to produce a more elegant dosage form or to protect the active ingredient from exposure to certain environmental conditions. Excipients in the coating formulation can aid in processing, facilitate film formation, and alter the mechanical or permeability properties of the film. In addition to the film coating formulation, the pharmaceutical scientist must consider the properties of the active ingredient and the design of the substrate to be coated as well as the processing parameters employed in the coating process to fully develop and optimize a coated drug product.

REFERENCES

1. Bechard SR, Quraishi O, Kwong E. Film coating: effect of titanium dioxide concentration and film thickness on the photostability of nifedipine. *Int. J. Pharm.* 1992;87:133–139.

2. Felton LA, Timmins GS. A nondestructive technique to determine the rate of oxygen permeation into solid dosage forms. *Pharm. Dev. Technol.* 2006;11:1–7.

3. Zheng W, Sauer D, McGinity JW. Influence of hydroxyethylcellulose on the drug release properties of theophylline pellets coated with Eudragit RS 30 D. *Eur. J. Pharm. Biopharm.* 2005;59:147–154.

4. Mwesigwa E, Buckton G, Basit AW. The hygroscopicity of moisture barrier film coatings. *Drug Dev. Ind. Pharm.* 2005; 31:959–968.

5. Stanley P, Rowe RC, Newton JM. Theoretical considerations of the influence of polymer film coatings on the mechanical strength of tablets. *J. Pharm. Pharmacol.* 1981;33:557–560.

6. Cerea M, Zheng W, Young CR, McGinity JW. A novel powder coating process for attaining taste masking and moisture protective films applied to tablets. *Int. J. Pharm.* 2004; 279:127–139.

7. Dashevsky A, Wagner K, Kolter K, Bodmeier R. Physicochemical and release properties of pellets coated with Kollicoat SR 30 D, a new aqueous polyvinyl acetate dispersion for extended release. *Int. J. Pharm.* 2005;290:15–23.

8. Siepmann F, Wahle C, Leclercq B, Carlin B, Siepmann J. pH-sensitive film coatings: towards a better understanding and facilitated optimization. *Eur. J. Pharm. Biopharm.* 2008;68:2–10.

9. Felton LA, Haase MM, Shah NH, Zhang G, Infeld MH, Malick AW, McGinity JW. Physical and enteric properties of soft gelatin capsules coated with Eudragit L 30 D-55. *Int. J. Pharm.* 1995;113:17–24.

10. Qussi B, Suess WG. Investigation of the effect of various shellac coating compositions containing different water-soluble polymers on *in vitro* drug release. *Drug Dev. Ind. Pharm.* 2005;1:99–108.

11. Theeuwes F. Elementary osmotic pump. *J. Pharm. Sci.* 1975;64:1987–1991.

12. Gazzaniga A, Iamartino P, Maffione G, Sangalli ME. Oral delayed-released system for colonic specific delivery. *Int. J. Pharm.* 1994;108:77–83.

13. Dashevsky A, Mohamad A. Development of pulsatile multiparticulate drug delivery system coated with aqueous dispersion Aquacoat® ECD. *Int. J. Pharm.* 2006;318:124–131.

14. Bussemer T, Peppas NA, Bodmeier R. Evaluation of the swelling, hydration and rupturing properties of the swelling layer of a rupturable pulsatile drug delivery system. *Eur. J. Pharm. Biopharm.* 2003;56:261–270.

15. Sungthongjeen S, Paeratakul O, Limmatvapirat S, Puttipipatkhachorn S. Preparation and *in vitro* evaluation of a multiple-unit floating drug delivery system based on gas formation technique. *Int. J. Pharm.* 2006;324:136–143.

16. Akhgari A. Sadeghi F, Garekani HA. Combination of time-dependent and pH-dependent polymethacrylates as a single coating formulation for colonic delivery of indomethacin pellets. *Int. J. Pharm.* 2006;320:137–142.

17. Guthmann C, Lipp R, Wagner T, Kranz H. Development of a multiple unit pellet formulation for a weakly basic drug. *Drug Dev. Ind. Pharm.* 2007;33:341–349.

18. Ramakrishna N, Mishra B. Plasticizer effect and comparative evaluation of cellulose acetate and ethylcellulose–HPMC combination coatings as semipermeable membranes for oral osmotic pumps of naproxen sodium. *Drug Dev. Ind. Pharm.* 2002;28:403–412.

19. Carlin B, Li JX, Felton LA. Pseudolatex dispersions for controlled drug delivery. In McGinity JW, Felton LA, editors. *Aqueous Polymeric Coatings for Pharmaceutical Dosage Forms,* 3rd edition, Informa Healthcare, New York, 2008, pp. 1–46.

20. Bodmeier R, Paeratakul O. The effect of curing on drug release and morphological properties of ethylcellulose pseudolatex-coated beads. *Drug Dev. Ind. Pharm.* 1994; 20:1517–1533.

21. Siepmann F, Siepmann J, Walther M, MacRae RJ, Bodmeier R. Aqueous HPMCAS coatings: effects of formulation and processing parameters on drug release and mass transport mechanisms. *Eur. J. Pharm. Biopharm.* 2006;63: 262–269.

22. Zheng W, McGinity JW. Influence of Eudragit NE 30 D blended with Eudragit L 30 D-55 on the release of phenyl-propanolamine hydrochloride from coated pellets. *Drug Dev. Ind. Pharm.* 2003;29:357–366.

23. Hamed E, Sakr A. Effect of curing conditions and plasticizer level on the release of highly lipophilic drug from coated multiparticulate drug delivery systems. *Pharm. Dev. Technol.* 2003;8:397–407.

24. Williams RO III, Liu J. The influence of plasticizer on heat–humidity curing of cellulose acetate phthalate coated beads. *Pharm. Dev. Technol.* 2001;6:607–619.

25. Williams RO III, Liu J. Influence of processing and curing conditions on beads coated with an aqueous dispersion of cellulose acetate phthalate. *Eur. J. Pharm. Biopharm.* 2000; 49:243–252.

26. Wesseling M, Bodmeier R. Influence of plasticization time, curing conditions, storage time, and core properties on the drug release from Aquacoat-coated pellets. *Pharm. Dev. Technol.* 2001;6:325–331.

27. Appel LE, Clair JH, Zentner GM. Formulation and optimization of a modified microporous cellulose acetate latex coating for osmotic pumps. *Pharm. Res.* 1992;9:1664–1667.

28. Rajabi-Siahboomi A, Farrell TP. The applications of formulated systems for the aqueous film coating of pharmaceutical oral solid dosage forms. In McGinity JW, Felton LA, editors. *Aqueous Polymeric Coatings for Pharmaceutical Dosage Forms,* 3rd edition, Informa Healthcare, New York, 2008, pp. 323–343.

29. Rohera BD, Parikh NH. Influence of plasticizer type and coat level on Surelease film properties. *Pharm. Dev. Technol.* 2002;7:407–420.

30. Lecomte F, Siepmann J, Walther M, MacRae RJ, Bodmeier R. pH-sensitive polymer blends used as coating materials to control drug release from spherical beads: elucidation of the underlying mass transport mechanisms. *Pharm. Res.* 2005;22:1129–1141.

31. Sadeghi F, Ford JL, Rubinstein MH, Rajabi-Siahboomi AR. Study of drug release from pellets coated with Surelease containing hydroxypropyl methylcellulose. *Drug Dev. Ind. Pharm.* 2001;27:419–430.

32. Verma RK, Mishra B, Garg S. Osmotically controlled oral drug delivery. *Drug Dev. Ind. Pharm.* 2000;26:695–708.

33. Binschaedler C, Gurny R, Doelker E. Osmotic water transport through cellulose acetate membranes produced from a latex system. *J. Pharm. Sci.* 1987;76:455–460.

34. Sastry SV, Khan MA. Aqueous based polymeric dispersion: Plackett–Burman design for screening of formulation variables of atenolol gastrointestinal therapeutic system. *Pharm. Acta Helv.* 1998;73:105–112.

35. Verma RK, Kaushal AM, Garg S. Development and evaluation of extended release formulations of isosorbide mononitrate based on osmotic technology. *Int. J. Pharm.* 2003;263:9–24.

36. Thombre AG, Cardinal JR, DeNoto AR, Herbigb SM, Smith KL. Asymmetric membrane capsules for osmotic drug delivery. I. Development of a manufacturing process. *J. Control. Release* 1999;57:55–64.

37. Cole ET, Scott RA, Connor AL, Wilding IR, Petereit HU, Schminke C, Beckert T, Cade D. Enteric coated HPMC

112 COATING SYSTEMS FOR ORAL CONTROLLED RELEASE FORMULATIONS

capsules designed to achieve intestinal targeting. *Int. J. Pharm.* 2002;231:83–95.

38. Thoma K, Bechtold K. Influence of aqueous coatings on the stability of enteric coated pellets and tablets. *Eur. J. Pharm. Biopharm.* 1999;47:39–50.

39. Amighi K, Moes A. Influence of plasticizer concentration and storage conditions on the drug release rate from Eudragit RD30D film-coated sustained-release theophylline pellets. *Eur. J. Pharm. Biopharm.* 1996;42:29–35.

40. Bodmeier R, Paeratakul O. Process and formulation variables affecting the drug release from chlorpheniramine maleate-loaded beads coated with commercial and self-prepared aqueous ethyl cellulose pseudolatexes. *Int. J. Pharm.* 1991; 70:59–68.

41. Wu T, Pan W, Chen J, Zhang R. Studies of the drug permeability and mechanical properties of free films prepared by cellulose acetate pseudolatex coating system. *Drug Dev. Ind. Pharm.* 2000;26:95–102.

42. Felton LA, Shah NH, Zhang G, Infeld MH, Malick AW, McGinity JW. Compaction properties of individual non-pareil beads coated with an acrylic resin copolymer. *STP Pharm. Sci.* 1997;7:457–462.

43. Felton LA, McGinity JW. Influence of plasticizers on the adhesive properties of an acrylic resin copolymer to hydrophilic and hydrophobic tablet compacts. *Int. J. Pharm.* 1997;154:167–178.

44. Wu C, McGinity JW. Non-traditional plasticization of polymeric films. *Int. J. Pharm.* 1999;177:15–27.

45. Lindholm T, Lindholm BA, Niskanen M. Koskiniemi J. Polysorbate 20 as a drug release regulator in ethyl cellulose film coatings. *J. Pharm. Pharmacol.* 1986;38:686–688.

46. Bodmeier R, Paeratakul O. Plasticizer uptake by aqueous colloidal polymer dispersions used for the coating of solid dosage forms. *Int. J. Pharm.* 1997;152:17–26.

47. Gutierrez-Rocca JC, McGinity JW. Influence of water soluble and insoluble plasticizers on the physical and mechanical properties of acrylic resin copolymers. *Int. J. Pharm.* 1994; 103:293–301.

48. Bodmeier R, Paeratakul O. The distribution of plasticizers between aqueous and polymer phases in aqueous colloidal polymer dispersions. *Int. J. Pharm.* 1994;103:47–54.

49. Siepmann J, Paeratakul O, Bodmeier R. Modeling plasticizer uptake in aqueous polymer dispersions. *Int. J. Pharm.* 1998; 165:191–200.

50. Felton LA, O'Donnell PB, McGinity JW. Mechanical properties of polymeric films prepared from aqueous dispersions. In McGinity JW, Felton LA, editors. *Aqueous Polymeric Coatings for Pharmaceutical Dosage Forms*, 3rd edition, Informa Healthcare, New York, 2008, pp. 105–128.

51. O'Donnell PB, McGinity JW. Mechanical properties of polymeric films prepared from aqueous polymeric dispersions. In McGinity JW, editors. *Aqueous Polymeric Coatings for Pharmaceutical Dosage Forms*, 2nd edition, Marcel Dekker, Inc., New York, 1997, pp. 517–548.

52. Sinko CM, Amidon GL. Plasticizer-induced changes in the mechanical rate of response of film coatings: an approach to quantitating plasticizer effectiveness. *Int. J. Pharm.* 1989;55: 247–256.

53. Wesseling M, Kuppler F, Bodmeier R. Tackiness of acrylic and cellulosic polymer films used in the coating of solid dosage forms. *Eur. J. Pharm. Biopharm.* 1999;47:73–78.

54. Fassihi RA, McPhillips AM, Uraizee SA, Sakr AM. Potential use of magnesium stearate and talc as dissolution retardants in the development of controlled drug delivery systems. *Pharm. Ind.* 1994;56:579–583.

55. Maejima T, McGinity JW. Influence of film additives on stabilizing drug release rates from pellets coated with acrylic polymers. *Pharm. Dev. Technol.* 2001;6:211–221.

56. Okhamafe AO, York P. Relationship between stress, interaction and the mechanical properties of some pigmented tablet coating films. *Drug Dev. Ind. Pharm.* 1985;11:131–146.

57. Gibson SHM, Rowe RC, White ET. Mechanical properties of pigmented tablet coating formulations and their resistance to cracking. I. Static mechanical measurement. *Int. J. Pharm.* 1988;48:63–77.

58. Okhamafe AO, York P. Effect of solids–polymer interactions on the properties of some aqueous-based tablet film coating formulations. II. Mechanical characteristics. *Int. J. Pharm.* 1984;22:273–281.

59. Nimkulrat S, Suchiva K, Phinyocheep P, Puttipipathachorn S. Influence of selected surfactants on the tackiness of acrylic polymer films. *Int. J. Pharm.* 2004;287:27–37.

60. Rowe RC, Sheskey PJ, Owen S. *Handbook of Pharmaceutical Excipients*, 5th edition, Pharmaceutical Press, London, 2006.

61. Felton LA, McGinity JW. Influence of insoluble excipients on film coating systems. *Drug Dev. Ind. Pharm.* 2002;28: 225–243.

62. Felton LA, McGinity JW. Influence of pigment concentration and particle size on adhesion of an acrylic resin copolymer to tablet compacts. *Drug Dev. Ind. Pharm.* 1999;25:599–606.

63. Maul KA, Schmidt PC. Influence of different-shaped pigments on bisacodyl release from Eudragit L 30 D. *Int. J. Pharm.* 1995;118:103–112.

64. Maul KA, Schmidt PC. Influence of different-shaped pigments and plasticizers on theophylline release from Eudragit RS30D and Aquacoat ECD30 coated pellets. *STP Pharm. Sci.* 1997;7:498–506.

65. Nyamweya N, Mehta KA, Hoag SW. Characterization of the interactions between polymethacrylate-based aqueous polymeric dispersions and aluminum lakes. *J. Pharm. Sci.* 2001; 90:1–11.

66. Rowe RC. Modulus enhancement in pigmented tablet film coating formulations. *Int. J. Pharm.* 1983;14:355–359.

67. Felton LA, Austin-Forbes T, Moore TA. Influence of surfactants in aqueous-based polymeric dispersions on the thermomechanical and adhesive properties of acrylic films. *Drug Dev. Ind. Pharm.* 2000;26:205–210.

68. Lindholm T, Juslin M, Lindholm BA, Poikala M, Tiilikainen S, Varis H. Properties of free ethyl cellulose films containing surfactant and particulate matter. *Pharm. Ind.* 1987;49: 740–746.

69. Bauer KH, Lehmann K, Osterwald HP, Rothgang G. *Coated Pharmaceutical Dosage Forms*, CRC Press, Boca Raton, FL, 1998.

70. Mehta AM. Processing and equipment considerations for aqueous coatings. In McGinity JW, Felton LA, editors. *Aqueous Polymeric Coatings for Pharmaceutical Dosage Forms*, 3rd edition, Informa Healthcare, New York, 2008, pp. 67–103.

71. Barbash D, Fulgham J, Yang J, Felton LA. A novel imaging technique to investigate the influence of atomization air pressure on film-tablet interfacial thickness. *Drug Dev. Ind. Pharm.* 2009;35:480–487.

72. Bodmeier R. Tableting of coated pellets. *Eur. J. Pharm. Biopharm.* 1997;43:1–8.

73. Dashevsky A, Kolter K, Bodmeier R. Compression of pellets coated with various aqueous polymer dispersions. *Int. J. Pharm.* 2004;279:19–26.

74. Sarisuta N, Punpreuk K. *In vitro* properties of film-coated diltiazem hydrochloride pellets compressed into tablets. *J. Control. Release* 1994;31:215–222.

75. Lopez-Rodriguez FJ, Torrado JJ, Torrado S, Escamilla C, Cadorniga R, Augsburger LL. Compression behavior of acetylsalicylic acid pellets. *Drug Dev. Ind. Pharm.* 1993;19:1369–1377.

76. Bruce LD, Koleng JJ, McGinity JW. The influence of polymeric subcoats and pellet formulation on the release of chlorpheniramine maleate from enteric coated pellets. *Drug Dev. Ind. Pharm.* 2003;29:909–924.

77. Felton LA, McGinity JW. Adhesion of polymeric films to pharmaceutical solids. *Eur. J. Pharm. Biopharm.* 1999;47:1–14.

78. Felton LA, Sturtevant S, Birkmire A. A novel capsule coating process for the application of enteric coatings to small batch sizes. In *American Association of Pharmaceutical Scientists Annual Meeting*, San Antonio, TX, 2006.

79. Felton LA, Friar AL. Enteric coating of gelatin and cellulosic capsules using an aqueous-based acrylic polymer. In *American Association of Pharmaceutical Scientists Annual Meeting*, Toronto, Canada, 2002.

80. Cunningham C, Kinsey BR, Scattergood LK, Turnbull N. The effect of tablet shape on the application of an enteric film coating. In *American Association of Pharmaceutical Scientists*, Toronto, Canada, 2002.

81. Okutgen E, Hogan JE, Aulton ME. Effects of tablet core dimensional instability on the generation of internal stresses within film coats. Part 1. Influence of temperature changes during the film coating process. *Drug Dev. Ind. Pharm.* 1991;17:1177–1189.

82. Bodmeier R, Paeratakul O. Evaluation of drug-containing polymer films prepared from aqueous latexes. *Pharm. Res.* 1989;6:725–730.

83. Felton LA, Shah NH, Zhang G, Infeld MH, Malick AW, McGinity JW. Physical–mechanical properties of film-coated soft gelatin capsules. *Int. J. Pharm.* 1996;127:203–211.

84. Bruce LD, Petereit HU, Beckert T, McGinity JW. Properties of enteric coated sodium valproate pellets. *Int. J. Pharm.* 2003;264:85–96.

85. Missaghi S, Fassihi R. A novel approach in the assessment of polymeric film formation and film adhesion on different pharmaceutical solid substrates. *AAPS PharmSciTech* 2004;5:article 29.

86. Felton LA, McGinity JW. Adhesion of polymeric films. In McGinity JW, Felton LA, editors. *Aqueous Polymeric Coatings for Pharmaceutical Dosage Forms*, 3rd edition, Informa Healthcare, New York, 2008, pp. 151–170.

87. Rowe RC. A reappraisal of the equations used to predict the internal stresses in film coatings applied to tablet substrates. *J. Pharm. Pharmacol.* 1983;35:112–113.

88. Croll SG. The origin of residual internal stress in solvent-cast thermoplastic coatings. *J. Appl. Polym. Sci.* 1979;23:847–858.

89. Sato K. The internal stress of coating films. *Prog. Org. Coat.* 1980;8:143–160.

90. Lehtola VM, Heinamaki JT, Nikupaavo P, Yliruusi JK. Effect of some excipients and compression pressure on the adhesion of aqueous-based hydroxypropyl methylcellulose film coatings to tablet surface. *Drug Dev. Ind. Pharm.* 1995;21:1365–1375.

91. Khan H, Fell JT, Macleod GS. The influence of additives on the spreading coefficient and adhesion of a film coating formulation to a model tablet surface. *Int. J. Pharm.* 2001;227:113–119.

92. Rowe RC. The measurement of the adhesion of film coatings to tablet surfaces: the effect of tablet porosity, surface roughness, and film thickness. *J. Pharm. Pharmacol.* 1978;30:343–346.

93. Felton LA, McGinity JW. Influence of tablet hardness and hydrophobicity on the adhesive properties of an acrylic resin copolymer. *Pharm. Dev. Technol.* 1996;1:381–389.

94. Johnson BA, Zografi G. Adhesion of hydroxypropyl cellulose films to low energy solid substrates. *J. Pharm. Sci.* 1986;75:529–533.

95. Rowe RC. The adhesion of film coatings to tablet surfaces—measurement on biconvex tablets. *J. Pharm. Pharmacol.* 1977;29:58–59.

96. Felton LA, Baca ML. Influence of curing on the adhesive and mechanical properties of an applied acrylic polymer. *Pharm. Dev. Technol.* 2001;6:1–9.

97. Felton LA, Yang J. Confocal laser scanning microscopy to investigate drug migration into a sustained release coating. In *Controlled Release Society Annual Meeting*, Miami, FL, 2005.

98. Felton LA, Perry WL. A novel technique to quantify film–tablet interfacial thickness. *Pharm. Dev. Technol.* 2002;7:1–5.

99. Houlton S. Electrostatics in continuous tablet coating. *Manuf. Chem.* 1998;69:13–16.

100. Porter SC. Coating of pharmaceutical dosage forms. In Dera-Marderosian A, et al., *Remington: the Science and Practice of Pharmacy*, 21st edition, University of the Sciences in Philadelphia, Philadelphia, PA, 2005, pp. 929–938.

101. Pearnchob N, Bodmeier R. Dry polymer powder coating and comparison with conventional liquid-based coatings for Eudragit RS, ethyl cellulose, and shellac. *Eur. J. Pharm. Biopharm.* 2003;56:363–369.

102. Bodmeier R, McGinity JW. Powder-coating systems. *Drug Deliv. Technol.* 2005;5:70–73.

103. Obara S, Maruyama N, Nishiyama Y, Kokubo H. Dry coating: an innovative enteric coating method using a cellulose derivative. *Eur. J. Pharm. Biopharm.* 1999;47:51–59.

104. Sauer D, Zheng W, Coots LB, McGinity JW. Influence of processing parameters and formulation factors on the drug release from tablets powder-coated with Eudragit L 100-55. *Eur. J. Pharm. Biopharm.* 2007;67:464–475.

105. Rowe RC. The cracking of film coatings on film-coated tablets—a theoretical approach with practical implications. *J. Pharm. Pharmacol.* 1981;33:423–426.

106. Rowe RC. Defects in aqueous film-coated tablets. In McGinity JW, Felton LA, editors. *Aqueous Polymeric Coatings for Pharmaceutical Dosage Forms*, 3rd edition, Informa Healthcare, New York, 2008, pp. 129–149.

107. Kucera SA, Felton LA, McGinity JW. Physical aging of polymers and its effect on the stability of solid oral dosage forms. In McGinity JW, Felton LA, editors. *Aqueous Polymeric Coatings for Pharmaceutical Dosage Forms*, 3rd edition, Informa Healthcare, New York, 2008, pp. 445–474.

108. Kucera SA, McGinity JW, Zheng WJ. Use of proteins to minimize the physical aging of Eudragit sustained release films. *Drug Dev. Ind. Pharm.* 2007;33:717–726.

8

FLUID BED COATING AND GRANULATION FOR CR DELIVERY

YUE TENG AND ZHIHUI QIU

Pharmaceutical Development Unit, Novartis Pharmaceuticals Corporation, East Hanover, NJ, USA

8.1 INTRODUCTION

Why is a controlled release (CR) solid dosage form needed? The main objectives of developing a controlled release dosage form are to decrease the frequency of administration, improve patient compliance, reduce side effects, deliver a compound to the desired site, and extend a drug product's market life. Target delivery, for example, colon delivery or enteric delivery, can improve therapeutic effect and reduce side effects. Taking pills less often means convenience and compliance for patients. With sustained release dosage forms, a steady drug plasma level can be maintained within a desired therapeutic range over a prolonged time period. Therefore, possible side effects caused by high C_{max} during multiple dosing can be reduced. Overall, the therapeutic effect can be improved.

Among all the systems used for controlled release, multiparticulate systems are the most commonly used and can offer many unique advantages. Compared to single-unit dosage forms, such as film-coated or matrix tablets, multiparticulate dosage forms distribute evenly in the GI track and can release drug substances in a uniform and continuous way. They offer both formulation flexibility and good *in vivo* performance [1].

Multiparticulate dosage forms can also

- avoid dose dumping;
- minimize unit to unit variation;
- include particles with different release rates in one drug product;
- provide dosing flexibility without extensive product development;
- coat incompatible active ingredients separately to deliver combination products;
- offer more reproducible release due to less variation in GI track transit time;
- reduce variation due to GI track condition or food effect;
- minimize GI track irritation.

There are several ways to prepare controlled release multiparticulate systems:

- a matrix core with or without polymer coating;
- inert beads with a layer of drug covered by a layer of polymer;
- immediate release core with functional coating to control release rate or site of release (enteric coating, colon delivery, etc.);
- a matrix core with a functional coating followed by an immediate release layer.

Fluidized bed processing systems have been widely used in the development of multiparticulate systems to achieve a desired release profile. They can be used for both preparing coating substrates and applying coating onto the substrates. In fluidized bed systems, solid particles are suspended by upward blowing air during granulation or coating processes. The bed looks like a vigorously boiling liquid, and has many of the properties of a fluid, hence being called "fluid bed."

Oral Controlled Release Formulation Design and Drug Delivery: Theory to Practice, Edited by Hong Wen and Kinam Park

The granulating or coating liquids are atomized and sprayed from the spray system. The droplets of the atomized solution or suspension are then spread onto the particles as the particles pass through the spray zone. For a granulation process, the particles form agglomerates. For a coating process, the spread droplets of the polymer coalescence and form a continuous film. The vigorous mixing of air with solids brings particles into intimate contact with each other, thus allowing removal of a large amount of liquid in a very short period of time. These features make fluid bed systems particularly suitable for providing cosmetic and functional coating of small particles.

Controlled release dosage form design covers a broad range of release forms: sustained release (SR), delayed release, targeted release, and so on. This chapter focuses on the development of SR forms: using the fluid bed technology to prepare coating substrates and apply a functional coating to multiparticulates in order to achieve a desired drug release profile.

The following will be discussed: release mechanism, formulation consideration, and process.

8.2 RELEASE MECHANISM FOR MULTIPARTICULATE SYSTEMS

For a SR dosage form, zero-order release is the ideal release profile, in which the release rate is constant within a specified period of time. The most commonly used formulation design strategies employ diffusion and/or dissolution as the rate-limiting step [2]. In a real SR dosage form, generally both mechanisms contribute to altering the release profile [3]. Theoretically, the drug release from a unit dose is governed by Fick's law of diffusion, where the flux of a drug across a surface is proportional to the concentration gradient of the drug along the diffusion path:

$$J = D \, dC/dx \qquad (8.1)$$

where J is the flux of diffusion, D is the diffusion coefficient in a given medium, and dC/dx is the concentration gradient of a drug across the diffusion path.

There are many ways to apply diffusion and dissolution as the means of rate control in formulation design, such as matrix dissolution systems, diffusion-controlled porous or nonporous reservoir systems, or diffusion-controlled monolithic systems (see Figures 8.1–8.4). Fick's law (Equation 8.1) can be used to derive release models for each type of formulation. The following assumptions are used in the subsequent controlled release models:

- drug particles are spherical with a smooth surface;
- both core and coating film are homogeneous;

- the dissolution medium is well mixed and in sink condition, that is, the diffusion layer is not rate limiting;
- diffusion through the membrane or dissolution of the matrix/coating is the rate-limiting step.

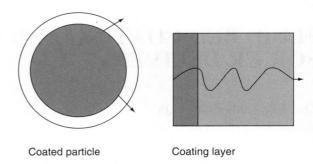

FIGURE 8.1   Illustration of drug release from insoluble polymer.

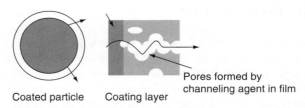

FIGURE 8.2   Illustration of drug release from porous polymer film.

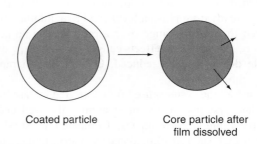

FIGURE 8.3   Illustration of drug release from drug dispersed in an insoluble matrix.

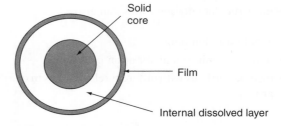

FIGURE 8.4   Illustration of drug release from film-coated core via dissolution mechanism.

8.2.1 Diffusion-Controlled Release

Diffusion is the most common mechanism controlling drug release from a core surrounded by a polymeric membrane. Drug substance is released by diffusion across the membrane to the dissolution media. For a water-insoluble polymer, either the solubility of the drug in the polymer drives the diffusion or a channeling agent in the film dissolves and allows the diffusion to occur. For a partially water-soluble polymer, the water-soluble portion of the polymer can dissolve and create small pores.

8.2.1.1 Diffusion-Controlled Nonporous Reservoir System
For a drug-loaded particle coated with a layer of insoluble polymer, the drug needs to penetrate the membrane from the high-concentration core to low-concentration medium. The release rate dM/dt is described in Equation 8.2.

$$dM/dt = ADK\Delta C/l \tag{8.2}$$

where A is the area, D is the diffusion coefficient, K is the partition coefficient of drug between membrane and core, l is the diffusion length, and ΔC is the concentration difference across the membrane.

To achieve a constant release rate, all the parameters on the right-hand side of Equation 8.2 should be held constant. Any changes in the above terms during release will lead to nonzero-order release profile. The release rate can be adjusted by changing the coating thickness (l), area (A) through particle size, partition coefficient (K), and/or diffusion coefficient (D) through polymer composition.

8.2.1.2 Diffusion-Controlled Porous Reservoir System
When a water-soluble pore former is in the film, the particle will be encapsulated by a partially soluble membrane. Upon exposure to water, the water-soluble polymer dissolves and forms channels. The drug diffusion can be described as

$$dM/dt = AD\Delta C/l \tag{8.3}$$

where A is the area, D is the diffusion coefficient, l is the diffusion length, ΔC is the concentration difference across the membrane.

8.2.1.3 Diffusion-Controlled Monolithic System
For a drug-loaded matrix core coated with water-soluble polymer film, polymer coating is quickly dissolved. The rate-limiting step for the release is the diffusion step through the insoluble solid matrix. Higuchi derived an equation to describe the drug release from such a system:

$$Q = \left[\frac{D\varepsilon}{\tau}(2A-\varepsilon C_S)C_S t\right]^{1/2} = k_H t^{1/2} \tag{8.4}$$

where Q is the amount of drug released in time t, D is the diffusion coefficient, C_s is the solubility of drug in the dissolution medium, ε is the porosity, τ is the matrix tortuosity, A is the drug content per cubic centimeter of matrix tablet, and k_H is the release rate constant for the Higuchi model.

8.2.2 Dissolution-Controlled Release

Sustained release profile can also be achieved by dissolution mechanism. On one hand, the drug can be incorporated into a coated or uncoated soluble matrix, in which the release rate depends on matrix erosion:

$$dC/dt = K_D A(C_s - C) \tag{8.5}$$

where K_D is the diffusion rate constant.

On the other hand, an immediate release core can be coated with a slowly dissolving layer to form an encapsulated dissolution system, or a reservoir system. The dissolution rate will depend on the dissolution rate constant of the coating polymer and the coating thickness. Once the coating is completely eroded, the encapsulated drug is released as a pulsed dose. Combinations of several discrete coating thicknesses lead to pulsatile release, while the inclusion of a spectrum of coating thickness leads to common sustained release.

8.3 FORMULATION CONSIDERATION

Generally, the following steps are used to prepare controlled release multiparticulate systems. The first step is to make substrates like pellets, granules, or minitablets with drug substances. The release of drug substances from the substrates can be either immediate release or sustained release. The second step is to coat the substrates with the desired coating formulations and coating level(s) to achieve the required drug release profile.

The formulation design strategy for a multiparticulate system should focus on two aspects: substrates and coating membranes. Depending on drug solubility, polymer films with different permeability can be selected for coating. For a water-soluble compound, the main challenge is to sustain the release. Thus, less permeable polymer films are often selected and sustained release matrix cores are used to provide further protection. For a poorly water-insoluble compound, the main challenge is to release most of the drug from the system. This can be achieved with an osmotically active drug core, coated with semipermeable membrane.

8.3.1 Substrates' Preparation

Ideal substrates for coating applications should be dense, have spherical and smooth surfaces with minimal porosity,

TABLE 8.1   Comparison of Pellets, Granules, and Minitablets as Coating Substrate

Substrate	Pellets			Granules	Minitablets
Technology	Extrusion spheronization	Direct pelletization	Drug layering	Granulation	Compaction
Pros	High drug loading, good flow, low friability, suited for coating	High drug loading, good flow	Smooth surface, narrower particle size distribution, good flow	Easy process	Easy process
Cons	Rough surface, not a very common pharmaceutical process (equipment might not be readily available at the intended manufacturing site)	Low yield, potentially hard to control particle size	Low drug loading, long processing time	Rough surface, porous structure, low density	Potential edge chipping, extra care and fine adjustments of tableting machines

and have narrow particle size distribution. The substrates should be hard enough to sustain fluidization, stable at elevated temperature and moisture level, and noncohesive. Pellets, granules, and minitablets are examples of substrates used in multiparticulate systems. Many different techniques, such as extrusion spheronization [4], drug layering, granulation, and compaction, can be used to make substrates for subsequent coating. Each of the techniques has its own advantages and disadvantages as listed in Table 8.1. Selection of a technique should be based on the physical–chemical characterization of the drug substance.

8.3.1.1   Pellets   Pellets are good substrates for coating because they have spherical shape, suitable particle size/density, and smooth surface. Drugs are incorporated into pellets either as drug layers or as matrix systems. For active layering, different inert cores, such as sugar spheres (nonpareil) and microcrystalline (MCC) spheres (Cellets), are commercially available. For matrix systems, pellets can be prepared by extrusion spheronization or rotary fluid bed pelletization process.

Active ingredients can be layered onto inert cores in two ways:

1. Spray a solution or suspension containing both drugs and binding agents onto inert cores.
2. Layer drug powders onto inert cores directly.

Solution layering is a good choice for drugs with reasonable solubility and viscosity. Suspension layering is more flexible due to its low viscosity and its capability for high drug loading. For both approaches, dissolving drug substances in liquids and spray layering onto pellets may cause transformation of the drug polymorphism state. Crystalline drugs may convert to a different crystalline state or even the amorphous state. This will change drug release profile on storage and cause stability concern. For both suspension layering and dry powder layering, it is important to control the particle size of the drug substance, which should be

sufficiently small compared to the inert cores, to ensure good adhesion and uniform coating. After the layering process, the drug layered particles maintain spherical shape and a narrow particle size distribution.

Pellets can also be made out of a mixture of active ingredients and other excipients via direct pelletization in rotary fluid bed systems or extrusion spheronization. The excipients include binders, fillers, and disintegrants. Various types of polyvinylpyrrolidone, starch, and so on can be used as binders. Fillers, such as MCC, lactose, and calcium phosphate, should be selected based on the desired properties of the core (such as solubility) and their compatibility with the drug substance. The composition of the core can influence the drug release profile significantly [5, 6]. The core can be immediate release or sustained release. It was reported that using waxy corn starch as a cofiller in the pellet cores could prevent the diffusion of water-soluble drugs from pellet core to the film layer [5].

Drug-loaded pellets made through drug layering and direct pelletization can be distinguished using scanning electronic microscopy (SEM). As shown in Figure 8.5, the cross section of the pellet is not uniform. The outer layer (drug) and core (inert) have different textures, which means the pellets are drug layered.

It is critical to control the size and size distribution of pellets prior to the coating process. The size of pellets has an impact on pellet's fluidization pattern and velocity, thus affecting coating thickness. Larger or heavier pellets, with a longer residence time in the spray zone, form a thicker film and exhibit a much slower release rate than smaller or lighter pellets [7]. Particles less than $100\,\mu m$ are very cohesive, thus leading to agglomeration and poor coating. Meanwhile, the particle surface area is directly related to the particle size (Table 8.2), whereas smaller particles have larger overall surface area and require a higher amount of coating than larger particles for the same release profile. For the same amount of coating material, smaller particles or higher surface areas lead to thinner films, thus a faster release. Therefore, both a desirable pellet size and a narrow particle

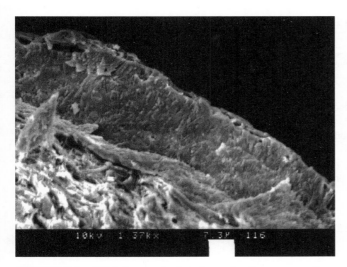

FIGURE 8.5 SEM picture of a drug-layered pellet, cross section.

TABLE 8.3 Melting Points for Binders Used in Melt Granulation Technique

Binder	Melting Point Range (°C)
Beeswax	62–65
Carnauba wax	82–86
Glyceryl behenate	65–67
Glyceryl monostearate	55–60
Glyceryl palmitostearate	52–55
Glyceryl stearate	54–63
Hydrogenated castor oil	61–69
Stearic acid	67–69

size distribution are important to achieve a tight and reproducible release profile.

Particles with a smooth surface are preferred over a rough surface. The rough surface can significantly increase the surface area and the coating materials trapped in the pores cannot function as a release barrier [8]. The preferred particle shape is spherical with an aspect ratio of 1.0. The influence of pellet shape on coating process in a fluid bed system had been investigated and found that the aspect ratio of a pellet should be less than 1.5 [8]. Therefore, cubic or plate-shaped drug substances that tend to compact well are the preferred substrates for drug layering and needle shape drug substances should be avoided. Particles with a high friability tend to break and create fines during fluidization and lead to uncontrolled release of drugs.

8.3.1.2 *Granules*

In general, granules are not the best substrates for coating due to their high porosity and irregular shapes. However, hot melt fluid bed granulation can overcome these disadvantages. Molten wax is often used to granulate drug powder in fluid bed systems. The wax congeals upon contact with drug particles at lower temperature, and serves as a retardant, hindering the premature release of drug from granules. This technology is advantageous

TABLE 8.2 Comparison of Particles Size and Relative Surface Area

Mesh Size	Diameter (mm)	Relative Surface Area
10	2.00	1.0
18	1.00	2.0
35	0.500	4.0
60	0.250	8.0
120	0.125	16.0

because there is no water involved in the whole process. Due to the inert nature of wax, formulations containing wax are stable and less sensitive to pH change. It should be noted that the melting points for commonly used binders in hot melt granulation range from 50 to 90°C, as listed in Table 8.3. This is because a binder with very low melting point may get melted or softened during handling and storage.

8.3.1.3 *Minitablets*

Minitablets refer to tablets with a diameter of 2–3 mm. Different doses can be achieved by adjusting the number of minitablets in each unit dose. Minitablets are more uniform in size, thus less unit-to-unit variation is expected. They are larger than pellets and granules; therefore, less amount of coating materials is needed to achieve the same release rate. A comparison study showed that, to obtain the same release rate of theophylline from minitablets, granules required about 2.5–3 times more coating materials [9].

Due to the vigorous movement of minitablets in fluid bed systems, it is very important to make tablets with a high hardness and low friability to avoid chipping during the subsequent coating step. Compared to layering drug substances onto pellets, it is relatively easier to produce minitablets containing drug substance. Zero-order kinetic release may be achieved by coating minitablets [10].

8.3.2 Subcoat

A subcoat can be defined as any protection layer underneath the outermost functional coating. It is needed if the drug substance is not compatible with the functional coating. A subcoat may also be applied to improve the spreading and adhesion of the functional coating to the substrate to provide more uniform coating. This is particularly true when the substrate has a sharp edge or abrupt curvature change like the edge of minitablets or logos.

Hydroxypropylmethylcellulose (HPMC) or hydroxypropylcellulose (HPC) based coating materials are common choices for a subcoat. For a fine pellet coating, less tacky materials like Metolose® SM-4 perform better than HPMC. A functional coating material can also be used as a subcoat.

TABLE 8.4   Comparison of Organic Solvent-Based and Aqueous-Based Coating System

Coating System	Pros	Cons
Organic solvent-based coating system	*Process*: low processing temperature; no water involved	*Process*: solvent recovery or disposal cost; special requirements for explosion proof equipment and rooms; special requirements for utility and ventilation
	Coating materials: less coating materials needed; less expensive material	*Coating materials*: limited choice of polymer
	Others: better film quality—dense and smooth	*Others*: residual solvent; environmental pollution
Aqueous-based coating system	*Process*: better safety; less expensive process and equipment	*Process*: high process sensitivity; longer processing time
	Coating materials: commercially available prepared dispersions	*Coating materials*: special handling requirement for shipping and storage of the aqueous coating materials
	Others: less environmental pollution	*Others*: residual moisture may cause stability issue and alter dissolution profile; thicker film required

It has been reported that, for a highly water-soluble compound, tartaric acid crystals, it was very hard to sustain its release rate with a single layer of Eudragit RS coating. With an addition of subcoat layer of Eudragit RSPM/S100, a significant retardation in the release of tartaric acid was achieved [11].

8.3.3   Functional Coating

Functional coatings are applied to multiparticulate systems to achieve desired drug release profiles. The release profiles of actives can be controlled by varying the composition and level of release barriers. As discussed earlier, there are several mechanisms by which drugs are released. Depending on the nature, thickness, and porosity of the coating film, the release of an active ingredient may take place by leaching, erosion, rupture, diffusion, or a combination of these mechanisms [12]. The active ingredient may be soluble in the polymer and diffuse from the core through the polymer layer to the bulk liquid medium. A pore former in the film may dissolve upon contact with liquid and form diffusion channels for the drug release. Drug release occurs by erosion if the polymer layer is erodible. Despite a constant drug release rate is highly desired, in reality, the drug release rate varies as a function of time due to the change of concentration gradient.

Coating can be applied as an organic solvent-based system, aqueous-based system, hot melt coating system, or direct powder coating system. Due to safety and environmental concerns, aqueous-based coatings are preferable over organic solvent-based coatings. Water-insoluble polymers can be applied as organic solvent solutions or as aqueous dispersions. Water-soluble polymers are applied as aqueous solutions. Typically, a coating system includes functional polymer(s), plasticizer(s), antitacking agent(s), and pigment(s).

8.3.3.1   *Organic Solvent-Based Polymer Coating Systems*
Organic solvent-based polymer coating systems have a long history of use; however, as already stated, aqueous-based

coatings are preferred in the pharmaceutical industry. The advantages and disadvantages of aqueous versus solvent coatings are listed in Table 8.4.

Typically, the composition of a solvent-based coating system may include

1. insoluble polymers such as Eudragit RS, Eudragit RL, and Ethylcellulose [13];
2. channeling agents, also called "pore formers," such as PEG, Eudragit L, CAP, and Polysorbate 20 [13];
3. plasticizers such as di-*n*-butyl-phthalate [14] and di-Bu sebacate (DBS);
4. solvent such as ethanol, methanol, acetone, ethanol, and methylene chloride mixture [14].

Special attentions to the residual solvent are needed. Residual solvent contents increase with coating levels. Moreover, the maximum spray rate is determined by the lower explosive limit of the organic solvent.

8.3.3.2   *Aqueous-Based Polymer Coating Systems*
Aqueous dispersions of water-insoluble polymers are the most widely used coatings in control release formulations. There are two major types of aqueous coating systems: water-soluble polymer solutions [15] and water-insoluble aqueous polymer dispersions. The pros and cons of water-soluble and -insoluble polymers are listed in Table 8.5. Coating systems containing soluble polymers are less expensive and do not need curing at the end of coating. But they tend to have high viscosity at low concentration, and tend to produce tacky films. On the contrary, aqueous coating dispersions can contain high polymer concentration at relatively low viscosity. However, the resulting process typically needs a curing step and tends to be more expensive.

The choice of polymer is a critical factor in achieving the desired drug release profile. Different polymers form films with different properties, including permeability and

TABLE 8.5   Comparison of Aqueous Solution and Aqueous Dispersion Polymers Coatings

	Aqueous Solution	Aqueous Dispersion
Viscosity	High	Low
Polymer concentration	Low	High
Curing requirement	No	Yes
Cost	Low	High
Tackiness	More likely	Less likely

TABLE 8.7   Solubility of Plasticizers

Plasticizer	Solubility
Triethyl citrate (TEC)	Water soluble
Tributyl citrate	Water insoluble
Stearic acid	Water insoluble
Dibutyl sebacate (DBS)	Water insoluble, soluble in ethanol and acetone, insoluble in propylene glycol
Diethyl phthalate (DEP)	Water insoluble
Propylene glycol	Soluble in water, ethanol, and acetone
Glycerol triacetate (triacetin)	Slightly soluble in water but soluble in alcohol and ether

hydrophilicity. The selected polymer has to be biocompatible and compatible with the drug as well. Both synthetic and semisynthetic polymers, such as HPMC [15], poly(acrylate, methacrylate) copolymer [9], and ethyl cellulose [3], are used in aqueous coating systems. Dispersions of fats and waxes, such as hydrogenated vegetable oil, carnauba wax, have also been reported [16]. Aqueous coating systems based on synthetic polymers and semisynthetic derivatives of naturally occurring polymers are commercially available as aqueous coating dispersions for pharmaceutical development (Table 8.6).

Polymers with different permeability can lead to different release profiles [9, 17]. The desired release profile can be obtained by varying the concentration of the polymer or by using mixtures of polymers. For example, Eudragit RL 30D films are readily permeable, while Eudragit RS 30D films are much less permeable. The two polymers can be used in various ratios to achieve the desired drug release profile [18, 19].

Since most of the polymers used in coating are brittle at room temperature, plasticizers are used to reduce the glass transition temperature (T_g) of the polymers and enhance film-forming qualities of the polymers. Plasticizers act as lubricants to reduce the cohesive intermolecular forces along the polymer chains, thus reduce the stiffness and improve the processability of polymers [20]. The solubility values of plasticizers are listed in Table 8.7.

TABLE 8.6   Examples of Commercially Available Water-Insoluble Polymers

Polymer	Trade Name
Poly(acrylate, methacrylate) copolymers	Eudragit RL pseudolatex disperson, Eudragit RS pseudolatex dispersion, Eudragit NE latex dispersion
Polyvinyl acetate/polyvinyl pyrrolidone	Kollicoat SR 30D is a 30% aq. dispersion of polyvinyl acetate
Polyacrylate	Carbopol 971P or 974P
Ethylcellulose	Surelease, Aquacoat ECD

The type of plasticizer plays an important role in modifying drug release profiles [21]. It is reported that for Kollicoat SR30D coating systems, the drug release profile decreases in the following order: no plasticizer > triethyl citrate > propylene glycol > dibutyl sebacate [22]. When selecting plasticizers, the following factors should be considered:

- compatibility with the polymer;
- capability to lower T_g;
- processability.

The distribution of plasticizers in films should be uniform at molecular level to avoid phase separation and ensure a consistent release profile. A mixture of plasticizers may be needed when more than one polymer are used in the coating system. To ensure a reproducible release profile, the process temperature should be 10–20°C higher than the minimum film-forming temperature of the system [23]. Plasticizers are used to reduce the T_g of the coating system. When adding plasticizers to coating systems, sufficient mixing and equilibration (plasticizing time) are the keys to achieve high coating quality and the desirable release profile. Therefore, plasticizers should be able to mix with polymers quickly. In addition, after mixing, the equilibration time should be reasonable.

The amount of plasticizers in a coating formulation should be optimized, as plasticizers play a critical role in controlling film quality and permeability, hence the release profile [24]. If the plasticizer level is too low, the coalescence of polymer particles is poor. This may lead to poor film formation and uncontrolled release. If the plasticizer level is too high, the tensile strength and elastic module of the film will be reduced, and tackiness may be an issue. The film may become so soft that poor fluidization and particle agglomeration may occur. At an optimal plasticizer level, better films with less permeability are formed and the release rate would be slower [24]. Schmidt [25] reported that a higher release rate was obtained by reducing the TEC (triethyl citrate) level in a coating system from 20% to 10%. The amount of plasticizer needed also depends on the processing temperature, as more plasticizer with low boiling point may be lost at higher temperature.

The presence of a pore former increases the porosity of a polymer film and facilitates drug diffusion. Upon exposure to water, the pore former dissolves and forms drug release channels in a less water-permeable film. For example, methylcellulose could be added to an ethylcellulose-based coating system as a pore former to modify the release rate of theophylline [4]. HPMC is another example of pore formers that can significantly increase the release rate [10, 26].

Another ingredient in a coating system is antifoaming agents. The existence of air bubbles is potentially detrimental to functional coating. The tiny bubbles may create pores in the film and cause uncontrolled leakage. Adding small amounts of simethicone as an antifoaming agent is a common method used to reduce foaming.

Film stickiness or tackiness is a common problem during coating. Film tackiness can also be reduced by

- reducing the plasticizer level;
- lowering coating temperature;
- lowering curing temperature;
- adding antitacky agents, such as talc.

Talc is widely used to eliminate stickiness during coating and curing. However, talc has the tendency to change the dissolution profile as it increases the hydrophobicity of the film. Many other antitacking agents, such as magnesium stearate, titanium dioxide, colloidal silicon dioxide [27], and glyceryl monostearate [28], have also been tested and proved to be efficient.

8.3.3.3 Direct Powder Coating Systems
Dry polymer powders can be applied directly onto the surface of core substrates, as opposed to being applied as solution or dispersion systems. Compared to liquid coating systems, powder coating requires higher amounts of plasticizer and coating material to obtain a similar drug release profile. In addition, a heat treatment, also known as curing, is needed to achieve a compact and uniform film. Eudragit® RS, ethylcellulose, and shellac had been applied onto drug-loaded pellets as dry powders [29]. A smaller polymer particle size could promote the particle coalescence, and thus form a better film [30].

8.3.3.4 Hot Melt Wax Coating Systems
Hot melt wax coating can be applied to substrates through dry powder coating or molten wax coating. For dry powder coating systems, curing is needed to form a dense layer to prolong the drug release. Different drug release profiles can be achieved by varying the thickness of the coat or film

composition, for example, incorporating inert substance or pore-forming agents. An inert substance, Aerosil R972, in the coating mixture was used to reduce the initial burst release, as well as the total drug release [31]. Pore-forming agents, like small water-soluble molecules and surfactants, were used to adjust the release rate. For example, the sustained release profile of a wax-coated drug substance could be modified by varying the ratio of Precirol® ATO5 wax coating to the pore former Gelucire® 50/02 [32]. Examples of waxes include carnauba wax, hydrogenated castor oil [33], polyglycolyzed glyceride [32], glycerol palmitostearate [32], Compritol® 888 ATO [34], and so on.

8.4 APPLICATION OF FLUID BED SYSTEMS IN CONTROLLED RELEASE DEVELOPMENT

In the development of controlled release coated multiparticulates, selections of equipment and process are equally as important as formulation design. It is critical to understand and control these factors to ensure the consistency of drug release profile.

In general, three kinds of fluidized bed configuration, top spray, bottom spray, and tangential spray, are used in the pharmaceutical industry. A schematic drawing of the three systems is described in Figure 8.6. All the three fluidized bed systems contain similar components, including

- product container;
- expansion chamber;
- filter housing;
- spray system;
- air handling system.

Typically, a cone-shaped expansion chamber is used for particle deceleration in order to minimize the entrapment of particles in the upper filter region. Two types of filters are commonly used: fabric filters and cartridges. The fabric filter housing can be divided into two sections as shown in Figure 8.7. This arrangement allows for continuous

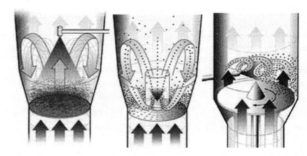

FIGURE 8.6 Schematic drawing of top spray/bottom spray/rotary spray system. Courtesy of Glatt Air Techniques.

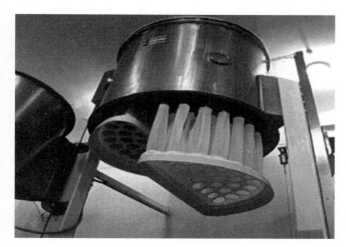

FIGURE 8.7 Picture of filter bags. Courtesy of Glatt Air Techniques.

TABLE 8.8 Selection of Fluid Bed Coating System Versus Processes

Process	Top Spray System	Bottom Spray System	Rotary Spray System
Hot melt granulation/coating	+	−	+
Solution/suspension layering	+	+	+
Powder layering	−	−	+
Solvent coating	−	+	+
Aqueous coating	+	+	+

−: not recommended; +: recommended.

fluidization even during shaking. When choosing a filter bag for a certain product, both of its permeability and porosity should be considered with equal importance. Permeability is a measure of air passing through a clean filter bag, while porosity means the size of the opening of the filter bag. Filters with poor permeability may not allow sufficient air through to provide enough fluidization. If the porosity is too high, small particles may escape from the system. Cartridge filters, even though not used as commonly as fabric filters, provide some unique advantages. Cartridge filters can be cleaned by automatic pulse, which permits continuous fluidization and cleaning in place. This makes cartridges a better option for highly potent materials.

For the spray system, peristaltic pumps are usually used to deliver the coating fluids. Lab-scale machines typically have single nozzles, while large-scale machines are commonly equipped with multinozzles. The nozzle position for each fluidized system is different and will be discussed separately.

The inlet air should be conditioned by an air handler before entering the system. Ideally, dehumidifier and humidifier are used to control the humidity of air. At the least, a dehumidifier should be installed, especially for a coating system used for functional coating processes.

As shown in Figure 8.6, the particle moving patterns in the three fluidized bed systems are not the same. The traveling direction and distance of sprayed droplets to particle surfaces also vary. These unique characteristics of the three systems determine that they have different coating efficiencies and applications. Selection of fluidized bed systems should be based on process conditions (see Table 8.8).

The coating efficiencies of bottom spray systems and tangential spray systems are much higher than that of top spray systems. For sustained release and enteric release coatings, bottom and tangential spray systems are used more

often than top spray fluidized bed systems. The unique features of each type of fluidized bed systems are discussed below.

8.4.1 Bottom Spray Systems

Bottom spray, also called Wurster coating, is the most commonly used process for controlled release products. This is because a bottom spray systems has a special designed air distribution plate, partition, and nozzle location.

The air distribution plate is designed with large holes in the center area under the partition, small holes outside the center area, and bigger holes close to the rim (see Figure 8.8). The partition refers to the column above the distribution plate, also known as Wurster column (see Figure 8.9). There is a gap, also known as partition height, between the partition and the air distribution plate. Because of the gap and the design of the perforated air distribution plate, particles are actually sucked through the gap, drawn to the center of the plate, and accelerated inside of the partition. The smaller the gap, the stronger the suction is. In addition, for smaller particles, smaller partition height is required.

FIGURE 8.8 Picture of air distribution plate. Courtesy of Glatt Air Techniques.

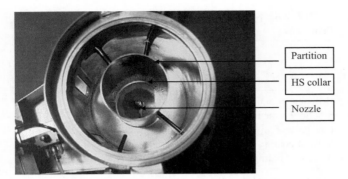

FIGURE 8.9   Picture of partition and HS collar. Courtesy of Glatt Air Techniques.

The spray nozzle is mounted on the air distribution plate. The coating solution or suspension is sprayed upward to form a spray zone, and the particles pass through the spray zone at a very high speed concurrently. As the particles continue moving upward, they are decelerated in the expansion chamber and then fall outside the Wurster column. The returned particles move downward until they are sucked back up to the loop again. This unique design induces a recurring flow of the particles to pass the spray zone in a concurrent mode; therefore, films produced from this process are much more uniform than films produced from top spray coating processes. This is why bottom spray systems are more suitable for functional coating.

For a successful coating in bottom spray systems, the batch size is an important factor. The maximum bowl charge is the volume of the container minus the volume of the partition(s). The minimum bowl charge is one-half of the maximum batch size.

During the coating process, particles to be coated come into close proximity to the spray nozzle. Agglomeration of small particles is not an unusual problem during coating process. Wurster HS is developed to keep the upward moving particles away from the nozzle tip (see Figure 8.9). The spray zone within the HS collar is the wettest portion of the spray zone and has the highest atomization air velocity. As a result, agglomeration can be significantly reduced. In addition, chances of attrition at the beginning of coating are reduced.

8.4.2   Rotary Spray Systems

Rotary tangential spray fluidized bed systems can be used for granules or core pellets preparation, and film coating processes with the same equipment setup [35]. The bowl charge can be 20–90% of the specified capacity. Four distinctive features differentiate fluid bed rotary systems from the other two types of fluid bed systems.

1. *Unique Powder Moving Pattern*:   Besides the gravity force and the upward force provided by the fluidization air, there is another force—the centrifugal force. The centrifugal force, generated by the rotating disc, drives the powder bed to move toward the wall of the chamber. All three forces act on the powder bed from different directions, causing it to move in a spiral pattern.

2. *Unique Inlet Air Pathway*:   The air enters the powder bed through a variable air gap along the edge of the rotation disc, instead of an air distribution plate.

3. *Different Spray Method*:   A nozzle is positioned inside the product and set tangentially to the rotation disc. A solution or suspension is sprayed in the direction concurrent to the rotating powder bed.

4. *Direct Powder Application Capacity*:   Rotary spray systems allow both liquid-based (solution or suspension) layering and powder layering. Loading large amounts of active ingredients onto pellets can be very time consuming. For these situations, powder layering in conjunction with a liquid binder is a better choice. Very thick and dense layers can be applied to the starting cores due to centrifugation. The increase in weight can be up to 10 times the original core weight.

During the pelletization process, the moistened powders are agglomerated, and spheronized into uniform and dense pellets due to the rotation of the rotor disc. The damp pellets can be dried either *in situ* or in another fluid bed dryer. Pellets produced with this technology are dense and spherical. The density of the pellets is greatly influenced by the rotation speed. The higher the speed, the denser the pellets are. Bouffard et al. [36] studied the optimal operating window for pellets production. Their results showed that the pellet mean particle size will increase with increased binder spray rate, while decrease with increased air flow rate and rotor plate speed. The impact on pellet flow properties is positive with increased air flow rate and is negative with increased binder spray rate.

8.4.3   Top Spray Systems

Top spray granulator/coater is the most basic unit of the three systems. It can be used for hot melt granulating/coating. The preferred batch size is usually 60–90% of the container capacity.

Besides the commonly shared system components, at the bottom of the product container, there is an air distribution plate and a retention screen. The air distribution plate ensures uniform fluidization of the product. The retention screen is then used to retain the product. For a hot melt coating process, the nozzle should be kept low in the expansion chamber, so that spray drying can be avoided before the coating solution or suspension arrives at the surface of fluidized particles. Also, the atomization air for the nozzle needs to be heated to avoid congealing before spraying.

8.4.4 Process Variables

For a fluid bed coating, the following process variables need to be monitored and controlled to produce a quality product with consistency:

- spray rate;
- atomization air pressure;
- inlet air temperature;
- inlet air volume;
- inlet air humidity;
- product and exhaust temperature.

Spray rate is the rate of applying liquid coating solutions/suspensions to solid substrates in a fluid bed system. Many factors, such as the viscosity and tackiness of the coating liquid, the droplet size, and the drying capacity of the fluidized bed system, can limit the maximum spray rate. If the spray rate is too high, irreversible agglomeration can occur, which may result in batch failure. Therefore, the spray rate should be set conservatively for initial trials. For any coating process, a short coating time and a high quality film must be balanced.

Atomization air pressure is an important key process parameter to control droplet size distribution and velocity. Optimal droplet size distribution and velocity are critical for an efficient process, which helps prevent overwetting and spray drying. Increasing the atomization air pressure will increase the volume and velocity of the air passing through the nozzle, thus affecting the spray pattern and reduce the droplet size. Too much atomization air pressure leads to spray drying, while too little atomization air pressure can result in agglomeration.

Inlet temperature is the temperature of the heated air taken right before it enters the product container. For coating, the inlet temperature is set relatively low compared to drying. This is because a high inlet temperature can lead to film softening, which may result in agglomeration during coating. An optimal range needs to be defined for every coating system.

Fluidization pattern and the upward velocity of the particles in a fluidized bed system are determined by the inlet air volume. The upward movement of the particles should be properly controlled. If the air volume is too high, the particles can be trapped in the filter housing. If the air volume is too low, the particles may get overwetted when passing through the spray zone.

Inlet air humidity is usually indicated by the dew point temperature, at which the air is saturated with moisture. With unconditioned air, the moisture carrying capacity of the air varies. The spray rate needs to be reduced if the inlet air humidity is high, otherwise overwetting may be an issue. High inlet air humidity also makes it difficult to remove residual moisture in films. Extra moisture remaining in the product may cause stability problem. The spray rate needs to be increased if the inlet air humidity is low, otherwise spray drying may occur. For a functional coating process, dew point control is even more important, because it is related to the moisture remaining in the film. Moisture in a film can often function as a plasticizer, and thus change the drug release profile during storage. With conditioned air, the dew point of the inlet air is consistent for each batch, which allows a reproducible drying rate and better film quality.

The combination of inlet temperature, air volume, and inlet air humidity determines the drying capacity of a fluid bed system, thus the maximum output of the system.

Product and exhaust temperatures are two commonly monitored parameters. The product temperature is more sensitive than exhaust temperature. The inlet temperature and inlet air volume have a positive impact, while the spray rate has a negative impact on the product temperature. In a coating process, if the product temperature is lower than the minimum film-forming temperature, it will result in the formation of poorly coalesced film and faster drug release rate. If the temperature is too high, agglomeration may occur.

8.4.5 Scale-Up Consideration

The scale-up from single to multiple nozzle units is quite straightforward. Ideally, the following parameters should be kept constant:

- the product temperature;
- the inlet air humidity;
- the ratio of air volume/cross-sectional area, if possible;
- the droplet size.

The air velocity needed to fluidize particles in a fluidized bed system depends on the size and density of the particles. For a given particle, during scale-up, the air velocity should be duplicated to achieve similar fluidization pattern. Typically, the air velocity can be calculated from the ratio of air volume to cross-sectional area.

$$V_s/A_s = V_l/A_l$$
$$V_l = A_l \times (V_s/A_s)$$

where V_l is the inlet air volume and A_l is the cross-sectional area in the larger scale system, V_s is the inlet air volume and A_s is the cross-sectional area in the small-scale system. This is a good starting point for setting air volume in large-scale equipment during scale-up.

The droplet size is an important factor during scale-up. It is determined by the viscosity of the coating liquid, the atomization condition, and the spray rate. The viscosity of the coating liquid will not change. During scale-up, changing atomizing conditions, such as nozzle type, ratio of air to coating liquid, and atomizing air pressure, may cause significant differences in the physical properties of the droplets. These differences will be particularly magnified when the film is to be used for a sustained release application because the coalescence process of the film is changed. Different types of nozzles have different optimal air to liquid ratios. Therefore, the same type of nozzle, ratio of air to liquid, and droplet size should be used during scale-up. The quantities of particles passing through the spray zone in each cycle are more relevant to the spray rate than the total batch size. In the same fluidized bed unit, increasing batch size only means longer spray time.

These general rules for scale-up apply to all of the three fluidized bed systems. Scale-up in rotary fluidized systems also involves the scale-up of rotor speed. A similar peripheral edge velocity (~rotor speed × plate radius) can be used to scale up rotor speed. Centrifugal force can also be used as the similarity factor for scale-up [37].

In summary, the ultimate goal of scale-up is to achieve a batch-to-batch consistency at different scales. Not only all the parameters must be understood and carefully defined but also the critical properties of polymers used in coating must be well controlled to ensure a reproducible product. Slight variations in the polymer may lead to major changes in the drug release profile, throwing off the batch-to-batch consistency.

REFERENCES

1. Lehmann KOR, Bossler HM, Dreher DK. Controlled drug release from small particles encapsulated with acrylic resin. *Niol Macromol. Monogr. Polym. Deliv. Syst.* 1979;5:111–119.

2. Robinson JR. *Sustained and Controlled Release Drug Delivery Systems*, Marcel Dekker, New York, 1978.

3. Ozturk AG, Ozturk SS, Palsson BO, Wheatley TA, Dressman JB. Mechanism of release from pellets coated with an ethylcellulose-based film. *J. Control. Release* 1990;14(3): 203–213.

4. Yuen KH, Deshmukh AA, Newton JM. Development and in-vitro evaluation of a multiparticulate sustained-release theophylline formulation. *Drug Dev. Ind. Pharm.* 1993;19(8): 855–874.

5. Guo HX, Heinämäki J, Yliruusi J. Diffusion of a freely water-soluble drug in aqueous enteric-coated pellets. *AAPS PharmSciTech* 2002;3(2):97–104.

6. Sousa JJ, Sousaa A, Mourab MJ, Podczeckc F, Newtonc JM. The influence of core materials and film coating on the drug release from coated pellets. *Int. J. Pharm.* 2002;233(1–2): 111–122.

7. Wesdyk R, Joshi YM, Jain NB, Morris K, Newman A. The effect of size and mass on the film thickness of beads coated in fluidized bed equipment. *Int. J. Pharm.* 1990;65(1–2):69–76.

8. Chopra R, Alderborn G, Newton JM, Podczeck F. The influence of film coating on pellet properties. *Pharm. Dev. Technol.* 2002;7(1):59–68.

9. Munday DL. A comparison of the dissolution characteristics of theophylline from film coated granules and mini-tablets. *Drug Dev. Ind. Pharm.* 1994;20(15):2369–2379.

10. Tomuta L, Leucuta SE. The influence of formulation factors on the kinetic release of metoprolol tartrate from prolong release coated minitablets. *Drug Dev. Ind. Pharm.* 2007;33(10): 1070–1077.

11. Venkatesh GM. Development of controlled-release SK&F 82526-J buffer bead formulations with tartaric acid as the buffer. *Pharm. Dev. Technol.* 1998;3(4):477–485.

12. Chen CM, Cheng XX. Controlled release oral dosage form of beta-adrenergic blocking agents. US Patent 7,022,342, 2002.

13. Munday DL, Fassihi AR. Controlled release delivery: effect of coating composition on release characteristics of mini-tablets. *Int. J. Pharm.* 1989;52(2):109–114.

14. Zeng HX, Cheng G, Pan WS, Zhong GP, Huang M. Preparation of codeine-resinate and chlorpheniramine-resinate sustained-release suspension and its pharmacokinetic evaluation in beagle dogs. *Drug Dev. Ind. Pharm.* 2007;33(6):649–665.

15. Sangalli ME, Maroni A, Foppoli A, Zema L, Giordano F, Gazzaniga A. Different HPMC viscosity grades as coating agents for an oral time and/or site-controlled delivery system: a study on process parameters and *in vitro* performances. *Eur. J. Pharm. Sci.* 2004;22(5):469–476.

16. Walia PS, Stout PJM, Turton R. Preliminary evaluation of an aqueous wax emulsion for controlled-release coating. *Pharm. Dev. Technol.* 1998;3(1):103–113.

17. Rafiee-Tehrani M, Sadegh-Shobeiri N. Effect of various polymers on formulation of controlled-release (CR) ibuprofen tablets by fluid bed technique. *Drug Dev. Ind. Pharm.* 1995;21(10):1193–1202.

18. Chang RK, Price JC, Hsiao C. Preparation and preliminary evaluation of Eudragit RL and RS pseudolatices for controlled drug release. *Drug Dev. Ind. Pharm.* 1989;15(3):361–372.

19. Kramar A, Turk S, Vrecer F. Statistical optimisation of diclofenac sustained release pellets coated with polymethacrylic films. *Int. J. Pharm.* 2003;256(1–2):43–52.

20. Gutierrez-Rocca JC, McGinity JW. Influence of water soluble and insoluble plasticizers on the physical and mechanical properties of acrylic resin copolymers. *Int. J. Pharm.* 1994; 103:293–301.

21. Kim TW, Ji CW, Shim SY, Lee BJ. Modified release of coated sugar spheres using drug-containing polymeric dispersions. *Arch. Pharm. Res.* 2007;30(1):124–130.

22. Shao ZJ, Moralesi L, Diaz S, Muhammadi Nouman A. Drug release from Kollicoat SR 30D-coated nonpareil beads: Evaluation of coating level, plasticizer type, and curing condition. *AAPS PharmSciTech* 2002;3(2):15.

23. Laicher A, Lorck CA, Grunenberg PC, Klemm H, Stanislaus F. Aqueous coating of pellets to sustained-release dosage forms in a fluid-bed coater. Influence of product temperature and polymer concentration on in vitro release. *Pharm. Ind.* 1993; 55 (12):1113–1116.

24. Lippold BH, Sutter BK, Lippold BC. Parameters controlling drug release from pellets coated with aqueous ethyl cellulose dispersion. *Int. J. Pharm.* 1989;54(1):15–25.

25. Schmidt PC, Niemann F. The MiniWiD-coater. III. Effect of application temperature on the dissolution profile of sustained-release theophylline pellets coated with Eudragit RS 30D. *Drug Dev. Ind. Pharm.* 1993;19(13):1603–1612.

26. ur Rahman N, Yuen KH, Khan N, Wong JW. Drug-polymer mixed coating: a new approach for controlling drug release rates in pellets. *Pharm. Dev. Technol.* 2006;11 (1):71–77.

27. Vecchio C, Fabiani F, Gazzaniga A. Use of colloidal silica as a separating agent in film forming processes performed with aqueous dispersion of acrylic resins. *Drug Dev. Ind. Pharm.* 1995;21(15):1781–1787.

28. Singh SK, Khan MA. Ibuprofen release from beads coated with an experimental latex: effect of certain variables. *Drug Dev. Ind. Pharm.* 1997;23(2):145–155.

29. Pearnchob N, Bodmeier R. Dry powder coating of pellets with micronized Eudragit® RS for extended drug release. *Pharm. Res.* 2003;20(12).

30. Pearnchob N, Bodmeier R. Dry polymer powder coating and comparison with conventional liquid-based coatings for Eudragit® RS, ethylcellulose and shellac. *Eur. J. Pharm. Biopharm.* 2003;56(3):363–369.

31. Chansanroj K, Betz G, Leuenberger H, Mitrevej A, Sinchaipanid N. Development of a multi-unit floating drug delivery system by hot melt coating technique with drug-lipid dispersion. *J. Drug Deliv. Sci. Technol.* 2007;17(5):333–338.

32. Sinchaipanid N, Junyaprasert V, Mitrevej A. Application of hot-melt coating for controlled release of propranolol hydrochloride pellets. *Powder Technol.* 2004;141(3):203–209.

33. Norihito S, Masumi U, Yasuhiko N. Design of controlled release system with multi-layers of powder. *Chem. Pharm. Bull.* 2002;50(9):1169–1175.

34. Barthelemy P, Laforet JP, Farah N, Joachim J. Compritol 888 ATO: an innovative hot-melt coating agent for prolonged-release drug formulations. *Eur. J. Pharm. Biopharm.* 1999; 47(1):87–90.

35. Vuppala MK, Parikh D, Bhagat HR. Application of powder-layering technology and film coating for manufacture of sustained-release pellets using a rotary fluid bed processor. *Drug Dev. Ind. Pharm.* 1997;23(7):687–694.

36. Bouffard J, Dumont H, Bertrand F, Legros R. Optimization and scale-up of a fluid bed tangential spray rotogranulation process. *Int. J. Pharm.* 2007;335(1):54–62.

37. Chukwumezie BN, Wojcik M, Malak P, Adeyeye M. Feasibility studies in spheronization and scale-up of ibuprofen microparticulates using the rotor disk fluid-bed technology. *AAPS PharmSciTech* 2002;3(1):2.

9

CONTROLLED RELEASE USING BILAYER OSMOTIC TABLET TECHNOLOGY: REDUCING THEORY TO PRACTICE

SHERI L. SHAMBLIN

Pfizer Global Research & Development, Groton, CT, USA

9.1 INTRODUCTION

The controlled delivery of an active pharmaceutical ingredient (API) using an osmotic pump is an elegant way to deliver drug at a constant rate. The ability to deliver drug with a zero-order release profile in a manner that is independent of external factors (e.g., hydrodynamics, media pH) offers distinct advantages over other controlled release technologies (e.g., matrix tablets). In particular, the ability to deliver a drug at a rate that is the same whether administered with or without food, and that does not depend on the pH or hydrodynamics of the external environment, makes the task of establishing an IVIVC much simpler. The use of osmotic pumps for oral drug delivery has evolved from very simple, monolithic tablets to more complex formulations that include multilayered tablet cores with different layers providing different functionalities. The first commercial osmotic tablets to reach the market in the late 1980s were Acutrim, developed by Alza as an OROS® and marketed by Ciba-Geigy Corp, and Procardia XL, an OROS® push–pull system developed by Alza Inc. and marketed by Pfizer. Following Procardia XL, a number of other commercial osmotic tablets were developed using Alza's push–pull system that incorporates a functional layer into the tablet core (e.g., Cardura XL®, Covera HS®, Glucotrol XL®, and Minipress XL®). The presence of a functional push layer in the tablet core offers particular advantages in that it yields a very robust technology that can deliver drugs with a wide range of solubilities, including very insoluble drugs for as long as 18 h.

This chapter will focus on the practical aspects of developing bilayer osmotic tablets that contain a functional swelling layer and a manufactured delivery port. Following a brief overview of all osmotic tablets, including a comparison of all types of osmotic tablets, the applicability of bilayer, swelling core osmotic tablet technology to specific delivery requirements formulation design will be discussed. In addition, the important aspects of formulation design, the processes used for manufacturing these types of osmotic tablets for both clinical evaluation and commercialization, and regulatory considerations are presented. The historical and theoretical aspects of osmotic pump delivery as they relate to the development of these dosage forms will be discussed only briefly, and as needed for context and background, since the theoretical basis for osmotic drug delivery has already been described in detail elsewhere [1–5].

9.2 OVERVIEW OF OSMOTIC PUMP DEVICES FOR DRUG DELIVERY

9.2.1 History

The first reported use of an osmotic pump for controlling the drug delivery to the gut in sheep and cattle was reported in 1955 by Rose and Nelson [6], whereby a chamber containing drug and excess salt was separated from a chamber containing water by a semipermeable membrane. The same principles that govern the control of drug in this primitive device

Oral Controlled Release Formulation Design and Drug Delivery: Theory to Practice, Edited by Hong Wen and Kinam Park

were used to develop a number of osmotic devices; many were complex mechanical systems that were used to deliver drugs to promote health in animals.

The osmotic pump delivery devices combine a semipermeable membrane and osmotic agents that move water across that membrane. These first devices were developed for veterinary applications for delivery of drugs, mainly hormones and antibiotics to animals [4], and were designed to be ingested and retained in the rumen. The invention of the elementary osmotic pump (EOP) [5] that came in the form of a conventional tablet brought osmotic delivery to the mainstream of pharmaceutical development and into the human health arena. The same theory and principles that are pertinent to controlling drug release with an osmotic tablet apply to other osmotic dosage forms, such as capsules and implants.

9.2.2 Basic Components in Any Osmotic Tablet

The "pump" in an osmotic tablet is created by the combination of water-soluble components contained inside a tablet core surrounded by a semipermeable membrane. The basic components of any osmotic tablet are illustrated in Figure 9.1. When drug solubility is low, the table core must include added osmogens to create an osmotic potential. Frequently, components that increase viscosity and maintain a uniform structure and composition within the tablet core as it hydrates are also included. The tablet core may also contain buffers or other agents to help modulate drug solubility within the core, polymers that swell upon hydration to help "push" or deliver the drug, and fillers and binders that help with compression and lead to a robust tablet core. The role of the semipermeable membrane is to control the ingress of water into the tablet and maintain water-soluble components (that provide osmotic potential) within the tablet core. Ultimately, it is the semipermeable membrane that controls the drug delivery rate. The semipermeable membrane consists of a water-insoluble film-forming polymer and a second component that moderates the permeability of water. This second component also reduces the brittleness of the film by acting as

a plasticizer. The drug can be delivered as solution through pores within the semipermeable membrane or through one or more delivery ports created in the film coating. For very insoluble drugs, the formation of a delivery port is required for osmotic delivery.

There are several categories of osmotic tablets that differ based on core construction, physical and chemical composition of the semipermeable membrane, and nature of the delivery port. The delivery of drug from an osmotic pump tablet requires a balance between the decrease in viscosity of core components, increase in hydrostatic pressure, and release of drug and other components as a result of water imbibitions. This balance is achieved through the appropriate selection of the semipermeable coating, osmotic potential, tablet core viscosity, and hydrostatic pressure within the tablet core. For each type of osmotic tablet technology (e.g., elementary, asymmetric membrane, or swelling core osmotic), this balancing act has been demonstrated only by a limited number of tablet core components (polymers, osmogens) and coating components (e.g., water-insoluble polymers and plasticizers).

9.2.3 Classification of Osmotic Tablet Technologies

While the focus of this chapter is on bilayer osmotic tablets containing a functional swelling layer, a brief discussion of other types of osmotic tablets is helpful in understanding the distinguishing features of bilayer core osmotic tablets. In Figure 9.2, a classification of osmotic tablets based on core structure (single or multiple layers), coating or membrane architecture (porosity), and delivery route through the membrane (membrane pores versus manufactured orifice) is illustrated.

In theory, an osmotic tablet can be engineered with nearly any combination of core structure, coating architecture, and delivery devices described in Figure 9.2. However, each of the four classes described below fulfill certain delivery needs and offers a distinct advantage over the others. In some cases, the drug solubility, desired PK profile, or manufacturing capabilities may limit the choice to a single type of osmotic tablet. In other cases, there may be several ways to design a tablet that achieve desirable performance, safety, and manufacturability using an osmotic pump. Some specific examples of how an osmotic pump has been used in commercial osmotic tablets are provided in Table 9.1.

9.2.3.1 Elementary Osmotic Pump and Liquid OROS
The first elementary osmotic tablets (EOP) were formed by compression of a soluble drug into a tablet using conventional equipment and then coating the tablet with a water-insoluble polymer, usually cellulose acetate, and forming one or more delivery ports in the coating. The soluble drug provides the driving force for water imbibitions in the EOP and OROS systems. Water is driven into the tablet core

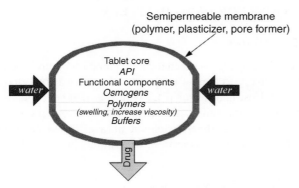

FIGURE 9.1 Basic structure of an osmotic tablet.

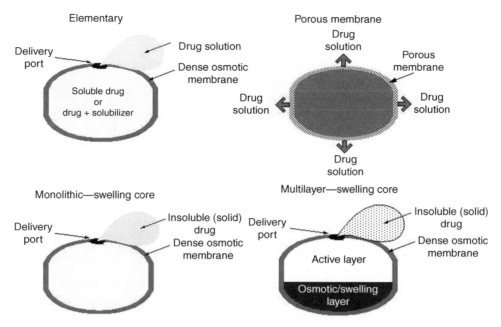

FIGURE 9.2 Schematic of different types of osmotic tablets.

through the semipermeable membrane by the soluble drug. A more detailed description of the history of the osmotic pump delivery leading to the invention of the elementary osmotic tablet and the advancements that followed this important invention is summarized in Ref. [4]. One of the limitations of these devices is the need for high water fluxes to achieve desirable drug release rates, which is usually achieved with very thin coatings that tend to rupture or by addition of the addition of plasticizers [1]. The other limitation of the elementary osmotic tablet is the decreasing release rate with time, particularly for very soluble drugs, for only 20% of the dose may be delivered as zero-order release. Many modifications to the first elementary osmotic tablets have resulted in improved manufacturability, better control of drug release rate, and increasing applicability by modification of core composition. Modifications to the core to increase solubility of the drug and the use of enteric polymers in the membrane coating composition have extended the range of applicability of the elementary osmotic tablets.

9.2.3.2 Porous Coating (Asymmetric Membrane) Tablets

Asymmetric membrane tablets, similar to other osmotic tablet technologies, consist of a core containing drug and osmogen that is coated by a semipermeable membrane. The imbibition of water through a membrane into the tablet core results in the dissolution of soluble components (including drug) in the core. The asymmetric semipermeable membrane coatings consist of a polymer film having a microporous structure formed by a phase separation during film formation [7] and are a proprietary membrane coating developed by Pfizer Inc. and Bend Research Inc. for application to tablets or other oral dosage forms [9]. The film is created by

TABLE 9.1 Summary of Marketed Osmotic Tablets

Type	Commercial Examples	Core Description	Membrane Description	Delivery Route	Drug Form delivered
Elementary (OROS, EOP)	Volmax® Sudafed 24® Accutrim®	Monolithic	Dense	Manufactured orifice(s)	Solution
Porous membrane	No commercial examples	Monolithic	Porous	Diffusion through membrane pores	Solution
Swelling core	Tegretol®	Monolithic	Dense	Manufactured orifice(s)	Suspension
Swelling core	Cardura XL® Procardia XL® Covera HS® Glucotrol XL® Minipress XL®	Bilayer	Dense	Manufactured orifice(s)	Suspension
Swelling core	Concerta®	Trilayer	Dense	Manufactured orifice(s)	Suspension

Information taken from U.S. Pharmacopeia, PDR, and Refs [3, 7, 8].

dissolving polymer (typically cellulose acetate) in a mixed-solvent system (comprised of a solvent, a nonsolvent, and a plasticizer for the polymer), which is sprayed onto tablet cores. During the coating process, the solvent (typically acetone) evaporates more rapidly than do the nonsolvent (typically water) and plasticizer (typically polyethylene glycol (PEG)). As the coating process continues, additional layers are formed in the same manner. Following drying, the coating consists of a porous solid composed of multilayered thin films or skins.

9.2.3.3 Swelling Core, Monolithic

The osmotic delivery of drugs with moderate and poor aqueous solubility requires some of the drugs to be delivered as a suspension. In such cases, using an elementary osmotic tablet can lead to a change in release rate that is more first order and governed by the solubility of the drug. This usually occurs once components within the core that keep the drug suspended are depleted. This leads to settling or aggregation of the drug. Polymers that swell upon hydration have been incorporated into monolithic tablets to keep the drug suspended as a way to deliver drugs with moderate and poor aqueous solubility.

The simplest approach to osmotic delivery of poorly soluble drugs is that of a swelling monolithic tablet [10]. However, the disadvantage of the single layer swelling core design is the incomplete release of drug, particularly in the delivery of moderate to low-solubility drugs in the form of a suspension through one or more delivery ports within the coating. The most robust approach to delivery of poorly water-soluble drugs in the form of a suspension has been to combine a second layer that swells upon hydration and pushes in a piston-like manner on the drug containing layer [11, 12], as described below.

9.2.3.4 Swelling Core, Bilayer Tablets

The first bilayer core tablets, osmotic tablets, were developed by Alza [13] and were classified as OROS® push–pull devices. The bilayer osmotic tablet technology developed by Alza Inc. was successfully commercialized in collaboration with Pfizer to produce Glucotrol XL, Cardura XL, and Procardia XL sustained release formulations. The bilayer tablet consists of an "active layer" and a functional layer, called an "osmotic layer" or "push layer," and coated with a semipermeable membrane. The second layer provides two functions: (1) to generate the osmotic potential within the tablet core and (2) to generate a hydrostatic pressure via a swelling force that "pushes" the extrudable material (drug layer) from the tablet. This layer primarily consists of a polymer that swells upon hydration (e.g., high molecular weight polyethylene oxide (PEO)) and an osmogen (e.g., sodium chloride). The "push" of the sweller layer against the drug layer forces the drug, which is suspended in the hydrated drug layer, through a delivery port that is formed into the membrane on the active layer side of the tablet.

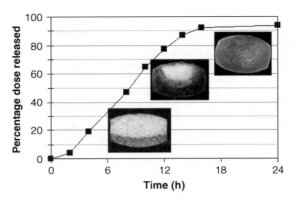

FIGURE 9.3 Typical release profile for a bilayer osmotic tablet. Photos show tablets removed at $t = 0$, 8, 24 h and bisected to illustrate core hydration, swelling, and extrusion of drug layer.

Figure 9.3 shows a typical release profile for a bilayer osmotic tablet and also includes photographs of tablets bisected at various stages of delivery. The profile nicely illustrates a zero-order release profile that is sustained for 90% of the dose, which is one of the important advantages of the bilayer (or multiple layer) architecture compared to other osmotic tablet technologies. The photographs illustrate pushing of the sweller layer against the drug layer as it is extruded through the delivery port. This back pressure provides a very robust mechanism and maintains a constant rate of delivery. The *in vitro* release profile also shows the initial lag time that is present in all osmotic tablets. This lag is associated with the time needed for water to diffuse through the coating into the tablet core and dissolution of the soluble components. The dissolution of soluble components creates the osmotic potential needed to drive additional water into the tablet at a constant rate. For bilayer osmotic tablets, this time lag includes the time it takes for sufficient water ingress so that there is enough hydrostatic pressure to "push" the drug through the hole. Typically, this time lag is 1–2 h and will depend on coating thickness and permeability, and to a lesser extent on the composition of the tablet core.

The rate of drug release achieved with a bilayer osmotic tablet depends on the size of the tablet, and drug loading within the core, but ultimately it is governed by the osmotic potential created by soluble components within the core and the permeability of the coatings. The practical working ranges for osmotic potentials and coating permeability based on compendial materials will be discussed later in this chapter. However, it is worthwhile to note some examples of release rates that have been achieved with bilayer osmotic tablets, including commercialized products and examples from the literature. These release estimates were calculated based on the duration of the delivery, taking into account time and dose per tablet (Table 9.2).

One characteristic of osmotic tablets is the potential for some fraction of the dose to remain trapped inside the tablet,

TABLE 9.2 Typical Release Rates Achieved with Bilayer Osmotic Tablets

Drug	Commercial Name	Dose (mg)	Rates (mg/h)
Doxazosin mesylate	Glucotrol XL	4, 8	0.2–0.6
Nifedipine	Procardia XL	30, 60, 90	1.5–6.6
Oxybutynin chloride	Ditropan XL	5	0.27
Glipizide	Glucotrol XL	5,10	0.27–0.53
Prozosin	Minipress XL	2.5, 5, 10	0.13–1.1
Verapamil	Covera HS	180, 240	18–27

leading to incomplete release of the dose. This is illustrated in Figure 9.3, which shows that only 94% of the drug is released from the tablet core. In bilayer osmotic tablets, this can occur because of "mixing" of the drug with the swelling layer at the final stages of the release, which results in trapping of the drug within this highly viscous layer. The amount of residual drug that is not released ranges from 2% to 10% for bilayer osmotic tablets and depends somewhat on the solubility of the drug. More soluble drugs will tend to diffuse through the hole even if they become "trapped" by the swelling layer as it pushes toward the delivery port, and result in lower residual drug values.

The remaining part of this chapter will focus on the bilayer osmotic tablet and provide more information about the applicability of this osmotic technology, the composition and manufacturing process, and important considerations for successful formulation development, such as the importance of *in vitro* characterization, manufacturing and scale-up, and regulatory considerations.

9.3 ADVANTAGES AND DISADVANTAGES OF BILAYER OSMOTIC TABLETS

9.3.1 Advantages

The most noted advantage of osmotic drug delivery is the ability to maintain a constant drug delivery rate. In reality, there are a limited number of situations when maintaining zero-order delivery is critical to achieving a PK profile that is both safe and efficacious. In most cases, it is the insensitivity of the drug release to solution pH, external hydrodynamic effects [14], and the physical properties (e.g., solubility, ionization, particle size) of the API that make osmotic tablets a desirable technology. These features can simplify dosage form design and development and increase the likelihood for establishing an IVIVC, even with poorly soluble drugs.

9.3.1.1 Independence of Release Rate on Hydrodynamics
One of the difficulties in developing a biorelevant test in matrix tablet formulations is the importance of hydrodynamics on the erosion in the matrix, particularly for moderately or poorly soluble drugs. This is challenging for two reasons. First, the ability to re-create the same hydrodynamics in the gastric environment in the laboratory is very challenging. Second, the hydrodynamics in the stomach change in response to food in a manner that is also difficult to mimic *in vitro*. With osmotic control, drug release rate depends only on the influx of water into the tablet, which is controlled by the osmotic potential and the properties of the coating (thickness and permeability). Thus, the rate of stirring in an *in vitro* test [15] and differences in the hydrodynamics in a fasted stomach and fed stomach will not alter drug release rate from an osmotic tablet. Thus, a target PK profile can be achieved with fewer iterations (or with fewer test formulations) because of the expectation of a good IVIVC.

9.3.1.2 Independence of Release on pH
The ability to maintain the intended delivery rate is challenging for ionizable drugs due to the changes in pH along the GI tract and inter- and intrasubject variability of pH. The differences in hydrodynamics and levels of surfactants associated with digestion can further exacerbate the effect of pH on the drug release and dissolution of ionizable drugs. These factors can make it very difficult to establish an IVIVC for ionizable drugs that are delivered using polymer-based matrix CR dosage forms. When control of release occurs using an osmotic pump, which includes a water-insoluble membrane barrier between the drug and the GI environment, the drug release rates in the acidic environment of the stomach and in the neutral pH distal to the stomach are the same. This is perhaps the most important advantage of an osmotic delivery system.

9.3.1.3 In Vitro and In Vivo Correlation
The development of controlled release dosage forms is much more challenging in the absence of an IVIVC. Often the lack of IVIVC is due to difference in the release rates *in vitro* and *in vivo*. Lack of an IVIVC is usually due to conditions *in vivo* that affect drug release and dissolution rates that were not reflected by the vitro test. For matrix tablets, a lack of IVIVC is often attributed to differences in release mechanism (erosion versus diffusion), and in turn release rate, as a result of external hydrodynamics, especially for low-solubility drugs. The uncertainty in how an *in vitro* release translates *in vivo* usually demands that multiple formulations provide a range of *in vitro* release rates to ensure that the right release rate is achieved *in vivo*. This, in turn, translates to additional time and resource requirements to successfully develop a controlled release dosage form.

The ability to establish an IVIVC for bilayer osmotic tablets is favored by the robustness of the technology and the lack of sensitivity to release rate to external factors such as the pH and ionic strength of the media, and to hydrodynamics. An excellent correlation between *in vitro* and *in vivo* release rates in dogs for a 150 mg allopruinol osmotic tablet is

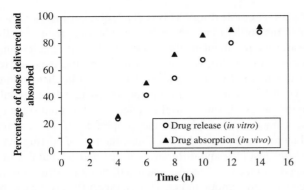

FIGURE 9.4 Percentage of allopruinol absorbed in dogs and cumulative percentage of drug released and dissolved during *in vitro* dissolution testing as a function of time. Data plotted with permission from Ref. [11].

shown in Figure 9.4 [11]. The cumulative drug absorption profile is zero order, with up to ~90% of the dose absorbed at nearly a constant rate. The drop in the *in vitro* drug release rate and apparent loss in osmotic control *in vitro* after ~6 h may be more of a reflection of the low solubility of the drug in gastric media (i.e., 0.48 mg/mL) and slowing of drug dissolution as the saturation solubility is approached. Similar correlations between *in vitro* and *in vivo* drug release rates in humans have been demonstrated for an OROS system that delivers metoprolol [16].

For a bilayer, swelling core osmotic tablet, a correlation between *in vitro* and *in vivo* performance has been demonstrated in humans in both the fasted and fed states, as shown in Figure 9.5. The *in vitro* release profile for a 50 mg CP-607,366 tablet was used to predict the PK profile with knowledge of the key pharmacokinetic properties of the drug, previously determined in PK studies with an immediate release formulation. The PK data in the graph on the right-hand side of Figure 9.5 shows an excellent correlation between the simulated PK profile and the actual PK data obtained in both the fed and fasted states. In this case, the

IVIVC was aided by the fact that the drug had reasonable dose volume (5000) and was well absorbed throughout the GI.

While the osmotic control of drug release does minimize differences in the hydrodynamics and changes in pH due to the effect of food, a food effect may still be observed due to the effect of food on gastric retention. For a drug that has differences in regional absorption, due to the role of transporters or the effects of bile salts on the dissolution rate or solubility *once* the drug is delivered from the osmotic tablet, increasing the residence time in the upper GI will likely change the pK profile. In such cases, the *in vivo* release profile is not predicted from *in vitro* performance not due to differences in release rate from the tablet, but rather to physiological factors that affect dissolution and/or absorption of drug once delivered.

9.3.2 Disadvantages

The advantages associated with osmotic delivery of drugs comes at the cost of a complex manufacturing process and a limited selection of excipients and polymers appropriate for membrane and core table design. While the process for manufacturing an osmotic tablet core is similar to that used for conventional tablets using standard equipment, when a bilayer tablet core is required, the use of a bilayer press, possibly needed for multiple granulations, makes the whole manufacturing process train more complex. All osmotic tablets require the application of a membrane coating to the tablet. The insolubility of the coating materials in water requires the use of nonaqueous solvent for preparation of the film coatings. The use of solvents requires safety and environmental considerations that add to the complexity and cost of manufacture. Current commercial osmotic tablets rely on the delivery of a drug through a manufactured port and also require laser drilling equipment. While most of the processes used are on standard equipment, the possible need for a multiple layer tablet press, solvent handling capability

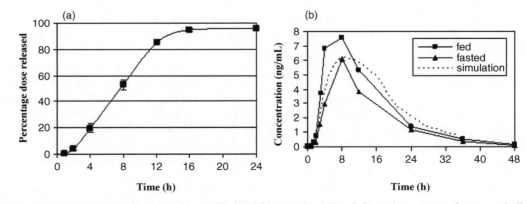

FIGURE 9.5 *In vitro* dissolution profile for a 50 mg CP-607,366 osmotic tablet (left) and *in vivo* performance indicated by plasma concentrations as a function of time.

for wet granulation and tablet film coating processes, and laser drilling may require capital investment and increase overall manufacturing costs.

Another disadvantage of osmotic tablets includes the potential problems associated with a limited selection of functional excipients. There are a limited number of excipients that can provide the functionality needed to drive the osmotic pump. Likewise, there are only a few compendial polymers that are suitable for the formation of a semipermeable membrane coating. These limitations in the choice of excipients may impede the ability to overcome physical and/or chemical incompatibilities, or inadequacies in the material properties, through a change in excipient.

The practical limitation on the drug loading, and therefore dose, can also be a potential downside of osmotic pump tablets. Generally, the upper limit in drug loading that can be achieved in an osmotic tablet is less than 25% of the tablet by weight, which is lower than what is possible with a matrix tablet. Thus, the administration of multiple tablets may require achieving an efficacious dose for low potency, high dose drugs.

9.3.3 Comparison of Osmotic Tablets to Other CR Tablet Technologies

The osmotic pump that controls drug release from a bilayer osmotic tablet is a distinguishing feature and leads to important differences in the mechanism for control compared to other more conventional tablet approaches. The mechanisms for controlling drug release and the external factors that can influence drug release rate in different types of controlled release tablet technologies (i.e., osmotic, hydrophilic matrix, coated tablet) are summarized in Table 9.3. This table illustrates how the osmotic pump mechanism translates to

the dependence of drug release on factors that can sometimes make the design and development of controlled release dosage forms very challenging.

9.4 APPLICABILITY OF BILAYER OSMOTIC TABLET TECHNOLOGY

The applicability of a bilayer osmotic tablet technology to achieving a particular clinical objective begins with the evaluation of dose and solubility, and later the release profile required to achieve a target PK profile. Even before considering the use of a bilayer osmotic tablet for controlled release, it is highly recommended that a more general feasibility assessment for controlled release [17] be conducted. The information collected during a more general controlled release feasibility assessment (e.g., drug solubility as a function of pH, regional absorption, target PK profile) will also be useful in the assessment of a bilayer osmotic tablet and likely save development time.

9.4.1 Controlled Release Feasibility

Before evaluating osmotic tablet technologies, or any controlled release technology for that matter, it is first important to conduct a feasibility assessment for controlled release that includes a clear definition of the clinical objectives for providing controlled release [12]. Defining the clinical objectives, such as improving safety by reducing the C_{max} by 25% while maintaining AUC in comparison to an immediate release dosage form, is essential to successful development of any controlled release dosage form. Once the clinical objective and target PK profile are identified, a review of the physical, chemical, and biopharmaceutical properties of the

TABLE 9.3 Comparison of the Delivery Mechanism for Osmotic Tablets to Other Controlled Release Tablet Technologies

	Osmotic	Polymer Matrix (Diffusion/ Swelling/Erosion)	Film[a]-Coated Tablet
Mechanism for rate control	Osmotic pump	Drug diffusion through viscous barrier (polymer matrix, hydrogel)	Drug diffusion through viscous barrier (polymer film coating)
Key formulation factors that control release	Membrane permeability	Polymer type[b]	Polymer type[b]
	Membrane thickness	Polymer molecular weight (viscosity)	Polymer molecular weight
	Osmotic potential	Polymer concentration	Coating thickness
Other factors that influence drug release			
Drug loading	Little or no effect	Moderate effect	Little or no effect
Tablets $\left(\dfrac{SA}{V}\right)^2$	Little or no effect	Moderate to large effect	Moderate to large effect
pH	No effect	Large effect for ionizable	Moderate effect for ionizable drugs
Hydrodynamics	No effect	Large effect	Little or moderate effect

[a] Film refers to functional film coating (e.g., enteric coatings).
[b] Type usually refers to the hydrophilic/hydrophobic nature of the polymer. For example, different grades of hydroxypropylmethylcellulose will achieve different release rates based on their ability to wet/interact with an aqueous environment.

drug provides the data required to evaluate the possible use of an osmotic tablet for achieving the target PK profile and meeting clinical objectives. The physical, chemical, and biopharmaceutical properties and the target PK profile are also important for choosing the right osmotic tablet technology.

The consideration of dose and solubility is a starting point when evaluating a drug candidate for controlled release using osmotic pump tablet technologies. The delivery volume, defined as the volume of water required to dissolve the dose, is a useful parameter for assessing which tablet technology is most appropriate and gives an indication of the challenge associated with successful development of a CR tablet. The delivery volume, Dv, is defined in Equation 9.3, where the solubility is simply the solubility in aqueous media. Clearly, this solubility can be different as a function of pH for ionizable drugs, and for some drug may depend on the presence of micelles and surfactants.

$$Dv = \frac{dose\ (mg)}{solubility\ (mg/mL)} \qquad (9.1)$$

For the purpose of selecting an osmotic tablet technology for CR delivery knowing the solubility of the drug in unbuffered water, intestinal media buffered between 6.5 and 7.5 or to a pH that can practically be achieved in a tablet core protected by a semipermeable coating are all useful measures of solubility. When the delivery volume is on the order of 1 mL, it is still possible to dissolve the dose within the osmotic tablet core, and it may allow other types of osmotic tablet technologies (e.g., asymmetric membrane and elementary). When the dose volume exceeds 1 mL, the entire dose cannot be dissolved in the 1–2 mL of water that is typically imbibed in tablets of an acceptable size (i.e., 1 g or less in weight). With increasing dose volume, the portion of the dose that

will be delivered as a suspension from an osmotic tablet increases.

The dose volume can also give an indication of how much of the dose can be expected to be absorbed in the lower GI based on the solubility. This is particularly important when the delivery of drug must be sustained more than 4–6 h and therefore requires some portion of the dose to be delivered to the colon where the amount of water is very limited. A high dose volume (i.e., >100 mL) in combination with a long-duration osmotic tablet (i.e., 16 h) indicates that absorption may be solubility limited since more than half of the dose will likely be delivered to the colon where the volume of available water is on the order of 50 mL or less [18, 19]. Some empirically based guidelines have been reported to suggest that as the dose volumes approach 1000 mL and higher, a means to enhance drug solubility will be required to promote absorption in the lower GI tract, even for relative short release durations (i.e., 4–6 h) [12]. In Figure 9.6, the applicability map for choosing an osmotic tablet technology based on dose and solubility is shown.

9.4.2 Physiologic Factors Important for Controlled Release

When developing a sustained release oral dosage form, the permeability of the drug, factors that affect absorption throughout the GI tract (e.g., efflux), and how the drug is metabolized must be considered. For example, if first-pass metabolism is high, there is a potential for loss of bioavailability as the input rate of the drug is slowed down by a controlled release dosage form. Likewise, if active transport is important to the absorption of the drug, the location of these transporters throughout the GI tract will affect the regional absorption from a dosage form that delivers drug throughout the GI tract (e.g., sustains delivery for greater

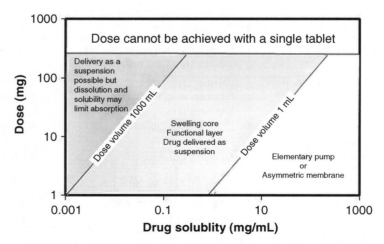

FIGURE 9.6 Applicability map for choosing osmotic tablet technologies based on the drug solubility and dose. Adapted with permission from Ref. [12].

than 4 h). As with any other orally administered controlled release dosage form, these factors should be known and understood as part of the feasibility process and prior to the development of an osmotic tablet. However, since one of the advantages of osmotic tablets is the ability to sustain drug delivery for as long as 16–18 h, it is especially important to understand the slowing down of the input of release in the context of metabolism and regional absorption.

The amount of water in various regions of the GI tract is an important consideration to both the performance of the osmotic pump. One advantage of bilayer osmotic tablets is the ability to deliver drug at the intended release rate, even when only small volumes of water are available. This is probably because the actual volume imbibed into the tablet and, in turn, extruded along with soluble drug or other water-soluble components is usually only a few milliliters. The advantage of this is the ability for osmotic tablets to continue to deliver drug even in the colon where the volume is as low as 11 mL [19].

9.5 COMPONENTS OF A BILAYER OSMOTIC TABLET

9.5.1 Overview

The push–pull design of a drug containing layer paired with a swelling, push layer can accommodate drugs with a range of physical–chemical properties (i.e., solubility, log P, ionization) [12]. A typical tablet core formulation for a bilayer osmotic tablet is shown in Table 9.4. The composition of the drug layer will depend mostly on the dose (drug loading) and the material properties of the drug. The composition of the swelling layer has some flexibility and can be standardized for most formulations. Typical ratios for drug layer:sweller

TABLE 9.4 Typical Tablet Compositions for a Bilayer Osmotic Tablet Formulation

Component	Weight Percentage/ Per Layer
Drug layer components	
API	1–30
Solubilizer (e.g., salts as buffers, cyclodextrin)	As needed
Entrainer, suspending agent (e.g., PEO MW 100,000–600,000)	40–90
Binder (e.g., hydroxypropyl cellulose)	0–5
Swelling/push layer components	
Swelling agent (e.g., PEO MW 4000–7000 K)	30–70
Osmogen (e.g., sodium chloride)	10–40
Binder (e.g., hydroxypropyl cellulose)	0–5
Drug layer:sweller layer ratio (by weight): 2:1–3:1	

layer are in the range from 1:2 to 1:3. The exact ratio will depend upon drug loading, tablet aspect ratio, and to some extent the rate of release. However, generally speaking, this ratio is not a critical parameter for typical bilayer osmotic tablet formulations described in Table 9.4. Some refinement in the ratio may be necessary once the tablets are compressed to ensure that the sweller layer extends into the tablet band upon compression.

In many cases, the same osmotic tablet formulation can be used to deliver drugs at approximately the same rate, although some minor changes in the coating composition or thickness may be required [20]. The robustness of the delivery mechanism results from a tablet core composition that hydrates and swells in a manner that is independent of the drug, provided the drug is maintained within a certain concentration in the core. This makes formulation development very rapid and reliable. The complexity in the composition and manufacturing process of bilayer osmotic tablet results in a robust technology, but there are associated costs in development and final cost of goods for the commercial dosage form. This is discussed in Section 9.5.2 in greater detail. However, this complexity may offer an advantage following loss of exclusivity if it provides a PK profile that is only achievable using an osmotic dosage form. The need to develop an osmotic tablet to achieve bioequivalence may prevent, or at least delay, the launch of a generic form of the drug getting to the market.

9.5.2 Tablet Core

9.5.2.1 Drug Layer The functional requirements of the drug layer include the ability to suspend drug particles and to be extruded through a delivery port as the tablet core becomes hydrated. The suspension of drug within the core as it is hydrated is achieved by formation of a concentrated polymer solution. The viscosity within the hydrating core is bound on the low side by the ability to suspend the drug and on the high side by the ability to flow through the delivery port. This viscosity range must be achieved with the addition of very small amount of water (0.5–2.0 mL) and in sufficient time so that relief of hydrostatic pressure from the influx of water and swelling of the functional layer is possible through extrusion of the drug layer. This functionality has been achieved by combining the drug with polyethylene oxide, a water-soluble polymer, and by limiting the amount of drug in this mixture so that the polymer dominates the hydration rate and viscosity of the layer.

Drug Properties In theory, there are few restrictions on the types of drugs that can be delivered in a controlled manner using bilayer osmotic tablets. Given the complexity and cost of manufacture of bilayer osmotic tablets, the use of this technology is usually justified only for poorly soluble drugs that require sustained delivery for greater than 6 h. For these drugs, the use of simpler osmotic tablets (e.g., elementary, AMT) that require dissolution of drug within the core is prohibited by the drug solubility. Likewise, controlled delivery using matrix tablet technology is challenging due to erosion controlled release mechanism that predominates when drug solubility is low.

The ability to suspend and deliver an insoluble drug through a manufactured delivery port is one of the advantages of delivery using a bilayer osmotic tablet technology. There are a few considerations, however, for drug particle size, solubility, and mechanical properties that are important, particular for higher doses and hence higher drug loadings within the tablet core. The drug particle size is bound at one extreme by the ability to "fit" through the delivery port within the membrane coating, which is typically 500–1000 μm. Since the justification for this complex controlled release technology is usually low solubility, the drug particle size should be engineered toward the low end to promote dissolution [21, 22]. Thus, engineering of drug particle size to help in dissolution of the drug will ensure that delivery of suspended drug through the coating orifice is also possible.

Drug Entraining/Suspending Agent The very specific requirements for hydration and viscosity in the drug layer of the tablet core have limited the choice of compendial excipients that can be used as the functional excipient in the drug layer. Water-soluble polymers such as polyethylene oxide, hydroxypropylmethylcellulose (HPMC), and polyvinylpyrrolidone (PVP) have been used for entrainment of drug; however, polyethylene oxide with average molecular weights ranging from 100,000 to 400,000 is most common [20, 23, 24]. The nearly exclusive use of PEO is due to its ability to wet and hydrate to the appropriate viscosity range to suspend drug particles and flow through the delivery port. Also, the rate of hydration and viscosity drop in the PEO-based drug layer works well in combination with the hydration and swelling of PEO-based functional push layers. When PEO is used as the primary component in the drug containing layer, typical amounts range 40–90% by weight of the total drug layer. Maintaining a high concentration of PEO in the hydrating layer helps to achieve a cohesive layer that suspends drug and maintains a uniform composition as the layer is extruded from the tablet.

The compendial grades of polyethylene oxide currently supplied by Dow® (Polyox™ NF) are available in a number of different molecular weight grades and ranges of particle sizes. Generally, the Polyox™ grades of PEO are free flowing powders and are very compressible, making them very amenable to conventional blending and compression processes. Thus, the inclusion of fillers and/or binders to the drug layer to help in powder handling or to help produce a robust tablet is generally not required. The importance of the material properties of the Polyox™ grades of PEO are

discussed later in the context of processing of the drug and swelling layers for compression.

9.5.2.2 *Granulation Aids and Binders* Even though the Polyox™ grades PEO that comprise most of the tablet core have very good flow and compression properties, there are situations when it may be necessary to include binders or fillers in the drug layer. For example, for low dose, low drug loading, it may be necessary to use a granulation step to ensure that uniform distribution of drug is achieved and maintained during processing. Alternatively, if the dose and drug loading are high, additional excipients may be required to improve flow or compression properties as a result of deficiencies in the material properties of the API. In these cases, the use of insoluble excipients should be minimized since they may cause delivery problems for the device. If insoluble excipients are required, they should be used in small amounts (e.g., <10% of formulation) to minimize the impact on the function of PEO in the drug layer.

9.5.3 Swelling Layer

The dual functionality required by the sweller layer is achieved by combining a component with a high osmotic potential with a high molecular weight water-soluble polymer that swells upon hydration. High molecular weight grades of PEO are very commonly used in bilayer osmotic tablets [23, 24], but the use of other hydrogels and superdisintegrants [20] has also been reported. The rate of swelling achieved with the imbibition of small amounts of water is part of a balance that must be achieved to work in concert with the extruding sweller layer. The hydrostatic pressure generated by the swelling layer is dictated by the physical and mechanical properties of the functional (swelling) polymers and the hydrating drug layer that creates resistance to the flow through the delivery orifice. If this hydrostatic pressure exceeds the mechanical strength of the semipermeable coating, a rupture of the coating membrane will occur. On the other hand, if the pressure is high enough to create a back pressure against the extruding drug layer, the "push–pull" mechanism of the two layers is lost, leading to entrapment of drug within the core and incomplete drug release.

9.5.3.1 *Osmogen* The osmotic pump is driven by the osmotic potential created by the presence of water-soluble components within the tablet core. The osmotic potential is the sum of the contribution from all water-soluble components within the tablet core, but the osmogen contributes the most to the osmotic potential. In a bilayer osmotic tablet, the primary source of osmotic potential is contained in the sweller layer. The most important consideration for selection of the osmogen is that it achieves sufficient osmotic pressure difference ($\Delta\pi$) across the membrane to drive water into the tablet and that enough osmogen is present to sustain

TABLE 9.5 Osmotic Potential for Commonly Used Osmogens

Osmogen	Osmotic Pressure (atm)
Sodium chloride	356
Potassium chloride	245
Sodium phosphate dibasic	31
Fructose	355
Sucrose	150
Dextrose	82

Osmotic pressure values taken from Ref. [26].

a saturated solution and $\Delta\pi$ over for the intended delivery duration. Thus, the osmolarity within the tablet core must be greater than that in the GI tract. The osmolarity of the GI tract is regulated to be on average ~270–290 mOsmol/L [18, 25] and can range from as low as 34 mOsm/Kg in the colon to 200–280 in the upper GI. In the stomach, the osmotic potential greatly fluctuates during digestion and can be as high as 1000 mOsmol/L [25] The ability for a water-soluble component to generate an osmotic potential (π) across the membrane depends on its solubility, molecular weight, and osmolarity as shown in Equation 9.2, where i is the dimensionless van't Hoff factor, M is the molarity of a saturated solution of the solute, R is the gas constant (0.08206 L atm/(mol K)), and T is temperature.

$$\pi = iMRT \qquad (9.2)$$

The best osmogens are soluble salts (e.g., NaCl) that dissociate in solution ($i = 2$ for NaCl) and sugars that are very soluble and low in molecular weight. As shown in Table 9.5, sodium chloride is the one of the best osmogens available from the list of compendial excipients. In addition, it is generally nonreactive and available in grades with a range of particle sizes, making it the osmogen of choice in most bilayer osmotic tablets. To maintain a constant osmotic pressure, the osmogen should be present at a level to maintain a saturated solution within the core for the entire duration of delivery. This level can easily be calculated with the knowledge that most osmotic tablets will imbibe approximately 1–2 mL of water and the solubility of the osmogen.

9.5.4 Membrane Coating

The function of the membrane coating is to isolate a highly water-soluble component that generates a high osmotic pressure and to create a rate-controlling barrier for diffusion for water into the core and delivery out of the pump. This barrier in an osmotic pump tablet is a semipermeable film formed using a water-insoluble, film-forming polymer, in combination with a plasticizer or a pore former. The cross-sectional appearance of a typical membrane coating on a bilayer osmotic tablet is shown in Figure 9.7. Another requirement for the membrane coating is sufficient mechanical strength so

FIGURE 9.7 A cross-sectional view of the membrane coating applied to a bilayer osmotic tablet.

that it can resist the hydrostatic pressure that is created from the influx of water and due to swelling of the push layer. Cellulose acetate is commonly used as the basis for the semipermeable membrane coating. The use of other water-insoluble polymers (e.g., ethylcellulose) as the primary component in the membrane coating has been described in the literature [27–29]. Enteric polymers have also been described for use as a membrane [30]. However, enteric polymers only provide osmotic delivery in gastric media since dissolution of the membrane occurs once the tablet reaches the intestine.

The second component in an osmotic film coating is a water-soluble component (usually polyethylene glycol) for the primary purpose of increasing membrane porosity, water permeability, and release rate. In addition, PEG also acts as plasticizer to reduce the brittleness of the cellulose acetate film and to improve film robustness [31, 32]. The amount of PEG directly affects the drug delivery rate through its effect on the permeability of water into the tablet core and must be optimized for each formulation. The critical aspects of designing a membrane coating to achieve a target release rate include selection of the ratio of the insoluble and soluble components (e.g., cellulose acetate:PEG), the membrane thickness, and coating solution composition (solvent system, solid content) use for spraying the film onto the tablets.

9.5.4.1 Controlling Release Rate with Membrane Composition and Thickness The semipermeable membrane controls the rate of water influx into the tablet core and, in turn, the rate of drug release. The resulting drug release rate is

a function of the osmotic potential across the membrane and the permeability and the thickness of the membrane coating. The influx of water dV/dt is described by the Equation 9.3,

$$\frac{dV}{dt} = \frac{Ak\Delta\pi}{l} \tag{9.3}$$

where A is the area of the membrane, k is the membrane permeability, $\Delta\pi$ is the difference in osmotic pressure across the membrane, and l is the diffusion length or coating thickness. Assuming that the osmotic potential remains constant, the volume of the tablet is maintained, drug diffusion out of the tablet is not rate limiting, and membrane permeability remains constant over time, the rate of drug delivered dM/dt can be described by the Equation 9.4,

$$\frac{dM}{dt} = \frac{Ak\Delta\pi c}{l} \tag{9.4}$$

where c is the concentration of drug in the tablet core, provided that drug release can be assumed to be dependent only on the rate of water influx as controlled by the coating thickness (l), permeability (k), area (A), and difference in osmotic pressure ($\Delta\pi$).

In a bilayer osmotic tablet, $\Delta\pi$ is due primarily to the osmogen, typically sodium chloride, in functional swelling layer. When present in sufficient quantities to maintain a saturated solution, the osmotic pressure can be assumed to be constant throughout the delivery. Bilayer osmotic tablets generally do not swell upon hydration, and thus the surface area of the membrane can be assumed to be constant. The permeability of the coating to water ingress, reflected in the value of the k, depends on the composition and to a lesser extent the porosity of the membrane. Typical permeabilities are 1×10^{-2} to 1×10^{-4} cm³ mL/(cm² h atm) [1]. Generally, the permeability of the membrane is assumed to be constant throughout the delivery. The drug concentration within the

core, c, remains constant provided the functional components, binders, and fillers are delivered at the same rate as the drug.

With the $\Delta\pi$, drug concentration in the core, and membrane coating surface area all remaining constant, the drug release rate is controlled only by the coating permeability (membrane composition) and thickness. The film coating composition and thickness are easily measured and controlled and make this a very elegant way to achieve a target drug release rate. The classic osmotic tablet profile includes an initial lag (0.5–2 h), followed by a constant rate of delivery of 80–90% of the dose as shown in Figure 9.8a, and the linear relationship between the coating thickness and the release rate for two different coating compositions. Figure 9.8b illustrates the coating control of drug release for tablets where all other variables in Equation 9.2 remain constant.

The assumptions that are made to allow the drug release rate to be controlled entirely by coating composition and thickness are generally valid ones. This allows the ability to make predictions of the zero-order release rate with knowledge of the $\Delta\pi$, drug loading within the tablet, and an empirical measure of coating permeability possible. There are situations when zero-order release is not observed, which can usually be attributed to one of the following reasons: (1) depletion of osmogen leading to a drop in osmotic pressure, (2) depletion of the plasticizer, pore former, or permeability enhancing agents in the film coating, (3) blockage of the manufactured delivery ports from swelling agents in bilayer or multilayer tablets, (4) and the osmotic pressure outside the tablets exceeding that within tablets. The first example occurs in very simple, single layer osmotic tablets that rely on the drug to act as the osmogen. The depletion of water-soluble component in the semipermeable membrane has been documented, including the loss of PEG in cellulose acetate-based coatings present in most commercial osmotic tablets. However, this loss in PEG does not lead to a change in permeability or loss in zero-order release rate in most cases.

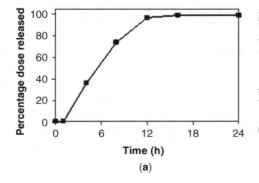

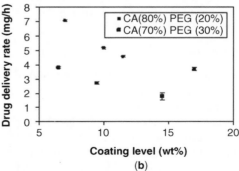

FIGURE 9.8 (a) Cumulative drug release as a function of time from an osmotic tablet with a cellulose acetate:PEG (7:3) coating and (b) the relationship between drug release rate as a function of semipermeable membrane composition (CA = cellulose acetate, PEG = polyethylene glycol) and coating level (weight percent relative to the tablet).

In the final example provided (4), the ability of a high osmotic pressure outside the tablet suppress release rate is rarely expected to be a problem. However, purposeful increases in the osmotic pressure in dissolution test media with the addition of salts and buffers to the point of "shutting down" the osmotic pump have been used to confirm osmotic control in tablets. One final and important comment about maintaining osmotic controlled release tablets is that an "apparent" loss in osmotic control can be suggested when using drug dissolution as an indirect measure of drug release. This is particularly true with osmotic systems that deliver insoluble drug through a delivery port that in turn must dissolve in a test media. In this case, an "apparent" first-order release profile could simply mean that release was constant but that the appearance of drug in solution is limited by drug dissolution. The importance of *in vitro* testing of insoluble drugs from osmotic tablets will be discussed in more detail later in this chapter.

9.5.4.2 *Selecting Membrane Composition*

In theory, the ratio of cellulose acetate and PEG in the membrane coating that is required to achieve a target drug release rate with a bilayer core tablet with known osmotic potential, drug loading, and tablet size (surface area) will depend on the target release rate that can be predicted if the permeability of the coating k in Equation 9.3 is known. In practice, however, the values of k for a range of CA:PEG ratios are difficult to measure since they can depend not only on composition, but also on the processing parameters used to apply the film. Thus, determination of the membrane coating composition is usually done through an iterative process that begins by trying two coatings with different CA:PEG ratios. Ideally, these compositions will bracket target release rate that can be achieved either by interpolation to ratio intermediate to those that were tried or by adjusting the thickness of the membrane. Typical ratios of CA:PEG in bilayer osmotic tablets range from 20:1 to 7:3, with high ratios leading to less permeable membranes and slower release of drug.

9.5.4.3 *Film Thickness*

Once a coating composition is selected based for the ability to achieve a target release rate range, refinement in that release rate is usually achieved by setting the film coating thickness. The average coating thickness ranges from 50 to 200 μm for bilayer osmotic membrane coatings. The average coating thickness is selected using an iterative process that involves choosing the thickness that achieves the desired release rate. This is usually done by selecting the coating composition (ratio of CA:PEG) and then coating to a range of coating thicknesses. If the desired release profile cannot be achieved with thickness values appropriate for the osmotic tablet technology, that is, 50–250 μm for dense coatings and 200–400 μm for porous coatings, it may be necessary to adjust the CA:PEG ratio. By staying within these boundaries, the release rates

can be adjusted by increasing or decreasing the PEG content to give a faster or slower release rate, or by applying a thinner or thicker coating.

Although the coating thickness is one of the key variables that controls the delivery rate, the process variable that is monitored during manufacturing is the weight gain during spray coating. During film coating, the weight of the film coat is used to target the desired release rate. The weight percentage of coating needed to achieve a particular coating thickness varies with tablet size. Because the relationship between coating weight percentage and coating thickness depends on tablet size, the thickness of the coating also should be monitored in relation to release rate, rather than coating weight percentage. Knowledge of coating thickness during coating process development and scale-up is invaluable since thickness reflects coating density and architecture that can affect the drug release rate even when the appropriate coating weight has been applied to the tablet surface.

9.5.4.4 *Solvent System and Coating Solution Composition*

Selection of the coating composition and solvent system for the membrane depends on the density of the membrane desired and also on the target release rate. In Figure 9.9, a schematic phase diagram is provided for a cellulose acetate and polyethylene glycol-based semipermeable membrane. Knowledge of the phase diagram for the components of the membrane film coating, including the components of the spray solution, is essential to select the membrane composition and development of the film coating process. The dense membranes used in most commercialized osmotic tablets are formed when components in the film coat are miscible and form a single phase when applied to the tablet surface. In the case of cellulose acetate and PEG membranes, higher CA:PEG ratios and the use of solvent high in acetone lead to denser films. With increasing PEG content, increasing solid content and higher amounts of water tend to lead to more porous coatings. As the phase boundary approaches a two-phase system, the films become more porous and should be avoided to achieve the dense membrane required for bilayer osmotic tablets.

The solid content of the film coating solution also plays a role in the resultant film porosity. For dense semipermeable membranes made with cellulose acetate and polyethylene glycol, the optimal solid content is 5–10% by weight. The upper limit in the solid content is bounded by potential for phase separation and spray drying that can lead to coatings that are less dense and having greater opacity. The lower limit in solids is driven by the practical considerations in spray time and spray efficiency.

9.5.5 Delivery Orifice

In bilayer osmotic tablets, drug is delivered through one or more delivery ports located on the face of the drug layer side

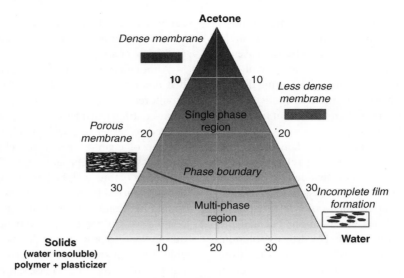

FIGURE 9.9   Schematic of the phase diagram for components of a semipermeable membrane containing a water-insoluble polymer and plasticizer and the solvent system used for spray coating the components.

of the tablet. The manufacture of a delivery orifices in a semipermeable coating has been accomplished by a number of ways, including the intentional formation of defects in coatings [33]. The most common way to manufacture delivery orifices in osmotic tablets is by using a laser to penetrate the semipermeable coating on one or both sides of the tablet face [26]. The disadvantage of this approach is that it is a costly and complex manufacturing step.

Generally speaking, the size and the number of holes are not critical to the delivery of drug in a bilayer osmotic tablet since the coating and the osmogen control the rate of release [20]. Thus, the most important requirements for a delivery port are (1) that it is large enough to allow extrusion of the viscous drug layer that includes suspended drug particles that tend to be less than 100 μm; (2) that it is not too large (or too many), or the integrity of the membrane coating may be compromised as the hydrostatic forces increase due to tablet core hydration and swelling. These two requirements in combination with the preference to make the laser drilling process as simple as possible usually result in a single hole that is on the order of 500–1000 μm in diameter.

9.6   PROCESSING CONSIDERATIONS FOR OSMOTIC TABLET FORMULATIONS

As with conventional IR tablet formulations, the material properties of the excipients and the API must be considered in the development of a bilayer osmotic tablet formulation. However, the ability to overcome material property challenges, such as an API that is cohesive and demonstrates poor powder flow, shifts the emphasis from excipient selection to the use of processing (e.g., granulation). The limitations on

the compositional approaches to improving the material properties result from a limited number of functional components that are suitable for the bilayer osmotic tablets.

9.6.1   Material Properties of Key Components in Osmotic Tablet Formulations

The NF grades of polyethylene oxide, Polyox™, are the primary components in bilayer osmotic tablets and are present in most of the osmotic tablets currently available on the market. The lower molecular weight grades (i.e., 200,000–400,000) serve as the primary component in the drug layer, while the higher molecular weight grades act as a swelling agent in the swelling layer (3,000,000–7,000,000). The high compressibility and flowability of two Polyox grades are reflected in the values of the compression stress and effective angle of internal friction shown in Table 9.6. The compression stress is a measure of the force required to densify a powder that for polyethylene oxide falls outside the 10% percentile compared to other excipients. Very low compression stress values reflect the ease with which the PEO powders can be compressed to form a compact. The low values of effective angle of internal friction, a measure of the ability of the powders to flow under compression, indicates good to excellent flow compared to other excipients. In many respects, the compressibility and ductility of PEO are attributes that are complementary to brittle tendencies demonstrated by many drugs. The nearly exclusive use of PEO in the drug layer is also a likely consequence of material properties that make them amenable to simple blending and compression processes.

The tensile strength values of PEO compacts are low compared to typical tablet fillers and binders, representing a deficiency as measured in conventional tablet formulations.

TABLE 9.6   Comparison of the Tensile Strength, Compressibility, and Effective Angle of Internal Friction (Measure of Flow) for Polyethylene Oxide

		Particle Size (mm)	Tensile Strength (MPa)[a]	Compression Stress (MPa)[a]	Effective Angle of Internal Friction[a,b]
Polyethylene oxide (MW 200,000)	Polox N80	131	1.4	16.4	32.6
Polyethylene oxide (MW 5,000,000)	Polox coagulant	177	1.0	13.4	32.3
	Typical range	NA	0.58–3.8	34–120	29–38
	Mean	NA	1.9	73	33

[a] Unpublished data generated in the authors' laboratory. The 10th–90th percentile and mean of over 100 excipients evaluated.
[b] A higher value indicates poorer flowability.

In the case tablets comprised mostly of PEO that tend to be very robust (low friability), the compressible and ductile nature seems to compensate for low tensile strength. A downside of the high compressibility and very ductile nature of PEO is the potential for crowning and smearing during processing. The importance of the material properties of PEO will be revisited during a discussion of the manufacture of bilayer osmotic tablets. The low tensile strength of polyethylene oxide compared to other excipients suggests that tablet strength and robustness may pose problems during manufacture, shipping, and storage. However, generally bilayer osmotic tablets that contain between 65% and 80% polyethylene oxide are very robust. Low friability and tendency for capping or flaking in these tablets may be due to both the compressibility and the high ductility. Generally, the API and the excipient used as osmogens (e.g., salts, sugars) are poorly compressible and brittle in nature. However, the high concentration of PEO in the drug layer tends to dominate the material attributes in most formulations.

The combination of high ductility and good compressibility of PEO can lead to a loss in compressibility with each compression cycle [34–36]. Thus, there is a tendency for tablet core formulations containing polyethylene oxide to exhibit low tensile strength upon compression, particularly if the formulation is processed using roller compaction prior to tableting [37]. In osmotic tablet formulations high in PEO concentration, there is a possibility for loss in compressibility that may lead to low tensile strength once granules are compressed into tablets [35]. Generally speaking, tensile strength values of 1 MPa or less are considered unacceptable for most tablets. However, given the unique material properties of PEO, these tensile strength values may be acceptable for PEO-rich tablet formulations.

9.6.2 Processing Osmotic Formulations for Compression

Considerations to processing of powder blends for tablet compression for osmotic tablets follow the same guiding principles as used in conventional immediate release or matrix tablet formulations. For bilayer osmotic tablets, the need for processing (e.g., granulation) of the drug layer is usually driven by the properties of the API, since the primary component of the drug layer (PEO) flows well and easily compressed. The need to granulate the drug layer usually stems from the need to ensure content uniformity when the API is present in low concentration (low dose) or is a cohesive powder, or to reduce the tendency for segregation when the API is very free flowing as is usually the case. In addition, an API with a high specific volume, low tensile strength, or high ductility may also require granulation to overcome these material deficiencies, especially when drug loading is higher.

Ideally, a direct compression of the drug layer blend with the sweller layer is the preferred process. The material attributes of the API that are most critical to evaluating its ability to be uniformly blended with Polyox, and for the blend to have suitable flow and compression properties, and that are easily measured are particle size, aspect ratio, and bulk density of the API. These attributes are very simple ways to assess API cohesivity and flow as well as potential for achieving a uniform blend that is maintained during powder transfer prior to and during compression. As the drug loading in the blend increases, it may be necessary to consider other material properties of the drug, such as tensile strength, compressibility, and ductility, that may compromise the material attributes of polyethylene oxide that in general favor good powder flow and compressibility.

For the swelling layer, wet granulation may be required to prevent segregation of the PEO and the osmogen since both components tend to have very good flow properties. The need for granulation in the sweller will be driven by the specific properties of the excipients, and the particular grade of each excipient must be evaluated to determine if direct compression of the sweller is possible, or if granulation is required. The ability to disperse granulation fluid into PEO-rich blends usually requires some amount of solvent in the fluid to promote wetting. The requirement for solvent to promote wetting and even distribution of granulation fluid through the bed may be reduced or eliminated by the use of pressurized spray nozzles in standard high-shear mixers.

Dry granulation may offer a solvent-free approach to reducing specific volume or achieving content uniformity in some cases. Dry granulation also may be a desirable approach to overcoming undesirable material properties for drugs that are chemically unstable and potentially labile during wet granulation and drying stages. An important consideration for dry granulation of blends rich in polymers (e.g., PEO), or other ductile materials, is to minimize the compression used during dry granulation. Minimizing the force needed to form a granule retains some of the compressibility that will be needed to form a compact during tablet compression.

Granules and ribbons formed by wet or dry granulation for compression can be milled using standard rotary, impact mills to produce free flowing particles with a size distribution that is appropriate for tableting. Impact mills may also be needed in direct compression formulations to remove clumps and improve content uniformity. The very ductile nature of PEO tends to cause PEO-rich granules to "flow" rather than "break" in response to an applied force. In some cases, this has led to difficulty in milling granules and ribbons manufactured by wet granulation or roller compaction because of the lack of brittleness. The presence of amorphous domains and low melting temperature of crystalline domains [38] in PEO can lead to smearing or melting during milling of granules. This can be minimized by taking care not to overwork the materials during granulation and/or precooling the blend.

9.6.3   Tablet Compression

In the manufacture of bilayer osmotic tablets, the drug layer is usually filled first and the functional swelling layer is filled second. Two of the main advantages of filling the drug layer include the ability to sample the compacted drug layer alone (to validate fill weight and potency in isolation of the sweller layer) and the ability to apply a light "tamping" or precompression force to the first layer. The drug layer, which is the first layer to be filled, is tamped with a very light precompression force prior to filling the second layer. The purpose of this step is to very lightly compact the drug layer and create an even bed for filling the second layer into the die. In some situations, the swelling or push layer thickness is too low for

a tamping or light precompression prior to the filling of the second layer. The final compression step for the multilayered tablets takes place during the final compression.

Inherent in all bilayer presses is the possibility for carryover of powders from one tablet station to the next. While in most cases the amount of carryover is unlikely to have an effect on tablet performance, it can lead to loss in product elegance. Furthermore, carryover observed in development, when the press is typically being run at speeds well below those used for large scale and commercial operations, may be an indicator that potency and uniformity problems will be observed on scale-up. This contamination problem results when fine particles in the sweller layer are carried under the scrapers and not removed by the vacuum system. This is particularly a problem for direct compression formulations. While both causes and fixes for carryover problems are likely to be specific to any particular press, recommendations for solving carryover problems include adjusting sweepers to improve contact with table (e.g., replacement of Teflon with metal) and adjusting (usually increasing) the vacuum at each of the fill stations. A word of caution: setting the vacuum to high can lead to loss in fines during direct compression of blends or granulations containing a large number of fines.

Osmotic tablet formulations containing high proportions of PEO are very ductile. This ductility can cause plastic deformation during tableting, which can result in crowning, as shown in Figure 9.10. Since crowning (flow around the punch) can lead to coating defects, it should be avoided. Crowning or flashing can be reduced by decreasing the compression force or increasing the dwell time. Lowering the tablet pressure will generally alleviate the problem, but may lead to low tablet hardness.

Tablets containing PEO tend to have very low friability even at hardness values that are lower than those typically observed from conventional tablet formulations. This behavior is consistent with high ductility and the tendency for PEO particles to yield rather than fracture under an applied load. The typical crushing strength values for PEO-based tablet cores tend to be lower than conventional immediate release

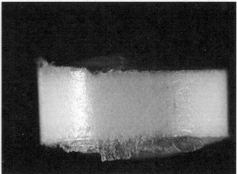

crowning

FIGURE 9.10   Example of tablet crowning or flashing for a PEO-based osmotic tablet.

TABLE 9.7 Typical Crushing Strength Values for Bilayer Osmotic Tablets Containing PEO in Both the Active and Swelling Layers

Tablet Weight (mg)	Tooling Size (SRC)	Typical Crushing Strength Osmotic Tablets (kP)	Typical Crushing Strength Values for IR Tablets (kP)
111	7/32″	3–5	4–6
332	11/32″	5–7	8–12
442	3/8″	5–7	9–12
617	7/16″	8–10	12–16
650	1/2″	9–11	14–18

tablet formulations containing excipients that are higher in tensile strength (e.g., microcrystalline cellulose, dicalcium phosphate) and having some brittle component (e.g., lactose). The crushing strength values for bilayer osmotic tablets containing PEO in both active and swelling layers are compared to conventional immediate release tablets of similar size and weight (Table 9.7). In all these cases, the tablets showed minimal friability and were successfully film coated.

9.6.4 Film Coating

9.6.4.1 *Preparation of Film Coating Solution*
The application of water-insoluble polymers such as cellulose acetate to the tablet surface to create a semipermeable membrane requires the use of organic solvent to dissolve the polymer in the spray solution. There has been some attempt to develop aqueous-based approaches to the formation of the cellulose acetate film coatings from latexes containing cellulose acetate and plasticizer [39, 40]. However, these approaches have resulted in rapid loss of the plasticizer from the film in the presence of aqueous media and show the polymers to be unstable during manufacture. Thus, the film coating process is a solvent-based process starting in development with even small-scale batches, all the way through to commercial manufacture.

Preparation of coating solutions for osmotic tablet systems requires some careful consideration since no *one* solvent can dissolve both components. Generally, the solvent system for preparing the most commonly used components in osmotic membranes, cellulose acetate, and PEG includes acetone and water. The process involves first completely dissolving the polyethylene glycol in water, and then slowly adding acetone while stirring to form the cosolvent. The cellulose acetate is the last component to be added and should be done slowly with stirring to minimize phase separation or precipitation of the PEG. Complete dissolution of both PEG and cellulose acetate is critical to the formation of the membrane coating.

The solvent-based semipermeable membranes used in current commercial osmotic tablets are applied using conventional pan coaters that are outfitted to handle the application of solvents. One of the challenges associated with development and commercialization of osmotic tablets is the ability to uniformly and reproducibly apply a film coating to the tablet core. An understanding of the phase diagram of the film coating solution described previously (see Figure 9.9) and how process variables such as spray rate, atomization pressure, outlet temperature, and dew point affect the nature of the film is essential. As with any coating process, the design and the setup of the pan coater are important to both appearance and performance attributes of the functional membrane coating. The movement and mixing of the tablets in the bed, the number of spray guns and their location relative to the tablet bed, and the increase in bed volume relative to the volume of the pan during the coating process contribute to the rate of film formation and the number and composition of phases in the film and film porosity. One of the challenges in scaling up a functional film coating process is understanding how the differences in pan coater design will translate to key attributes of the film. A thermodynamic model that provides a precise relation of key variables such as the inlet air (temperature, humidity, and flow rate), coating solution, and outlet air (temperature, humidity, and flow rate) can enable prediction of target process conditions on scale-up [41].

The selection, setup, and maintenance of spray nozzle are extremely important to the successful and reproducible application of the film coating to the tablets. The potential for bearding at the nozzle surface can complicate the coating process and in some cases require that the process be interrupted to clean or replace the nozzle. Recently, modifications in the nozzle design have been made to reduce the tendency for bearding. Small differences in the assembly of the nozzle and the setup of the nozzle in the spray gun (orientation) can affect the droplet size and extent of spray drying during film coating.

The measurement and characterization of coating thickness on osmotic tablets may seem trivial; however, the technology for rapid online or even at-line measurement of coating thickness has not been fully developed. The coating thickness may be measured using calipers to measure coated versus uncoated tablets or using microscopy to look at the coatings on bisected cores. This approach, which is slow and tedious, requires some subjectivity, particularly for very porous coatings. The use of vibrational spectroscopy to measure coating thickness and tablet to tablet variability has been explored [42, 43]. However, these approaches will likely require further refinement before they are ready for implementation into development and commercial manufacture on a routine basis. Monitoring thickness is an important process control since the thickness plays an important role in release rate. At the same time, it is the coating thickness at the tablet edge, which is more difficult to measure, that is important for robustness.

9.6.5 Delivery Port Manufacture

In bilayer osmotic tablets, the holes are drilled on the drug layer side of the tablet as close as possible to the center of the face. It has been shown that having the holes slightly off-center does not affect performance, and many of the commercial osmotic tablets have delivery ports that are slightly off-center. It is important, however, to make sure that the hole is not too close to the edge of the tablet face where the coating is the thinnest and most prone to failures. In principle, the delivery port should penetrate completely through the coating to the surface of the tablet core. The diameter of the delivery port does not need to be tightly controlled, since the coating thickness and osmotic potential control drug release rate. However, the diameter, including the "roundness" of the hole, is usually validated as part of the hole drilling process.

For commercial manufacture, the delivery ports are usually drilled using a laser. This has proved to be a fast and reliable method to reproducibly form a hole in the membrane film coating. Commercially available laser drilling systems for both development and commercial manufacture are fully automated with respect to tablet handling, side recognition, and inspection. Usually, laser drilling equipment designed for large-scale, commercial manufacture have more sophisticated automation and operate at much higher output rates.

The process of laser drilling is complex and includes tablet handling, establishing the duration and intensity of the laser needed to penetrate the film coating, and validation of the process. The tablet handling process involves presentation of the tablet to the laser, side recognition (i.e., drilling on the active layer side of the tablet), and/or tablet orientation. Recently, the incorporation of infrared spectroscopy into tablet handling and laser drilling equipment has been used for side recognition based on spectral difference in the drug layer and the swelling layer of a coated bilayer osmotic tablet [44, 45]. The validation of the laser drilling process includes confirmation that the correct side of the tablet has been drilled, the number and the location of the holes are correct, the penetration depth is adequate, and every tablet has been drilled. Most commercial lasers incorporate the ability to "accept" or "reject" tablets that have been correctly drilled as part of the laser drilling process using vision systems. For example, the laser drilling process may include verification that the hole completely penetrates the membrane coating. Because of the typical variation in the thickness of the coating from tablet to tablet, this will require penetration of the laser into the tablet cores for some tablets. Validation that the proper hole depth has been achieved can be done by verification that the tablet core is revealed by

drilling the delivery hole. This is not a trivial exercise, since drilling into the tablet core can results in browning of the tablet core and may impact API stability (purity). Thus, the extent that the laser penetrates into the core should be minimized. This can be aided by good control of tablet compression (i.e., tablet thickness) and a well-designed and well-controlled film coating process that minimizes batch-to-batch and tablet-to-tablet differences in film coating membrane thickness to a minimum.

It is worth noting that for bilayer, swelling osmotic tablets that have components that swell upon hydration, complete penetration through the coating may not be required to allow delivery. In these systems, the hydrostatic pressure associated with swelling of polymer upon hydration can break any residual coating that remains intact after drilling. Thus, the laser drilling process needs to only partially penetrate the coating and thereby form a defect or weak point in the coating. If stability as a result of laser penetration into the core is a concern, it is better to err on the side of drilling just short of the core surface.

Generally, the laser drilling process does not promote drug degradation, presumably due the short duration of exposure to the laser. There have been a few examples of discoloration near the hole when the laser drilled hole penetrated well into the tablet core. The discoloration is attributed to the presence of BHT in Polyox and other impurities in other excipients that are chemically labile when subjected to the local heating that can result from laser drilling. To minimize discoloration and other physical/chemical changes due to laser drilling, it is best to avoid penetration of the laser into the table core.

During small-scale development, holes can be drilled manually using a benchtop mechanical drill. Provided the diameters of the delivery holes are approximately the same size, the release rate and the tablet performance are the same when the holes are hand drilled and laser drilled. Development and small-scale laser drilling equipment used for the manufacture of clinical supplies are manually operated and require both loading tablets into trays and 100% visual inspection. While these types of lasers are labor intensive, they offer a rapid approach to drilling small-scale clinical supplies (less than 5000 tablets) and the greatest flexibility.

9.7   *IN VITRO* PERFORMANCE TESTING

Measures of tablet release rates primarily involves *in vitro* characterization using dissolution testing. The objectives of the dissolution testing will change at different stages of development and manufacture of osmotic tablets for clinical evaluation and commercial manufacture. Early on the dissolution testing is done to determine correlations between the film coating thickness and the composition, to verify osmotic pressure as the mechanism of control and release, and to establish tablet robustness under conditions representative of fed and fasted states as well as accelerated storage conditions. At later stages, the *in vitro* characterization is used to verify successful scale-up of tablet manufacture, particularly the film coating process, and validate successful manufacture of commercial tablets.

9.7.1 Measuring *In Vitro* Drug Release Rate Versus *In Vitro* Drug Dissolution Rate

As stated earlier, one of the advantages of osmotic delivery is the insensitivity of drug release to the pH, ionic strength, and hydrodynamics of the media surrounding the tablet. While this is certainly the case for the release of drug from bilayer osmotic tablets, it may not be true for the *dissolution* of insoluble drug delivered as a suspension. If the drug is delivered from the tablet at rate or extend that approaches or exceeds the solubility of the drug, the dissolution rate of the drug may be affected by hydrodynamics. Likewise, the dissolution rate of poorly soluble ionizable drugs from bilayer osmotic tablets may vary according to the pH of the media. This has important implications for both *in vitro* and *in vivo* performance. If dissolution of the drug once delivered from the GI tract is retarded due to the effect of pH on solubility, this may be a barrier for absorption. Likewise, the dissolution rate of an ionizable drug delivered as a suspension may differ according to the pH of the dissolution medium. For example, a weak basic drug may dissolve very quickly in gastric media, making a dissolution test an accurate measure of drug release. In contrast, dissolution of the same tablet formulation measured in intestinal buffer may be limited by the solubility of drug in the unionized form. In this case, the "apparent release" profile reflects slower *drug dissolution* once released from the tablet, not slower *drug release* (Figure 9.11).

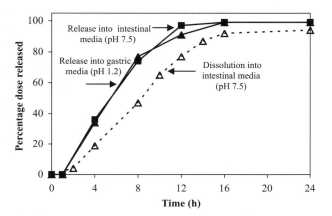

FIGURE 9.11 Comparison of cumulative percentage of release of a weakly basic drug in gastric media, pH 1.2, (▲), intestinal buffer, pH 7.5 (■), and dissolution in intestinal buffer, pH 7.5 (△).

9.7.2 Dissolution Testing to Support Formulation Development

Dissolution studies are first used to determine the relationship between release rate and coating composition and coating thickness. These tests are easily done using a standard USP apparatus II (paddles) dissolution test in a media that will provide sink conditions for the API dose per tablet. It is important that there should be rapid and complete dissolution of the drug at this stage so that the measure of drug dissolution rate can be used as a surrogate for drug release. This ensures that differences in dissolution are due only to the coating. Selection of the media for this purpose is based only on API solubility that should take into account the pK_a of ionizable drugs. The aim of these dissolution tests is to determine the drug release rate as a function of coating composition and thickness, such as that shown in Figure 9.8, and ultimately to select the appropriate coating composition thickness to achieve the target *in vivo* drug release rate.

Once the coating composition and coating thickness are selected, *in vitro* dissolution testing is used to verify true osmotic control, independence of release on media pH and stir rate, and to establish variability of release rates within the batch, the effect of key coating process variables on release rate. Provided that the drug core formulation is typical with regard to API loading and the use of typical molecular weight grades of PEO in the drug and swelling layers includes an osmogen (NaCl) that is coated with a cellulose acetate-based coating, the drug release rate will be controlled by the osmotic pump generated by the tablet core and the coating. In cases where drug loading is high, or when nonstandard formulations are used because of stability or manufacturing considerations, it may be worthwhile to verify that drug release is osmotically controlled. This is done by conducting dissolution testing in media with increasing molality to demonstrate a decrease in the release rate that eventually stops when the osmotic potential across the membrane is eliminated. *In vitro* studies to confirm the independence to hydrodynamic effects and pH in theory are not necessary. However, these additional dissolution experiments can build experience with both dosage form and analytical methods that may be worth having.

One word of caution when looking at the effect of pH on the dissolution is to ensure that changing pH does not lead to loss in sink conditions for ionizable drugs. This would invalidate the assumption that dissolution is also a measure of release. In such cases, the addition of surfactant may be helpful in achieving sink conditions at pH ranges where the drug in unionized. When dissolution is required at pH where the dose per tablet is not soluble in a reasonable media volume, direct release testing can be conducted using multiple tablets, where each tablet is pulled at the designated time, rinsed, cut open, and assayed for residual drug remaining in the tablet. This approach for measuring drug release as a function of time based on residual drug content within the tablet is more laborious and requires multiple tablets, but is the best approach for determination of release rate when drug solubility is extremely low.

9.7.3   *In Vitro* Testing to Support Clinical and Commercial Manufacture

The purpose of dissolution testing of osmotic tablets manufactured for clinical evaluation or clinical manufacturing is usually to ensure process control and quality assurance, ensuring performance stability (release characteristics do not change with time), and to support regulatory documentation [46]. This testing may also provide the basis for an IVIVC that may be helpful for widening the specifications on *in vitro* testing when variability in release rates is problematic.

Provided that a formulation has been confirmed to be a well-behaved bilayer swelling core osmotic tablet (i.e., osmotic control of release), the drug release rate is expected to be the same in different media, and not a function of stirring rate. Thus, the design of a dissolution test to confirm the target drug release profile for clinical use and to support regulatory documentation for clinical and commercial usually is fairly straightforward. Typically, all that is needed is a single test that is robust and that ensures the drug has good solubility. Since many drugs incorporated into bilayer swelling osmotic tablets have low aqueous solubility, it may be important to be aware of the typical particle size ranges (specifications) for the API and how that range of particle sizes may contribute to variability in the release profile. It may be possible that the API dissolution contributes more to the variability in the apparent release rate (measured indirectly using a dissolution test) than the drug itself.

The recommendation for dissolution testing of modified release formulations includes a minimum of three points. The first point demonstrates the absence of dose dumping and is usually taken within the first 1–2 h. The last time point is taken when at least 80% of the dose has been released and dissolved, and for an osmotic tablet ensures what percentage of the dose is completely released from the tablet. The middle point is often taken at a time corresponding to the delivery of 40–80% of the dose. Since a bilayer, swelling core osmotic tablet gives a true zero-order release profile, the selection of the midpoint value is not critical, but is usually the time corresponding to ~50% release. The specifications for target cumulative percentage release at the designated time points are obtained from the mean values for lots used to establish the bioavailability (or bioequivalence). These specifications usually require that the cumulative drug release is ±10% of the label claim.

Usually the most variability in the release profile is observed for the middle point. Sometimes, meeting the label claim for the second time point within ±10% of the mean can

be difficult. Sources of variability in the dissolution profile are most likely due to process-related changes in the semipermeable membrane such as coating thickness variability within the batch. In batches that have correct target coating thickness and minimal tablet-to-tablet variability, qualitative changes in the film coating (e.g., porosity, phase behavior) associated with more subtle differences in the film coating process may lead to variability in the *in vitro* release profile. Variability in the excipients, such as the degree of substitution on derivatized polymers such as cellulose acetate, can also cause batch-to-batch variability in release rate. An important part of developing a robust commercial dosage form is collecting the data that correlate typical process variation and coating thickness and release rate, and includes monitoring tablet-to-tablet and batch-to-batch variabilities. Included in this process understanding is knowledge of how typical variation in API particle size and excipient specifications (e.g., molecular weight, impurities, degree of substitution) can affect the *in vitro* performance. In some cases, specification on the API or key functional excipients may be required.

9.7.4   Preclinical *In Vivo* Testing

The insensitivity of drug release rate to hydrodynamics and media pH affords an easy correlation between *in vitro* and *in vivo* performance. Thus, there is little benefit in testing prototype formulations in animal models for the sake of establishing *in vivo* release rate. Furthermore, the differences in GI residence time and regional absorption in different species may be misleading with respect to drug release rate.

In lieu of testing of osmotic tablet formulations in animal models, the best approach for achieving *in vivo* success with an osmotic tablet formulation in humans is to understand the potential for differences in regional absorption due to permeability, metabolism, or efflux based on preclinical *in vitro* models prior to development of any osmotic tablets. An assessment of the solubility of the drug, rate of delivery, and expected dissolution rate as it is delivered into different regions of the human GI tract is another preclinical activity that is extremely valuable and more likely to determine the likelihood of success in achieving a target PK profile.

9.7.5   Consideration for Clinical Study Design

While the drug release rate from an osmotic tablet *in vivo* is expected to be the same as that measured from *in vitro* studies, differences in dissolution rate, absorption, and metabolism at different locations along the GI tract may lead to *in vivo* results that are not expected based on drug release rate and release duration. Some drugs are known to increase the GI mobility, leading to faster transit of osmotic tablets. A faster transit time may lead to lower absorption, especially if the tablet passes through the GI before the entire drug has been delivered. In the event that bioavailability is low, it is helpful if tablet recovered following defecation are collected and assayed for residual drug [20]. Even knowing the transit time for the tablet, based on the dosing and recovery times, may help to explain low bioavailability base on a transit time that is fast relative to the delivery duration of the tablet. A study that measures the effect of food on the *in vivo* performance is also highly recommended, and earlier in development rather than later. Even though the release rate is not affected by the physiological differences between fasted and fed GI tracts, food may affect residence time in the stomach, transit time, and drug solubility (e.g., due to presence of bile salts). These effects can lead to the dissolution and absorption of a drug delivered from the osmotic tablet.

9.8 STABILITY TESTING

During the standard testing for drug potency and purity, and release performance required to set shelf life and packaging requirements, the high osmotic potential within the tablet core and the presence of polymers that swell upon hydration must be considered. This presents some challenges for using accelerated stability testing to determine real-time storage and setting drug product shelf life. In addition, high temperature and high relative humidity can lead to changes in the coating that results in changes in the delivery profile over time.

9.8.1 Accelerated Stability Testing

The disadvantage to the need for only very small amounts of water to activate and drive the osmotic mechanism in tablets is the potential for "preactivation" when stored at high water activities during accelerated stability testing. Sodium chloride deliquesces at 75% RH, and thus exposure to high RH values can lead to nonreversible change in the physical properties of the tablet, associated with water vapor sorption and subsequent swelling of the core. At the more extreme accelerated stability conditions (e.g., 50°C/20% RH and 40°C/75% RH), osmotic tablets have been shown to begin the delivery of drug (extrusion) when stored without desiccant or in open containers. In most cases, storage in high-density polyethylene (HDPE) bottles with desiccant provides adequate protection from water vapor allowing physical stability of the osmotic tablet.

9.8.2 Excipient Impurities and API Compatibility Considerations

The addition of butylated hydroxytoluene (BHT) to some grades of polyethylene oxide prevents oxidative degradation to lower molecular weight chains and in turn loss in viscosity/swelling functionalities. Its presence is more critical for the higher molecular weight grades that are used for their swelling properties upon hydration. There are grades of Polyox with lower molecular weights that do not have added BHT and can be used as entrainers if BHT and API are reactive. It should be noted in bilayer osmotic tablets that BHT from Polyox grade in one layer can migrate into another layer containing drug, which may also contribute to stability problems even if the BHT is not originally present in the drug layer.

Oxidative degradation of high molecular weight grades of PEO during long-term storage or purposefully done to manufacture lower molecular weight grades can lead to the formation of potentially reactive functional groups, as shown in Figure 9.12 [47–49]. One such species is formate ion that is generated during oxidative degradation of high molecular weight grades of PEO to manufacture lower molecular weight grades. The levels of formate ion are usually highest in the lower molecular weight grades that have been irradiated for the longest period of time. Formate has the potential to react with drug containing secondary amines to form an *n*-formyl adduct as shown in Figure 9.12. When such a reaction does occur, the reaction can be minimized by selection of grades with lower levels of formate ion that may require the use of a higher molecular weight grade. It is also possible to form a drug–polymer adduct with drugs containing a secondary amine; however, this reaction is less commonly observed. There are some peroxides present in PEO, but generally the levels are lower in the higher molecular weight grades used in the tablet core compared low molecular weight grades (i.e. <4000). For drugs that can chemically degrade in the presence of peroxides, the compatibility with the PEO in the tablet core and in the PEG used for the coating should be determined.

Peroxides are present in PEG as impurities and have been shown to migrate in the tablet core upon storage. Therefore, it should not be assumed that for APIs that have the potential to react with peroxides the reaction will be limited to the interface between the core and the coating. The addition of the antioxidant BHT to the tablet core has been used to stabilize drugs that degrade in the presence of peroxide impurities present in PEG that migrated into the tablet core [50].

9.8.3 Potential for Drug–Polymer Interactions

The potential for drug–polymer interactions in the solid and solution states cannot be overlooked in systems containing so many polymers. The amphiphilic properties of PEO allow it to act as a nonionic surfactant that can both promote wetting and increase the concentration of drugs in solution [51]. In

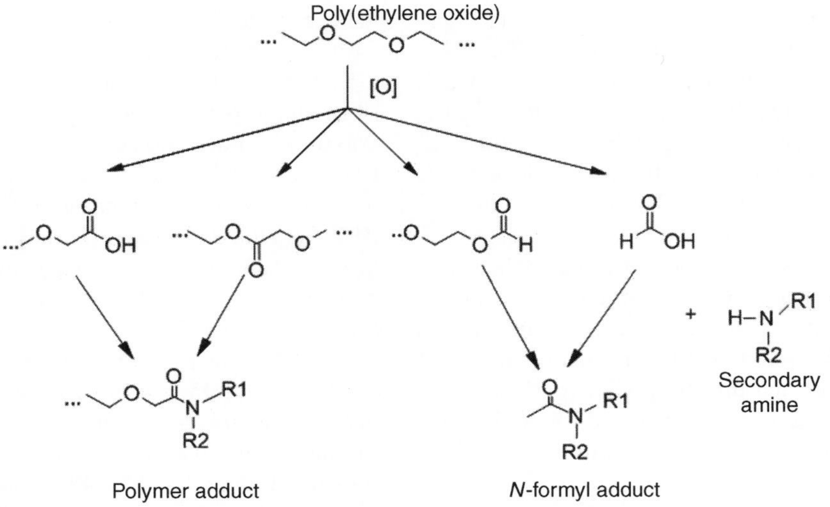

FIGURE 9.12   Oxidative degradation of polyethylene oxide and possible reactions with secondary amines.

the solid state, the formation of solid dispersions between drugs and PEO has been used to promote dissolution of poorly soluble drugs by formation of amorphous drug, a eutectic mixture, or by promoting wetting at the drug surface [51–56]. For example, a hot melt extrusion of ketoprofen and PEO resulted in the formation of a eutectic mixture that had improved dissolution compared to crystalline drug alone [53]. Polyethylene oxide has been shown to promote conversion of crystalline indomethacin to an amorphous form in the presence of high levels of moisture [57]. Schmidt et al. studied the crystallization tendency of amorphous indomethacin when compressed with PEO in comparison to other excipients (e.g., microcrystalline cellulose) [58]. The stabilization of amorphous drug form was attributed to indomethacin–PEO interactions that were detected using Raman spectroscopy.

The potential for PEO to form intermolecular interactions leading to changes in the physical form of the drug that have the potential to reduce shelf life stability, pose regulatory risks, or alter dissolution performance should be considered. Thus, it is strongly advised that the effect of PEO and other functional components (e.g., surfactants, buffers) on the solid-state form, the chemical stability profile, and the behavior in solution of the drug are examined as a part of formulation design, processing, accelerated stability testing, and package selection.

9.9   SOLUBILIZATION OF DRUGS FOR OSMOTIC DELIVERY

One of the advantages of the bilayer osmotic technology is the ability to deliver insoluble drug. However, when the dose:solubility ratio is very high (>1000 mL), the dissolution

of drug once delivered precludes drug absorption. In such cases, approaches to help promote dissolution may need to be considered. This may be as "simple" as using particle size reduction (when appropriate), but may require more sophisticated approaches such as forming a complex with cyclodextrin (CD) to overcome barriers to dissolution. Approaches to improving solubility of drug within the tablet core or promoting dissolution once delivered are considered below.

For very insoluble drugs, the most obvious way to promote dissolution once delivered from the tablet is the use of particle size reductions, or when salt formation is possible to consider a very soluble salt. The development of a prodrug with higher solubility can be considered. The dissolution of a drug can also be improved by preparing it as an amorphous form, which may also achieve a higher solubility, although temporary, compared to crystalline drug [59, 60]. Such an approach may be explored provided the amorphous form has adequate physical stability over the expected shelf life of the tablet and during the hydration and extrusion of the tablet core.

When changes to API form are not possible, or do not offer significant increases in solubility or dissolution rate, the use of added excipients to improve solubility is the next obvious approach. Several workers have reported the use of megulamine for increasing solubility within osmotic tablets [61, 62]. The inclusion of buffers may also help to promote solubility of ionizable drugs within the core. However, the ability of buffers to maintain the effect on drug ionization and solubility in the GI tract is not well understood. For those drugs that can form an inclusion complex with cyclodextrins, the incorporation of cyclodextrins into the tablet core can help improve the delivery of low-solubility drug using osmotic tablet technologies [63]. Derivatives of cyclodextrins, such as hydroxypropyl β-cyclodextrin and sulfobutyl ether β-cyclodextrin, having higher solubilities that β-CD have been shown to have some osmotic potential, which is an advantage since the addition of solubilizers to the components necessary for creating an osmotic pump and/or the "push–pull" mechanism can limit the dose that can be delivered [64]. For example, the ability to deliver testosterone from an elementary osmotic tablet was achieved by using sulfobutyl ether-beta-cyclodextrin sodium salt that acted as both an osmotic agent and a solubilizer for the drug [65].

Except when the dose is very low, attempts to increase drug solubility within the core will only slightly increase the proportion of the drug delivered as a solution. The potential for additives that increase drug solubility to bilayer osmotic tablets to disrupt the hydration rate and viscosity balance between the two layers in the drug, which in turn can lead to incomplete release or rupture of coatings, should be considered. Another potential problem is preferential dissolution and diffusion of the solubilizing agent through the delivery orifice before an insoluble drug that is delivered more slowly with osmotic control. It attempts to keep the drug and solubilizer retained within the core. Thombre has coated the solubilizer with a hydrophobic polymer [66]. This approach adds additional processing steps to an already complex process train and may be a problem for higher dose drugs since it is increasing the loading of functional components within the tablet core.

9.10 SUMMARY

Significant advancements over the first osmotic pumps intended for delivery of biologically active agents have lead to the successful commercialization of osmotic tablets for improved human health. Bilayer osmotic tablets, which include a functional swelling layer, are very versatile and extremely robust ways to achieve controlled release, especially for low-soluble drugs. The robustness of these tablets with respect to delivery, including the ability to deliver drugs with a wide range of physical properties, and the development of formulations can be done in a time frame that is comparable to development of other types of controlled release tablet formulations. As is true for most osmotic dosage forms, the challenge to developing and manufacturing bilayer osmotic tablets is having access to standard pan coaters that are solvent read, a bilayer press, and laser drilling capabilities. With access to these process capabilities, either in-house or through external vendors, the ability to deliver insoluble drug in a manner that is independent of GI location (e.g., pH), hydrodynamics (effects of food), and can easily be correlated to *in vitro* performance, the complexity and the costs of the manufacture of these dosage forms may be justified and even recouped for candidates with more challenging biopharmaceutical requirements.

ACKNOWLEDGMENTS

The author acknowledges Beth Langdon and Danni Supplee for their help in measuring the material properties of polyethylene oxide and to Matt Mullarney for providing guidance in processing of blends containing polyethylene oxide for compression. Thanks to Steev Sutton for simulations of a PK profile from *in vitro* data to demonstrate IVIVC. The author is also grateful to Al Berchielli, Jeremy Bartlett, Alan Carmody, Bruno Hancock, Avi Thombre, and Ken Waterman for helpful scientific discussions and to Scott Herbig for continued support for research in this area.

REFERENCES

1. Baker RW. *Controlled Release of Biologically Active Agents*, John Wiley & Sons, New York, 1987.

2. Cardinal JR. In Amidon GL, Lee PI, Topp EM, editors. *Transport Processes in Pharmaceutical Systems*, Marcel Dekkar, New York, 2000, pp. 411–444.

3. Kim C. *Controlled Release Dosage Form Design*, Technomic Publishing Company, Lancaster, PA, 2000.

4. Santus G, Baker RW. Osmotic drug delivery: a review of the patent literature. *J. Control. Release* 1995;35:1–21.

5. Theeuwes F. Elementary osmotic pump. *J. Pharm. Sci.* 1975; 64:1987–1991.

6. Rose S, Nelson JF. A continuous long-term injector. *Aust. J. Exp. Biol.* 1955;33:415–420.

7. am Ende MT, Herbig SM, Korsmeyer RW, Chidlaw MB. In Wise DL, editor. *Handbook of Pharmaceutical Controlled Release Technology*, Marcel Dekker, New York, 2000, pp. 751–785.

8. Verma RK, Mishra B, Garg S. Osmotically controlled oral drug delivery. *Drug Dev. Ind. Pharm.* 2000;26(7):695–708.

9. Cardinal JR, Herbig SM, Korsmeyer RW, Lo J, Smith KL, Thombre AG. The use of asymmetric membranes in pharmaceutical delivery devices. EP 357369 A2, 19890829, 1990.

10. Nokhodchi A, Momin MN, Shokri J, Shahsavari M, Rashidi PA. Factors affecting the release of nifedipine from a swellable elementary osmotic pump. *Drug Deliv.* 2008;15(1): 43–48.

11. Wang X, Nie S-F, Li W, Luan L, Pan W. Studies on bi-layer osmotic pump tablets of water-insoluble allopurinol with large dose: *in vitro* and *in vivo*. *Drug Dev. Ind. Pharm.* 2007;33(9): 1024–1029.

12. Thombre AG. Assessment of the feasibility of oral controlled release in an exploratory development setting. *Drug Discov. Today* 2005;10(17):1159–1166.

13. Cortese R, Theeuwes F. Osmotic device with hydrogel driving member. EP Patent 52917, 19820602, 1982.

14. Zentner GM, Rork GS, Himmelstein KJ. The controlled porosity osmotic pump. *J. Control. Release* 1985;1(4):269–282.

15. Jedras Z, Janicki S. *In vitro* study of gastrointestinal diffusion system using disopyramide phosphate. *Pharmazie* 1987;42 (12): 842–844.

16. Theeuwes F. Oral dosage forms design: status and goals of oral osmotic systems technology. *Pharm. Int.* 1984;5:293–296.

17. Thombre AG. Feasibility assessment and rapid development of oral controlled release prototypes. *ACS Symp. Ser.;* 752 (Controlled Drug Delivery): 2000, 69–77.

18. McConnell EL, Fadda HM, Basit AW. Gut instincts: explorations in intestinal physiology and drug delivery. *Int. J. Pharm.* 2008;364(2):213–226.

19. Schiller C, Frohlich C-P, Giessmann T, Siegmund W, Monnikes H, Hosten N, Weitschies W. Intestinal fluid volumes and transit of dosage forms as assessed by magnetic resonance imaging. *Aliment. Pharmacol. Ther.* 2005;22(10):971–979.

20. Thombre AG, Appel LE, Chidlaw MB, Daugherity PD, Dumont F, Evans LAF, Sutton SC. Osmotic drug delivery using swellable-core technology. *J. Control. Release* 2004; 94:75–89.

21. Johnson KC. Dissolution: fundamentals of *in vitro* release and the biopharmaceutics classification system. *Drugs Pharm. Sci.* 2007;165(Pharmaceutical Product Development):1–28.

22. Lu ATK, Frisella ME, Johnson KC. Dissolution modeling: factors affecting the dissolution rates of polydisperse powders. *Pharm. Res.* 1993;10(9):1308–1314.

23. Nie S-F, Li W, Luan L, Pan W, Wang X. Studies on bi-layer osmotic pump tablets of water-insoluble allopurinol with large dose: *in vitro* and *in vivo. Drug Dev. Ind. Pharm.* 2007;33 (9):1024–1029.

24. Sastry SV, DeGennaro MD, Reddy IK, Khan MA. Atenolol gastrointestinal therapeutic system. I. Screening of formulation variables. *Drug Dev. Ind. Pharm.* 1997;23(2):157–165.

25. Poon CY. *Remington: The Science and Practice of Pharmacy,* Lippincott Williams and Wilkins, New York, 2000.

26. Theeuwes F, Ayer AD. Osmotic devices having composite walls US Patent 4,077,407, 1978.

27. Savastano L, Carr J, Quadros E, Shah S, Khanna SC. Controlled-release osmotic drug delivery devices. EP Patent Application 94-810212 621032, 19940414, 1994.

28. Zaffaroni A, Michaels AS, Theeuwes F. Method for administering drug to the gastrointestinal tract. US Patent 4,096,238, 19780620, 1978.

29. Edgren D, Wong PSL, Theeuwes F. Oral device for osmotic delivery of drugs. US Patent 4503030, 19850305, 1985.

30. Stevens RE, Limsakun T, Evans G, Mason DH, Jr. Controlled, multidose, pharmacokinetic evaluation of two extended-release carbamazepine formulations (Carbatrol and Tegretol-XR). *J. Pharm. Sci.* 1998;87(12):1531–1534.

31. Guo JH. Effects of plasticizers on water permeation and mechanical properties of cellulose acetate: antiplasticization in slightly plasticized polymer film. *Drug Dev. Ind. Pharm.* 1993;19(13):1541–1545.

32. Yuan J, Shang PP, Wu SH. Effects of polyethylene glycol on morphology, thermomechanical properties, and water vapor permeability of cellulose acetate-free films. *Pharm. Technol. North Am.* 2001;25(10):62, 64, 66, 68, 70, 72, 74.

33. Liu L, Xu X. Preparation of bilayer-core osmotic pump tablet by coating the indented core tablet. *Int. J. Pharm.* 2008; 352 (1–2):225–230.

34. Lin C-W, Cham T-M. Compression behavior and tensile strength of heat-treated polyethylene glycols. *Int. J. Pharm.* 1995;118:169–179.

35. Yang L, Venkatesh G, Fassihi R. Characterization of compressibility and compactibility of poly(ethylene oxide) polymers for modified release application by compaction simulator. *J. Pharm. Sci.* 1996;85(10):1085–1090.

36. Yang L, Venkatesh G, Fassihi R. Compaction simulator study of a novel triple-layer tablet matrix for industrial tableting. *Int. J. Pharm.* 1997;152(1):45–52.

37. Jaminet F, Hess H. Compression and dry granulation. *Pharm. Acta Helv.* 1966;41(1):39–58.

38. Sanchez-Soto PJ, Gines JM, Arias MJ, Novak C, Ruiz-Conde A. Effect of molecular mass on the melting temperature, enthalpy and entropy of hydroxy-terminated PEO. *J. Therm. Anal. Calorim.* 2002;67(1):189–197.

39. Bindschaedler C, Gurny R, Doelker E. Osmotically controlled drug delivery systems produced from organic solutions and aqueous dispersions of cellulose acetate. *J. Control. Release* 1986;4(3):203–212.

40. Kelbert M, Bechard SR. Evaluation of a cellulose acetate (ca) latex as coating material for controlled release products. *Drug Dev. Ind. Pharm.* 1992;18(5):519–538.

41. am Ende MT, Berchielli A. A thermodynamic model for organic and aqueous tablet film coating. *Pharm. Dev. Technol.* 2005;10(1):47–58.

42. Romero-Torres S, Perez-Ramos JD, Morris KR, Grant ER. Raman spectroscopic measurement of tablet-to-tablet coating variability. *J. Pharm. Biomed. Anal.* 2005;38(2): 270–274.

43. Romero-Torres S, Perez-Ramos JD, Morris KR, Grant ER. Raman spectroscopy for tablet coating thickness quantification and coating characterization in the presence of strong fluorescent interference. *J. Pharm. Biomed. Anal.* 2006;41(3): 811–819.

44. Geerke JH. Method and apparatus for semipermeable membrane detection in osmotic tablets by near-infrared spectroscopy. US Patent 2007201024, 20070830, 2007.

45. Geerke JH. Method and apparatus for drilling orifices in osmotic tablets incorporating near-infrared spectroscopy. US Patent 2007196487, 20070823, 2007.

46. Malinowski HJ, Morroum PJ. In Mathiowitz E, editor. *Encyclopedia of Controlled Drug Delivery*, Vol. 1, John Wiley & Sons, Inc., New York, 1999, pp. 381–395.

47. Donbrow M. Stability of the polyoxyethylene chain. *Surfactant Sci. Ser.* 1987;23(Nonionic Surfactants):1011–1072.

48. Han S, Kim C, Kwon D. Thermal/oxidative degradation and stabilization of polyethylene glycol. *Polymer* 1997;38: 317–323.

49. Marchal J. Oxidative degradation of polymers and organic compounds via unimolecular decomposition of peroxy radicals. *Cent. Rech. Macromol., CNRS* 1972.

50. Puz MJ, Johnson BA, Murphy BJ. Use of the antioxidant BHT in asymmetric membrane tablet coatings to stabilize the core to the acid catalyzed peroxide oxidation of a thioether drug. *Pharm. Dev. Technol.* 2005;10(1):115–125.

51. Elworthy PH, Lipscomb FJ. Solubilization of griseofulvin by nonionic surfactants. *J. Pharm. Pharmacol.* 1968;20(11): 817–824.

52. Kanagale P, Patel V, Venkatesan N, Jain M, Patel P, Misra A. Pharmaceutical development of solid dispersion based osmotic drug delivery system for nifedipine. *Curr. Drug Deliv.* 2008;5 (4):306–311.

53. Margarit MV, Rodriguez IC, Cerezo A. Physical characteristics and dissolution kinetics of solid dispersions of ketoprofen and polyethylene glycol 6000. *Int. J. Pharm.* 1994;108(2): 101–107.

54. Law D, Schmitt EA, Marsh KC, Everitt EA, Wang W, Fort JJ, Krill SL, Qiu Y. Ritonavir-PEG 8000 amorphous solid dispersions: *in vitro* and *in vivo* evaluations. *J. Pharm. Sci.* 2004;93 (3):563–570.

55. Stachurek I, Pielichowski K. Preparation and thermal characterization of poly(ethylene oxide)/griseofulvin solid dispersions for biomedical applications. *J. Appl. Polym. Sci.* 2009; 111(4):1690–1696.

56. Schachter DM, Xiong J, Tirol GC. Solid state NMR perspective of drug-polymer solid solutions: a model system based on poly(ethylene oxide). *Int. J. Pharm.* 2004;281(1–2): 89–101.

57. Marsac PJ, Romary DP, Shamblin SL, Baird JA, Taylor LS. Spontaneous crystallinity loss of drugs in the disordered regions of poly(ethylene oxide) in the presence of water. *J. Pharm. Sci.* 2008;97(8):3182–3194.

58. Schmidt AG, Wartewig S, Picker KM. Polyethylene oxides: protection potential against polymorphic transitions of drugs?. *J. Raman Spectrosc.* 2004;35(5):360–367.

59. Hancock BC, Parks M. What is the true solubility advantage for amorphous pharmaceuticals?. *Pharm. Res.* 2000;17(4): 397–404.

60. Hancock BC, Zografi G. Characteristics and significance of the amorphous state in pharmaceutical systems. *J. Pharm. Sci.* 1997;86:1–12.

61. Kidane A, Ray SK, Bhatt PP, Bryan JW. Osmotic delivery of drugs by solubility enhancement. US Patent Application 2003-655725, 2005053653, 2005.

62. Thombre AG. Delivery device having encapsulated excipients. WO Patent 9412152, 19940609, 1994.

63. Mehramizi A, Asgari ME, Pourfarzib M, Bayati K, Dorkoosh FA, Rafiee-Tehrani M. Influence of β-cyclodextrin complexation on lovastatin release from osmotic pump tablets. *DARU* 2007;15(2):71–78.

64. Sotthivirat S, Haslam JL, Stella VJ. Evaluation of various properties of alternative salt forms of sulfobutylether-β-cyclodextrin, (SBE)7M-β-CD. *Int. J. Pharm.* 2007;330(1–2):73–81.

65. Okimoto K, Rajewski RA, Stella VJ. Release of testosterone from an osmotic pump tablet utilizing (SBE)7M-beta-cyclodextrin as both a solubilizing and an osmotic pump agent. *J. Control. Release* 1999;58(1):29–38.

66. Thombre AG. Delivery device having encapsulated excipients. US Patent 5697922A, 1995.

10

FAST DISINTEGRATING TABLETS

SEONG HOON JEONG
College of Pharmacy, Pusan National University, Busan, South Korea

JAEHWI LEE
College of Pharmacy, Chung-Ang University, Seoul, South Korea

JONG SOO WOO
College of Pharmacy, Yeungnam University, Gyongsan, Gyeongbuk, South Korea

10.1 INTRODUCTION

Oral drug delivery system has been regarded as the most convenient and economical method of drug administration with good patient compliance. Among the oral dosage forms, a tablet dosage form has been the most acceptable one [1–3]. However, approximately one-third of the population, mainly geriatric and pediatric population, has difficulty in swallowing oral solid dosage forms including tablets and capsules [4]. It may cause poor patient compliance leading to reduced overall therapeutic efficacy. A novel tablet technology, which can offer ease of oral administration and benefits of increased patient compliance, needs to be developed and fast disintegrating tablets (FDTs) might be one of the perfect fits for the purpose.

FDT technology makes tablets dissolve or disintegrate quickly on the tongue without additional water intake [5]. It is also referred to as orally disintegrating tablet (ODT), fast melting tablet, fast dissolving/dispersing tablet, rapid dissolve tablet, rapid melt tablet, and quick disintegrating tablet. In April 2007, based on the FDA Guidance for Industry regarding ODT [6], the dosage form was defined as "A solid dosage form containing medicinal substances that disintegrates rapidly, usually within a matter of seconds, when placed upon the tongue." However, the United States Pharmacopeia (USP) does not have a published definition for the ODT yet. Even though the FDA recommends that ODTs be kept to a maximum tablet size of 500 mg in the guidance [6], there may be many ODT products in the pipeline that exceed the limit due to their various advantages.

Administration of FDTs is different from conventional tablets and they should have several unique properties to accommodate the rapid disintegration time [7]. They should dissolve or disintegrate into smaller particles or melt in the mouth with the limited amount of the patient's saliva. The disintegrated tablet should become a soft paste, gel, or liquid suspension, which provides good mouth feel and enables easy swallowing in the oral cavity. These tablets are designed to disintegrate rapidly in the oral cavity generally within 1 min (optimally less than 30 s). Ideal properties of FDT would be

- no or minimum water requirement, yet fast disintegration in the mouth;
- pleasant or acceptable taste;
- minimal or no residue in the mouth after administration;
- low cost in manufacturing;
- strong enough to be resistant to the conventional manufacturing process;
- low sensitivity to temperature and humidity change.

FDT has advantages over conventional solid dosage forms with regard to the swallowing issue. It has advantages over oral solutions or suspensions with regard to stability, accurate

Oral Controlled Release Formulation Design and Drug Delivery: Theory to Practice, Edited by Hong Wen and Kinam Park
Copyright © 2010 John Wiley & Sons, Inc.

dosing, and portability. Compared to the solid dosage forms, any suffocation issue or choking can also be avoided, because there is no physical obstruction when swallowed as a soft paste or liquid. Moreover, good mouth feel and different administration concept may help to change the children's perception on bitter and hard-to-swallow pills to be more fun to take them. The dosage form is now extending to more general patient population with daily medication.

Since the tablets disintegrate in the mouth, active pharmaceutical ingredients can be absorbed in the buccal, pharyngeal, and pregastric regions, which can circumvent first-pass metabolism. Therefore, rapid drug action and increased bioavailability might be expected [8]. However, it is hard to control the amount of swallowed drug, so inconsistent bioavailability is highly probable. The claimed increase in bioavailability might be open to debate. FDTs can provide a new dosage form strategy for drugs nearing the end of their patent life in the pharmaceutical industry. The companies can extend market exclusivity by reformulating an existing product as an FDT [9].

With the above background information, this chapter will introduce the current FDT technologies, products in the market, and also the properties that need to be evaluated for the dosage form development.

10.2 CURRENT FDT PRODUCTS IN THE MARKET

A number of FDTs are commercially available for human use using technologies developed by pharmaceutical companies such as Cardinal Healthcare, Janssen Pharmaceutical (Johnson & Johnson), Cima (Cephalon), Bioavail, and Yamanouchi (Astellas Pharma). These technologies either use expensive processing technology producing fragile tablets that require costly specialized packaging or use conventional compression procedures that give longer than desired disintegration time. Some compressed tablets still need specialized packaging. Table 10.1 shows typical examples of FDT products in the market and their applied fast disintegrating technologies. Figure 10.1 shows the scanning electron microscopic pictures of the selected tablet's inner matrix.

10.3 FDT FORMULATION METHODS

It is important to keep in mind that the critical properties of the FDTs are instant wetting or absorption of water into the tablet matrix and disintegration of associated components into individual particles for easy swallowing. Regardless of the preparation methods, freeze drying or compression, one of the common strategies to achieve rapid disintegration or dissolution is to maintain a highly porous tablet structure, which will ensure fast water absorption into the matrix.

Moreover, suitable tablet excipients need to have high "wetting properties" to improve the water absorption. When compressed, the porosity of the resulting tablets is inversely related to compression pressure. Therefore, it is also important to find the optimum porosity that allows both fast water absorption and high mechanical strength. Since low compression pressure resulted in fragile tablets, a formulation and/or processing strategy will be necessary to increase tablet mechanical strength without compromising porosity or necessitating special packaging. Pharmaceutical scientists have been trying to develop new and improved fast disintegrating technologies and examples will be introduced briefly.

10.3.1 Lyophilization (Freeze Drying)

Lyophilization is a pharmaceutical process in which water is sublimated from the product after freezing. It allows drying of drugs and biologicals, which are heat sensitive, at low temperature under conditions that can remove water by sublimation. Lyophilization can result in highly porous pharmaceutical preparations with a very high specific surface area, which dissolve instantly and show improved absorption (Figure 10.1a). It can also remove solvents from a frozen solution or suspension of drug with structure-forming excipients. The process imparts amorphous structure resulting in highly porous and lightweight product, which can be exploited in formulating FDTs. The freeze-dried unit dissolves instantly to release the drug when placed on the tongue.

There are some limitations to the lyophilization technology. The process of freeze drying is a relatively expensive process and also the formulation is very lightweight and fragile. Moreover, the formulation readily absorbs water and is very sensitive to degradation at high humidity and temperature. If there is any damage to the package, moisture can pass through freely and the lyophilized product will be collapsed due to absorption of the moisture. Patients should be advised to pay attention when storing the freeze-dried tablets [7, 8]. Moreover, patients need to be advised not to push the tablets through the foil film, instead peel the film back to release the tablet. Several technologies including Zydis, Lyoc, and QuickSolv have been patented as the lyophilization process, and they will be covered later. Desired characteristics of the lyophilization technology are as follows [9]:

- Drug should be chemically stable and water insoluble.
- Dose for water-soluble drugs is limited (about 60 mg).
- Particle size is smaller than 50 μm.

10.3.2 Molding Method

Molding process includes moistening, dispersing, or dissolving the active ingredient with a hydroalcoholic solvent,

TABLE 10.1   Typical Examples of FDT Products on the Market and Their Applied Fast Disintegrating Technologies

Drug Products	Active Ingredient	Indication	Company	Technology	Technology Company
Claritin RediTabs	Loratadine	Allergy	Schering-Plough		
Clarinex RediTabs	Desloratadine	Allergy	Schering-Plough		
Dimetapp	Brompheniramine, phenylpropanolymin	Cold, allergy	Wyeth		
Feldene Melt	Piroxicam	NSAID, pain	Pfizer		
Imodium Lingual	Loperamide	Antidiarrhea	J&J		
Innovace Melt	Enalapril maleate	Hypertension	Merck	Zydis	Cardinal Health
Maxalt-MLT	Rizatriptan benzoate	Migraine	Merck		
Motilium	Domperidone maleate	Motion sickness	J&J		
Pepcid RPD	Famotidine	Antiulcer	Merck		
Seresta Expidet	Oxazepam	Tranquilizer	Wyeth		
Zofran ODT	Ondansetron	Nausea (cancer chemo induced)	GSK		
Zyprexa VeloTab®	Olanzapine	Antipsychotic	Eli Lilly		
Seglor Lyoc	Dihydroergotamine mesylate	Migraine	Sanofi	Lyoc	Cephalon
Propulsid QuickSolv	Cisapride monohydrate	Gastroesophageal reflux disease	Janssen (J&J)		
Risperdal M-TAB/ Quicklet	Risperidone	Schizophrenia, bipolar mania	Janssen (J&J)	QuickSolv	Janssen (J&J)
Fazalco	Clozapine	Schizophrenia	Alamo		
Remeron SolTab	Mirtazapine	Antidepressant	Organon, Solvay	OraSolv	
Tempra® Qicklets/ FirsTabs	Acetaminophen	Analgesic	BMS		Cima (Cephalon)
Alavert	Loratadine	Allergy	Wyeth		
Nulev	Hyoscyamine sulfate	Irritable bowel	Schwarz Pharma	DuraSolv	
Zomig ZMT	Zolmitriptan	Antimigraine	AstraZeneca		
Benadryl Fastmelt	Diphenhydramine pseudoephedrine	Allergy, cold, sinus	Johnson & Johnson	WOW Tab	Yamanouchi (Astellas)
Paroxetine FD	Fluoxetine	Depression, obsessive compulsive disorder	Biovail	Flashdose	Biovail
Zolpidem FD	Zolpidem tartrate	Insomnia	Biovail		
Excedrin Quicktabs	Acetaminophen, caffeine	Pain	BMS		
Febrectol & DolFlash	Acetaminophen	Pain, antipyretic	Ethypharm	FlashTab	Ethypharm
Prevacid SoluTab	Lansoprazole	Duodenal ulcer	TAP		

molding the moist mixture into tablets under pressure lower than that of the conventional tablet (compression molding), and then removing the solvent by air drying [10]. As the compression force employed is lower than conventional tablets, the molded tablet results in highly porous structure, which can decrease the disintegration time of the product. Usually, molding process is employed with soluble ingredients (saccharides) for improved mouth feel and disintegration of the tablets. However, molded tablets still have low mechanical strength, which can be an issue during delivery and handling [11]. Moreover, evaporation may occur before compression. There are some issues related to compression including stickiness and adhesiveness because of the high moisture content of granules [12].

10.3.3   Compression Method

While the lyophilization and molding methods are effective, they are time consuming and technically difficult, often requiring special processing equipments. Moreover, the resulting tablets are usually very weak and friable, so it may not withstand general packaging, transportation, and patient handling. Therefore, it is necessary to develop a new formulation technology to obtain FDTs by using conventional compression method, which is attractive because of the low manufacturing cost and ease of technology transfer. Many strategies have been investigated to adapt the traditional tablet press to FDT preparation, and to acquire both high porosity and adequate tablet strength [13–15]. Typical examples will be introduced in this section. Most of them are

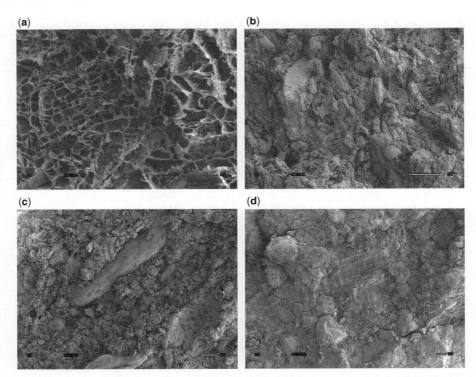

FIGURE 10.1   Scanning electron microscopic pictures of typical fast disintegrating tablets: Claritin® RediTabs (a), Alavert™ (b), Benadryl® Fastmelt (c), and Excedrin® Quicktab (d).

making use of physical or chemical changes of specific excipients and many formulations may include two or more examples together to achieve the desirable properties [16].

10.3.3.1 Addition of Tablet Disintegrants

Disintegrants help a tablet to break up into individual pieces upon contact with aqueous solution, which is called disintegration, and every compressed tablet has a certain level of them. Disintegration starts when a small amount of water or saliva wets the dosage form and penetrates the tablet matrix by capillary action. Especially, fast disintegration of a tablet matrix in the oral cavity facilitates swallowing and increases the surface area of the tablet particles, which enhances the rate of absorption of the active ingredients [17].

Most disintegrants swell to some extent and the swelling pressure is generally considered as the major factor for tablet disintegration [18, 19]. One of the most desirable properties of the disintegrants is rapid swelling without gel formation (no viscosity increase), since high viscosity on the surface of the tablet will prevent water penetration into the tablet matrix. High swelling capacity along with high water penetration leads to fast tablet disintegration. Disintegrants are usually water-insoluble polymers and remain as granules in the mouth, so the addition of excess disintegrants may lead to "sandy feeling grittiness" on the tongue after tablet disintegration.

There are a lot of disintegrants and superdisintegrants in the market. Typical examples include crospovidone (cross-linked polyvinylpyrrolidone), croscarmellose (cross-linked cellulose), sodium starch glycolate (cross-linked starch), and low-substituted hydroxypropylcellulose [18, 19]. Kollidon and Polyplasdone products are crospovidone of BASF and ISP, respectively. Ac-Di-Sol® (croscarmellose sodium) is an internally cross-linked sodium carboxymethylcellulose and Primojel® is sodium starch glycolate produced by cross-linking and carboxymethylation of potato starch (Table 10.2).

Selection of the right disintegrants depends on the formulation and preparation procedures applied. For example, when FDTs were prepared using lactose as a diluent, *in vivo* disintegration time was dependent on the type of disintegrants including Ac-Di-Sol, L-HPC (low-substituted hydroxylpropylcellulose), ECG-505 (carmellose calcium), Polyplasdone XL-10 (crospovidone), and carmellose (NS-300) [20]. NS-300 gave the fastest disintegration of the five evaluated, and disintegration time was not affected by tablet hardness at the tested levels. In another example, croscarmellose sodium was superior to crospovidone, Indion 414, and sodium starch glycolate [21]. As mannitol and crospovidone were used together for direct compression method, an optimum tablet formulation with mannitol (34%) and crospovidone (13%) was disclosed to give a short wetting time and a sufficient crushing strength [22].

When direct compression was utilized with microcrystalline cellulose (MCC) and low-substituted hydroxypropyl

TABLE 10.2   Overview of Typical Disintegrants with Their Particle Size and Swelling Pressure [19]

Disintegrants		Particle Size (μm)	Swelling Pressure (kPa)	Company
Crospovidone	Kollidon CL	110–130	170	
	Kollidon CL-F	20–40	30	BASF
	Kollidon CL-SF	10–30	25	
	Kollidon CL-M	3–10	70	
	Polyplasdone XL	100–130	110	ISP
	Polyplasdone XL-10	30–50	95	
Croscarmellose sodium	Ac-Di-Sol	49	271	FMC
Sodium starch glycolate	Primojel	41	158	DMV

cellulose (L-HPC) as disintegrants, the MCC/L-HPC ratio in the range of 8:2–9:1 showed the quick disintegration time [23, 24]. Poly(acrylic acid) superporous hydrogel microparticles were also reported to have a unique porous structure and they were added as a wicking agent to decrease disintegration time [25].

10.3.3.2   Crystalline Transition   The crystalline transition method (CTM) utilizes the phase transition of pharmaceutical excipients, especially sugars, from the amorphous to crystalline state to increase tablet mechanical strength while keeping porosity [16]. Freeze drying or spray drying method can be used to prepare the amorphous state. As amorphous forms of sugars have higher compressibility than crystalline forms, they can contribute to high mechanical strength [26, 27]. However, amorphous sugars have a tendency to absorb more moisture than crystalline ones. Moreover, an amorphous state is metastable and tends to convert to a thermodynamically stable crystalline state over time [28]. The absorbed moisture may act as a plasticizer for the amorphous sugars and facilitate the rate of crystalline transition [28]. For example, an FDT was prepared by compressing a mixture of mannitol and amorphous sucrose (Figure 10.2) [29]. The amorphous sucrose can be regarded as a binder and the mannitol as a diluent. The blend of the two was compressed and exposed to various conditions of temperature and humidity to induce the phase transition [29]. Higher storage

temperature or humidity caused a faster moisture uptake and also crystallization of the amorphous material resulting in the faster increase of the tablet mechanical strength [29–31]. It also showed the faster the crystalline transition, the faster the rate of increase in tablet tensile strength [29–31].

The mechanism of the CTM can be understood by using the moisture sorption model [32]. Amorphous sucrose is hygroscopic, so it can absorb ambient moisture leading to the formation of hydrated amorphous material. The absorbed water can play a role as a plasticizer, which can influence the free volume due to breakage of hydrogen bonds between the molecules in the solid [28]. The above can lower the glass transition temperature to or below the operation temperature changing it from a glassy to a rubbery state [33]. The hydrated amorphous material with increased physical reactivity may lose a relative amount of moisture and the crystallization of the amorphous sucrose may be induced with the release of the absorbed water. In summary, the hydrated amorphous sucrose in an FDT can be converted into the crystalline form and the crystalline sucrose forms new internal contact points in the tablet [29, 30]. The tablets' mechanical strength increases with an increasing percentage of amorphous sucrose due to its good compressibility [29, 31]. However, the higher amorphous content causes a longer disintegration time in the mouth. Moreover, it affected the internal structure of the tablets; the tablets with higher initial porosity shrank, whereas the

FIGURE 10.2   Schematic view of the crystalline transition method to prepare FDT using mannitol and amorphous sucrose (modified from Ref. 29).

tablets with lower porosity expanded due to the recrystallization of the sucrose [31].

Conditioning of tablets at a certain temperature and humidity was also investigated, and involved different kinds of pharmaceutical polymers, such as polyvinylpyrrolidone (PVP), or other excipients [34–36]. Although crystal transition may not occur in the case of polymers [33], the mechanical strength of the tablets can be increased significantly with humidity conditioning. Highly water-soluble polymers absorb moisture and can form new contact points as the amorphous sugars do.

10.3.3.3   Phase Transition   Saccharides and sugar alcohols can be divided into two groups, low and high melting sugars, based on their melting points. The mechanical strength of tablets can be improved with relevant mixing level of both and also the processing conditions [37]. For example, xylitol is a low melting (93–95°C) and erythritol is a high melting (122°C) sugar alcohol (Figure 10.3). Xylitol and erythritol were utilized as a binder and a diluent, respectively, for fluid bed granulation. After compaction, the resulting tablets were put in a drying oven and heated to a temperature close to the melting point of the lower one (xylitol, approximately 93°C). After maintaining for a certain period of time, the tablets were allowed to cool to room temperature. The mechanical strength of the processed tablets increased with higher low melting (xylitol) content [37].

The heating process may affect tablet hardness and disintegration time, as the content of the saccharides or sugar alcohols does [37]. Heating increased pore size within the tablets and it was suggested that the diffusion of xylitol in the tablet resulted in the increased pore size. Moreover, as xylitol melted, diffused, and solidified in the heated tablets, it caused greater bonding surface area between the particles and also increased hardness. For example, tablets containing about 5% xylitol showed hardness of 4 kp and the oral disintegration time of less than 30 s [38]. The increasing tablet hardness by heating and storage was not dependent on the crystal state of the sugar alcohols, but related to the formation of interparticle bonds or the increased bonding surface area induced by the melting of xylitol particles and their subsequent solidification upon cooling [37]. Other pharmaceutical ingredients, such as polyethylene glycol, and wax, have been also utilized for the PTM [39, 40].

10.3.3.4   Sublimation   When volatile solids are compressed into tablets together with other tablet excipients using a conventional method, they can be removed by sublimation resulting in highly porous structures. Fast dis-

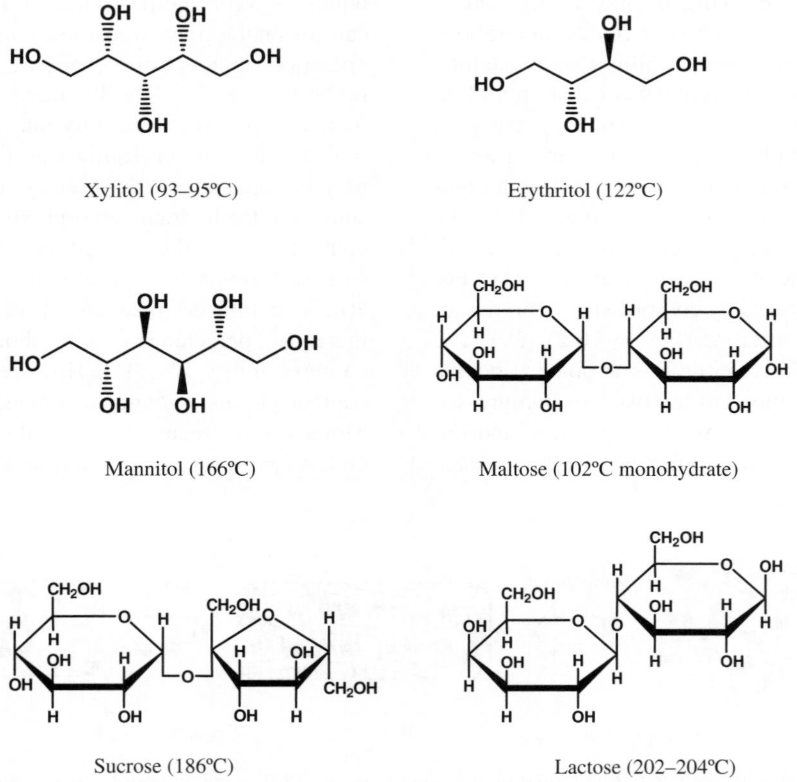

FIGURE 10.3   Molecular structure and melting point of typical pharmaceutical sugars.

integration of compressed tablets can be facilitated as more water penetrates into the tablet matrix. Therefore, giving higher porosity on the compressed tablets is one of the methods to increase the water penetration. Typical materials may include camphor, menthol, thymol, urea, ammonium carbonate, and ammonium bicarbonate [41–43]. As an example, camphor was incorporated into FDTs and then sublimated in a vacuum oven resulting in highly porous tablets with a porosity of up to 40% [44]. However, the vacuum process is still costly and also time consuming.

10.3.3.5 *Effervescence*

Effervescent tablets are making use of a chemical reaction between a soluble acid and an alkali metal carbonate with the help of water to make CO_2 gas, which can give a "fizzy taste" in the mouth. The CO_2 gas generated in the tablet can also improve the overall taste. The reaction is spontaneous as saliva is supplied to the materials and the small amount of saliva will be enough to start the reaction. Moreover, the process is a chain reaction with water, which is one of the reaction products as following [16]:

$$R\text{-COOH} + NaHCO_3 + H_2O \rightarrow R\text{-COONa} + H_2CO_3$$

$$+ H_2O \rightarrow R\text{-COONa} + CO_2 + 2H_2O$$

Effervescence reaction also can occur with moisture from the atmosphere during storage. Therefore, special packaging will be necessary to protect them from the moisture. Another limitation is that only chemically stable drugs under acid and alkali conditions could be incorporated. Typical acidic components are food acids (citric acid, tartaric acid, and malic acid), acid anhydride (succinic anhydride), and acid salts (NaH_2PO_4, $Na_2H_2P_2O_7$). Carbonate sources are bicarbonate and carbonate forms. Among the various acids, citric acid is the most frequently used acidic component. Typical carbonates are sodium bicarbonate, sodium carbonate, potassium bicarbonate, and potassium carbonate.

The effervescent carbon dioxide can facilitate tablet disintegration quickly, so the effervescent couples can be regarded as a disintegrant in the FDT formulation. OraSolv® technology is a good example of using the effervescent excipients [45, 46].

10.3.3.6 *Spray Drying*

Highly porous, fine powders can be obtained by a spray drying method. An FDT was prepared utilizing this process and the formulation consisted of hydrolyzed/unhydrolyzed gelatin as a supporting agent for matrix, mannitol as a bulking agent, and sodium starch glycolate or croscarmellose sodium as a disintegrating agent [47]. The formulation was spray dried to yield a porous powder and the resulting tablets showed fast disintegration, less than 20 s [48, 49]. Disintegration and dissolution were improved further by adding effervescent components, citric acid and sodium bicarbonate.

10.4 PATENTED TECHNOLOGIES

10.4.1 Zydis

Zydis technology was patented by R. P. Scherer and now Cardinal Health (Dublin, Ohio). Zydis is the first marketed fast disintegrating dosage form. The tablet literally dissolves in the mouth within seconds after placement on the tongue [7, 9, 50]. A Zydis tablet is prepared by lyophilizing or freeze drying the drug in a matrix composed of a saccharide and a polymer [51, 52]. Polymers can be partially hydrolyzed gelatin, hydrolyzed dextran, dextrin, alginates, polyvinyl alcohol, polyvinylpyrrolidine, and/or acacia. Solution or dispersion of tablet components is prepared and filled into blister cavities and then frozen in a liquid nitrogen environment. The frozen solvent is freeze dried to produce porous wafers and then peelable backing foil is used to pack the tablets. Since the product is very lightweight and fragile, it must be dispensed in a special blister pack. Moreover, the tablets are sensitive to moisture and may degrade at higher humidity. Due to the low moisture content in the final freeze-dried product, the Zydis formulation does not allow microbial growth and hence self-preserving [9]. Many products are currently available using the Zydis technology [7] including Claritin Reditab, Dimetapp Quick Dissolve, Feldene Melt, Maxalt-MLT, Pepcid RPD, Zofran ODT, Zyprexa Zydis, and so on (Table 10.1) [53].

10.4.2 Lyoc and QuickSolv

Other technological examples for manufacturing FDTs by lyophilization are Lyoc (Farmalyoc, now Cephalon, Franzer, PA) and QuickSolv (Janssen Pharmaceutica, Beerse, Belgium). Lyoc tablet is a porous, solid wafer prepared by lyophilizing an oil-in-water emulsion placed directly in a blister [54]. The wafer can accommodate high drug dosing and disintegrates rapidly. QuickSolv is patented by Janssen Pharmaceuticals (Beerse, Belgium). QuickSolv tablets are made with a similar technology that makes a porous solid matrix by freeze drying an aqueous dispersion or solution of the matrix formulation [55]. The process works by removing water using an excess of alcohol (solvent extraction).

10.4.3 OraSolv

OraSolv is Cima's (now Cephalon) first fast disintegrating dosage form [56, 57]. As mentioned earlier, the OraSolv technology is making use of the effervescence to facilitate the disintegration in the saliva. Moreover, the unpleasant flavor of a drug might be compromised by the effervescence together with sweeteners or flavors, even though coating the drug powder is the major tool of taste masking. The major limitation of this technology is its low mechanical strength as tablets are prepared at low compression pressure. It yields weak and brittle tablets compared to conventional tablets, so Cima developed a special handling and packaging system, PakSolv [58]. It is to protect the OraSolv tablets from breaking during storage and delivery. PakSolv is a dome-shaped blister package preventing vertical movement of the tablet and offering moisture, light, and child resistance. With the low compression pressure, the particle coating for the taste masking is not compromised by fracture during processing.

10.4.4 DuraSolv

DuraSolv is Cima's second-generation fast disintegrating tablet formulation [59]. The tablets are prepared similar to OraSolv, but it utilizes the conventional compression process. Therefore, DuraSolv has much higher mechanical strength. Tablets are formulated by using a drug, nondirect compression fillers, and lubricants. Nondirect compressible fillers are generally directly compressible sugars, which can be dextrose, mannitol, sorbitol, lactose, and sucrose. They have advantage of quick dissolution and avoiding gritty texture on the tongue. The tablets are strong enough to be packed in bottles and blisters. However, pharmacists are advised to take care when dispensing such formulations from the bottles to ensure that they are not exposed to high levels of moisture or humidity. Excess handling of tablets can introduce enough moisture to initiate dissolution of the tablet matrix [60]. Another limitation of DuraSolv is that the technology is not compatible with larger doses of active ingredients. Since the high compaction pressures are applied, the structural integrity of any taste masking powder coating in DuraSolv may become fractured during compaction, exposing the bitter-tasting drug to a patient's taste buds. The technology might be suitable for formulations including relatively small doses of active ingredients [61].

10.4.5 WOW Tab

The WOW Tab technology was developed by Yamanouchi (now Astellas Pharma, Japan) and WOW stands for without water. The technology was in the Japanese market first and it was introduced into the U.S. market later. This technology utilizes conventional granulation and compression methods to prepare FDTs with low- and high-moldable saccharides (Table 10.3) [62, 63]. The two different types of saccharides are combined to obtain a tablet formulation with adequate hardness and fast disintegration rate. Low-moldable saccharides are lactose, mannitol, glucose, sucrose, and xylitol. High-moldable saccharides are maltose, maltitol, and sorbitol. When these saccharides are utilized alone, the resulting tablets do not have desired properties of fast disintegration and hardness. Tablets were compressed with granules prepared using low-moldable saccharides with high-moldable saccharides as a binder and the tablets were processed further

TABLE 10.3 Compression Characteristics of Various Saccharides with Their Disintegration Time [62, 63]

Moldability	Saccharides	Hardness (kp)	Disintegration Time (*in vivo*, s)
Low	Mannitol	0	<10
	Lactose	0	<10
	Erythritol	0	<10
	Xylitol	0	<10
	Sucrose	0.5	<10
High	Sorbitol	2.2	>30
	Maltitol	2.5	>30
	Trehalose	3.4	>30
	Maltose	6.8	>30

by moisture treatment. They showed adequate hardness and rapid disintegration, so they are suitable for both conventional bottle and blister packaging. The taste masking technology utilized in the WOW Tab is proprietary, but claims to offer superior feeling due to the patented SmoothMelt action (Yamanouchi Pharma Technologies, Inc., June 2001, www.ypharma.com). Typical formulations currently in the U.S. market are Benadryl Allergy & Sinus FastMelt and Children's Benadryl Allergy & Cold FastMelt.

10.4.6 Flashdose

Fuisz (Fuisz Technologies, Chantilly, VA) developed Flashdose technology, which utilizes a unique spinning mechanism to produce a floss-like crystalline structure called "cotton candy" [64, 65]. Active ingredients can be added to this crystalline sugar and then compressed into a tablet. This procedure is also known as Shearform termed as floss. The final product has a very high surface area and disperses or dissolves quickly once placed on the tongue [5]. Nurofen Meltlet is the first commercial product launched by Bioavail Corporation using Flashdose technology. Moreover, the properties of the product can be altered significantly by changing the temperature and other conditions during the manufacturing. Instead of a floss-like structure, small spheres of sugar can be produced to carry the drug and the process of making the microspheres has been patented and is known as Ceform [5]. It can be an alternative way of taste masking.

10.4.7 Frosta

Frosta technology was developed by Akina (West Lafayette, IN) and it utilizes conventional wet granulation and compression processing for cost-effective production of FDTs [66]. The technology used the concept of formulating plastic granules and compressing at relatively low pressure to prepare strong tablets with high porosity (Figure 10.4) [67]. Plastic granules composed of porous plastic material, water

FIGURE 10.4 Scanning electron microscopic picture of Frosta tablet's inner matrix showing high porosity.

penetration enhancer, and binder. The tablets obtained have good hardness and rapid disintegration time ranging from 15 to 30 s depending on the size of the tablet. Frosta tablets are claimed to be strong with friability much less than 1% and stable in open air environment. Conventional tablet machines are used for the production, and no other special instruments are necessary. The suggested benefits are [66]

- fast melting in the mouth (melting within 5–20 s for easy swallowing);
- extremely cost effective (the same as making conventional tablets);
- simple processing (one-step wet granulation processing before making tablets);
- strong mechanical property (friability less than 1%);
- possible of multitablet packaging (dozens of tablets in one bottle).

10.4.8 FlashTab Technology

FlashTab technology was developed by Ethypharm (Saint-Cloud, France) and it combines two key proprietary technologies: ease of administration and taste masking [68]. Moreover, the technology uses regular granulation and compression process. FlashTab tablets consist of microparticles that disperse quickly in the mouth thereby allowing the rapid and safe release of the active ingredients into the body. Two types of excipients are used in this technology: one is disintegrating agents including reticulated polyvinylpyrrolidine or carboxymethylcellulose and the other is swelling agents such as carboxymethylcellulose, starch, modified starch, microcrystalline cellulose, and carboxymethylated starch. The resulting tablets show decent physical resistance with disintegration time within 1 min.

10.4.9 Pharmaburst Technology

The Pharmaburst was patented by SPI Pharma (New Castle, DE) and the technology uses the coprocessed excipients to prepare FDTs. The manufacturing process involves dry blending of active ingredients with other excipients including flavor and lubricant, which is followed by compression into tablets [50]. The process is claimed to be carried out under normal temperature and humidity. Resulting tablets have sufficient strength so they can be packed in blister packs or bottles.

10.5 SELECTION OF ACTIVE INGREDIENTS

The major issue of incorporating a drug into FDT is taste and the drug needs to have no or acceptable taste for the patients. Otherwise, taste masking should be utilized. If the drug has short half-life with frequent dosing, it may not be a good candidate for the formulation development. Moreover, the drug needs to have good stability in saliva (mainly stable at the saliva pH and aqueous environment) and decent permeability. Sustained or controlled release might be another issue.

The FDT technology should be able to accommodate a wide range of drugs' physicochemical properties [16]. However, a drug's properties affect tablet performance: solubility, crystal morphology, particle size, and bulk density of a drug can affect tablet characteristics, such as strength and disintegration. FDTs should have low sensitivity to humidity, but many water-soluble excipients are utilized to enhance fast dissolving/disintegrating properties as well as to create good mouth feel. Some of those highly soluble excipients may attract moisture and may cause an effect on water-sensitive active ingredients.

10.6 TASTE MASKING

Taste masking is one of the most important areas in the preparation of the FDTs [10]. During the disintegration process in the patient's mouth, the drug will be partially dissolved in close proximity to the taste buds. Unless the drug is tasteless or does not have an undesirable taste, taste masking techniques should be used [10]. An ideal taste masking technology should not give grittiness and should produce good mouth feel. The amount of taste masking materials used in the dosage forms should be minimized to avoid excessive tablet size. The technology should also be compatible with other components and properties of the formulation.

Approaching methods for the taste masking may include addition of sweeteners and flavors, encapsulation of the unpleasant drug, and adjustment of pH. Typical sweeteners are sugar-based excipients, which are highly water soluble, dissolve quickly in saliva, and provide pleasant taste and mouth feel [5]. Sugars can be used as diluents and binders, as well as taste improving agents. For example, mannitol is one of the most common excipients for the dosage form because it is water soluble, nonhygroscopic. It also produces a unique cooling sensation in the mouth and has a pleasant taste when chewed or dissolved [69]. Aspartame and citric acid are commonly used along with various flavors such as mint, orange, strawberry flavor, and peppermint flavor to impart pleasant taste. Some of the unpleasant active ingredients cannot be masked by incorporation of sweeteners and flavors. In those cases, an alternative method of taste masking is to encapsulate or coat the active ingredient with acceptable polymers or excipients. This process retards or prevents dissolution of the coated drug before taste is perceived in the mouth [70]. Various techniques for this purpose were introduced already in the patented technologies.

10.7   EVALUATION OF FDTS

Some of the tablet evaluation parameters introduced in the pharmacopoeias can be assessed, along with some specific "in-house" tests. Tablet mechanical strength can be measured simply by hardness tester or using the tablet tensile strength (T), which can be calculated from the following equation:

$$T = \frac{2F}{\pi \times d \times t}$$

where d and t represent the diameter and thickness of the tablet, respectively. The crushing strength of the tablet may be measured using Texture analyzer (Texture Technologies Corp., Scarsdale, NY). The force that causes a breakage of a tablet in the radial direction can be taken as the crushing load (F) for the tablet [62]. Enough mechanical strength of FDTs is difficult to achieve due to the specialized processes and ingredients used for the preparation. Therefore, the limit of crushing strength is usually kept in a lower range to facilitate quick disintegration in the mouth.

Friability is quite a challenging task to propose a guideline because many preparation methods of FDT are responsible for increasing the percentage (%) of friability. Therefore, it is necessary that this parameter should be evaluated carefully and be proposed with the limits. Tablet porosity ε can be calculated using the following equation:

$$\varepsilon = 1 - \frac{m}{\rho_t V}$$

where ρ_t is the true density, and m and V are the weight and volume of the tablet, respectively. True density can be measured using helium pycnometer. Tablet porosity may

give valuable information regarding the mechanical strength and friability, and also the tablet wetting time.

Tablet wetting time is another important parameter, which can be assessed to give an insight into the disintegration properties of the tablet. Faster wetting implies a quicker disintegration of the tablet. Wetting time of the dosage form is related to the contact angle. This parameter can be measured using various "in-house" methods. As an example, a tablet can be put on a prewetted filter paper placed in a Petri dish with water and the time for complete wetting can be measured using a stopwatch [71]. In order to differentiate the time more easily, relevant dyes might be added to the wetting solution. For the evaluation of water absorption ratio, the weight of the dry tablet before putting in the Petri dish (W_b) and the weight of the wet tablet from the Petri dish (W_a) can be measured and the water absorption ratio R can be calculated according to the following equation:

$$R = \frac{W_a - W_b}{W_b} \times 100$$

The time for disintegration of FDTs is generally less than 1 min, ideally less than 30 s, and actual disintegration time that patients could experience ranges from 5 to 30 s depending on the formulations. The standard procedure for the disintegration has limitations and it is not ideal to measure very short disintegration times. The disintegration test should mimic disintegration in mouth with limited amount of saliva and its contents. FDTs fall under the general guidance of the USP method 701 for disintegration, which is the current regulatory condition for meeting the "definition" of ODT. This is currently under review by the FDA, since this dissolution test is too rigorous for ODTs due to their fast disintegration time. Alternatively, texture analyzer instrument (TA) (Stable Micro Systems, UK) could be utilized to measure the beginning and ending time of disintegration [72]. A tablet was adhered to the bottom of a probe, which is attached to the load cell with a double-sided scotch tape. The tablet under a constant force was immersed in a defined volume of water. The time for the tablet to disintegrate was determined by measuring the distance the probe traveled into the tablet. Typical time–distance profiles generated by the TA software enabled the calculation of beginning and ending of the disintegration time (Figure 10.5).

Dissolution methods for FDTs are comparable and practically identical to the ones adopted for conventional tablets. USP dissolution apparatus 1 and 2 can be used, which are basket and paddle apparatus, respectively. Generally, the dissolution of FDT is very fast when using USP monograph conditions, so slow paddle speeds may be used, such as a paddle speed of 50 rpm.

Additionally, it is necessary to study the moisture uptake for FDT to evaluate the stability of the formulation. A specific

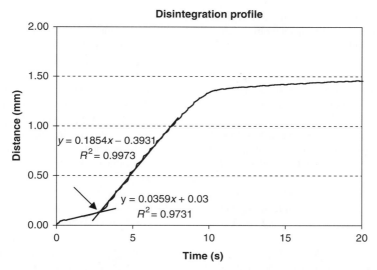

FIGURE 10.5 Interpretation of disintegration profile. The curve has three different areas. The first straight line illustrates the physical resistance of the tablet before the beginning of disintegration. When the tablet starts to disintegrate, in order to maintain a constant force, the probe needs to travel a certain distance, which is represented by the second line. Finally, the profile approaches to a constant distance and it represents the time where disintegration of the tablet is finished. Using a simple linear regression model, the best fitting straight lines can be defined and by projecting the two slope intercepts, the *in vitro* disintegration time can be calculated (2.8 s in the graph).

number of tablets from each formulation, which are dried already, are weighed and exposed to high relative humidity (typically 75%) at room temperature for weeks. Tablets are weighed and the percentage increase in weight can be monitored.

10.8 CONCLUSION

Many people, especially the geriatric and pediatric populations, have swallowing issues resulting in poor compliance with oral solid dosage forms, which may lead to reduced overall therapeutic efficacy. A new dosage form, the fast disintegrating tablet has been developed, which can offer the advantages of both solid and liquid dosage forms. The tablets remain solid during storage, maintaining good stability of dosage forms, and transform into liquid within few seconds after their administration, which facilitates swallowing of the dosage form in the absence of water. The dosage forms have drawn the attention of pharmaceutical industry over a decade because they have potential advantages over conventional dosage forms, such as improved patient compliance, convenience, and bioavailability with rapid onset of action [73–75]. Some of the technologies can offer FDT formulations with sufficient mechanical strength and quick disintegration/ dissolution in the mouth without water. Due to the constraints of the current FDT technologies as mentioned so far, there are still needs for improved formulations and manufacturing processes for the FDTs that are mechanically strong, allowing ease of handling and packaging with competitive production cost similar to that of conventional tablets.

REFERENCES

1. Sastry SV, Degennaro MD, Reddy LK, Khan MA. Atenolol gastrointestinal therapeutic system. I. Screening of formulation variables. *Drug Dev. Ind. Pharm.* 1997;23:157–165.
2. Li VHK, Lee VHL, Robinson JR. Influence of drug properties and routes of drug administration on the design of sustained and controlled release systems. In *Controlled Drug Delivery— Fundamentals and Application*, Marcel Dekker, New York, 1987, pp. 3–94.
3. Fasano A. Novel approaches in oral delivery of macromolecules. *J. Pharm. Sci.* 1998;87:1351–1356.
4. Lindgren S, Janzon L. Dysphagia: prevalence of swallowing complaints and clinical findings. *Med. Clin. North Am.* 1993;77:3–5.
5. Chang RK, Guo X, Burnside BA, Couch RA. Fast-dissolving tablets. *Pharm. Technol.* 2000;24:52–58.
6. FDA Guidance for Industry Orally Disintegrating Tablets, April 2007.
7. Habib W, Khankari R, Hontz J. Fast-dissolve drug delivery systems. *Crit. Rev. Ther. Drug Carrier Syst.* 2000;17: 61–72.
8. Bogner RH, Wolpsz MF. Fast-dissolving tablets. *US Pharm.* 2002;27:34–43.
9. Seager H. Drug-delivery products and the Zydis fast-dissolving dosage form. *J. Pharm. Pharmacol.* 1998;50:375–382.
10. Dobetti L. Fast-melting tablets: developments and technologies. *Pharm. Technol. North Am.* 2001;44–50.
11. Van Scoik KG. Solid Pharmaceutical dosage in tablet triturate form and method of producing the same. US Patent 5,082,667, 1992.

12. Abdelbary G, Prinderre P, Eouani C, Joachim J, Reynier JP, Piccerelle P. The preparation of orally disintegrating tablets using a hydrophilic waxy binder. *Int. J. Pharm.* 2004;278: 423–433.

13. Shimizu T, Nakano Y, Morimoto S, Tabata T, Hamaguchi N, Igari Y. Formulation study for lansoprazole fast-disintegrating tablet. I. Effect of compression on dissolution behavior. *Chem. Pharm. Bull.* 2003;51:942–947.

14. Shimizu T, Kameoka N, Iki H, Tabata T, Hamaguchi N, Igari Y. Formulation study for lansoprazole fast-disintegrating tablet. II. Effect of triethyl citrate on the product. *Chem. Pharm. Bull.* 2003;51:1029–1035.

15. Shimizu T, Sugaya M, Nakano Y, Izutsu D, Mizukami Y, et al. Formulation study for lansoprazole fast-disintegrating tablet. III. Design of rapidly disintegrating tablets. *Chem. Pharm. Bull.* 2003;51:1121–1127.

16. Jeong SH, Takaishi Y, Fu Y, Park K. Materials properties for making fast dissolving tablets by a compression method. *J. Mater. Chem.* 2008;18:3527–3535.

17. Shanraw R, Mitrevej A, Shah M. A new era of tablet disintegrants. *Pham. Technol.* 1980;4:48–57.

18. Zhao N, Augsburger LL. The influence of granulation on super disintegrant performance. *Pharm. Dev. Technol.* 2006;11:47–53.

19. Quadir A, Kolter K. A comparative study of current super-disintegrants. *Pharm. Technol.* 2006;30:s38–s42.

20. Fukami J, Yonemochi E, Yoshihashi Y, Terada K. Evaluation of rapidly disintegrating tablets containing glycine and carboxymethylcellulose. *Int. J. Pharm.* 2006;310:101–109.

21. Amrutkar JR, Pawar SP, Nakath PD, Khan SA, Yeole PG. Comparative evaluation of disintegrants by formulating famotidine dispersible tablets. *Indian Pharm.* 2007;6:85–89.

22. Schiermeier S, Schmidt PC. Fast dispersible ibuprofen tablets. *Eur. J. Pharm. Sci.* 2002;15:295–305.

23. Bi YX, Sunada H, Yonezawa Y, Danjo K, Otsuka A, et al. Preparation and evaluation of a compressed tablet rapidly disintegrating in the oral cavity. *Chem. Pharm. Bull.* 1996;44:2121–2127.

24. Watanabe Y, Koizumi KI, Zama Y, Kiriyama M, Matsumoto Y, et al. New compressed tablet rapidly disintegrating in saliva in the mouth using crystalline cellulose and a disintegrant. *Biol. Pharm. Bull.* 1995;18:1308–1310.

25. Yang S, Fu Y, Jeong SH, Park K. Application of poly(acrylic acid) superporous hydrogel microparticles as a super-disintegrant in fast-disintegrating tablets. *J. Pharm. Pharmacol.* 2004;56:429–436.

26. Vromans H, Bolhuis GK, Lerk CF, Kussendrager KD, Bosch H. Studies on tableting properties of lactose. IV. Consolidation and compaction of spray dried amorphous lactose. *Acta Pharm. Suec.* 1986;23:231–240.

27. Sebhatu T, Ahlneck C, Alderborn G. The effect of moisture content on the compression and bond-formation properties of amorphous lactose particles. *Int. J. Pharm.* 1997;146:101–114.

28. van Campen L, Amidon GL, Zografi G. Moisture sorption kinetics of water soluble substances. I. Theoretical consideration of heat transport control. *J. Pharm. Sci.* 1983;72:1381–1388.

29. Sugimoto M, Matsubara K, Koida Y, Kobayashi M. The preparation of rapidly disintegrating tablets in the mouth. *Pharm. Dev. Technol.* 2001;6:487–493.

30. Sugimoto M, Narisawa S, Matsubara K, Yoshino H, Nakano M, et al. Development of manufacturing method for rapidly disintegrating oral tablets using the crystalline transition of amorphous sucrose. *Int. J. Pharm.* 2006;320:71–78.

31. Sugimoto M, Maejima T, Narisawa S, Matsubara K, Yoshino H. Factors affecting the characteristics of rapidly disintegrating tablets in the mouth prepared by the crystalline transition of amorphous sucrose. *Int. J. Pharm.* 2005;296: 64–72.

32. Makower B, Dye WB. Equilibrium moisture content and crystallization of amorphous sucrose and glucose. *J. Agric. Food Chem.* 1956;4:72–77.

33. Ahlneck C, Zografi G. The molecular basis of moisture effects on the physical and chemical stability of drugs in the solid state. *Int. J. Pharm.* 1990;62:87–95.

34. Tatara M, Matsunaga K, Shimizu T, Maeda S. JPO Patent JP8291051, 1996.

35. Chowhan ZT. Role of binders in moisture-induced hardness increase in compressed tablets and its effect on *in vitro* disintegration and dissolution. *J. Pharm. Sci.* 1980;69:1–4.

36. Chowhan ZT, Palagyi L. Hardness increase induced by partial moisture loss in compressed tablets and its effect on *in vitro* dissolution naproxen, naproxen sodium 2% HPMC. *J. Pharm. Sci.* 1978;67:1385–1389.

37. Kuno Y, Kojima M, Ando S, Nakagami H. Evaluation of rapidly disintegrating tablets manufactured by phase transition of sugar alcohols. *J. Control. Release* 2005;105:16–22.

38. Nystrom C, Karehill PG. Use of tablet tensile strength adjusted for surface area and mean interparticulate distance to evaluate dominating bonding mechanisms. *Powder Technol.* 1986;47:201–209.

39. Lo JB. Preparation of tablets of increased strength. PCT Patent WO013758, 1993.

40. Masuda Y, Mizumoto T, Fukui M. JPO Patent JP11033084, 1999.

41. Heinemann H, Rothe W. Preparation of porous tablets. US Patent 3,885,026, 1975.

42. Roser B, Blair J. Rapidly soluble oral solid dosage forms, methods of making same, and compositions thereof. US Patent 5,762,961, 1998.

43. Lee CH, Woo JS, Chang HC. Rapidly disintegrating tablet and process for the manufacture thereof. US Patent 0,001,617, 2002.

44. Koizumi K, Watanabe Y, Morita K, Utoguchi N, Matsumoto M. New method of repairing high-porosity rapidly saliva soluble compressed tablets using mannitol with camphor, a subliming material. *Int. J. Pharm.* 1997;152:127–131.

45. Wehling F, Schuehle S. Effervescent formulations of base coated acid particles. PCT Patent WO021239, 1994.

46. Wehling F, Schuehle S, Madamala N. Effervescent dosage form and method of administering same. PCT Patent 004757, 1991.

47. Allen LV, Wang B. Process for making a particulate support matrix for making a rapidly dissolving dosage form. US Patent 6,207,199, 2001.

48. Allen LV, Wang B. Process for making a particulate support matrix for making a rapidly dissolving tablet. US Patent 5,587,180, 1996.

49. Allen LV, Wang B, Davis LD. Rapidly dissolving tablet. US Patent 5,807,576, 1998.

50. Fu Y, Yang S, Jeong SH, Kimura S, Park K. Orally fast disintegrating tablets: developments, technologies, taste-masking and clinical studies. *Crit. Rev. Ther. Drug Carrier Syst.* 2004;21:433–476.

51. Gregory GK, Ho DS. Pharmaceutical dosage form packages. US Patent 4,305,502, 1981.

52. Yarwood R, Kearnery P, Thompson A. Process for preparing solid pharmaceutical dosage form. US Patent 5,738,875, 1998.

53. Green R, Kearney P. Process for preparing fast dispersing solid oral dosage form. PCT Patent WO002140, 1999.

54. Lafon L. Galenic form for oral administration and its method of preparation by lyophilization of an oil-in-water emulsion. US Patent 4,616,047, 1986.

55. Gole DJ, Levinson RS, Carbone J, Davies JD. Preparation of pharmaceutical and other matrix systems by solid-state dissolution. US Patent 5,215,756, 1993.

56. Wehling F, Schuehle S. Base coated acid particles and effervescent formulation incorporating same. US Patent 5,503,846, 1996.

57. Wehling F, Schuehle S, Madamala N. Effervescent dosage form with microparticles. US Patent 5,178,878, 1993.

58. Amborn J, Tiger V. Apparatus for handling and packaging friable tablets. US Patent 6,311,462, 2001.

59. Khankari RK, Hontz J, Chastain SJ, Katzner L. Rapidly dissolving robust dosage form. US Patent 6,024,981, 2000.

60. Bi YX, Yonezawa Y, Sunada H. Rapidly disintegrating tablets prepared by the wet compression method: mechanism and optimization. *J. Pharm. Sci.* 1999;88:1004–1010.

61. Sunada H, Bi YX, Preparation evaluation and optimization of rapidly disintegrating tablets. *Powder Technol.* 2002;122:188–198.

62. Mizumoto T, Masuda Y, Fukui M. Intrabuccally dissolving compressed moldings and production process thereof. US Patent 5,576,014, 1996.

63. Mizumoto T, Masuda Y, Kajiyama A, Yanagisawa M, Nyshadham JR. Tablets quickly disintegrating in the oral cavity and process for producing the same. US Patent 6,589,554, 2003.

64. Myers GL, Battist GE, Fuisz RC. Process and apparatus for making rapidly dissolving dosage units and product there from. PCT Patent WO34293-A1, 1995.

65. Cherukuri SR, et al. Quickly dispersing comestible unit and product. PCT Patent WO34290-A1, 1995.

66. Jeong SH, Fu Y, Park K. Frosta®: a new technology for making fast-melting tablets. *Expert Opin. Drug Deliv.* 2005;2:1107–1116.

67. Fu Y, Jeong SH, Park K. Fast-melting tablets based on highly plastic granules. *J. Control. Release* 2005;109:203–210.

68. Cousin G, Bruna E, Gendrot E. Rapidly disintegratable multiparticular tablet. US Patent 5,464,632, 1995.

69. Elshattawy HH, Kildsig DO, Peck GE. Aspartame-mannitol resolidified fused mixture: characterization studies by differential scanning calorimetry, thermomicroscopy, photomicrography and X-ray diffractometry. *Drug Dev. Ind. Pharm.* 1984;10:1–17.

70. Morella AM, Pitman IH, Heinicke GW. Taste masked liquid suspensions. US Patent 6,197,348, 2001.

71. Fu Y, Jeong SH, Park K. Preparation of fast dissolving tablets based on mannose. *PMSE Preprints* 2003;89:821–822.

72. Dor PJ.M, Fix JA. *In vitro* determination of disintegration time of quick-dissolve tablets using a new method. *Pharm. Dev. Technol.* 2000;5:575–577.

73. Wilson CG, et al. The behavior of a fast dissolving dosage form (Expidet) followed by g-scintigraphy. *Int. J. Pharm.* 1987;40:119–123.

74. Fix JA. Advances in quick-dissolving tablets technology employing Wowtab. In IIR Conference on Drug Delivery Systems, 1998.

75. Virely P, Yarwood R. Zydis—a novel, fast dissolving dosage form. *Manuf. Chem.* 1990;61:36–37.

11

BUCCAL DRUG DELIVERY SYSTEMS

JOHN D. SMART

School of Pharmacy and Biomolecular Sciences, University of Brighton, Brighton, UK

GEMMA KEEGAN

Vectura Group plc, Chippenham, Wiltshire, UK

11.1 INTRODUCTION

A major challenge for the pharmaceutical sciences are strategies for the more effective delivery of conventional drugs, improving patient acceptability and therapeutic outcomes, as well as designing formulations to deliver the new products arising from innovations in molecular biology ("biopharmaceutical" products, typically large, hydrophilic and unstable proteins, oligonucleotides, and polysaccharides). The buccal route offers one approach to address these issues. Buccal formulations are typically those delivered to the buccal pouch, the area in the oral cavity between the upper gingivae (gums) and cheek (lined by the buccal mucosa) (Figure 11.1). In some texts, however, the whole oral cavity is referred to as the buccal cavity. In this chapter, drug delivery via the buccal/gingival mucosa will be considered. Sublingual drug delivery, placing a dosage form under the tongue to achieve a rapid onset of action, or orally disintegrating tablets, designed for absorption lower down the gastrointestinal tract for patients who have difficulty in swallowing, will not be discussed.

11.2 ANATOMY AND PHYSIOLOGY OF THE ORAL CAVITY

To appreciate the formulation challenges presented by buccal drug delivery, it is important to understand the environment of the oral cavity. The anatomy and physiology of the oral cavity have been reviewed in many texts [1–3] and will be described briefly below.

The oral cavity is the first functional compartment of the alimentary tract, and has numerous functions: mastication, taste, swallowing, lubrication, digestion, speech, the signaling of thirst, and protection of the body from harmful ingested substances.

The main task of the oral cavity, the fragmentation of large food masses, is performed by the teeth. The teeth are arranged in two arcs, with the free surface of those embedded in the mandible oppose and contact those in the maxilla, and are characterized by a layer of hard material called enamel, which is composed of hydroxyapatite $(Ca_5(PO_4)_3OH)$ arranged in tightly packed hexagonal rods or prisms.

The oral mucosal is covered by a stratified squamous epithelium (Figure 11.2), which is keratinized in region(s) subjected to abrasion, such as the gingivae. This type of epithelium is characterized by a variable number of cell layers, which undergo a morphological and functional transition from the cuboidal basal layer to the flattened surface layers. At the surface, degenerate cells are sloughed off and replaced by cells from the deeper layers. The thickness of the mucosae varies from 500–800 μm for the buccal mucosa (40–50 cell layers) to 100–200 μm for the gingival mucosa. The rate at which the entire thickness of the epithelium can be replaced by this process is known as the turnover time [4], and this has been reported to be 3–8 days [3] and 14–24 days [5]. The nature of the epithelial lining varies according to the tissue function, but generally it behaves as a lipophilic barrier and is considered relatively impermeable.

Oral Controlled Release Formulation Design and Drug Delivery: Theory to Practice, Edited by Hong Wen and Kinam Park

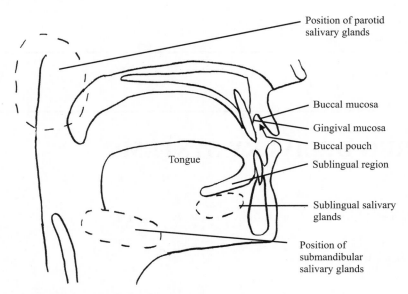

FIGURE 11.1   A schematic diagram of a section through the oral cavity, indicating the buccal mucosa and the position of the major salivary glands.

The keratinization of the oral epithelium occurs as a result of both intrinsic and extrinsic factors such as the expression and regulation of differentiation markers and the heavy abrasion, humidity, and temperature of the oral cavity, respectively [6]. The differentiation of cells within a keratinizing squamous epithelium leads to the production of a layer analogous to the stratum corneum of the skin—a layer of flat, hexagonal cells bounded by a thickened cell envelope and surrounded by an external lipid matrix [4]. This layer can be described as a nonliving proteinaceous material, which is mechanically strong, but flexible, and acts as a physical barrier, particularly to microorganisms and small molecules such as drugs [7]. Occasionally, nonkeratinizing epithelium can become keratinized, for example, persistent cheek biting can cause keratinization of the buccal mucosa. In general, keratinizing epithelium has a slower turnover time than nonkeratinizing epithelium.

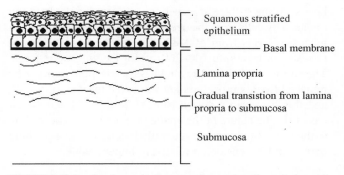

FIGURE 11.2   A schematic diagram of a section through the oral mucosa.

Three different types of oral mucosa are recognized within the oral cavity [4, 8], of which two are relevant for buccal drug delivery.

The lining mucosa covers the soft palate, ventral (lower) surface of the tongue, the floor of the mouth, the alveolar processes excluding the gingivae (gums), and the internal surfaces of the lips and cheeks. It is characterized by nonkeratinizing epithelium, underneath which a loose lamina propria provides an elastic, deformable surface capable of stretching with movement.

The masticatory mucosa covers the gingivae and hard palate and is characterized by keratinized epithelium. These regions are subject to the mechanical forces of mastication, such as abrasion and shearing. The lamina propria is firmly attached to the underlying bone via a meshwork of collagen fibers. The masticatory mucosa is consequently rigid and cannot move freely above the structures it is tethered to.

Whole saliva is the collective product of pairs of parotid, submandibular, and sublingual salivary glands, which are situated near the oral cavity (Figure 11.1) and whose secretions are carried via salivary ducts. In addition, there are numerous minor glands located in the buccal mucosa and palate. The parotid gland duct drains into the *buccal cavity* (the inside of the cheek) opposite the upper second molar, the other two glands drain into the sublingual region. Generally, salivary glands can be considered in terms of the type of secretory cells located in the acini: serous, mucous, or mixed [9–12]. Serous secretions are watery in nature due to the lack of mucin, but contain salts, proteins, and the enzyme amylase. The acini of the parotid salivary glands are composed of serous cells. Mucous secretions are viscid and consist almost entirely of salivary mucins. The minor glands located in the buccal, lingual, and palatine mucosa are composed of mucous cells and although the secretions from these glands account for only 10% of the whole saliva, they can contribute up to 70% of the mucin [10]. Both the submandibular and the sublingual glands are composed of a mixture of secretory cells.

Humans typically produce 1 L of saliva per day [3, 10]. The composition of whole saliva is highly variable and is dependent on a number of factors, including the time of day and the degree and type of stimulation [13]. The secretions from each gland can also vary significantly according to the salivary flow rate, which is regulated by nervous stimulation. During unstimulated states, saliva is secreted at a basal level (about 0.5 mL/min), termed the resting flow rate. It follows a circadian pattern with reduced production in the morning and increased production in the afternoon [10]. During sleep, the salivary flow rate is significantly reduced due to the lack of physical stimulation. Nervous stimulation accounts for 80–90% of the daily production of saliva [14] and flow rates of up to 7 mL/min can be achieved [3, 10].

The constituents that are found within whole saliva can be divided into two major categories: water and electrolytes (inorganic components) and macromolecules (organic components). The ions present within the oral cavity include sodium, potassium, chloride, and bicarbonate, amongst others. Bicarbonate is the major buffer in saliva and its concentration increases during nervous stimulation. The buffering capacity of saliva is important in maintaining the oral pH at around neutral. Saliva is supersaturated with calcium phosphates, which help to prevent the demineralization of the teeth. Water and the ionic constituents are derived from blood plasma; however, saliva is hypotonic when compared to plasma due to the active transport mechanisms within salivary gland ducts. Nonelectrolytes such as urea, uric acid, and ammonia are passively diffused from plasma into saliva and increase the buffering capacity of saliva [11].

Saliva contains many organic components, which contribute to functions such as, digestion (through enzymatic activity), lubrication, and the protection of dental tissues. These include amylase, lipase, mucin, proline-rich protein (PRP), tyrosine-rich protein, histadine-rich protein, peroxidase, lysozyme, secretory IgA, and protease inhibitors [11, 15, 16].

The protective film that covers oral surfaces is known as the salivary pellicle; however, the components that adsorb to different locations are not identical [17]. The pellicle that forms on mucosal surfaces is referred to as the mucus coat, while the pellicle that forms on enamel is referred to as the acquired enamel pellicle. The mucous coat is predominantly composed of salivary mucin and is approximately 53 μm thick [15]. Mucins have been shown to interact with buccal mucosa via receptor proteins located within the epithelial cell membranes. A function of the mucus coat is to preserve oral

mucosal integrity through the modulation of intracellular calcium levels [15]. The mucus coat also limits the diffusion of various compounds, neutralizes weak acids, and resists proteolysis, protecting the mucosa underneath [10]. The mucus layer has an antiadherent function, preventing materials adhering within the oral cavity, which has clear implications for the retention of buccal formulations.

11.2.1 Microbial Ecology of the Oral Cavity

The oral cavity is ecologically distinct from all other surfaces of the body; these properties dictate the types of microorganisms able to persist [17]. Several different habitats that each supports a characteristic community of microorganisms can be identified within the oral cavity [17–20]. Within this ecosystem, a variety of factors can influence the selection of microorganisms and help maintain the equilibrium among bacterial populations.

A wide variety of organisms have been identified in the oral cavity, 300 different species of bacteria have been isolated and identified, including *Streptococcus*, *Enterococcus*, *Staphylococcus*, and *Micrococcus* species, as well as some fungi (*Candida* species), mycoplasmas, and protazoa (e.g., *Entamoeba* species) and viruses (e.g., herpes simplex) [20]. The density of microorganisms in the oral fluids is high; saliva, which derives its flora from oral surfaces, contains 10^7–10^8 bacteria/mL [21].

The oral mucosa represents a soft tissue site for the colonization of oral bacteria. This surface is characterized by the desquamation of surface epithelial cells, and consequently the removal of surface-bound microorganisms [20]. The gingival, palatine, buccal, and lingual mucosa are colonized by a few microorganisms [19]. Frequent disruption accounts for the lack of complex microbial communities at these sites.

The teeth represent a hard, nonshedding surface for microbial colonization, which enables the accumulation of microorganisms [17]. Microorganisms exist in a complex matrix composed of microbial extracellular products and salivary compounds known as dental plaque. Multiple habitats are associated with the tooth surface, each supporting the growth of different populations of oral bacteria.

The growth of oral microorganisms is influenced by a variety of physicochemical, host, microbial, and external factors. These include temperature, pH, oxidation–reduction potential, nutrients (both endogenous and exogenous), microbial adherence and agglutination, antimicrobial agents, host defences, and host genetics [17, 19].

The importance of these microorganisms are that not only are they potentially pathogenic and therefore the target for localized buccal drug delivery but they may also colonize a surface or infiltrate a drug delivery system, causing potential irritation, infection, or instability. The mechanism by which oral mucosae prevents colonization by microorganisms, a constant shedding of cells, also has implications for development of retentive buccal systems.

11.3 LOCAL DISEASES OF THE ORAL CAVITY

Disease conditions not only provide the target for localized drug delivery but can also alter the environment of the oral cavity, effecting systemic drug delivery by, for example, altering the barrier function of oral mucosa.

Dental caries can be defined as the localized destruction of the tissues of the tooth by bacterial fermentation of dietary carbohydrates [17]. It is one of the most prevalent infectious diseases; in the United Kingdom alone, 55% of dentate adults had one or more teeth with visual or cavitated caries [22]. The caries process is dependant upon (1) the interaction of protective and deleterious factors in saliva and plaque, (2) the balance between the cariogenic and noncariogenic microbial population within saliva and in particular plaque, and (3) the physicochemical characteristics of enamel, dentin, and cementum that make the dental hydroxyapatite more or less vulnerable to an acidogenic challenge [23].

The term "periodontal disease" describes a group of conditions in which the supporting structures of the teeth are attacked [17]. Periodontal diseases can be divided into two main categories: gingivitis and periodontitis [24]. Gingivitis is a nonspecific, inflammatory response to dental plaque involving the gingival margin. It is characterized by swelling, redness, and bleeding of the gingiva and is often linked with the onset of periodontitis. Gingivitis is associated with poor oral hygiene, which results in the development of more organized supragingival plaque [25]. Periodontitis usually follows gingivitis and is characterized by periodontal pockets formed by the migration of the junctional epithelial tissue at the base of the gingival sulcus down the root of the tooth. It involves the loss of attachment between the root surface, the gingivae, and the alveolar bone, and bone loss itself may occur.

In the United Kingdom, 54% of dentate adults had some periodontal pocketing while 43% had some loss of attachment between the tooth and periodontal tissue in 1998 [22]. The development of periodontal disease is the result of direct action by plaque-associated microorganisms and indirect inflammatory response of the human immune system. Treatment is usually by the introduction of enhanced oral hygiene procedures, and the use of antibiotics or antimicrobial agents, although gaining access to deep pockets is a significant challenge in their treatment. Calculus, or tartar, results from the calcification of dental plaque and is linked to the retention of bacterial antigens and further plaque accumulation [26]. Consequently, calculus at the gingival margin is associated with an increased risk of developing periodontal disease. Calculus can be removed by mechanical means by a dentist, and reduced by enhanced dental hygiene procedures.

Aphthous ulceration (common mouth ulcers) are erosions of the oral mucosa of largely unknown cause that are usually self-limiting, but can be treated with steroids, antibacterials to reduce secondary infections and analgesics. Acute bacterial infections arising from endogenous opportunistic pathogens are the most common types of oral infection after periodontal diseases and dental caries [17]. These infections occur when the number of pathogens exceeds the minimum infective dose, normally as a result of predisposing factors, for example, antibiotic chemotherapy, affecting colonization resistance and selecting resistant species. Chronic bacterial infections can develop from acute infections and commonly affect medically compromised individuals [17]. These are treated by the local application of antimicrobial agents or appropriate systemic antibiotics. Oral fungal infections such as candidosis are generally the result of a range of predisposing factors, but more commonly affect the medically compromised. In particular, the incidence of oral fungal infections has increased with the emergence of acquired immunodeficiency syndrome (AIDS). These are treated by local and/or systemic antifungal agents. Oral viral infections are common in the population and are generally not associated with predisposing factors. Regularly encountered viruses include herpes I–VIII (although herpes I is by far the most common), measles, and papilloma. Oral carcinomas, such as squamous cell carcinomas of the mucosa, are linked to smoking and alcohol consumption.

Xerostomia is defined as the perception of oral dryness and is often due to a reduction in salivary flow rate of the major and minor salivary glands [10, 11]. Saliva has a range of important functions that are essential to the physiological processes that occur within the oral cavity. Xerostomia is therefore implicated as a predisposing factor for many oral diseases including dental caries, candidosis, and mucositis [10–12]. Saliva substitutes can be used to help relieve the symptoms of this condition.

11.4 PERMEABILITY OF THE BUCCAL AND GINGIVAL MUCOSA

For localized drug delivery, the buccal mucosal barrier is useful in helping retain the drug at the site of action, while for systemic drug delivery, the barrier properties present one of the major challenges. One advantage, however, of systemic drug delivery via this route is that it avoids the hepatic first-pass elimination of drugs. Drugs are absorbed into the reticulated vein, jugular vein, and then drained into the systemic circulation.

The superficial layers (approximately the outermost quarter) of the oral mucosa represent the primary barrier to the entry of substances from the exterior, although the lower layers have also been proposed to provide a significant barrier [27, 28]. There are two possible routes of drug absorption through the squamous stratified epithelium: transcellular (intracellular, passing through the cell) and paracellular (intercellular, passing around the cell). Permeation across the buccal mucosa has been reported to be mainly by the paracellular route [4, 29, 30], which is analogous to the situation for skin. Other researchers have proposed that in the nonkeratinized buccal and sublingual mucosae, the hydrophilic nature of the intercellular lipids produced by membrane coating granules means that this is the predominant route for the absorption of hydrophilic molecules, while lipophilic molecules pass through the cell membranes and are absorbed by the transcellular route [31, 32]. The amphiphilic nature of the intercellular lipids indicates the possibility of both a hydrophobic and hydrophilic pathway through the paracellular route [27].

Lower molecular weight, more lipophilic (log P of 1.6–3.3) molecules show greatest absorption via this route; however, for highly lipophilic molecules, limited water solubility restricts their absorption [33]. As most drugs are either weak acids or bases, the salivary pH will have a pronounced effect on molecular charge, and thus the relative hydrophilic/lipophilic nature of the drug, and its absorption from this route. Although passive diffusion is the main mechanism of drug absorption [31, 32], specialized transport mechanisms have been reported to exist in other oral mucosa (that of the tongue) for a few drugs and nutrients, such as glucose and cefadroxil [34, 35]. The buccal mucosa is challenging for the delivery of hydrophilic macromolecular therapeutic agents such as peptides, oligonucleotides, and polysaccharides, these having low permeability and bioavailability [36], and permeation enhancers may be required to overcome this [37]. The presence of proteases within the mucosa and salivary enzymes may also reduce stability.

Damage to the oral mucosa, by mechanical abrasion or diseases such as lichen planus, pemphigus, viral infections, and allergic reaction, would also be expected to increase its permeability.

11.4.1 Permeation Enhancers

To overcome the oral mucosal barrier, permeation (absorption) enhancers have been included in buccal formulations. These act by a number of mechanisms, such as increasing the fluidity of the cell membrane, extracting inter/intracellular lipids, altering cellular proteins, or altering surface mucin [37]. Permeation enhancers have been found to significantly increase absorption, for example, for some proteins from 1–3% *in vitro* to up to 10% [27].

The most common permeation enhancers are fatty acids, bile salts, and surfactants (such as sodium dodecyl sulfate). Oleic acid has been reported to be a good permeation enhancer for insulin [38]. The fatty acids in cod liver oil have also been reported to enhance the buccal delivery of ergotamine [39]. Bile salts have been used extensively as

penetration enhancers, and are believed to act by the extraction of lipids or proteins from the cell wall, membrane fluidization, and reverse membrane micellation without causing major damage to the mucosa. Sodium dodecyl sulfate is reported to have a significant permeation enhancing effect, but may also produce mucosal damage [27].

Other materials have been reported to have a permeation enhancing effect. Solutions/gels of chitosan were found to promote the transport of mannitol and fluorescent-labeled dextrans across a tissue culture model of the buccal epithelium [40], chitosan glutamate being particularly effective. Glyceryl monooleates were reported to enhance peptide absorption by a cotransport mechanism [41]. Lipophilic skin penetration enhancers such as Azone™ have been shown to enhance the buccal absorption of triamcinolone acetonide *in vitro*, while reducing the penetration of oestradiol and caffeine [42]. The enzyme inhibitors aprotinin [43] and puromycin [44] have been added to buccal formulations to reduce peptide degradation and thus aid absorption of the intact molecule.

11.5 DRUG DELIVERY SYSTEMS FOR THE ORAL CAVITY

Local therapy is used to treat conditions of the oral cavity outlined in Section 11.3, while systemic delivery carries the drug into the main circulation avoiding the pH and digestive enzymes of the middle gastrointestinal tract as well as hepatic "first-pass metabolism." A relatively rapid onset of action can be achieved compared to the oral route although sublingual is considered more rapid of the two, and the formulation can be removed if therapy is required to be discontinued. Delivery systems used include mouthwashes, aerosol sprays, chewing gums, bioadhesive tablets, gels, and patches [45, 46]. The challenges facing buccal drug delivery are typically related to whether local or systemic action is required [47–50].

For local action, the rapid elimination of drugs due to the flushing action of saliva or the ingestion of foodstuffs may lead to the requirement for frequent dosing. The nonuniform distribution of drugs within saliva on release from a solid or semisolid delivery system [51–53] could mean that some areas of the oral cavity might not receive effective levels. Typically saliva flows from the opening of the salivary ducts and pools at the floor of the mouth prior to swallowing. The effect of saliva may therefore depend on the location of a dosage form within the buccal region. If placed in the upper buccal pouch near the incisors or canine teeth, then the formulation will typically be exposed to relatively little saliva. If placed near the opening of the parotid salivary gland ducts (adjacent to the upper second molar on either side of the mouth), or in the lower region of the buccal pouch where saliva will pool, then a much greater exposure to saliva would be expected. For both local and systemic action patient acceptability in terms of taste, irritancy and "mouth feel" is an issue. For systemic delivery, the relative impermeability of oral cavity mucosa with regard to drug absorption, especially for large hydrophilic biopharmaceuticals, is a major concern.

11.6 FORMULATIONS

11.6.1 Solid Dosage Forms

A formulation with extended action placed into the oral cavity is challenged regularly by food and drink, mouth movements, and saliva. Parameters influencing the clearance of a product introduced in the oral cavity [54] include the following:

Physiological and/or Anatomical Parameters.
- Salivary flow rate
- Swallowing frequency
- Residual volume of saliva
- Anatomical position of the salivary glands
- Anatomical factors (space between teeth, tongue position, etc.)

Oral Muscular Movements.
- Factors relating to the product and its administration
- Dose and concentration of the product
- Duration and frequency of the administration
- Association between products and/or buccal substrates
- Dilution (drinking habits, rinsing, etc.)
- Time of administration (day/night, proximity of a meal, etc.)

There are some nonattached formulations for buccal drug delivery [55]. Commercial examples include Actiq® (Cephalon) that is "lozenge on a stick" containing the analgesic fentanyl, which is rubbed along the inside of the cheek, or OraVescent® (Cima), an effervescent tablet that rapidly releases fentanyl in the buccal pouch.

One interesting option for delivering drugs is described as being a "mechatronic" approach, where attaching a device to a denture or two adjacent prosthetic tooth crowns allows modulated drug release aimed at buccal absorption [56]. This device is clearly limited currently in terms of the limited number of potential users. The majority of buccal formulations use bioadhesive polymers to allow prolonged drug therapy.

11.6.1.1 Bioadhesive Formulations Bioadhesion within the pharmaceutical sciences usually refers to the adhesion of a pharmaceutical formulation or material to a biological tissue; when that tissue is mucus or a mucous membrane,

the more specific term mucoadhesion is often used. The aim is usually to enhance drug delivery by retaining a dosage form at the site of action or site of optimum absorption for extended periods, and this is clearly applicable to buccal drug delivery. Bioadhesive formulations typically contain an adhesive material into which the drug is incorporated. There are a wide range of bioadhesives available for medical applications [57], the most widely investigated group used in buccal delivery systems being the hydrophilic macromolecule "wet" adhesives [58–60]. The mechanism of action of these materials is discussed briefly below.

Conventional bioadhesive materials are hydrophilic, have a high molecular weight and a high density of hydroxyl, carboxyl, or amine groups, and many polymeric materials that satisfy this description will show adhesive properties. They are called "wet" adhesives in that they are activated by moistening and will adhere nonspecifically to many surfaces [60]. Unless water uptake is restricted, they may overhydrate to form a slippery mucilage. Typical examples are the poly(acrylic acid)s (e.g., carbomer) and the polysaccharides such as chitosan, the alginates, and cellulose derivatives (Figure 11.3). These were used initially as they were available "off-the-shelf" with regulatory approval, but in the past few years, new enhanced materials have been developed with more favorable properties. An ideal bioadhesive would be

- nontoxic and nonabsorbable from the gastrointestinal tract (both polymer and its degradation products);
- nonirritant to the mucous membrane;
- able to form strong but nonpermanent bonds with the mucin/epithelial surface;
- able to adhere quickly to moist tissue and possess some site specificity;
- compatible with the active ingredient, allowing easy incorporation and desired *in vivo* release;
- stable;
- cost effective.

For dry or partially hydrated dosage forms, two basic steps in mucoadhesion have been identified [60]. Step one is the "contact stage" where intimate contact is formed between the mucoadhesive and mucous membrane: Within the oral cavity, the formulation can usually be readily placed into contact with the required mucosa and held in place to allow adhesion to occur. Step two is the "consolidation" stage where various physicochemical interactions occur to consolidate and strengthen the adhesive joint, leading to prolonged adhesion. Mucus gel dehydration by the bioadhesive polymer and macromolecular interpenetration have both been proposed as ways that strengthening of the mucus layer can occur [60, 61]. It is worth noting that the attached mucus coat on buccal and gingival mucosa is thin relative to the rest

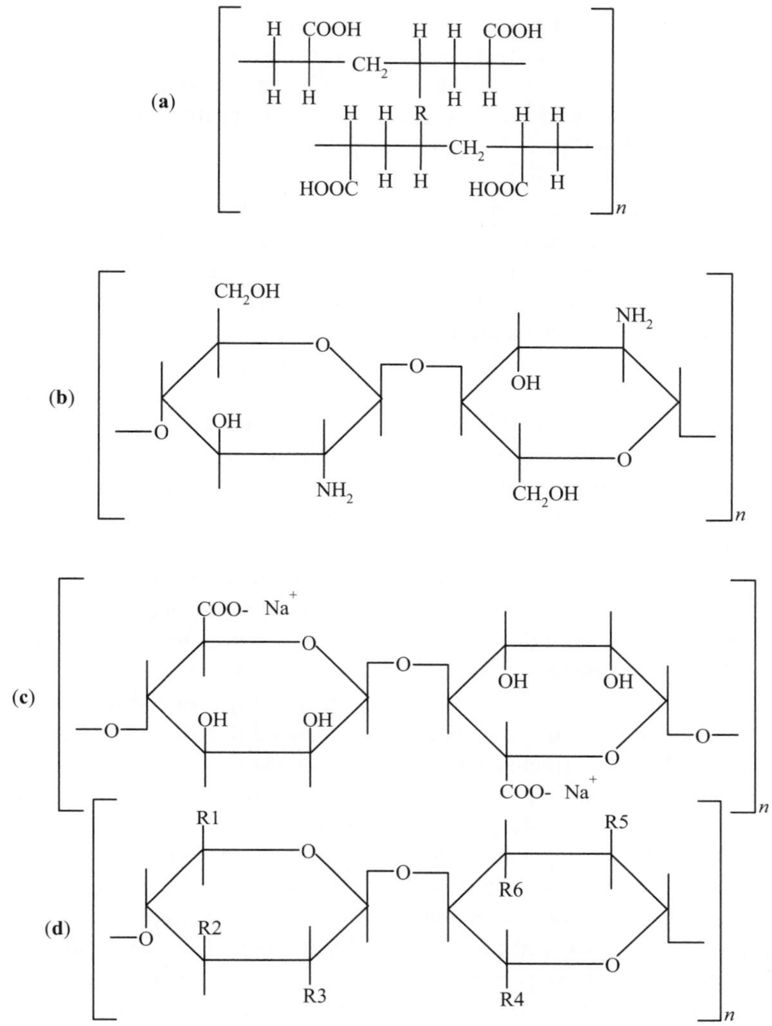

FIGURE 11.3 The structure of some common mucoadhesive polymers: (a) polyacrylic acid, R = allyl sucrose (Carbopol) or divinyl glycol (polycarbophil); (b) chitosan; (c) sodium alginate; and (d) cellulose derivatives, for example, SCMC, R1, R4 = CH$_2$OH; R2, R3, R5 = OH; R6 = OCH$_2$CO$_2$-Na +, or HPMC, R1 = CH$_2$OCH$_3$; R2 = OH; R3 = OCH$_2$CHOHCH$_3$; R4 = CH$_2$OH; R5, R6 = OCH$_3$.

of the gastrointestinal tract, so would readily dehydrate and "collapse" in contact with a swelling mucoadhesive. Adhesive joint failure will normally occur within the weakest region of the bioadhesive joint. For weak adhesives this would be the adhesive interface, for weak gel-forming materials this would be the hydrating layer, while for stronger adhesives this would initially be the mucus layer, but later the hydrating adhesive material. The strength of the adhesive joint will therefore depend on the cohesive nature of the weakest region. In the buccal cavity, this is unlikely to be thin and dehydrated mucus coat.

To strengthen the adhesive bond, thiol groups (by coupling cysteine, thioglycolic acid, cysteamine) have been placed onto several mucoadhesive polymers such as the carbomers, chitosans, and alginates by Bernkop-Schnurch and coworkers [62–67]. *In situ* they form disulfide links between the polymers themselves, but also with the mucin layer/mucosa itself [68].

Glyceryl monooleate/water (8–20%) liquid crystalline phases have also been used as bioadhesives in the buccal cavity. They differ significantly from the materials used in most bioadhesive studies [69]. Mucoadhesion is said to occur after the uptake of water, allowing dehydration of the substrate, and an inverse relationship between water content and mucoadhesion was obtained.

Control of water uptake into mucoadhesive tablets is important. Poly(acrylic acid)s and cellulose derivatives, which are commonly used in tablet formulation, hydrate to form an outer gel layer that increases the size of the tablets with the potential of negatively affecting patient compliance, reducing the duration of adhesion, and in some cases the drug bioavailability. To address this, new low-swellable polymethylmethacrylate (PMMA) sodium salts were prepared by adding sodium hydroxide solution to Eudragit® S100 aqueous suspension and spray drying [70, 71]. These materials exhibited enhanced *in vitro* and *in vivo* mucoadhesion relative to Eudragit alone but dissolved rather quickly. Modification (cross-linking) of these PMMA sodium salts with bivalent inorganic salts, principally magnesium and calcium salts, yielded a reduction of at least 25% of the dissolution rates of tablets, increasing the *in vivo* adhesion time.

11.6.1.2 Buccal Adhesive Tablets

These are typically prepared by mixing the powdered drug, adhesive material (such as a poly(acrylic acid) or cellulose derivative alone or in combination), and any excipient, then compressing into tablets, perhaps after a granulation process. These can then be placed into contact with the oral mucosa and allowed to adhere. Drug release from tablets can either be multidirectional or unidirectional (Figure 11.4). In multidirectional tablets, the drug is incorporated into a bioadhesive polymer matrix, which hydrates and releases the drug upon contact with the mucosa. These are often used for local therapy, with drug being released into the saliva and then distributed within the oral cavity. In unidirectional tablets, a second impermeable layer, often consisting of an insoluble polymer covers the underlying drug/polymer adhesive layer. On placing into the buccal pouch, the drug release is directed only toward the mucosal surface, and the high local concentration, protection of the drug from the environment of the oral cavity and intimate contact with the mucosa favors absorption. The impermeable layer may also be helpful in restricting the rate of hydration of the formulation, by reducing the access of saliva.

The amount of drug incorporated into a tablet matrix is often limited due to its possible interference with the mucoadhesive characteristics.

Disadvantages of buccal tablets include patient acceptability (mouth feel, taste, and irritation) and the nonubiqui-

tous distribution of drug within saliva for local therapy and the nature of the buccal mucosa [51–53].

Commercially available examples of buccal tablets are Suscard Buccal® (a matrix containing modified hydroxypropylmethylcellulose for the delivery of glyceryl trinitrate), Buccastem® (a matrix containing xanthan and locust bean gums for the delivery of prochlorperazine) [50], Aphtach® (a double-layered tablet with the drug triamcinolone incorporated into a hydroxypropylcellulose and carbomer adhesive matrix) [72], and Striant® (Columbia Laboratories, a buccal tablet containing the bioadhesives polycarbophil, carbomer, and hydroxypropylmethylcellulose for the delivery of testosterone). An interesting recent product based on the Lauriad® drug delivery technology is the Loramyc®, a miconazole 50 mg mucoadhesive gingival tablet, developed by BioAlliance Pharma, for the treatment of oropharyngeal candidiasis in immunocompromised patients, including HIV patients and head or neck cancer patients undergoing radiotherapy [73]. Salivary miconazole pharmacokinetics of the once-daily 50 mg bioadhesive eroding tablet were compared to those of a gel directly applied three times daily with a total dose of 375 mg. Results demonstrated that salivary drug concentrations using the tablet were higher with prolonged duration above the minimum inhibitory concentration (MIC) required for therapy of some *Candida* species. The company intends to introduce other actives such as acyclovir and fentanyl into this technology.

A U.S. patent assigned to Jenapharm GmbH & Co. KG [74] describes an interesting bioadhesive tablet with one surface having concentric or parallel, straight, and/or curved depressions. This was said on swelling not to inhibit the passage of the active agents to the mucosa allowing high bioavailability, the device was also said to protect its active agents against microbial degradation.

An *in vivo* study showed that a 10 mg once-daily mucoadhesive buccal tablet of miconazole was as effective as 400 mg ketoconazole administered once daily in the treatment of HIV-positive patients with oropharyngeal candidiasis, with less side effects [75], indicating the potential usefulness of

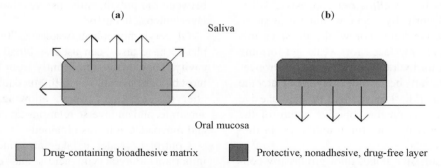

FIGURE 11.4 Typical design of a mucoadhesive tablet formulation: (a) the drug is incorporated throughout the tablet matrix, allowing multidirectional drug release and (b) the drug-containing polymer matrix is protected by the nonadhesive backing layer, favoring unidirectional drug release toward the mucosa.

TABLE 11.1   Some Recently Described Bioadhesive Buccal Tablet Formulations

Type of Formulation	Bioadhesives	Active (and Other Major Components)	Details
Drug reservoir incorporated into hollow in the surface of adhesive tablet [76]	Carbomer and HPC	Ergotamine	Absorption from buccal tablet more effective than oral administration in guinea pigs
Matrix [77]	HPMC, SCMC	Cetyl pyridinium chloride	Antimicrobial levels maintained for over 8 h *in vivo*
Matrix [78]	Hydroxyethyl-cellulose, carbomer	Metronidazole	Salivary antimicrobial levels maintained for 12 h
Matrix [79]	Carbomer and SCMC	Secnidazole	Showed anaerobic activity with planktonic bacteria in *in situ* studies
Matrix [80]	Carbomer	Testosterone	Could include up to 60% active, *in vivo* studies gave elevated drug levels over 20 h
Matrix [81]	Carbomer, HPMC	Nystatin	Biphasic tablet, immediate release layer (lactose)
Matrix [82, 83]	Freeze-dried palmitoyl glycol chitosan	FITC dextran, denbufylline	Could be loaded with high levels (27.5%) of a model macromolecule and allowed delivery of a small molecule in an animal model for over 5 h
Matrix [84]	Carbomer and corn starch	Chlorhexidine	Less effective that mouthwash in reducing plaque formation
Matrix [85, 86]	Starch–poly(acrylic acid) grafted copolymers	Testosterone/ theophylline	Allowed testosterone delivery for 5–13.5 h *in vivo* in dogs, a slow release of theophylline *in vitro* over 3–6 h and retained *in vivo* in dogs for 2–3 days
Matrix [68]	Thiolated polycarbophil		More bioadhesive *in vitro* and disintegrated less rapidly than the original polymer. It was also able to reduce enzymatic degradation
Matrix [87]	HPMC/carbomer	Danazol	Cyclodextrin added to promote drug release. A 25% bioavailability found *in vivo*
Matrix [88]	Ethyl cellulose or hydroxypropylcellulose	Iodine	Complex formed, gave slow release iodine
Bilayer tablet [89]	PVP plus others if required	Range—mouth freshners to peptides	Patent that puts active in one layer with a hydrophilic polymer and, adhesive in second
Matrix [90]	Polyacrylic acid PVP	Plant extracts	Patent recommends range of excipients to produce hard flat adhesive tablets

HPMC: hydroxypropylmethylcellulose; HPC: hydroxypropylcellulose; PVP: poly(vinylpyrrolidone); SCMC: sodium carboxymethylcellulose.

buccal tablets for localized therapies relative to systemic treatments.

Over the past 20 years, a range of such adhesive tablet formulations have been described in the literature and some more recent examples are given in Table 11.1.

11.6.1.3   Films and Patches

Patches are usually prepared by casting a solution of polymer, drug, and any excipients (such as a plasticizer) onto a (low-adhesive) surface and allowing it to dry. Patches can be made up to 10–15 cm^2 in size but are more usually 1–3 cm^2 with perhaps an ellipsoid shape to fit comfortably into the center of the buccal mucosa [91]. In a similar fashion to buccal tablets, they can be made multidirectional or unidirectional, for example, by the application of an impermeable backing layer. They have many of the advantages and disadvantages of buccal tablets but by being thin and flexible, they tend be less obtrusive and more acceptable to the patient. The relative thinness of the films, however, means that they are more susceptible to overhydration and loss of the adhesive properties. Dentipatch® (Noven), a modified transdermal patch using "Dot Matrix®" technology and silicone as the adhesive for the delivery of lidocaine, OraDisc® (Access Pharmaceuticals), a multilayered patch for the delivery of amlexanox, and BEMA® technologies (Atrix), a bioerodible mucoadhesive disk, are examples of commercial products. Some other recent examples of bioadhesive patches described in the literature are given in Table 11.2.

11.6.1.4   Particulates

These are typically delivered as an aqueous suspension, but can also be applied by aerosol or incorporated into a paste or ointment. Particulates have the advantage of being small relative to tablets and films, and therefore more likely to be acceptable to the patient, but the rheological properties, rigid (gritty) or more gel like, will also have an effect on patient perception. The dose of drug

TABLE 11.2 Some Recently Described Bioadhesive Patches and Films

Bioadhesives	Active (and Other Major Components)	Details
PVP/SCMC [92]	Ibuprofen	Films tolerable and comfortable relative to tablets
Polymethylvinylether-*co*-maleic anhydride [93]	Toluidine blue O	*In vitro* study to consider its use in photodynamic treatment of fungal infections
Poly(acrylic acid), PVP, and carmellose with a cellulose derivative (HPMC) [94]	Anti-inflammatory or steroid	Two layered device, bioadhesive layer caste onto hydrophobic layer
Carbomer, HPC, and HPMC [95]	Doxycycline	Film, giving *in situ* drug release and effective concentrations for up to 6 h
Carbomer and chitosan complexes [96]	Acyclovir	Intermacromolecular complexes were advantageous in retention and drug release studies
Chitosan, poly(vinylalcohol), and HPMC and PVP [97]	Miconazole	The poly(vinylalcohol)/PVP patch gave the best release profile, and allowed drug release over 6 h in human volunteers

PVP: poly(vinylpyrrolidone); HPMC: hydroxypropylmethylcellulose; SCMC: sodium carboxymethylcellulose; HPC: hydroxypropylcellulose.

delivered to the buccal mucosa is likely to be more variable than a single-unit dosage form such as a patch or buccal tablet.

Polymeric microparticles (23–38 μm) of Carbopol, polycarbophil, chitosan, or Gantrez [98] were capable of adhering to porcine oesophageal mucosa, with particles prepared from the poly(acrylic acid)s exhibiting greater mucoadhesive strength during tensile testing studies while in "elution" studies, particles of chitosan or Gantrez were seen to persist on mucosal tissue for longer time periods [99].

Govender et al. [100] investigated the optimization of a formulation of tetracycline-loaded chitosan microparticles prepared by cross-linking using tripolyphosphate. Increasing the concentration of tripolyphosphate increased the *in vitro* adhesive strength of the microparticles. The optimal formulation, 3% w/w chitosan, 10% w/w tetracycline hydrochloride, 9% w/w tripolyphosphate, was characterized in terms of hydration, release kinetics, antimicrobial activity, thermal properties, morphology, and surface pH. Tetracycline was released at concentrations above the minimum inhibitory concentration of *Staphylococcus aureus* for more than 8 h.

Giunchedi et al. [101] prepared chitosan microparticles loaded with chlorhexidine to achieve prolonged therapeutic concentrations within the oral cavity. The microparticles were prepared by spray drying solutions of chlorhexidine and chitosan in 1:2 and 1:4 weight ratios, respectively. Particle characterization showed drug incorporation and small particle sizes were achieved using this method. *In vitro* drug release from the microparticles significantly improved the dissolution of chlorhexidine when compared to chlorhexidine diacetate powder. These particles were compressed with sodium alginate, mannitol, and saccharin to form tablets that were then assessed *in vivo*, where chlorhexidine was detected in the saliva for more than 3 h, longer than for a chlorhexidine mouthwash (2 h).

11.6.2 Semisolids

With the exceptions of some commercially available teething gels, the majority of formulations are described as bioadhesive in nature, because of their resistance to dislodgement. They typically contain a mucoadhesive polymer and drug plus any excipient dissolved or suspended as a fine powder in an aqueous or nonaqueous base, depending on their solubility and concentration [50]. Hydrogels have become a commonly investigated semisolid dosage form, typically formed from a cross-linked polymer, which upon application, either topically with the finger or via a syringe, swell slowly in the presence of water releasing any drug entrapped within.

The ability of a viscous liquid or gel to cling to, and be retained on, a mucosal surface is influenced by the thermodynamic affinity of the semisolid for the mucous membrane, the rheological properties, and the formulation stability within the biological system [60]. It could be argued that it would be perhaps better to describe these formulations as "retentive" rather than "bioadhesive," as the process of bioadhesive bond formation described for solids may not take place in the same way. They may also deliver variable amounts of active ingredients in comparison to a unit dosage form.

Commercial examples are Adcortyl in Orabase® (Squibb), an ointment containing triamcinolone acetonide in a pectin, gelatin, carboxymethylcellulose-containing ointment base, and "Bioral gel" (Merck), carbenoxolone in a carboxymethylcellulose-containing ointment base.

A method for making oil-in-water emulsions having mucoadhesive properties has been described by Friedman et al. [102]. The emulsion includes a hydrophobic core, a surfactant, and a mucoadhesive polymer such as carbomer or chitosan in the external phase sufficient to confer bioadhesive properties. Some other recent formulations are given in Table 11.3.

TABLE 11.3   Some Recently Described Semisolid Formulations for Buccal Drug Delivery

Types of Formulation	Bioadhesives	Active (and Other Major Components)	Details
Gel [103]	Chitosan	Nystatin	*In vivo* effective concentrations maintained for over 90 min
Gel [104]	Poly(methylvinyl-ether-*c*-maleic anhydride) and PVP		Polymers were bioadhesive showing rheological synergy
Gel [105]	Eudispert hv	rgEGF	Increased healing rate of ulcers in hamster buccal mucosa
Ointment [106]	Poly(methacrylamide)	Benzyl nicotinate	Effective drug delivery to rat mucosa
Gels [107]	Hexadimethrine	Triclosan	Anionic cyclodextrin complexed with bioadhesive retention on buccal mucosa allowing prolonged release

PVP: poly(vinylpyrrolidone).

Portero et al. [108] describe the novel formulation of chitosan sponges for the delivery of insulin. The sponge was required to allow the porosity necessary to promote the release of a macromolecule. The insulin was incorporated into the chitosan adhesive layer, on top of which was placed an impermeable ethylcellulose outer layer to promote unidirectional delivery to the mucosa. *In vitro* studies indicated that these sponges adhered to mucus and could give a controlled release of insulin, but their utility needs to be evaluated *in vivo*.

11.6.3   Liquid Dosage Forms

The use of liquid dosage forms for the delivery of therapeutic agents increases the likelihood that the formulation will reach all areas of the oral cavity, including the distal, hard to reach crevices, and surfaces. However, few liquid bioadhesive dosage forms have been investigated. For all but the most viscous liquids, retention is an issue, and the solutions are usually readily removed. Retention is usually achieved by the adsorption of components of the formulation such as dissolved polymers onto a mucosal surface [109–111]. A sufficient amount of adsorption from solution is therefore required within a short period of time (at most 60 s) to achieve prolonged local therapeutic concentrations. It is interesting to note that the same materials that are widely reported as being bioadhesive as solid formulations, such as chitosan and carbomer, also readily adsorb onto oral mucosa from solution [109, 110] despite there clearly being a different mechanism. The polymer absorption principal has been investigated with regard to the prolonged delivery of zinc within the oral cavity [112].

Ungphaiboon and Maitani [113] prepared mucoadhesive mouthwash formulations containing the corticosteroid triamcinolone acetonide for the treatment of oral lichen planus, a chronic inflammatory condition that manifests itself as lesions on the oral mucosa. Carbopol 934 was incorporated into a cosolvent system and the *in vitro* permeation of drug across hamster cheek pouch was analyzed using a Franz type diffusion cell. Only formulations containing Carbopol 934 resulted in the permeation of drug and it was proposed that

the inclusion of this polymer resulted in increased tissue penetration rather than mucosal accumulation; it was also noted that the rheology of such formulations probably precluded their clinical application.

Lectins are proteins or glycoproteins that bind to specific sugar residues and can therefore interact with the glycoconjugates including those present on cell surfaces or the mucins in the salivary pellicle. This therefore provides the opportunity for their use to "anchor" a delivery system on the oral mucosa or a mucosal lesion [114]. Smart and coworkers [115] found that the lectins from *Arachis hypogaea*, *Canavalia ensiformis*, and *Triticum vulgaris* bound to oral mucosal cells, the latter showing the greatest binding (calculated to be about 7×10^9 molecules per cell). In an *in vivo* study in rats, the *T. vulgaris* lectin showed the greatest levels of retained lectin after 30 min (about 30 μg) on the buccal mucosal tissue and was still detected at similar levels after 2 h.

An interesting example of the therapeutic application of a nonretentive solution is that of the use of buccal midazolam for the treatment of acute status epilepticus and serious tonic–clonic seizures in children [116]. The intravenous preparation of midazolam hydrochloride was administered into the buccal cavity between the gum and cheeks, the dose being titrated to the size of the child, and was found to be more rapid and effective than diazepam suppositories in preventing further seizures. Although administered into the buccal pouch area, it is likely that the solution will spread throughout the oral cavity, and may indeed be swallowed, so drug absorption may occur from a variety of sites.

11.6.4   Toxicity and Irritancy

Formulations that produce local damage at the site of application, such as ulceration of the mucosa, would preclude their widespread usage as a result of the associated pain and discomfort. This is particularly important in buccal drug delivery where the formulation is in contact with the same section of mucosa for extended periods [117].

Toxic effects can arise from the drug itself, the bioadhesive, or from other components of the formulation. For example, carbomers have been reported to produce mucosal irritation [118] believed to result from a localized low pH [119]. To overcome this, Llabot et al. [120] used a mixture of Carbopol and lyophilized Carbopol sodium salt to prepare mucoadhesive tablets for the delivery of nystatin to avoid the acidity of Carbopol-only preparations. These tablets showed good *in vitro* mucoadhesive properties, high water uptake, and were able to modulate the release of nystatin. Lyophilized Carbopol induced a higher water uptake and a greater uptake capacity due to the porous structure obtained as a result of the lyophilization process.

11.7 CONCLUSIONS

The buccal route represents a useful alternative for the local delivery of therapeutic agents, while the opportunities of systemic delivery are more limited. The advantage of this route is that it is readily accessible, avoids first-pass metabolism, can deliver drugs directly to produce a localized targeted effect, and is noninvasive. However, the fact that there are comparatively few formulations available commercially that exploit this method of administration indicates that there are still some major barriers to overcome, such as

- patient acceptability, the drug and formulation has to have an acceptable flavor and "mouthfeel," and must be nonirritant (most patients would still prefer to swallow a dosage form);
- the stresses within the oral cavity, including salivary washout, eating and drinking, and mouth movements;
- swallowing the drug and dosage form.

In addition, for local action,

- a therapeutic agent may not reach the area of the oral cavity requiring therapy.

For systemic effects,

- the drug must be able to cross the relatively impermeable buccal or gingival mucosa.

There have been a plethora of formulations for buccal drug delivery described in the literature over the past 20 years, but the conversion of these to actual commercial products remains limited. With existing technology, localized delivery and the delivery of small potent molecules into the systemic circulation remain the most attainable target for buccal drug delivery. To make this route appropriate for a wider range of drugs, notably biopharmaceutical products, the technology will need to advance significantly, in terms of formulation design, enhancing mucosal permeability, taste masking/acceptability, and the use of more efficient bioadhesives. A closer consideration of the environment of the oral cavity,

and the challenges facing a formulation *in situ*, should also be a primary consideration when formulating a buccal delivery system.

REFERENCES

1. Smith ME, Morton DG. *The Digestive System: Basic Science and Principles*, Churchill Livingstone, Edinburgh, 2001, pp. 22–37.
2. Stevens A, Lowe J. *Human Histology*, 2nd edition, Mosby, Barcelona, 1997, pp. 177–190.
3. Gandhi RB, Robinson JR. Oral cavity as a site for bioadhesive drug delivery. *Adv. Drug Deliv. Rev.* 1994;13:43–74.
4. Squier CA, Wertz PW. Structure and function of the oral mucosa and implications for drug delivery. In Rathbone MJ. editor. *Oral Mucosal Drug Delivery*, Marcel Dekker, New York, 1996, pp. 1–26.
5. Squier CA, Wertz PW. Permeability and pathophysiology of the oral mucosa. *Adv. Drug Deliv. Rev.* 1993;12:13–24.
6. Gibbs S, Ponec M. Intrinsic regulation of differentiation markers in human epidermis, hard palate and buccal mucosa. *Arch. Oral Biol.* 2000;45:149–158.
7. Rathbone MJ, Tucker IG. Mechanisms, barriers and pathways of oral mucosal drug permeation. *Adv. Drug Deliv. Rev.* 1993;12:41–60.
8. Johnson DR, Moore WJ. *Anatomy for Dental Students*, 3rd edition, Oxford University Press, Oxford, 1997, pp. 181–193.
9. Bloom W, Fawcett DW. *A Textbook of Histology*, WB Saunders Co, Philadelphia, 1975, pp. 598–613.
10. Herrera JL, Lyons MF, Johnson LF. Saliva: its role in health and disease. *J. Clin. Gastroenterol.* 1988;10:569–578.
11. Dowd FJ. Saliva and dental caries. *Dent. Clin. North Am.* 1999;43:579–597.
12. Hand AR, Pathmanathan D, Field RB. Morphological features of the minor salivary glands. *Arch. Oral Biol.* 1999;44: S3–S10.
13. Bradley RM. *Essentials of Oral Physiology*, Mosby, St. Louis, 1995, pp. 161–186.
14. Mandel ID, Wotman S. The salivary secretions in health and disease. *Oral Sci. Rev.* 1976;8:25–47.
15. Slomiany BL, Murty VLN, Piotrowski J, Slomiany A. Salivary mucins in oral mucosal defense. *Gen. Pharmacol.* 1996;27:761–771.
16. Humphrey SP, Williamson RT. A review of saliva: normal composition, flow and function. *J. Prosthet. Dent.* 2001;85: 162–169.
17. Marsh PD, Martin MV. *Oral Microbiology*, 4th edition, Wright, Edinburgh, 1999.
18. Mager DL, Ximenez-Fyvie LA, Haffajee AD, Socransky SS. Distribution of selected bacterial species on intraoral surfaces. *J. Clin. Periodontol.* 2003;30:644–654.
19. Marcotte H, Lavoie MC. Oral microbial ecology and the role of salivary immunoglobulin A. *Microbiol. Mol. Biol. Rev.* 1998;62:71–109.

20. Theilade E. Factors controlling the microflora of the healthy mouth. In Hill MJ, Marsh PD. editors. *Human Microbial Ecology*, CRC Press, London, 1990, pp. 1–9.

21. Addy M. Local delivery of antimicrobial agents to the oral cavity. *Adv. Drug Deliv. Rev.* 1994;13:123–134.

22. Kelly M, Walker A. Adult Dental Health survey: oral health in the United Kingdom, Office of National Statistics, 1998. Available at http://www.statistics.gov.uk/downloads/theme_health/DHBulletinNew.pdf.

23. Hicks J, Garcia-Goidoy F, Flaitz C. Biological factors in dental caries: role of saliva and dental plaque in the dynamic process of demineralisation and remineralisation. *Part 1. J. Clin. Pediatr. Dent.* 2003;28:47–52.

24. Liebana J, Castillo AM, Alvarez M. Periodontal diseases: microbiological considerations. *Med. Oral Patol. Oral Cir. Buccal* 2004;9:75–91.

25. Sbordone L, Borkolaia C. Oral microbial biofilms and plaque related diseases: microbial communities and their role in the shift from health to disease. *Clin. Oral Invest.* 2003;7:181–188.

26. Listgarten MA. The structure of dental plaque. *Periodontology 2000* 1994;5:52–65.

27. Veuillez F, Kalia YN, Jacques Y, Deshusses J, Buri P. Factors and strategies for improving buccal absorption of peptides. *Eur. J. Pharm. Biopharm.* 2001;51:93–109.

28. Kurosaki Y, Kimura T. Regional variation in oral mucosal drug permeability. *Crit. Rev. Ther. Drug Carrier Syst.* 2000;17:467–508.

29. Squier CA, Lesch CA. Penetration pathways of different compounds through epidermis and oral epithelia. *J. Oral Pathol.* 1988;17:512–516.

30. Junginger H, Hoogstraate J, Verhoef J. Recent advances in buccal drug delivery and absorption—*in vitro* and *in vivo* studies. *J. Control. Release* 1999;62:149–159.

31. Song Y, Wang Y, Thakur R, Meiden VM, Michniak B. Mucosal drug delivery: membranes, methodologies and applications. *Crit. Rev. Ther. Drug Carrier Syst.* 2004;21:195–256.

32. Shojaei AM. Buccal mucosa as a route for systemic drug delivery, a review. *J. Pharm. Pharm. Sci.* 1998;1:15–30.

33. Florence AT Attwood DA. Buccal sublingual absorption. In *Physicochemical Principals of Pharmacy*, Palgrave, Basingstoke, 1998, p. 392.

34. Kurosaki Y, Nishimura H, Terao K, Nakayama T, Kimura T. Existence of a specialised absorption mechanism for cefadroxil, an aminocephalosporin antibiotic, in the human oral cavity. *Int. J. Pharm.* 1992;82:165–169.

35. Kurosaki Y, Yano K, Kimura T. Perfusion cells for studying regional variation in oral mucosal permeability in humans. 2. A specialized transport mechanism in D-glucose absorption exists in dorsum of tongue. *J. Pharm. Sci.* 1998;87:613–615.

36. Yamamoto A, Iseki T, Ochi-Sugiyama M, Okada N, Fujita T, Muranishi S. Absorption of water-soluble compounds with different molecular weights and [Asu$^{1.7}$]–eel calcitonin from various mucosal administration sites. *J. Control. Release* 2001;76:363–374.

37. Senel S, Hincal AA. Drug permeation enhancement via the buccal route: possibilities and limitations. *J. Control. Release* 2001;72:133–144.

38. Morishita M, Barichello JM, Takayama K, Chiba Y, Tokiwa S, Nagai T. Pluronic F-127 gels incorporating highly purified unsaturated fatty acids for buccal delivery of insulin. *Int. J. Pharm.* 2001;212:289–293.

39. Tsutsumi K, Obata Y, Nagai T, Loftsson T, Takayama K. Buccal absorption of ergotamine tartrate using the bioadhesive tablet system in guinea pigs. *Int. J. Pharm.* 2002;238:161–170.

40. Portero A, Remunan-Lopez C, Nielsen HM. The potential of chitosan in enhancing peptide and protein absorption across the TR146 cell culture model-an *in vitro* model of the buccal epithelium. *Pharm. Res.* 2002;19:169–174.

41. Lee J, Kellaway IW. Buccal permeation of [D-Ala^{2}D-Leu^5] enkephalin from liquid crystalline phases of glyceryl mono-oleate. *Int. J. Pharm.* 2000;195:3538.

42. Nicolazzo JA, Reed BL, Finnin BC. Modification of buccal delivery following pre-treatment with skin penetration enhancers. *J. Pharm. Sci.* 2004;93:2054–2063.

43. Yamamoto A, Hayakawa E, Lee VHL. Insulin and proinsulin proteolysis in mucosal homogenates of the albino rabbit,: implications in peptide drug delivery from non-oral routes. *Life Sci.* 1990;47:2465–2474.

44. Tavakoli-Saberi MR, Williams A, Audus KL. Aminopeptidase activity in human buccal epithelium and primary cultures of hamster buccal epithelium. *Pharm. Res.* 1991;6:S197.

45. Wilson CG, Washington N. Drug delivery to the oral cavity. In Wilson CG, Washington N. editors. *Physiological Pharmaceutics: Biological Barriers to Drug Absorption*, Ellis Horwood, Chichester, 1989, pp. 21–36.

46. Sudhakar Y, Kuotso K, Bandyopadhyay AK. Buccal bioadhesive drug delivery—a promising option for orally less efficient drugs. *J. Control. Release* 2006;114:15–40.

47. Samaranayake LP, Ferguson MM. Delivery of antifungals to the oral cavity. *Adv. Drug Deliv. Rev.* 1994;13:161–179.

48. Harris D, Robinson JR. Drug delivery via the mucous membranes of the oral cavity. *J. Pharm. Sci.* 1992;81:1–10.

49. Rathbone MJ, Drummond BK, Tucker IG. The oral cavity as a site for systemic drug delivery. *Adv. Drug Deliv. Rev.* 1994;13:1–22.

50. Smart JD. Buccal drug delivery. *Expert Opin. Drug Deliv.* 2005;2:507–517.

51. Wheatherell JA, Robinson C, Rathbone MJ. Site specific differences in the salivary concentrations of substances in the oral cavity—implications for the aetiology of oral disease and local drug delivery. *Adv. Drug Deliv. Rev.* 1994;13:23–42.

52. McPherson LMD, Dawes C. Distribution of sucrose around the mouth and clearance after a sucrose mouthrinse or consumption of three different foods. *Caries Res.* 1994;28:150–155.

53. Vivien-Castioni N, Gurny R, Baenhi P, Kalsatos V. Salivary fluoride concentrations following applications of bioadhesive

tablets and mouthrinses. *Eur. J. Pharm. Biopharm.* 2000;49: 27–33.

54. Vivien-Castioni N, Baehni PC, Gurny R. Current status in oral fluoride pharmacokinetics and implications for the prophylaxis against dental caries. *Eur. J. Pharm. Biopharm.* 1998; 45:101–111.

55. Burnside BA, Keith AD, Snipes W. Microporous hollow fibres as a peptide delivery system for the buccal cavity. *Proc. Int. Symp. Control. Release Bioact. Mater.* 1989;16:93–94.

56. Scholz OA, Wolff A, Schumacher A, Giannola LL, Campisi G, Ciach T, Velten T. Drug delivery from the oral cavity: focus on novel mechatronic delivery device. *Drug Discov. Today* 2008; 13:247–252.

57. Smart JD. Bioadhesion. In Wnek GE, Bowlin GL. editors. *Encyclopaedia of Biomaterials and Biomedical Engineering*, Marcel Dekker, New York, 2004, pp. 62–71.

58. Harding SE, Davis SS, Deacon MP, Fiebrig I. Biopolymer mucoadhesives. *Biotechnol. Genet. Eng. Rev.* 1999;6:41–85.

59. Lee JW, Park JH, Robinson JR. Bioadhesive-based dosage forms: the next generation. *J. Pharm. Sci.* 2000;89:850–866.

60. Smart JD. The basics and underlying mechanisms of mucoadhesion. *Adv. Drug Deliv. Rev.* 2005;57:1556–1568.

61. Ahuja A, Khar RP, Ali J. Mucoadhesive drug delivery systems. *Drug Dev. Ind. Pharm.* 1997;23:489–515.

62. Bernkop-Schnurch A, Scholler S, Biebel RG. Development of controlled release systems based on thiolated polymers. *J. Control. Release* 2000;66:39–48.

63. Bernkop-Schnurch A, Kast CE, Richter MF. Improvement in the mucoadhesive properties of alginate by the covalent attachment of cysteine. *J. Control. Release* 2001;71: 277–285.

64. Bernkop-Schnurch A, Clausen AE, Hnatyszyn M. Thiolated polymers, synthesis and *in vitro* evaluation of polymer-cysteamine conjugates. *Int. J. Pharm.* 2001;226:185–194.

65. Kast CE, Bernkop-Schnurch A. Thiolated polymers–thiomers: development and in vitro evaluation of chitosan–thioglycolic acid conjugates. *Biomaterials* 2001;22:2345–2352.

66. Guggi D, Marschultz MK, Bernkop-Schnurch A. Matrix tablets based on thiolated poly(acrylic acid): pH dependent variation in disintegration and mucoadhesion. *Int. J. Pharm.* 2004;274:97–105.

67. Langoth N, Kalbe J, Bernkop-Schnurch A. Development of buccal drug delivery systems based on a thiolated polymer. *Int. J. Pharm.* 2003;252:141–148.

68. Leitner VM, Walker GF, Bernkop-Schnurch A. Thiolated polymers: evidence for the formation of disulphide bonds with mucus glycoproteins. *Eur. J. Pharm. Biopharm.* 2003; 56:207–214.

69. Lee J, Young SA, Kellaway IW. Water quantitatively induces the mucoadhesion of liquid crystalline phases of glyceryl monooleate. *J. Pharm. Pharmacol.* 2001;53:629–636.

70. Cilurzo F, Selmin F, Minghetti P, Montanari L. The effects of bivalent inorganic salts on the mucoadhesive performance of a polymethylmethacrylate sodium salt. *Int. J. Pharm.* 2005; 301:62–70.

71. Cilurzo F, Minghetti P, Selmin F, Casiraghi A, Montanari L. Polymethacrylate salts as new low-swellable mucoadhesive materials. *J. Control. Release* 2003;88:43–53.

72. Nagai T. Topical mucosal adhesive dosage forms. *Med. Res. Rev.* 1986;6:227–242.

73. Cardot JM, Chaumont C, Dubray C, Costantini D, Aiache JM. Comparison of the pharmacokinetics of miconazole after administration via a bioadhesive slow release tablet and an oral gel to healthy male and female subjects. *Br. J. Clin. Pharmacol.* 2004;58(4):345–351.

74. Timpe C, Dittgen M, Grawe D, Schumacher J, Zimmermann H, Hoffmann H.Bioadhesive tablet. US Patent 6,063,404, 2000.

75. Van RJ, Haxaire M, Kamya M, Lwanga I, Katabira E. Comparative efficacy of topical therapy with a slow-release mucoadhesive buccal tablet containing miconazole nitrate versus systemic therapy with ketoconazole in HIV-positive patients with oropharyngeal candidiasis. *J. Acquir. Immune Defic. Syndr.* 2004;35:144–150.

76. Tsutsumi K, Obata Y, Nagai T, Loftsson T, Takayama K. Buccal absorption of ergotamine tartrate using the bioadhesive tablet system in guinea pigs. *Int. J. Pharm.* 2002;238:161–170.

77. Ali J, Khar R, Ahuja A, Kalra R. Buccoadhesive erodible disc for the treatment of oro-dental infections: design and characterisation. *Int. J. Pharm.* 2002;283:93–103.

78. Perioli L, Ambrogi A, Angelici F, Ricci M, Giovagnoli S, Capuccella M, Rossi C. Development of mucoadhesive patches for buccal administration of ibuprofen. *J. Control. Release* 2004;99:73–82.

79. Arora S, Kohli K, Ali A. Microbiological evaluation of mucoadhesive buccal tablets of secnidazole against anaerobic strains for the treatment of periodontal diseases. Presented at Controlled Release Society 30th Annual Meeting, 2003, Abstract #376.

80. Ameye D, Mus D, Foreman P, Remon JP. Spray-dried Amioca® starch/Carbopol® 974P mixtures as buccal bioadhesive carriers. *Int. J. Pharm.* 2005;301:170–180.

81. Llabot JM, Manzo RH, Allemandi DA. Double layered mucoadhesive tablets containing nystatin. *AAPS PharmSciTech* 2002;3:1–6.

82. Martin L, Wilson CG, Koosha F, Tetley L, Gray AL, Sevda S, Uchegbu IF. The release of model macromolecules may be controlled by the hydrophobicity of palmitoyl glycol chitosan hydrogels. *J. Control. Release* 2002;80:87–100.

83. Martin L, Wilson CG, Koosha F, Uchegbu IF. Sustained buccal delivery of the hydrophobic drug denbufylline using physically cross-linked palmitoyl glycol chitosan hydrogels. *Eur. J. Pharm. Biopharm.* 2003;55:35–45.

84. Coessens P, Herrebout F, De Boever JA, Voorspoels J, Remon JP. Plaque inhibiting effect of bioadhesive mucosal tablets containing chlorhexidine in a 4-day plaque regrowth model. *Clin. Oral Invest.* 2002;6:217–222.

85. Ameye D, Voorspoels J, Foreman P, Tsai J, Richardson P, Geresh S, Remon JP. *Ex vivo* bioadhesion and *in vivo* testosterone bioavailability study of different bioadhesive

formulations based on starch-*g*-poly(acrylic acid) copolymers and starch/poly(acrylic acid) mixtures. *J. Control. Release* 2002;79:173–182.

86. Geresh S, Gdalevsky GY, Gilboa I, Voorspoels J, Remon JP, Kost J. Bioadhesive grafted copolymers as platforms for peroral drug delivery: a study of theophylline release. *J. Control. Release* 2004;94:391–399.

87. Jain AC, Aungst BJ, Adeyeye MC. Development and *in vitro* evaluation of buccal tablets prepared using danazol-sulfobutylether 7β-cyclodextrin (SBE 7) complexes. *J. Pharm. Sci.* 2002;91:1659–1668.

88. Boaz Mizrahi AJD. Mucoadhesive tablet releasing iodine for treating oral infections. *J. Pharm. Sci.* 2007;96:3144–3150.

89. Acharya RN, Baker JL.Oral transmucosal delivery of drugs or any other ingredients via the inner buccal cavity. US Patent 6,210,699, 2001.

90. Levine W, Saffer A.Transmucosal oral delivery device. US Patent 7,285,295, 2007.

91. Merkle HP, Anders R, Wermerskirchen A. Mucoadhesive buccal patches for peptide delivery. In Lenaerts V, Gurny R. editors. *Bioadhesive Drug Delivery Systems*, CRC Press, Boca Raton, 1990, pp. 105–136.

92. Perioli L, Ambrogi A, Angelici F, Ricci M, Giovagnoli S, Capuccella M, Rossi C Development of mucoadhesive patches for buccal administration of ibuprofen. *J. Control. Release* 2004;99:73–82.

93. Donnelly RF, McCarron PA, Tunney MM, Woolfson AD. Potential of photodynamic therapy in treatment of fungal infections of the mouth design and characterisation of a mucoadhesive patch containing toluidine blue O. *J. Photochem. Photobiol. B: Biol.* 2007;86:59–69.

94. Moro DG, Callahan H, Nowotnik D.Mucoadhesive erodible drug delivery device for controlled administration of pharmaceuticals and other active compounds. US Patent 6,585,997, 2003.

95. Kholi K, Puri A, Arora A, Baboota S. Design and characterisation of novel buccoadhesive film for doxycycline for the treatment of pyorrhea. Presented at Controlled Release Society 30th Annual Meeting, Glasgow, UK, Abstract 445, 2003.

96. Rossi S, Sandri G, Ferrari F, Bonferoni MC, Caramella C. Buccal delivery of acyclovir from films based on chitosan and polyacrylic acid. *Pharm. Dev. Technol.* 2003;8:199–208.

97. Nafee NA, Ismail FA, Boraie NA, Mortada LM. Mucoadhesive buccal patches of miconazole nitrate: *in vitro/in vivo* performance and effect of ageing. *Int. J. Pharm.* 2003;264:1–14.

98. Kockisch S, Rees GD, Young SA, Tsibouklis J, Smart JD. Polymeric microspheres for drug delivery to the oral cavity: an *in vitro* evaluation of mucoadhesive potential. *J. Pharm. Sci.* 2003;92:1614–1623.

99. Kockisch S, Rees GD, Young SA, Tsibouklis J, Smart JD. *In situ* evaluation of drug-loaded microspheres on a mucosal surface under dynamic test conditions. *Int. J. Pharm.* 2004;276:51–58.

100. Govender S, Pillay V, Chetty DJ, Essack SY, Dangor CM, Govender T. Optimisation and characterisation of bioadhesive controlled release tetracycline microspheres. *Int. J. Pharm.* 2005;306:24–40.

101. Giunchedi P, Juliano C, Gavini E, Cossu M, Sorrenti M. Formulation and *in vivo* evaluation of chlorhexidine buccal tablets prepared using drug-loaded chitosan microspheres. *Eur. J. Pharm. Biopharm.* 2002;53:233–239.

102. Friedman D, Schwarz J, Amselem S.Bioadhesive emulsion preparations for enhanced drug delivery. US Patent 5,993,846, 1999.

103. Aksungur P, Sungur A, Unal S, Iskit AB, Squier CA, Senel S. Chitosan delivery systems for the treatment of oral mucositis: *in vitro* and *in vivo* studies. *J. Control. Release* 2004;98:269–279.

104. Jones DS, Lawlor MS, Woolfson AD. Rheological and mucoadhesive characterisation of polymeric systems composed of poly(methylvinylether-*co*-maleic anhydride) and poly(vinylpyrrolidone), designed as platforms for topical drug delivery. *J. Pharm. Sci.* 2003;92:995–1007.

105. Park JS, Yoon JI, Li H, Moon DC, Han K. Buccal mucosal ulcer healing effect of rhEGF/Eudispert hv hydrogel. *Arch. Pharm. Res.* 2003;26:659–665.

106. Petelin M, Pavlica Z, Bizimoska S, Sentjurc M. *In vivo* study of different ointments for drug delivery into oral mucosa by EPR oximetry. *Int. J. Pharm.* 2004;270:83–91.

107. Sigardsson HH, Knudsen E, Loftsson T, Leeves N, Sigurjonsdottir JF, Masson M. Mucoadhesive sustained drug delivery system based on cationic polymer and anionic cyclodextrin/triclosan complex. *J. Incl. Phenom. Macrocycl. Chem.* 2002;44:169–172.

108. Portero A, Tiejeiro-Osorio D, Alonso MJ, Remunan-Lopez C. Development of chitosan sponges for buccal administration of Insulin. *Carbohydr. Polym.* 2007;68:617–625.

109. Patel D, Smith AW, Grist N, Barnett P, Smart JD. *In vitro* mucosal model predictive of bioadhesive agents in the oral cavity. *J. Control. Release* 1999;61:175–183.

110. Patel D, Smith JR, Smith AW, Grist NW, Barnett P, Smart JD. An atomic force microscopy investigation of bioadhesive polymer adsorption onto human buccal cells. *Int. J. Pharm.* 2000;200:271–277.

111. Kockisch S, Rees GD, Young SA, Tsibouklis J, Smart JD. A direct-staining method to evaluate the mucoadhesion of polymers from aqueous dispersion. *J. Control. Release* 2001;77:1–6.

112. Keegan G, Smart JD, Ingram M, Barnes L, Rees G, Burnett G. An *in vitro* assessment of bioadhesive zinc/carbomer complexes for antimicrobial therapy within the oral cavity. *Int. J. Pharm.* 2007;340:92–96.

113. Ungphaiboon S, Maitani Y. *In vitro* permeation studies of triamcinolone acetonide mouthwashes. *Int. J. Pharm.* 2001;220:111–117.

114. Smart JD. Lectin-mediated drug delivery in the oral cavity. *Adv. Drug Deliv. Rev.* 2004;56:481–489.

115. Smart JD, Nantwi PKK, Rogers DJ, Green KL. A quantitative evaluation of radiolabelled lectin retention on oral mucosa *in vitro* and *in vivo*. *Eur. J. Pharm. Biopharm.* 2002;53:289–292.

116. Macintyre J, Robinson S, Norris E, Appleton R, Whitehouse WP, Phillips B, Martland T, Berry K, Collier J, Smith S, Choonara I. Safety and efficacy of buccal midazolam versus rectal diazepam for emergency treatment of seizures in children: a randomized controlled trial. *Lancet* 2005;366:205–210.

117. Hoogstraate JA, Wertz PW. Drug delivery via the buccal mucosa. *Pharm. Sci. Technol. Today* 1998;1:309–316.

118. Bottenburg P, Claymaet R, De Muynck C, Remon JP, Coomans D, Michotte Y, Slop D. Development and testing of bioadhesive fluoride-containing slow-release tablets for oral use. *J. Pharm. Pharmacol.* 1991;43:457–464.

119. Smart JD, Mortazavi SA. An investigation of the pH within the hydrating gel layer of a poly(acrylic acid) compact. *J. Pharm. Pharmacol.* 1995;47:1099.

120. Llabot JM, Manzo RH, Allemandi DA. Drug release from carbomer:carbomer sodium salt matrices with potential use as a mucoadhesive drug delivery system. *Int. J. Pharm.* 2004; 276:59–66.

12

ORAL TARGETED DRUG DELIVERY SYSTEMS: GASTRIC RETENTION DEVICES

HOSSEIN OMIDIAN

Department of Pharmaceutical Sciences, Health Professions Division, Nova Southeastern University, Fort Lauderdale, FL, USA

KINAM PARK

Departments of Biomedical Engineering and Pharmaceutics, Purdue University, School of Pharmacy, West Lafayette, IN, USA

12.1 INTRODUCTION

Drug delivery in general is simply a two-step process: entrapment and release. In controlled drug delivery, a given drug needs to be entrapped in a platform, device, or any matrix that can later on be released in a controlled way. If the drug is intended to be released at a specific site or organ and taken orally, the whole system is called oral targeted drug delivery system. One of the targets in oral controlled delivery is the stomach area. The stomach can be either a target organ or can serve as a reservoir to release the drug at the specific site. In either case, the drug is required to stay in the stomach area and the challenge will be to find ways to do the task.

Maintaining a drug delivery system in the stomach is beneficial for most drugs, especially for those of which absorption is not homogeneous throughout the gastrointestinal (GI) tract. Some drugs are absorbed only from the upper small intestine, that is, duodenum and jejunum, and in most cases, absorption decreases in the lower small intestine and large intestine. Such a decrease in drug absorption is known to be an absorption window. Those drugs with absorption windows usually result in poor bioavailability. When the drug has an absorption window in the upper small intestine, it cannot be efficiently formulated into conventional dosage forms. These drugs require longer retention in the stomach to offer better bioavailability. Since the residence time of the conventional dosage form is short, the amount of drug release

in the stomach and upper intestine will be small, which significantly decreases the bioavailability [1].

Gastric retention (GR) is beneficial not only for prolonging the residence time of an oral dosage form in the stomach, but also for local drug delivery in both the stomach and the upper small intestine. GR may be ideal for delivery of a drug for treating peptic ulcers. GR, however, is not for some drugs. Those drugs that can cause gastric legions, for example, aspirin and nonsteroidal anti-inflammatory drugs, are not good candidates for sustained delivery by GR. In addition, drugs that are unstable in acidic condition are not suitable for formulation in a gastroretentive system. The drugs that are absorbed equally well throughout the GI tract, such as isosorbide dinitrate, also do not benefit from a gastroretentive system. In general, drugs that can benefit from a longer retention in the stomach area are those that act locally in the stomach, are primarily absorbed in the stomach, are poorly soluble at an alkaline pH, have a narrow window of absorption, are absorbed rapidly from the GI tract, and degrade in the colon.

Achieving GR requires a platform that can withstand the contractile forces resulting from peristaltic waves in the stomach as well as constant grinding and churning processes. A GR platform requires two properties that seem to be contradicting each other: resisting gastric emptying during drug release and easy emptying after its purpose is served. The approaches used to develop GR platform include low-density floating systems, bioadhesives, and expandable and

Oral Controlled Release Formulation Design and Drug Delivery: Theory to Practice, Edited by Hong Wen and Kinam Park

swellable systems [1, 2]. Practically, floating and expanding approaches have reached to a better position of acceptability to be exploited in gastroretentive devices. The floating approach relies on natural GI physiology and provides retention through delayed emptying from the fed stomach. On the other hand, larger size dosage forms with or without floatability or bioadhesion properties are designed to fight the stomach physiology and emptying in the fasted state [3]. An alternative to gastric retention devices is to use multiple-unit systems, by which some, not all, units will gain a better chance of retention in the gastric medium. Typically, this method is employed to enhance the retention of a floating gastroretentive system. Another enhancement that is often used is the mucoadhesion. This is achieved by coating a GR system with a material that can stick to the stomach wall. Many GR systems combine the two main methods, in the hope that if one method does not work, the other will. Davis [4] has reviewed different gastric retention systems with the focus on bioadhesive microspheres and large single-unit systems. Although gastric retention technologies are intended to enhance local drug delivery, they all suffer from lack of specificity, mucus turnover, and low resistance to gastric forces. The muco- or bioadhesive system is claimed to be promising to address those concerns because it can penetrate into the mucus layer and can prolong the drug activity at the mucus–epithelial layer [5].

Discovery of *Helicobater pylori* by Warren and Marshall about 20 years ago triggered the development of efficient gastroretentive dosage forms. *Helicobacter* is a Gram-negative bacterium with the ability to grow in a very acidic environment of the human stomach by producing urease enzyme that significantly increases the stomach pH. Bardonnet et al. [6] reviewed the state of the art of the gastric retention with the particular coverage on GR dosage forms designed for *Helicobacter pylori* infection. The idea is that *H. pylori* infection can be better treated if the therapeutic agent (antibiotic) can stay local and for a longer period of time at the site of infection. Atherton et al. [7] studied intragastric distribution and gastric retention of the therapeutic agent in the presence of food and omeprazole, which lowers the acid secretion. The study shows that food has a strong positive effect on the local delivery of the drug and omeprazol can improve the effect even more. Since the concept of gastric retention is perceived promising in increasing the drug bioavailability and treating local diseases, a number of specialty pharmaceutical companies have focused their efforts to better understand the concept and to innovate more efficient technologies and platforms. Among GR technologies, mucoadhesion, expansion, and density systems have been taken to the clinical trials more frequently. To the same extent as a new technology, it has been found that calorie intake can have a notable effect on gastric retention as retention time was found to be proportional to the calorie intake. Even simple non-disintegrating controlled release tablets have displayed sig-

nificant retention in fed state when a high-calorie meal is taken. Nevertheless, the goal is still valuable, especially to achieve retention in the fasted state [8].

Although many research activities have been conducted in the GR area by using individual and combined mechanisms of retention, implementation and prediction of *in vivo* behavior still remains very challenging for a successful clinical practice [9, 10]. In studying gastroretentive systems, the first step will be the proof-of-principle in animals. To be able to extend the results from animal models to human subjects, the selected animal model should have GI characteristics similar to humans. Since the issue is very challenging, the animal data on gastrointestinal transit should be interpreted with considerable caution [11]. In fact, gastric retention is one of those areas for which a reliable proof-of-principle requires a human trial.

12.1.1 Gastrointestinal Absorption

It is needless to say that oral delivery is the most preferred route of drug administration. After entering into the mouth, the drugs start their journey across the GI tract from the mouth to the anus. To be therapeutically effective, the drug should ideally have a long half-life with even absorption across the GI tract. Drugs with a narrow absorption window require frequent dosing, which is perceived as being patient unfriendly. To help the drug become absorbed in its absorption area, care must be taken to maintain the drug in and around that particular segment. In this way, the drug will be released over a longer period of time and will be absorbed at the right place. The outcome of this approach will be an increased bioavailability of the drug and patient compliance. The majority of drugs are absorbed in the upper intestine (duodenum), and hence they need to be maintained in the stomach. The approaches that will help the drug to be retained in the stomach are called gastric retention technologies and the devices designed based on these approaches are called gastric retention devices.

12.1.2 Stomach Physiology and Function

Anatomically, the stomach is composed of three compartments: fundus, body, and antrum. The fundus and body (proximal stomach) serve as a reservoir for the ingested material, whereas the distal stomach (antrum) is a mixing site and acts as a drain pump to the duodenum. The fundus generates a positive pressure (due to gas accumulation) that helps the stomach contents to move along. The stomach serves as a food reservoir, food processor, food sterilizer, and food delivery port to the intestine. Food will enter and exit the stomach through the cardiac and pylorus, respectively. When the stomach starts to process the food, the cardiac is closed to prevent the food from entering the esophagus. The pylorus, on the other hand, sieves the processed food to prevent the large objects and undigested foods from entering into the upper intestine. The stomach is not a fixed size container and its size and volume vary with the size and type of its contents. After ingestion, the food and other ingested materials (including drug) are mixed with the gastric juice containing water, HCl, electrolytes, pepsin, protein, and mucus. Depending on the gastric content, HCl secretion will be different, and hence the pH of the gastric medium. Food, on the other hand, may buffer and neutralize the gastric acid, which increases the pH.

While the absorption of drug molecules from the stomach is generally not great compared to the intestinal absorption (due to the surface area of the epithelium), gastric emptying is an important physiological event that significantly influences the uptake of drug substances from the intestine [12, 13].

12.1.3   Gastric Retention and Emptying

If a drug can stay in the gastric medium longer, this means it will be emptied from the stomach with delay. Therefore, factors affecting gastric emptying will be affecting retention status of a given drug in the gastric medium. The stomach itself can be considered a critical variable as it provides a dynamic service medium that varies on an intrasubject and intersubject basis. Subjects take different foods at different times, with or without drink, and the duration of chewing process is notably subjective. A single subject does not follow the same eating habits everyday. When a drug starts its journey throughout the GI tract, its fate will be quite unpredictable because many variables are involved. These include type of food (digestible, indigestible, oil content, calorie content, solid content, temperature), stomach physiology (stomach size, pH, contents), stomach mode (fasted, fed), eating habit, mental behavior and status (level of stress) during eating, so on. Foods have different calories depending on the amount of protein, carbohydrate, and fat. High-calorie food provides predictable emptying time of around 2–3 h. Water, tea, coffee, soda, and juices all have different characteristics; they provide different pH mediums and promote gastric emptying in most of the cases. Solids follow a zero-order emptying kinetics, while fluids are emptied from the stomach following a first-order kinetics. Indigestible components and larger objects have longer retention in the stomach. Body postures such as lying on right, left or back, sitting upright, and standing during and after eating are also very important factors. The drug may be taken with or without food. With food, drug–food interaction requires consideration. Without food, unpredictable gastric emptying time will be a concern. In a fasted state, chyme can leave the stomach anytime from a few minutes to 3 h, which offers absolutely unpredictable retention time.

12.1.4   Gastric Motility

The stomach can assume a fasted state when it is empty and a fed state when it is full. Followed by emptying chyme into the upper intestine, the stomach will be left with the residues of mucus and undigested solids. These remain in the stomach until absorption in the intestine is completed, which takes about 2 h. At this point, the digestive phase of stomach activity stops and is replaced by interdigestive phase. All unprocessed residues left in the stomach are removed by migrating myoelectric complex (MMC) forces or the so-called housekeeper waves [13], which are the strongest stomach contractions of the digestive and interdigestive phases. In designing a gastric retention device, this is the force that requires careful consideration because of its strength and suddenness.

12.1.5   Drugs and Gastric Retention

Generally, drugs that act locally in the stomach (antacids), are primarily absorbed in the stomach and upper intestine (albuterol), are poorly soluble at an alkaline pH, have a narrow absorption window (riboflavin and levodopa), absorb rapidly from the GI tract (amoxicillin) and degrade in the colon (metoprolol), can be good candidates for gastric retention technologies. On the other hand, drugs that irritate the stomach lining, are unstable in the acidic environment, and have equal absorption throughout the GI tract would definitely not be the beneficiary. Drugs for viral infection especially herpes infections (acyclovir, zovirax®), hypertension (captopril, capoten®), diuretic and antihypertensive (furosemide, lasix®), type 2 diabetes (metformin, glucophage®), controlling seizure (gabapentin, Neurontin®), Parkinson's disease (levodopa, dopar®), muscle relaxation (baclofen, lioresal®), and antibiotic (ciprofloxacin, cipro®) are good candidates for gastric retention [4]. The market for the drugs that would potentially benefit from gastric retention technologies amounts to over $20 billion. Drugs studied in this research area are compiled in Table 12.1.

12.2   DESIGN OF A GASTRIC RETENTION DEVICE

Designing successful gastroretentive devices requires overcoming many important challenges ranging from small size for easy administration to removal from the stomach after the drug release is complete. They are listed in Table 12.2.

12.3   REVIEW OF GASTRIC RETENTION TECHNOLOGIES

The performance of the floating GR systems is critically dependent on the size of the platform, fasted or fed state, and the subject posture, upright or supine (lying on the back). In an upright position, floating platforms can stay continuously buoyant irrespective of their size, whereas nonfloating systems rapidly and irreversibly sink after administration. Therefore, the floating forms show prolonged and more reproducible gastric retention times compared to the nonfloating unit. The significance and extent of prolongation

TABLE 12.1 Drugs Suitable for Delivery by Gastroretentive Systems

Drug	Effect	Brand Name	References
Acetohydroxamic acid	Ureas inhibitor to treat urinary infections	Lithostat®	[14–16]
Acyclovir	Treats viral infections	Zovirax®	[17]
Amoxicillin	A penicillin antibiotic	Amoxil®	[18, 19]
Cefuroxime Axetil	Antibiotic	Ceftin®	[20]
Celecoxib	Reduces inflammation	Celebrex®	[21]
Ciprofloxacin	Antibiotic	Cipro®	[22, 23]
Clarithromycin	Antibiotic	Biaxin®	[24]
Diltiazem HCl	Calcium channel blocker	Tiazac®	[25, 26]
Famotidine	Treats stomach and intestinal ulcers	Pepcid®	[27]
Furosemide	A diuretic	Lasix®	[28]
Gabapentin	Treats partial seizures in adults with epilepsy; relieve nerve pain	Neurontin®	[29]
Ibuprofen	Anti-inflammatory	Motrin®	[30, 31]
Levodopa	Treats Parkinson's disease	Dopar®	[32–35]
Meloxicam	Pain reducer in osteoarthritis or rheumatoid arthritis	Mobic®	[36]
Metformin HCl	Treats type 2 diabetes	Glumetza™, Fortamet®	[37–40]
Metoclopramide	Relieve symptoms of gastroesophageal reflux, helps stomach motility	Reglan®	[31, 41]
Metoprolol tartrate	A beta-blocker, reduces the heart workload	Lopressor®	[42, 43]
Metronidazole	Kills or prevents the growth of certain bacteria and protozoa	Flagyl®	[44, 45]
Ofloxacin	Antibiotic	Floxin®	[46–48]
Omeprazole	Prevents stomach acid production	Prilosec®	[49]
Ranitidine HCl	Blocks release of stomach acid; treats stomach or intestinal ulcers	Zantac®	[50, 51]
Repaglinide	Treats type 2 diabetes mellitus	Prandin®	[52]
Riboflavin	Treats vitamin B_2 deficiency	Vitamin B_2	[53–58]
Tetracycline	Antibiotic	Sumycin®	[45]
Theophylline	Bronchodilator	Theo-Dur®	[59]
Verapamil HCl	Calcium channel blocker	Calan®	[26, 60, 61]

however is size dependent: as the platform becomes smaller, the effect becomes bigger. Moreover, there is no significant difference between the mean GRTs of the small, medium, and large floating units ($p > 0.05$). To the contrary, the retention of the nonfloating units in the stomach is only size dependent. In supine subjects, both floating and nonfloating units show better retention as they become larger. Generally, upright posture provides better and reproducible retention

TABLE 12.2 Challenges to Overcome for Designing Gastroretentive Devices

Retention	Device is required to stay in the stomach for at least 6 h, preferably in the fasted state; this will be the gold standard; DepoMed's metformin HCl ER for type 2 diabetes and ciprofloxacin HCl for urinary infections both require high-calorie meal to display gastric retention; Intecpharma's Accordion Pill displays a gastric retention of 6 h after low-calorie meal
Administration	Platform needs to be administered orally and has to be in tablet or encapsulated form
Design	Should have a simple design for a feasible scale-up
Drug loading	Should house a sensible amount of the drug
Controlled drug release	Should release the drug in a controlled manner
Removal	Has to be removed from the body after its service is completed, can be biodegradable or disintegrable
Effect on gastric motility	Should not affect gastric emptying pattern
Adverse effects	Should not impose local irritation and sensitization
Safety and toxicity	Should not be physically threatening like esophagus obstruction
Strong	Should be mechanically strong to resist stomach pressures especially during housekeeping contractions in fasted state
Animal models	Proper animal model should be selected for the proof-of-principle studies; dogs and pigs are not good models for gastric retention studies; GR proof of principle should be conducted in humans
Study design	Clinical study design would be very challenging because of numerous influential factors
Feasibility	Design should be feasible in terms of materials, equipments, and technology

compared to the supine position. As far as the food content of the stomach is concerned, all floating units are emptied much later in the frequently fed subjects than those who were fed a single meal [62]. The gastric residence times and transit behavior of floating and nonfloating hydrophilic matrices in different sizes have been examined *in vivo* with changing the subject position. In upright subjects, the floating forms were protected against postprandial emptying. The study shows high variability of gastric residence time with nonfloating forms and its strong dependency on the dosage size [63].

12.3.1 Density-Based Systems

Density-based systems can be classified as high and low density. Theoretically, a dense material will sit longer at the bottom of the greater curvature of the stomach. Although ruminants showed very promising results, the system has not proved to be effective in man. On the other hand, many systems reported in the literature are based on low-density floatable units and their retention requires the presence of fluid (for floatation) and food (food effect). These systems are generally emptied from the stomach at the end of the digestive phase or at the beginning of the interdigestive phase when the stomach applies its housekeeping sweeping forces. Hydrodynamically balanced systems (HBS) are the first generation of low-density formulations. The system contains a gel-forming high molecular weight hydrophilic polymer. The polymer in contact with the stomach fluid starts to swell, forms a loose gelatinous layer, and finally floats. Contraction and friction forces inside the stomach erode the swollen gel layer by layer and drug will be released via diffusion and erosion mechanisms. To increase the buoyancy of these systems, a gas-generating agent such as bicarbonate is added to the formulation. As bicarbonate reacts with the stomach acid, carbon dioxide gases are generated and trapped in the gelling polymer, helping the whole system to float. The system can be made as a single unit (one piece) or multiple units (many pieces). Another approach will be the use of low-density polymers (polyethylene, polypropylene) as a core or the use of hollow microcapsules or microspheres. Floating is a very complex concept and not all drugs respond similarly to this approach. Single- and multiple-unit hydrodynamically balanced systems, single- and multiple-unit gas-generating systems, hollow microspheres, and raft-forming systems are typical floatable GR systems. The floating system is generally multifaceted as its efficacy is dependent on so many factors including gastrointestinal physiology, dosage form characteristics, and patient-related factors [64]. Singh and Kim [65] have a detailed review on patents and marketed products related to the floating delivery system, and their advantages and potential for controlled oral delivery. Arora et al. [66] also have a complete review on floating-based gastric retention systems with the focus on principles of the floatable systems, physiological and formulation variables,

and single and multiple-unit systems. This review also summarizes the *in vitro* techniques and *in vivo* studies to evaluate the performance and application of floating systems. Floating and nonfloating versions of lyophilized calcium alginate multiple-unit systems have been studied *in vivo*. After a standard breakfast, radiolabeled formulations were administered to seven healthy volunteers and monitored by gamma scintigraphy. With all subjects, floating formulations offered a prolonged gastric retention time of over 5.5 h, while the mean emptying time for the nonfloating units were found to be about 1 h [67]. The study shows that floating systems can have significant positive effect on gastric retention in the fed state.

12.3.1.1 Conventional Floatable Systems Once daily sustained release ofloxacin tablet has been formulated into a gastroretentive platform using psyllium husk, HPMC K100M, and crospovidone. Different formulations were evaluated for buoyancy lag time, duration of buoyancy, dimensional stability, drug content, and *in vitro* drug release profile. While psyllium husk increases the dimensional stability of the dosage form, an increased rate of drug release is due to the crospovidone content. The optimized formulation was tested in 24 healthy human subjects and the pharmacokinetic data showed bioavailability similar to the marketed zanocin [46]. Dorozynski et al. [68] prepared HBS by filling hard gelatin capsules with different polymers including chitosan, sodium alginate, and hydroxypropyl methylcellulose (HPMC). They measured density, hydration, erosion, and floating force of the platforms and found that the erosion and swelling play a dominant role in flotation of the dosage forms. The floating properties of the dosage were found to be reliant on the polymer type and the fasted/fed state. Besides, the size of the HBS influenced the floating force value. Whitehead et al. [69] prepared floatable freeze-dried calcium alginate beads to release amoxicillin in solution and suspension forms. The study shows that the type of dosage form can play a significant role in drug release sustainability. They showed that the solution drug has better sustained release properties at the expense of loading capacity. On the other hand, beads containing suspension drug allowed higher loading capacity, but at the expense of lower buoyancy and hence faster release. A 23 factorial design has been employed to optimize a floating gastroretentive delivery system containing metoprolol tartrate. A study shows that the total polymer to drug ratio and polymer to polymer ratio significantly affect the floating time and the release properties, but the effect of different grades of HPMC was found to be insignificant [43].

12.3.1.2 Gas-Driven Floatable Systems In an earlier study, Desai and Bolton [70] formulated a floatable tablet by entrapping air and oil into the agar gel network. A 300 mg

theophylline tablet with the density of 0.67 was prepared and compared to the commercial Theo-Dur® *in vitro* and *in vivo*. The human study in fasted and fed conditions showed that the effect of food was much more pronounced in increasing gastric retention time than floatability. A floating multiple-unit system composed of a calcium alginate core, air, and calcium alginate/polyvinyl alcohol membrane has been developed. A clinical study was designed based on three regimens (fasted state, fed state after single meal, and fed state after frequent meals) using both floating and control dosage forms. Gastric retention was monitored by X-ray technique by loading dosage forms with barium sulfate. Unlike the control, the floating system remained buoyant on the gastric content under both fasted and fed states. In the fasted state, the buoyancy did not help the gastric residence time, but the floating units could stay buoyant for 5 h and could prolong the GRT by about 2 h in fed state after a single meal. With the regimen three, most floating units showed floatation of about 6 h and GRT was prolonged about 9 h compared to the control [71]. Atyabi et al. [59, 72] developed beads of the ion exchange resin coated with a semipermeable membrane and loaded with bicarbonate and the active drug. In the stomach, the bicarbonate component reacts with the gastric acid and generates carbon dioxide gases. The semipermeable membrane will act as a barrier and gases will be entrapped inside the beads. The approach has been tried with theophylline as a model drug. Authors used gamma scintigraphy and showed that coated resins could significantly ($P < 0.002$) improve the gastric residence time of the model drug. Bajpai and Tankhiwale [73] prepared a floatable GR platform based on Ca cross-linked alginate/dextran beads loaded with sodium bicarbonate. The beads containing higher concentrations of bicarbonate and calcium displayed 100% buoyancy and remained buoyant for longer than 20 h. Choi et al. [74] prepared floating beads from a sodium alginate solution containing calcium carbonate and sodium bicarbonate as porogens. The alginate solution was dropped into 1% calcium chloride solution (to promote alginate gelation) containing 10% acetic acid (to generate gas). The study shows that the size and floating properties of the beads increase with an increase in the amount of gas-forming agent. The authors used mercury porosimetry and scanning electron microscopy to examine the porosity, pore size, and morphology of the beads. Porosity and pore size were significantly increased with $NaHCO_3$ than with $CaCO_3$. On the other hand, smoother beads were obtained with $CaCO_3$. With no surprise, the gel strength was found to be decreasing with an increase in gas-forming agent. The effect has been more pronounced with $NaHCO_3$ than with $CaCO_3$.

Singh et al. [41] formulated metoclopramide hydrochloride into the floating matrix tablets to improve its bioavailability. Tablets were composed of guar gum, karaya gum, HPMC E15 without and with HPMC K15M, calcium carbonate, and citric acid. Study shows that gas porogens contribute to initial floating of the tablets and faster drug release, while HPMC K15 is in charge of integrity of the platform and sustainability of the drug release. Gastroretentive delivery systems based on guar gum, xanthan gum, and hydroxypropyl methylcellulose have been suggested for ranitidine hydrochloride. Sodium bicarbonate was added as porogen to make the platform floatable. With this research, the effects of citric and stearic acids on the drug release profile and floating properties were investigated. This study shows that the GR formulations containing low amounts of citric acid and high amounts of stearic acid favor sustained release of ranitidine hydrochloride [50]. The sustained release floating systems using different hydrocolloid polymers including HPMC, hydroxypropylcellulose (HPC), ethylcellulose (EC), and Carbopol (CP) have been reported for verapamil hydrochloride. Floating was achieved by adding an effervescent mixture of sodium bicarbonate and anhydrous citric acid [60]. Ranitidine hydrochloride, a water-soluble drug, has been formulated as a floating tablet composed of the drug, HPMC and sodium bicarbonate. The study shows that floating properties and drug release were strongly dependent on the tablet composition and its mechanical strength. The time taken for tablets to emerge on the water surface (floating lag time), or buoyant time, and the time during which the tablets constantly float on the water surface (duration of floating) were evaluated in a dissolution vessel. Drug release from the tablets was found to be sufficiently sustained (12 h) compared to conventional tablets. The dosage form has been tested in fasted rabbits, and the pharmacokinetic parameters and hence relative bioavailability of floating and conventional tablets were reported [51]. Metformin hydrochloride with better absorption in the upper intestine was formulated into a floating matrix tablet containing sodium bicarbonate (porogen) and hydroxypropyl methylcellulose (gelling agent). The formulation was optimized on the basis of floating ability and *in vitro* drug release, as well as hardness, weight uniformity, and friability. A formulation containing 500 mg metformin, 75 mg sodium bicarbonate, 170–180 mg HPMC K4M, 15–20 mg citric acid, and 32–40 mg polyvinyl pyrrolidone K90 with the hardness between 6.8 and 7.5 kg/cm^2 stayed afloat for more than 8 h with promising drug release results [39]. Citric acid within the formulation might also be partially responsible for some increase in gastric retention as previously reported. Diltiazem hydrochloride (DTZ) floatable tablets were prepared by direct compression technique, using hydroxypropylmethylcellulose (HPMC, Methocel K100M CR), Compritol 888, sodium bicarbonate, and succinic acid. The study indicates that floating and controlled release features are favored at high concentrations of HPMC and compritol [25]. Famotidine floatable tablets were prepared from methocel K100 and methocel K15M, sodium bicarbonate, and citric acid. The floating tablets were evaluated for weight uniformity, hardness, friability, drug content, *in vitro* buoyancy, and dissolution properties. The swollen tablets remained buoyant for 6–10 h. A longer floating period was favored by using methocel K100 and lower amounts of citric acid. The drug release from the tablets was found to be sufficiently sustained [27]. A coated bilayer multiple-unit floating delivery system was prepared by an extrusion–spheronization process. The inner layer is the effervescent layer composed of sodium bicarbonate and the outer layer is an emulsion-borne polymeric membrane. With no surprise, the authors found that the time to float decreases with an increase in the amount of the effervescent agent and with a decrease in the coating or membrane thickness. The system provided a rapid floating but sustained release properties [75]. Utilizing melt granulation and compression processes, sustained release minitablets with floating property have been prepared. Formulations were composed of a meltable binder, a swellable hydrocolloid, and a mixture of gas-generating agents. A small-scale clinical study on riboflavin-loaded samples showed that riboflavin concentration in urine was higher in floating than nonfloating samples. Since riboflavin has a narrow window of absorption in the upper part of GI tract, this observation was attributed to the gastric retention ability of the floating samples [57]. Floating pellets were prepared by melt granulation of a mixture of Compritol® and Precirolo as meltable binders and sodium bicarbonate and tartaric acid as gas-generating agents. Riboflavin-loaded floating pellets (FRF) were administered orally to nine healthy volunteers and compared with nonfloating pellets. An increase of urinary excretion of riboflavin with floating pellets was related to the gastric retention property of the floatable pellets [58]. Hot melt extruded floating tablets containing Eudragit RS PO and/or Eudragit EPO, as well as sodium bicarbonate, were prepared and evaluated for the controlled release of acetohydroxamic acid (AHA) and chlorpheniramine maleate. Prepared tablets were compared with direct compression tablets. The hot melt tablets could stay afloat for 24 h and displayed sustained release properties, while DC tablets showed no buoyancy but fast drug release [16].

12.3.1.3 Hollow or Low Density Cores
Low-density porous carriers are widely used in pharmaceutical applications. Absorption and release of ibuprofen from microporous polypropylene carriers have been studied. Ibuprofen was adsorbed on the polymer using methanol and dichloromethane [30]. Repaglinide was successfully formulated into hollow microspheres of cellulose acetate butyrate (CAB) and poly(ethylene oxide) (PEO) prepared by emulsion–solvent evaporation technique. Various formulations were prepared by varying the ratio of CAB and PEO, drug loading, and concentration of poly(vinyl alcohol) (PVA) solution. Microspheres (160–600 μm in size) were found to be porous at the surface and displayed an encapsulation efficiency of 95%, floatability longer than 10 h, an excellent flow and good packing properties. *In vitro* release studies in SGF indicated the dependence of release rate on the extent of drug loading and the amount of PEO [52]. Verapamil HCl was formulated into multiparticulate floating GR system based on polypropylene foam powder, Eudragit RS, ethyl cellulose, or poly(methyl methacrylate) utilizing an O/W solvent evaporation method. The microparticles were irregular in shape, highly porous, and had high encapsulation efficiency with good floating behavior. The release rate increased with increasing drug loading and decreasing polymer content [61]. In recent years, different multiparticulate floatable drug delivery systems have been developed that have clear advantages over single-unit dosage forms. The floatable hollow microspheres are developed for peroral drug administration using solvent evaporation technique [76]. Model drugs including chlorpheniramine maleate, diltiazem HCl, theophylline, or verapamil HCl were formulated into floating microparticles consisting of polypropylene foam powder, Eudragit RS, or poly(methyl methacrylate) by soaking the microporous foam carrier with an organic solution of drug and polymer, followed by drying. Major advantages of this technique are reported to be short processing times, no exposure to high temperatures, the possibility to avoid toxic organic solvents, high encapsulation efficiencies close to 100%, good *in vitro* floating, and a wide range of release patterns. The low-density microparticles were compressed into rapidly disintegrating tablets to provide an orally administrable dosage form [26]. A multiparticulate floating and pulsatile delivery system was developed using porous calcium silicate and sodium alginate for the time- and site-specific delivery of meloxicam. Meloxicam was adsorbed onto the silicate surface and calcium alginate beads were prepared by ionotropic gelation. Prepared beads were spherical and 2.0–2.7 mm in diameter, with crushing strength of 182–1073 g and drug entrapment efficiency of 70–94%. Floating time for the beads was controlled by the bead density and hydrophobic character of the drug. Formulations show a lag period ranging from 1.9 to 7.8 h in acidic medium followed by a rapid meloxicam release in simulated intestinal fluid. The authors claim that pulsatile release of meloxicam can be useful in chronopharmacotherapy of rheumatoid arthritis [36]. Floating microcapsules containing melatonin were prepared by the ionic interaction of chitosan and a negatively charged surfactant, sodium dioctyl sulfosuccinate (DOS). The infrared spectroscopy, differential scanning calorimetry (DSC), solubility, and X-ray diffraction were used to verify complex formation. The floating microcapsules were compared with the conventional nonfloating microspheres made of chitosan and sodium tripolyphosphate (TPP). The DOS/chitosan spheres were hollow in the core and provided drug loading capacity of 31–59%. Moreover, floating

microspheres could stay afloat for longer than 12 h and provided sustained release of several hours, while the release from a nonfloating unit was almost instant. DOS/chitosan microcapsules did not dissolve and swell in the SGF in a 3 days period, while TPP/chitosan units swelled remarkably in SGF and lost their integrity in 5 h [77].

12.3.1.4 Miscellaneous Floating Systems

Metronidazole (MZ), a common antibacterial drug used in the treatment of *H. pylori*, was loaded into chitosan-treated alginate beads prepared by ionotropic gelation. To prepare beads, methylcellulose, carbopol 934P, and kappa carrageenan were used as a viscosifying agent. A low-density magnesium stearate was also used to impart floatability. The study shows beads containing 0.5% κ-carrageenan, 0.4% chitosan, and 5% magnesium stearate offer immediate buoyancy, optimum drug entrapment efficiency, and extended drug release. The MZ-loaded floating beads and MZ suspension administered orally to *H. pylori* infected mice and the *H. pylori* clearance was tested in both cases *in vivo*. The study shows that a 15 mg/kg dose of MZ floating beads provided 100% clearance rate, whereas a 33.33% clearance rate was achieved using a 20 mg/kg dose of MZ suspension [44]. A floating dosage form based on freeze-dried calcium alginate beads was developed to increase riboflavin bioavailability. Gamma scintigraphy was used to monitor the movement of the calcium alginate beads *in vivo*. Urine riboflavin concentrations was measured by HPLC and used for bioavailability evaluations. The interesting outcome of this study is the effect of citric acid on delaying gastric emptying. The study finds that the bioavailability of riboflavin can be improved in the fasted state because citric acid is included into the dosage form [53]. An in situ foaming and gelling system has been developed for controlled delivery of amoxicillin for the treatment of peptic ulcer disease caused by *H. pylori*. The formulation is composed of gellan, water, sodium citrate, drug, and calcium carbonate. The study finds that the concentration of gellan gum and calcium carbonate significantly affected the *in vitro* drug release from the floating platform. The *in vivo* study conducted on infected Mongolian gerbils shows that the amount of amoxicillin needed to eliminate *H. pylori* is 10 times less than that of the suspension dosage form when amoxicillin is formulated into a gellan-based floating system [19]. Gellan-based in situ gelling system has been studied for potential treatment of gastric ulcers, associated with *H. pylori*, using clarithromycin. The formulation was prepared by dissolving gellan in DI water to which drug and sucralfate are added and dispersed well. The concentrations of gellan gum and sucralfate significantly influenced the rate and extent of the drug release *in vitro*. The addition of sucralfate to the formulation significantly suppressed the degradation of clarithromycin at low pH. Moreover, the amount of clarithromycin needed for elimination of *H. pylori* was found to be less from the gelling platform than from the clarithromycin suspension [24]. Lipid-coated pellets have been studied as a floatable delivery system. Metoprolol tartrate and hydrogenated soybean oil have been used as a water-soluble drug and lipid carrier, respectively. The floatable pellets were prepared by hot spraying the drug-in-lipid dispersion onto the pellets. Regardless of the coating amount, the lipid-coated pellets displayed a good floating property *in vitro* [42]. Multi-unit floating beads were developed based on calcium alginate, sunflower oil, and a drug of interest through an emulsification/gelation process. As opposed to oil-free beads, oil-loaded beads could continuously float over 24 h periods under constant agitation. Ibuprofen, niacinamide, and metoclopramide HCI that have different water solubility could be released in sustained form from these platforms [31]. Celecoxib has been formulated as microspheres to increase its bioavailability through increase in its gastric retention. The microspheres were prepared through solvent diffusion technique (dichloromethane and ethanol) using poly(ethylene oxide), Eudragit S, cellulose acetate, and Eudragit RL. Although formulations were found to be very different in terms of topography, diffractometry, and DSC data, the formulation containing PEO–Eudragit S–Celecoxib (2 : 2 : 1) offered the best *in vitro* release [21]. Ofloxacin has been formulated into a floatable system for prolonged local delivery to infected *H. pylori* area. The hydrodynamically balanced capsules were prepared by physical mixing of various grades of HPMC and PEO. Cellulose acetate phthalate, liquid paraffin, and ethyl cellulose were used as release modifiers. Floating microspheres were also prepared based on HPMC and PEO using solvent diffusion technique. The use of dichloromethane and ethanol with different diffusion rates was partially responsible for the buoyancy of microspheres. The *in vitro* release of the floating capsules and microspheres was found to be very similar (around 96%) [48]. Cefuroxime axetil is absorbed from the upper GI tract and is hydrolyzed to cefuroxime once it goes to the intestine. Unabsorbed drug causes high concentration of antibiotic to enter the colon, which contributes to colitis. A bilayer tablet was designed in which each layer contains the same amount of the drug, one layer provides immediate release, and the other layer has floating characteristics to provide sustained release property. The floating layer showed buoyancy lag time of 12–35 min and floating time of 8–24 h [20]. A coated multiple-unit sustained release floating unit was developed and tested with levodopa as a model drug. The unit consisted of a 3 mm drug-loaded, gas-generating core, prepared by melt granulation and compression and coated with a flexible membrane. For the membrane, the Eudragit RL30D and ATEC were used as a film former and a plasticizer, respectively. The formulation was optimized to achieve a floating lag time of 20 min and floating period over 13 h. Levodopa could be released from this unit over a 20 h period [32].

12.3.2 Assisted Floating Systems

An assisted floating system is a floatable system in which its retention performance is enhanced utilizing other gastric retention technologies. Cholestyramine (an ion exchange resin) floatable microcapsules coated with cellulose acetate butyrate were developed as an intragastric floating drug delivery system. The particles were loaded with bicarbonate followed by AHA and coated with CAB by emulsion solvent evaporation method. The X-ray diffraction was used to monitor drug/matrix interaction and the fluorescent probe method was used for *in vivo* mucoadhesion studies using rhodamine isothiocyanate. Cholestyramine microcapsules were found to be distributed throughout the stomach with prolonged gastric residence via mucoadhesion, which makes them a potential candidate in *H. pylori* treatment [14]. Combined swelling, floating, and bioadhesive mechanisms have been utilized to develop a GR platform to release ofloxacin. Various release retarding polymers including psyllium husk, HPMC K100M, and crosspovidone (superdisintegrant) were used to obtain a 24 h release profile. The developed formulation showed similar release to the marketed product zanocin OD [47]. Floating-bioadhesive tablets were developed to help retention of ciprofloxacin in the proximal area of the GI tract. Effervescent tablets were made using sodium carboxymethyl cellulose (CMC), HPMC, polyacrylic acid (AA), polymethacrylic acid (MAA), citric acid, and sodium bicarbonate. All tablets stayed floating for about 24 h. Tablets with 10% effervescent base, CMC/HPMC ratio of 4/1, or AAc/MAA ratio of 4/1 sound desirable [22]. Combined swelling and floating mechanisms of gastric retention were utilized in designing a triple-layer tablet to treat *H. pylori*. Hydroxypropyl methylcellulose and poly(ethylene oxide) were the rate-controlling excipients. Triple-layer construction also offers a combined immediate and sustained release feature to the delivery system. Tetracycline and metronidazole were incorporated into the core layer for controlled delivery, while bismuth salt was included in the outer layer for instant release. While the tablet remained afloat, sustained delivery of tetracycline and metronidazole could be easily achieved over a 6–8 h period [45]. Combined floating and mucoadhesion has been explored for the treatment of *H. pylori*. Floating microspheres containing the antiurease drug AHA were coated with polycarbophil. The bioadhesive property of the floating microspheres was investigated by the detachment force measurement method. *In vitro* growth inhibition studies were performed in isolated *H. pylori* culture. Although ureas help with colonization of *H. pylori*, the study shows that AHA-loaded floating microspheres are potent urease inhibitors [15]. Combined floating and pulsatile mechanisms

were used to design a floatable platform with delayed release feature. Pulsatile capsule was prepared by sealing the drug tablet and the buoyant filler inside the impermeable capsule body with an erodible plug. The lag time prior to the pulsatile drug release correlated well with the erosion properties of plugs. While the platform buoyancy was dependent on the bulk density of the dosage form, the delayed release was well correlated to the erodibility of the plug [78].

12.3.3 Mucoadhesion Systems

Dosage form will adhere to the mucosal surface of the stomach wall. Once attached, it will remain there until the mucus turns over. Polycarbophil and carbopol have been well tried as mucoadhesive matrices. Although these materials show great adhesion to the stomach wall, studies in animals and humans are rather disappointing. These acrylic-based polymers are able to stick to almost anything including the gelatin of the capsule and water-soluble proteins in the stomach and mucus. This nonspecific adhesion property severely reduces the amount of bioadhesive available to stick to the epithelial mucus. Microcrystalline chitosan granules were evaluated *in vivo* to see if they can provide a mucoadhesive sort of gastric retention to furosemide. Although *in vitro* results showed that furosemide release could be prolonged by increasing the molecular weight and concentration of chitosan in the granules, no *in vivo* evidence was found to suggest this. In fact, the *in vivo* study shows that a large amount of the drug passes its absorption window (upper intestine) before being released. This kind of pharmacokinetic profile of furosemide suggests that the gastric retention time of the granules is too short in relation to the release rate and a large amount of the drug passes its "absorption window" before being released. This example clearly shows that the *in vitro* results cannot be extended or used as evidence for *in vivo* trials [28]. Ion exchange resins have the ability to prolong gastric retention and also to distribute evenly over the gastric mucosa. Factors responsible for gastric retention status of ion exchange resins are the amounts of the administered resin, fasted or fed states, and also the surface charge of the resin. A single-blind, three-way (25, 250, and 250 mg cholestyramine tablet coated with ethyl cellulose) crossover study was performed in 12 healthy volunteers using gamma scintigraphy to visualize the distribution of the ion exchange resin (cholestyramine) in the stomach. The study concludes that dose size is not a factor in prolonged retention and even distribution of the resin, but it is the binding between the resin and the mucosa that provides retention [79]. It has been demonstrated that ion exchange resin cholestyramine can provide gastric retention via mucoadhesion and distribution. Gastric emptying and retention of this resin was compared with two other common mucoadhesive resins carbopol 934P and sucralfate using gamma scintigraphy. Sucralfate is a sucrose aluminum hydroxide compound that forms a gel layer over ulcerated or eroded tissues. The clinical study shows that 25% of cholestyramine could remain in the stomach after 6 h, while carbopol and sucralfate could remain up to 3.84% and 2.65%, respectively. The study also shows wide distribution of cholestyramine across the stomach, whereas carbopol and sucralfate were concentrated in the body and the antrum. This feature makes cholestyramine a good candidate for local treatment conditions such as *H. pylori* infections [80]. The rate of turnover of the intestinal mucus gel layer is the main reason for poor performance of the mucoadhesive GR systems.

12.3.4 Size-Dependent Technologies

Size-dependent GR technologies are classified into two categories, expandable and swellable. Both become large enough to be a size larger than pylorus, which assures their retention in the gastric medium. Expandable platforms are compact before administration, and their structure will be unfolded by physical means. Swellable platforms are also compact before administration, but their structure will be unfolded as a result of physicochemical interactions. There is a common risk with both platforms as they might obstruct the subject's esophagus as a result of expansion and swelling. This can be life threatening or at least unpleasant.

12.3.4.1 Swellable Platforms Leung et al. [81] studied the gastric retention property of a polyanionic hydrogel, polycarbophil, from the canine stomach using a duodenal cannulation technique. The basis of the study was to convert a fasted stomach to its fed state by employing a swelling hydrogel that could grow to a big size and could also increase the viscosity of the stomach content. The gastric emptying lag time was found to increase with the viscosity of the administered dose. The addition of sodium bicarbonate led to increased gastric retention due to increased apparent viscosity. The platform started to leave the stomach after a sufficient amount of stomach acid was secreted and protonated the anionic polymer. Riboflavin has been formulated in a gastric retention platform based on carbohydrate polymers. Proof of principle of gastric retention has been tested in fasted dogs and fasted healthy humans and compared with an immediate release formulation. The platform in dry state swells rapidly in the stomach juice and provides a zero-order release over a 24 h period. *In vivo* dog studies in fasted state showed that a rectangular platform could stay in the stomach for longer than 9 h, then disintegrated, and finally reached the colon in 24 h. Pharmacokinetic data obtained from urinary excretion of six fasted human subjects showed that the bioavailability depends on the platform size. Compared to the immediate release formulation, riboflavin in a large size platform showed three times more bioavailability because the platform could be retained in the stomach for about 15 h [56]. Groning et al. [82] developed a new collagen gastroretentive

dosage forms (GRDFs) with the ability to expand in the stomach after contacting gastric fluids. The dosage forms were prepared by freeze drying a collagen solution containing riboflavin. Within a few minutes of exposure to the artificial gastric juice, the collagen tablets expanded to a size of about $8 \times 18 \times 60$ mm and released riboflavin over a 16 h period. Chitosan–poly(acrylic) acid-based controlled release system has also been developed for gastric delivery of antibiotic amoxicillin. The study shows that the complex offers a significantly delayed gastric emptying compared to the chitosan alone [18].

For drug delivery applications, the use of conventional hydrogels is limited due to their slow kinetics of swelling. One way to improve the swelling rate is to make the hydrogel porous or even superporous. The word superporous means that the pore structure of the hydrogel is open and connected. In this way, water or aqueous fluid will be absorbed into the hydrogel structure due to combined diffusion and capillary mechanisms. The drawback of this approach is that the hydrogel will gain better swelling kinetics at the expense of its mechanical property. The more porous the structure is, the weaker the hydrogel structure will be. Chen et al. [83] first developed conventional superporous hydrogels and later superporous hydrogel composites with better mechanical properties. They improved the mechanical property by adding a pharmaceutically acceptable superdisintegrant into the hydrogel formulation. For gastric retention applications, a good mechanical property is a basic requirement. With no drug included, the retention status of these hydrogels was evaluated in dogs in both fasted and fed conditions. While under fasted condition, the hydrogels stayed in the stomach for only 2–3 h, the fed conditions prolonged the retention to about 24 h.

Omidian and Park [84] used a Taguchi experimental design to find out optimized foaming and gelation processes to prepare mechanically strong hydrogels. In another attempt to improve mechanical properties of conventional superporous hydrogels, Qiu and Park [85] utilized interpenetrating polymer networks. They treated the conventional SPH polymer with polyacrylonitrile as a second synthetic polymer and could improve compressive strength and elasticity of the SPHs. The enhanced mechanical properties were attributed to the formation of a scaffold-like fiber network structure within the cell walls of SPHs. Omidian et al. [86] developed superporous hydrogel hybrids by utilizing the gelation ability of polysaccharides in the presence of ions. They used this technique to prepare elastic superporous hydrogels with rubber-like properties [87]. It has also been shown that mechanical properties of SPH hybrids can be tailor-made by utilizing mixed ions having different activity and valence [88].

For administration purposes, a superporous hydrogel-based gastric retention system needs to be encapsulated into an oral dosage form like a 00 gelatin or a HPMC capsule by compression. Compression in both longitudinal and transverse directions may change the SPH swelling properties. The SPH should not be compressed to the extent that its porosity is affected drastically. Because it absorbs moisture very quickly, the SPH structure is generally very soft and the pores can readily lose their integrity. A study performed by Gemeinhart et al. [89] shows that SPH can maintain its swelling property if it is compressed parallel to the direction the pores are formed. Omidian et al. [90, 91] provided a detailed review on formulation, characterization, properties, and applications of superporous hydrogels. They highlighted how the synthetic features and properties of the SPH materials have been modified and improved over the years to meet the requirements for gastric retention application [92]. Feasibility of chitosan and glycol chitosan superporous hydrogels as gastroretentive platform has been studied in acidic solutions by Park et al. [93]. They optimized the SPH preparation by adjusting the gelation and foaming processes at different acidic conditions. The SPH swelling was found to depend on foaming/drying methods and also cross-link density.

12.3.4.2 *Expandable Platforms*

Geometric shapes such as a continuous solid stick, a ring, and a planar membrane have been disclosed. While these possess unfolding or expansion properties, they have to be made based on biodegradable materials for safe elimination. They are usually made of water-insoluble biomaterial polymers such as PLGA. Since they are not water soluble or water swellable, they are mechanically strong and can resist severe stomach contractions. Nevertheless, they lose their integrity and mechanical properties once they are digested by enzymes. Expandable gastroretentive dosage forms were originally designed for veterinary use; the design however was modified later for enhanced therapy in humans. The platform is easily swallowed and expands in the stomach to a very large size due to unfolding. After drug release, the platform dimensions are reduced to a size that can easily be removed from the stomach [94]. Klausner et al. [34] have developed a foldable gastroretentive platform to increase levodopa bioavailability in human subjects. The platform was a multilayer dosage form made of rigid polymeric matrices with different mechanical properties, folded into gelatin capsules, and administered to healthy volunteers with a light breakfast. They used gastroscopy and radiology to monitor the unfolding and integrity of the platform *in vivo*. Based on the pharmacokinetic data, levodopa in GR platform showed a notable increase in absorption compared to the controlled release Sinemet® as control. They attributed this feature to the size and mechanical rigidity of the GR platform. This finding is consistent with what Shalaby et al. [95, 96] found much earlier. To maintain a sustained therapeutic level, levodopa has to be administered continuously to the upper part of the intestine. With this in mind, Hoffman et al. [35] developed an expandable system with high rigidity and dimension.

Levodopa-loaded platforms were dosed to beagle dogs pretreated with carbidopa. The platform location in the GI tract was determined by X-ray, and blood samples were taken for levodopa concentration. Compared to non-GR particles and oral solution as control, mean absorption time for levodopa was found to be significantly extended. Groning et al. [97] developed oblong tablets based on collagen that expand to 4–6 cm within a few minutes of exposure to the GI fluids. Collagen sponge was loaded with riboflavin, and tablets (3.5 mm × 9 mm × 18 mm) were prepared by a computer controlled single punch tablet machine. An *in vivo* study with 12 healthy male and female subjects was conducted using oblong tablets and small sustained release hydrocolloid tablets as reference. The amount of riboflavin excreted into the urine from the testing and reference units were statistically different. Kagan et al. [54] examined the unfolding IntecPharma platform (Accordion Pill) to see whether they can improve riboflavin bioavailability in humans. The immediate release formulation and the GR formulation were administered with a low-calorie meal. Magnetic resonance imaging (MRI) was used to monitor gastric retention time and bioavailability data were obtained by taking blood and urine samples. The study showed prolonged gastric retention of about 10.5 h and increased bioavailability of $209 \pm 37\%$. Klausner et al. [55] administered different unfolding platforms with different dimensions and mechanical properties to beagle dogs and monitored them by X-ray. The absorption of riboflavin from a GR prototype was compared with a nongastroretentive controlled release dosage form and an oral solution of the drug. The extended absorption phase (>48 h) of riboflavin administered in a gastroretentive platform led to fourfold increased bioavailability.

12.3.5 Magnetic Systems

The GR platform usually contains a magnetic material such as a ferrite at the center of the dosage form. An external magnet will serve to anchor the dosage form within the body. Groning et al. [17] developed peroral depot tablets containing 200 mg acyclovir with gastric retention properties by incorporating internal magnets and using an extracorporeal magnet device. The device could improve the plasma concentration up to 170%. Other drugs including acetaminophen and riboflavin have shown better bioavailability when they were administered from a magnetic device. Apparently, the approach is difficult to perform and unfriendly and these have been the major obstacles in developing such products.

12.4 ANIMAL MODELS

Dogs and pigs are the two common animal models to study gastric retention. Although beagle dogs have been widely used as animal models to study gastric retention, the drug absorption profiles for man and dog are significantly different.

Even though they are similar in anatomy and size, the fed pattern lasts differently, much longer in dog than in man followed by a normal meal. GR studies with pigs have also been shown to be misleading, as a simple enteric-coated nondisintegrating magnesium hydroxide caplet could display a gastric residence time of 5 days [13]. Since the results of an animal GR study cannot be extended to man, companies are advised to consider a small-scale clinical trial for the proof-of-principle studies. This has to be considered as an important step when outlining the plan for the formulation and product development.

12.5   GASTRIC RETENTION MONITORING

For the proof-of-concept studies, the transit of a gastric retention platform inside the gastrointestinal tract needs to be monitored. Imaging techniques from simple to complex, cheap to expensive, and fairly reliable to very reliable are available and have been used for different clinical and proof of concept studies. These techniques include endoscopy (using fiber optic), gastroscopy (upper body endoscopy), X-ray (radiology using X-rays), fluoroscopy (X-ray radiology using a contract medium such as barium and iodine), magnetic marker monitoring (using iron oxide, ferrous–ferric oxides), magnetic resonance imaging (using strong magnetic fields), scintigraphy (using radionuclide isotopes including Tc-99, In-111, and iodine-123), and ultrasound (using high-frequency sound waves). Medical I.D. Systems, Inc. supplies barium impregnated polyethylene spheres (BIPS) to diagnose gastrointestinal obstructions and motility patterns for veterinary applications. These are gelatin capsules that can be administered easily in food or by mouth and claim to provide a safe, accurate diagnostic procedure with reducing the need for exploratory surgery. Shalaby et al. [95] utilized radiography, fluoroscopy, and ultrasound imaging techniques for full monitoring of swellable hydrogels including movement in the GI tract, swelling, and tissue–gel interaction. The authors found that the size and the integrity of the gels are prime controlling factors to prolong gastric retention. The study shows that hydrogels with less deformation under peristaltic gastric contractions have a better chance to be retained longer [96]. Table 12.3 displays recent examples of imaging studies with different platforms in animal and human clinical studies.

12.6   VETERINARY APPLICATIONS

Gastric retention with the intention of controlled drug release (to provide convenience, compliance, controlled rate, maintain plasma concentration or pulse dosing) might be useful for veterinary applications. The review covers potential new therapeutic areas where oral controlled release products can be applied in veterinary medicine [115]. The review also examines different species with the focus on gastrointestinal differences, which provides useful information to select animal models for gastric retention studies.

12.7   CORPORATE TECHNOLOGIES

There are several patents held for floating GR systems. West Pharmaceutical Services, Drug Delivery, and Clinical Research Centre in the United Kingdom have a patent for a system [116] that uses mucoadhesive microspheres. Eisai of Japan [117] has a gas-generating multiple-unit floating device comprised of a core (drug), an intermediate foaming layer, and an outer expandable coating. E.R.Squibb and Sons, Inc. has a powder [118] that floats on gastric fluid and becomes a gel. With many other works, a single-unit system having a special flotation chamber [119], a bilayer floating tablet [120], and a film [121] has been introduced. Although the number of patents and research activities are significant on the floatable devices, the size-dependent technologies (expandable and swellable) have been commercially more attractive and a few specialty pharmaceutical companies set up an infrastructure to pursue swellable and expandable GR platforms.

DepoMed (Menlo Park, CA, USA) utilized swelling feature of a cross-linked water-swellable polymer to develop a sustained release tablet or capsule to release a drug at a controlled rate that is dictated by the solubility of the drug. Upon ingestion, the polymer particles swell, maintain their integrity in the swollen form, and promote gastric retention, by which gastric fluid can penetrate into the particles and extract the drug in solution form [122]. Alkyl cellulose polymers including hydroxyethyl cellulose and hydroxypropyl cellulose have been used for this purpose [123]. DepoMed has a swellable and erodible system made of poly(ethylene oxide) that potentially releases the drug in a controlled rate to the fed stomach for local gastric disorders. Drug carriers including liposomes, nanoparticles, or enteric-coated drug particles are claimed to be delivered utilizing such GR platforms [124]. Controlled and sustained release GR platforms have been developed utilizing swellable hydrophilic polymers that expand to a large size after water absorption. The swollen polymer remains in the gastric cavity of the fed stomach for several hours and releases the drug in a controlled manner before it loses its integrity [125, 126]. The latest advancement in DepoMed's GR technology is the use of two polymers, poly(ethylene oxide) and hydroxypropyl methylcellulose, that offer controlled release, reproducibility, gastric retention, and disintegrability for safe elimination [127]. DepoMed has also manipulated the shape of the swellable polymers to promote gastric retention [128]. Metformin HCl has been the company's drug candidate for its GR technology [40]. Metformin extended release is a proprietary once daily formulation of metformin hydrochloride (glumetza) that DepoMed is developing for the treatment

TABLE 12.3 Imaging Studies on Various Gastroretentive Platforms

Study Topic	Technique	Subject	References
Bioavailability of riboflavin from a carbohydrate polymer-based GR platform	Endoscopy	Dog	[56]
Gastrointestinal transport and disintegration of oral dosage form	MMM	Human	[98]
Esophageal, gastric, and duodenal transit of nondisintegrating capsules	MMM	Human	[99, 100]
Tablet disintegration study using ferromagnetic black iron oxide	MMM		[101]
Fate and behavior of disintegrable dosage forms	MMM	Human	[102]
Gastrointestinal transit and release of felodipine	MMM	Human	[103]
A novel magnetic method to monitor esophagus, stomach, duodenum, and bowl	MMM	Human	[104]
Real-time monitoring of gastrointestinal tract	MMM		[105]
Solvent penetration into different hydrodynamically balanced systems based on alginate, chitosan, and HPMC	MRI		[68]
Gastric retention of unfolding Accordion Pill to increase riboflavin bioavailability	MRI	Human	[54]
Gastric emptying time of floatable tablets coated with a cross-linked polymer	Scintigraphy	Human	[106]
Gastric movement of floatable freeze-dried calcium alginate gels	Scintigraphy	Human	[53,67, 107]
Gastric transit of floating and nonfloating dosage forms	Scintigraphy	Human	[108]
Delayed gastric emptying of proton pump inhibitors like omeprazole	Scintigraphy	Human	[49]
Floating and pulsatile drug delivery system	Scintigraphy	Human	[78]
Ion exchange resins loaded with bicarbonate and coated with semipermeable membrane	Scintigraphy	Human	[59]
Improve retention of two radionuclides, indium (In-111) and samarium (Sm-153), using charcoal and cellulose acetate	Scintigraphy	*In vitro* and human	[109]
Monitor and compare gastric retention of raft-forming alginate-based Gaviscon and Gastrocote tablets	Scintigraphy	Human	[110]
Radionuclide gastric emptying study using technetium (Tc-99) in patients with Parkinson's disease	Scintigraphy	Human	[111]
Gastric retention evaluation of sucralfate gel and suspension dosage forms using Technetium (Tc-99)	Scintigraphy	Human	[112, 113]
Gastric distribution, emptying, and retention of cholestyramine ion exchange resin	Scintigraphy	Human	[79, 80]
Gastric emptying of oral silicone dosage forms using two isotopes iodine-123 and indium-111	Scintigraphy	Human	[114]
Hydrogel swelling and gel–gastric tissue interaction	Ultrasound	Dog	[95]
Gastric retention of enzyme-digestible hydrogels	Ultrasound, fluoroscopy	Canine	[96]
Intragastric behavior of floatable-coated calcium alginate gels having air compartment loaded with barium sulfate	X-Ray	Human	[71]
Gastric retention of unfolding polymer membrane to increase levodopa bioavailability	X-Ray	Beagle dog	[33]
Gastric retention of unfolding polymeric membrane to increase riboflavin bioavailability	X-Ray	Beagle dog	[55]
Gastrointestinal transit of enzyme-digestible hydrogels loaded with diatrizoate/diatrozoate meglumine	X-Ray, fluoroscopy	Dog	[95]
Unfolding thin multilayer dosage form to increase levodopa bioavailability	X-Ray, gastroscopy	Human	[34]

of diabetes. Glumetza tablets are modified release dosage forms that contain 500 or 1000 mg of metformin HCl. Each 500 mg tablet contains colorant, hypromellose, magnesium stearate, microcrystalline cellulose, and poly(ethylene oxide). Each 1000 mg tablet contains crospovidone, dibutyl sebacate, ethyl cellulose, glyceryl behenate, polyvinyl alcohol, polyvinylpyrrolidone, and silicon dioxide. Referring to a clinical study with more than 1000 diabetic patients, the metformin extended release has shown to significantly decrease the glycosylated hemoglobin level similar to that of metformin immediate release [37, 38]. DepoMed is also developing an oral formulation of extended release gabapentin to treat epilepsy and seizures, neuropathic pain, and hot flushes. Gabapentin ER is based on DepoMed's AcuForm gastric retention drug delivery technology. DepoMed's AcuForm is a polymeric platform that provides targeted drug delivery for a variety of compounds. Following ingestion, AcuForm tablet swells and is retained for 6–8 h in the stomach, enabling controlled and prolonged release of gabapentin to the upper intestinal tract. Compared to the immediate release formulation (neurontin), gabapentin displays better absorption and bioavailability and as such is suitable for twice-daily administration [29].

Merck & Co., Inc. (Rahway, NJ, USA) has a system that can be any of a number of folded geometric stick shapes, rings, planes, and so on based on erodible polymers [129, 130]. Bioerodible, thermoset, covalently cross-linked, elastomeric poly(ortho esters) with dimensional stability have been disclosed as a matrix for company's drug delivery devices [131, 132].

Alza Corporation (Mountain View, CA, USA) has a swellable system (a bilayer tablet) surrounded portionwise by a band of an insoluble material. The bands are intended to prevent the polymer from swelling in those sections. Because of less or no swelling, the banded area provides better physical integrity to the swollen platform and withstands the stomach contractions [133–137]. Low and high molecular weight HPMC polymers have been used in Alza's developments [138].

Pfizer, Inc. (New York, NY, USA) has a system with retractable arms that open once the device is in the stomach [139]. The system offers a delayed gastrointestinal transit [140].

West Pharmaceutical (Nottingham, UK) has a microsphere system comprised of an active ingredient in the inner core of the microsphere and a rate-controlling layer of a water-insoluble polymer and an outer layer of a bioadhesive cationic polymer that is claimed to be gastroretentive [116].

Teva Pharmaceutical Industries Ltd. (Petah Tiqva, Israel) has a system composed of a nonhydrated hydrogel, a superdisintegrant, and tannic acid with the potential to deliver a therapeutic bisphosphonate such as alendronate to the stomach of a patient over an extended period of time [141]. Gelling and expansion of the GR dosage forms have been claimed to

potentially improve bioavailability of antineoplastic agents, such as irinotecan, etoposide, paclitaxel, doxorubicin, and vincristine, whose oral effectiveness is limited by presystemic and systemic deactivation in the GI tract [142].

Kos Pharmaceuticals, Inc. (Miami, FL, USA) has a technology based on superporous hydrogels having simultaneous desirable swelling and mechanical properties. The system is strong but partially elastic in its swollen state and can withstand compression forces. The system utilizes ion equilibration concept, as swelling and mechanical properties are dependent on the amount and type of the ions used in gelation process. As with interpenetrating networks, one polymer is swellable and the other offers mechanical properties. It has been found that by properly adjusting the cations and the sequence in which they are added, superporous hydrogels can be formed that are highly absorbent with favorable structural properties, including strength, ruggedness, and resiliency [88].

Ranbaxy Laboratories Ltd. (New Delhi, India) have a product whose swelling results from the presence of a superdisintegrant. The system is comprised of a drug, a gas-generating component, a swelling agent (superdisintegrant, a cross-linked carboxymethyl cellulose, or polyvinyl pyrrolidone), a viscosifying agent, and optionally a gel-forming polymer. While the swelling agent provides a diffusion path for the sustained drug release, the latter two excipients help to entrap the gas and retain the tablet in the stomach or upper intestine. Two once daily formulations based on sodium alginate, xanthan gum, sodium bicarbonate, and cross-linked polyvinyl pyrrolidone containing ciprofloxacin base [23] and profloxacin base [143] have been suggested.

Bristol-Myers Squibb Co. (Princeton, NJ, USA) has a system that is claimed to provide prolonged gastric retention to highly water-soluble drugs like metformin HCl. The system has a solid core containing drug and one or more polymers surrounded by another solid layer composed of one or more polymers. The core and the shell may contain hydrophobic excipients including waxes, fatty alcohols, or fatty acid esters [144, 145].

Perio Products Ltd. (Jerusalem, Israel) has a core–shell system composed of drug surrounded by a water-insoluble or relatively water-insoluble coating material in which a particulate material is embedded. When the delivery device enters the gastrointestinal tract, the particulate matter takes up liquid, thus forming channels interconnecting the drug-containing core with the shell of the delivery device. These channels allow the release of drug from the core into the gastrointestinal tract. By controlling parameters in the device, such as the core material, carrier material in the coating, and particulate matter, the location of release of the drug can be carefully controlled [146].

LTS Lohmann Therapie-Systeme AG (Andernach, Germany) has a device composed of an expandable component and a liquid-permeable polymer coating. The release of the drug is dependent more on the drug than on the polymer membrane. The device can be rolled, folded, or filled into a capsule [147].

Purdue Research Foundation (West Lafayette, IN, USA) developed different generations of SPHs for general and specific applications in pharmaceutical and biomedical industries. Park and Park [148] developed a superabsorbent polymer with an interconnected pore structure, in which its swelling properties is independent of its size. The interpenetrating network of two swellable polymers (superporous hydrogel composites), one with very high swelling capacity (polyacrylamide or polyacrylic acid) and the other with low swelling capacity (pharmaceutical superdisintegrants), were used to develop superporous hydrogels with improved mechanical properties [149]. Purdue Research Foundation has developed superporous hydrogel hybrids with much improved elasticity and mechanical properties that can reversibly resist high mechanical forces and some behave like rubber [87]. Interpenetrating polymer networks have been utilized in this development. One polymer provides swelling property and the other acts as a strengthening agent. The strengthening agent is a polysaccharide polymer whose dissolution and swelling properties can be tailor-made by the amount and type of a complexing agent like metal cations. These hydrogels have a huge potential in developing successful gastric retention platforms [86]. Omidian et al. [90–92] have discussed the recent advances and developments in superporous hydrogels with the focus on gastric retention applications.

12.8   CONCLUSION

Gastric retention is mainly intended for drugs with a narrow absorption window and for local treatment of gastric disorders. Drugs degrading in the colon and instable at high pH can also benefit from this concept. Although many approaches have been tried to design gastric retention devices, two have received more attention in the pharmaceutical industry: size dependent and floatable devices. Gastric retention is still a "Holy Grail" in oral drug delivery, and there are plenty of opportunities for researchers to come up with a robust and therapeutically effective and commercially feasible system. This requires a constructive collaboration between academia and industry to work effectively on the area of formulation, design, technology, and assessment.

REFERENCES

1. Streubel A, Siepmann J, Bodmeier R. Drug delivery to the upper small intestine window using gastroretentive technologies. *Curr. Opin. Pharmacol.* 2006;6(5):501–508.

2. Hwang SJ, Park H, Park K. Gastric retentive drug-delivery systems. *Crit. Rev. Ther. Drug Carrier Syst.* 1998;15(3): 243–284.

3. Hou SYE, Cowles VE, Berner B. Gastric retentive dosage forms: a review. *Crit. Rev. Ther. Drug Carrier Syst.* 2003;20 (6):461–497.

4. Davis SS. Formulation strategies for absorption windows. *Drug Discov. Today* 2005;10(4):249–257.

5. Conway BR. Drug delivery strategies for the treatment of *Helicobacter pylori* infections. *Curr. Pharm. Des.* 2005;11 (6):775–790.

6. Bardonnet PL, et al. Gastroretentive dosage forms: overview and special case of *Helicobacter pylori*. *J. Control. Release* 2006;111(1–2):1–18.

7. Atherton JC, et al. Scintigraphic assessment of the intragastric distribution and gastric-emptying of an encapsulated drug: the effect of feeding and of a proton pump inhibitor. *Aliment. Pharmacol. Ther.* 1994;8(5):489–494.

8. Waterman KC. A critical review of gastric retentive controlled drug delivery. *Pharm. Dev. Technol.* 2007;12(1):1–10.

9. Talukder R, Fassihi R. Gastroretentive delivery systems: a mini review. *Drug Dev. Ind. Pharm.* 2004;30(10):1019–1028.

10. Kiss D, Zelko R. Gastroretentive dosage forms. *Acta Pharm. Hung.* 2005;75(3):169–176.

11. Davis SS, Wilding EA, Wilding IR. Gastrointestinal transit of a matrix tablet formulation: comparison of canine and human data. *Int. J. Pharm.* 1993;94(1–3):235–238.

12. Karali TT. Gastrointestinal absorption of drugs. *Crit. Rev. Ther. Drug Carrier Syst.* 1989;6:39.

13. Washinton N, Washington C, Wilson CG. *Physiological Pharmaceutics: Barriers to Drug Absorption*, 2nd edition, Taylor & Francis, New York, 2001, p 312.

14. Umamaheshwari RB, Jain S, Jain NK. A new approach in gastroretentive drug delivery system using cholestyramine. *Drug Deliv.* 2003;10(3):151–160.

15. Umamaheswari RB, et al. Floating-bioadhesive microspheres containing acetohydroxamic acid for clearance of *Helicobacter pylori*. *Drug Deliv.* 2002;9(4):223–231.

16. Fukuda M, Peppas NA, McGinity JW. *Floating hot-melt extruded tablets for gastroretentive controlled drug release system*. *J. Control. Release* 2006;115(2):121–129.

17. Groning R, Berntgen M, Georgarakis M. Acyclovir serum concentrations following peroral administration of magnetic depot tablets and the influence of extracorporal magnets to control gastrointestinal transit. *Eur. J. Pharm. Biopharm.* 1998;46(3):285–291.

18. Torrado S, et al. Chitosan-poly(acrylic) acid polyionic complex: *in vivo* study to demonstrate prolonged gastric retention. *Biomaterials* 2004;25(5):917–923.

19. Rajinikanth PS, Balasubramaniam J, Mishra B. Development and evaluation of a novel floating in situ gelling system of amoxicillin for eradication of *Helicobacter pylori*. *Int. J. Pharm.* 2007;335(1–2):114–122.

20. Dhumal RS, et al. Design and evaluation of bilayer floating tablets of cefuroxime axetil for bimodal release. *J. Sci. Ind. Res.* 2006;65(10):812–816.

21. Ali J, et al. Development and evaluation of a gastroretentive drug delivery system for the low-absorption-window drug

celecoxib. *PDA J. Pharm. Sci. Technol.* 2007;61(2): 88–96.

22. Varshosaz J, Tavakoli N, Roozbahani F. Formulation and *in vitro* characterization of ciprofloxacin floating and bioadhesive extended-release tablets. *Drug Deliv.* 2006;13(4): 277–285.

23. Talwar N, Sen H, Staniforth JN. Orally administered controlled drug delivery system providing temporal and spatial control. US Patent 6,261,601, 2001, Ranbaxy Laboratories Limited (New Delhi, India).

24. Rajinikanth PS, Mishra B. Floating in situ gelling system for stomach site-specific delivery of clarithromycin to eradicate *H-pylori*. *J. Control. Release* 2008;125:33–41.

25. Gambhire MN, et al. Development and *in vitro* evaluation of an oral floating matrix tablet formulation of diltiazem hydrochloride. *AAPS PharmSciTech* 2007;8:E73.

26. Streubel A, Siepmann J, Bodmeier R. Multiple unit gastroretentive drug delivery systems: a new preparation method for low density microparticles. *J. Microencapsul.* 2003;20(3): 329–347.

27. Jaimini M, Rana AC, Tanwar YS. Formulation and evaluation of famotidine floating tablets. *Curr. Drug Deliv.* 2007;4(1):51–55.

28. Sakkinen M, et al. Evaluation of microcrystalline chitosans for gastro-retentive drug delivery. *Eur. J. Pharm. Sci.* 2003;19 (5):345–353.

29. Gabapentin extended release—DepoMed: gabapentin ER, gabapentin gastric retention, gabapentin GR. *Drugs R D* 2007;8(5):317–320.

30. Sher P, et al. Low density porous carrier: drug adsorption and release study by response surface methodology using different solvents. *Int. J. Pharm.* 2007;331(1):72–83.

31. Tang YD, et al. Sustained release of hydrophobic and hydrophilic drugs from a floating dosage form. *Int. J. Pharm.* 2007;336(1):159–165.

32. Goole J, et al. New levodopa sustained-release floating minitablets coated with insoluble acrylic polymer. *Eur. J. Pharm. Biopharm.* 2008;68:310–318.

33. Klausner EA, et al. Novel levodopa gastroretentive dosage form: *in-vivo* evaluation in dogs. *J. Control. Release* 2003;88 (1):117–126.

34. Klausner EA, et al. Novel gastroretentive dosage forms: evaluation of gastroretentivity and its effect on levodopa absorption in humans. *Pharm. Res.* 2003;20(9):1466–1473.

35. Hoffman A, et al. Pharmacokinetic and pharmacodynamic aspects of gastroretentive dosage forms. *Int. J. Pharm.* 2004; 277(1–2):141–153.

36. Sharma S, Pawar A. Low density multiparticulate system for pulsatile release of meloxicam. *Int. J. Pharm.* 2006;313 (1–2):150–158.

37. Metformin extended release—DepoMed: metformin, metformin gastric retention, metformin GR. *Drugs R D* 2004;5(4): 231–233.

38. Metformin extended release: metformin gastric retention, metformin GR, metformin XR. *Drugs R D* 2005;6(5): 316–319.

39. Basak SC, Rahman J, Ramalingam M. Design and *in vitro* testing of a floatable gastroretentive tablet of metformin hydrochloride. *Pharmazie* 2007;62(2):145–148.

40. Cowles VE. Inhibition of emetic effect of metformin with 5-HT3 receptor antagonists. US Patent 6,451,808, 2002 (DepoMed, Inc., Menlo Park, CA).

41. Singh S, et al. Gastroretentive drug delivery system of metoclopramide hydrochloride: formulation and *in vitro* evaluation. *Curr. Drug Deliv.* 2007;4(4):269–275.

42. Chansanroj K, et al. Development of a multi-unit floating drug delivery system by hot melt coating technique with drug-lipid dispersion. *J. Drug Deliv. Sci. Technol.* 2007;17: 333–338.

43. Narendra C, Srinath MS, Babu G. Optimization of bilayer floating tablet containing metoprolol tartrate as a model drug for gastric retention. *AAPS PharmSciTech* 2006;7(2):E34.

44. Ishak RAH, et al. Preparation, *in vitro* and *in vivo* evaluation of stomach-specific metronidazole-loaded alginate beads as local anti-*Helicobacter pylori* therapy. *J. Control. Release* 2007; 119(2):207–214.

45. Yang LB, Eshraghi J, Fassihi R. A new intragastric delivery system for the treatment of *Helicobacter pylori* associated gastric ulcer: *in vitro* evaluation. *J. Control. Release* 1999;57 (3):215–222.

46. Chavanpatil M, et al. Development of sustained release gastroretentive drug delivery system for ofloxacin: *in vitro* and *in vivo* evaluation. *Int. J. Pharm.* 2005;304(1–2):178–184.

47. Chavanpatil MD, et al. Novel sustained release, swellable and bioadhesive gastroretentive drug delivery system for ofloxacin. *Int. J. Pharm.* 2006;316(1–2):86–92.

48. Ali J, Hasan S, Ali M. Formulation and development of gastroretentive drug delivery system for ofloxacin. *Methods Find. Exp. Clin. Pharmacol.* 2006;28(7):433–439.

49. Tougas G, et al. Omeprazole delays gastric emptying in healthy volunteers: an effect prevented by tegaserod. *Aliment. Pharmacol. Ther.* 2005;22(1):59–65.

50. Dave BS, Amin AF, Patel MM. Gastroretentive drug delivery system of ranitidine hydrochloride: formulation and *in vitro* evaluation. *AAPS PharmSciTech* 2004;5(2):

51. Hassan MA. Design and evaluation of a floating ranitidine tablet as a drug delivery system for oral application. *J. Drug Deliv. Sci. Technol.* 2007;17(2):125–128.

52. Rokhade AP, et al. Preparation and evaluation of cellulose acetate butyrate and poly(ethylene oxide) blend microspheres for gastroretentive floating delivery of repaglinide. *J. Appl. Polym. Sci.* 2007;105(5):2764–2771.

53. Stops F, et al. Citric acid prolongs the gastro-retention of a floating dosage form and increases bioavailability of riboflavin in the fasted state. *Int. J. Pharm.* 2006;308(1–2):14–24.

54. Kagan L, et al. Gastroretentive Accordion Pill: enhancement of riboflavin bioavailability in humans. *J. Control. Release* 2006;113(3):208–215.

55. Klausner EA, et al. Novel gastroretentive dosage forms: evaluation of gastroretentivity and its effect on riboflavin absorption in dogs. *Pharm. Res.* 2002;19(10):1516–1523.

56. Ahmed IS, Ayres JW. Bioavailability of riboflavin from a gastric retention formulation. *Int. J. Pharm.* 2007;330(1–2): 146–154.

57. Goole J, et al. *In vitro* and *in vivo* evaluation in healthy human volunteers of floating riboflavin minitablets. *J. Drug Deliv. Sci. Technol.* 2006;16(5):351–356.

58. Hamdani J, et al. *In vitro* and *in vivo* evaluation of floating riboflavin pellets developed using the melt pelletization process. *Int. J. Pharm.* 2006;323(1–2):86–92.

59. Atyabi F, et al. Controlled drug release from coated floating ion exchange resin beads. *J. Control. Release* 1996;42(1): 25–28.

60. Elkheshen SA, et al. *In vitro* and *in vivo* evaluation of floating controlled release dosage forms of verapamil hydrochloride. *Pharm. Ind.* 2004;66(11):1364–1372.

61. Streubel A, Siepmann J, Bodmeier R. Floating microparticles based on low density foam powder. *Int. J. Pharm.* 2002;241 (2):279–292.

62. Moes AJ. Gastroretentive dosage forms. *Crit. Rev. Ther. Drug Carrier Syst.* 1993;10(2):143–195.

63. Jacques Timmermans AJM. Factors controlling the buoyancy and gastric retention capabilities of floating matrix capsules: new data for reconsidering the controversy. *J. Pharm. Sci.* 1994;83(1):18–24.

64. Reddy LHV, Murthy RSR. Floating dosage systems in drug delivery. *Crit. Rev. Ther. Drug Carrier Syst.* 2002;19(6): 553–585.

65. Singh BN, Kim KH. Floating drug delivery systems: an approach to oral controlled drug delivery via gastric retention. *J. Control. Release* 2000;63(3):235–259.

66. Arora S, et al. Floating drug delivery systems: a review. *AAPS PharmSciTech* 2005;6(3):E372–E390.

67. Whitehead L, et al. Floating dosage forms: an *in vivo* study demonstrating prolonged gastric retention. *J. Control. Release* 1998;55(1):3–12.

68. Dorozynski P, et al. The macromolecular polymers for the preparation of hydrodynamically balanced systems—methods of evaluation. *Drug Dev. Ind. Pharm.* 2004;30(9): 947–957.

69. Whitehead L, Collett JH, Fell JT. Amoxycillin release from a floating dosage form based on alginates. *Int. J. Pharm.* 2000;210(1–2):45–49.

70. Desai S, Bolton S. A floating controlled-release drug-delivery system: *in vitro–in vivo* evaluation. *Pharm. Res.* 1993;10(9): 1321–1325.

71. Iannuccelli V, et al. Air compartment multiple-unit system for prolonged gastric residence. Part II. *In vivo* evaluation. *Int. J. Pharm.* 1998;174(1–2):55–62.

72. Atyabi F, et al. *In vivo* evaluation of a novel gastric retentive formulation based on ion exchange resins. *J. Control. Release* 1996;42(2):105–113.

73. Bajpai SK, Tankhiwale R. Preparation, characterization and preliminary calcium release study of floating sodium alginate/ dextran-based hydrogel beads: part I. *Polym. Int.* 2008; 57:57–65.

74. Choi BY, et al. Preparation of alginate beads for floating drug delivery system: effects of CO_2 gas-forming agents. *Int. J. Pharm.* 2002;239(1–2):81–91.

75. Sungthongjeen S, et al. Preparation and *in vitro* evaluation of a multiple-unit floating drug delivery system based on gas formation technique. *Int. J. Pharm.* 2006;324(2):136–143.

76. Soppimath KS, et al. Microspheres as floating drug-delivery systems to increase gastric retention of drugs. *Drug Metab. Rev.* 2001;33(2):149–160.

77. El-Gibaly I. Development and *in vitro* evaluation of novel floating chitosan microcapsules for oral use: comparison with non-floating chitosan microspheres. *Int. J. Pharm.* 2002;249 (1–2):7–21.

78. Zou H, et al. Design and gamma-scintigraphic evaluation of a floating and pulsatile drug delivery system based on an impermeable cylinder. *Chem. Pharm. Bull.* 2007;55(4): 580–585.

79. Thairs S, et al. Effect of dose size, food and surface coating on the gastric residence and distribution of an ion exchange resin. *Int. J. Pharm.* 1998;176(1):47–53.

80. Jackson SJ, Bush D, Perkins AC. Comparative scintigraphic assessment of the intragastric distribution and residence of cholestyramine, Carbopol 934P and sucralfate. *Int. J. Pharm.* 2001;212(1):55–62.

81. Leung SHS, Irons BK, Robinson JR. Polyanionic hydrogel as a gastric retentive system. *J. Biomater. Sci. Polym. Ed.* 1993;4 (5):483–492.

82. Groning R, Cloer C, Muller RS. Development and *in vitro* evaluation of expandable gastroretentive dosage forms based on compressed collagen sponges. *Pharmazie* 2006;61(7): 608–612.

83. Chen J, et al. Gastric retention properties of superporous hydrogel composites. *J. Control. Release* 2000;64(1–3): 39–51.

84. Omidian H, Park K. Experimental design for the synthesis of polyacrylamide superporous hydrogels. *J. Bioact. Compat. Polym.* 2002;17(6):433–450.

85. Qiu Y, Park K. Superporous IPN hydrogels having enhanced mechanical properties. *AAPS PharmSciTech* 2003;4(4): E51.

86. Omidian H, et al. Hydrogels having enhanced elasticity and mechanical strength properties. US Patent 6,960,617, 2005 (Purdue Research Foundation, West Lafayette, IN).

87. Omidian H, Rocca JG, Park K. Elastic, superporous hydrogel hybrids of polyacrylamide and sodium alginate. *Macromol. Biosci.* 2006;6(9):703–710.

88. Omidian H, Rocca JG. Formation of strong superporous hydrogels. US Patent 7,056,957, 2006 (Kos Pharmaceuticals, Inc., Miami, FL).

89. Gemeinhart RA, Park H, Park K. Effect of compression on fast swelling of poly(acrylamide-*co*-acrylic acid) superporous hydrogels. *J. Biomed. Mater. Res.* 55(1): 2001; 54–62.

90. Omidian H, Park K, Rocca JG. Recent developments in superporous hydrogels. *J. Pharm. Pharmacol.* 2007;59 (3):317–327.

91. Omidian H, Park K. Swelling agents and devices in oral drug delivery. *J. Drug Deliv. Sci. Technol.* 2008;18(2):82–93.

92. Omidian H, Rocca JG, Park K. Advances in superporous hydrogels. *J. Control. Release* 2005;102(1):3–12.

93. Park H, Park K, Kim D. Preparation and swelling behavior of chitosan-based superporous hydrogels for gastric retention application. *J. Biomed. Mater. Res. A* 2006;76(1): 144–150.

94. Klausner EA, et al. Expandable gastroretentive dosage forms. *J. Control. Release* 2003;90(2):143–162.

95. Shalaby WSW, Blevins WE, Park K. Gastric retention of enzyme-digestible hydrogels in the canine stomach under fasted and fed conditions: preliminary analysis using new analytical techniques. *ACS Symp. Ser.* 1991;469:237–248.

96. Shalaby WSW, Blevins WE, Park K. Use of ultrasound imaging and fluoroscopic imaging to study gastric retention of enzyme-digestible hydrogels. *Biomaterials* 1992;13(5): 289–296.

97. Groning R, et al. Compressed collagen sponges as gastro-retentive dosage forms: *in vitro* and *in vivo* studies. *Eur. J. Pharm. Sci.* 2007;30(1):1–6.

98. Kosch O, et al. Investigation of gastrointestinal transport by magnetic marker localization. *Biomed. Tech. (Berl)* 2002;47 (Suppl 1 Pt 2):506–509.

99. Weitschies W, et al. Magnetic marker monitoring of esophageal, gastric and duodenal transit of non-disintegrating capsules. *Pharmazie* 1999;54(6):426–430.

100. Weitschies W, et al. Magnetic marker monitoring: an application of biomagnetic measurement instrumentation and principles for the determination of the gastrointestinal behavior of magnetically marked solid dosage forms. *Adv. Drug Deliv. Rev.* 2005;57(8):1210–1222.

101. Weitschies W, et al. Determination of the disintegration behavior of magnetically marked tablets. *Eur. J. Pharm. Biopharm.* 2001;52(2):221–226.

102. Weitschies W, et al. Magnetic marker monitoring of disintegrating capsules. *Eur. J. Pharm. Sci.* 2001;13(4):411–416.

103. Weitschies W, et al. Impact of the intragastric location of extended release tablets on food interactions. *J. Control. Release* 2005;108(2–3):375–385.

104. Andra W, et al. A novel magnetic method for examination of bowel motility. *Med. Phys.* 2005;32(9):2942–2944.

105. Andra W, et al. A novel method for real-time magnetic marker monitoring in the gastrointestinal tract. *Phys. Med. Biol.* 2000;45(10):3081–3093.

106. Agyilirah GA, et al. Evaluation of the gastric retention properties of a cross-linked polymer coated tablet versus those of a nondisintegrating tablet. *Int. J. Pharm.* 1991;75(2–3): 241–247.

107. Stops F, et al. The use of citric acid to prolong the *in vivo* gastro-retention of a floating dosage form in the fasted state. *Int. J. Pharm.* 2006;308(1–2):8–13.

108. Timmermans J. Comparative evaluation of the gastric transit of floating and non-floating matrix dosage forms. *Bull. Mem. Acad. R. Med. Belg.* 1990;145(8–9):365–375.

109. Burke MD, et al. A novel method to radiolabel gastric retentive formulations for gamma scintigraphy assessment. *Pharm. Res.* 2007;24:695–704.

110. Davies NM, et al. A comparison of the gastric retention of alginate containing tablet formulations with and without the inclusion of excipient calcium ions. *Int. J. Pharm.* 1994;105 (2):97–101.

111. Djaldetti R, et al. Gastric emptying in Parkinson's disease: patients with and without response fluctuations. *Neurology* 1996;46(4):1051–1054.

112. Hardy JG, et al. A comparison of the gastric retention of a sucralfate gel and a sucralfate suspension. *Eur. J. Pharm. Biopharm.* 1993;39(2):70–74.

113. Vaira D, et al. Gastric retention of sucralfate gel and suspension in upper gastrointestinal diseases. *Aliment. Pharmacol. Ther.* 1993;7(5):531–535.

114. Kedzierewicz F, et al. Evaluation of peroral silicone dosage forms in humans by gamma-scintigraphy. *J. Control. Release* 1999;58(2):195–205.

115. Lavy E, Steinman A, Soback S. Oral controlled-release formulation in veterinary medicine. *Crit. Rev. Ther. Drug Carrier Syst.* 2006;23(3):165–204.

116. Illum L, Ping H. Gastroretentive controlled release microspheres for improved drug delivery. US Patent 6,207,197, 2001 (West Pharmaceutical Services Drug Delivery & Clinical Research Centre Limited, Nottingham, UK).

117. Ichikawa M, Watanabe S, Miyake Y. Granule remaining in stomach. US Patent 4,844,905, 1989 (Eisai Co., Ltd., Tokyo, Japan).

118. Dennis A, Timmins P, Lee K. Buoyant controlled release powder formulation. US Patent 5,169,638, 1992 (E. R. Squibb & Sons, Inc., Princeton, NJ).

119. Harrigan RM. Drug delivery device for preventing contact of undissolved drug with the stomach lining. US Patent 4,055,178, 1977.

120. Franz MR, Oth MP. Sustained release, bilayer buoyant dosage form. US Patent 5,232,704, 1993 (G. D. Searle & Co., Chicago, IL).

121. Mitra SB. Sustained release oral medicinal delivery device. US Patent 4,451,260, 1984 (Minnesota Mining and Manufacturing Company, St. Paul, MN).

122. Shell JW. Sustained-release oral drug dosage form. US Patent 5,007,790, 1991 (DepoMed Systems, Inc., Hillsborough, CA).

123. Shell JW. Alkyl-substituted cellulose-based sustained-release oral drug dosage forms. US Patent 5,582,837, 1996 (DepoMed, Inc., Foster City, CA).

124. Shell JW. Gastric-retentive, oral drug dosage forms for the controlled-release of sparingly soluble drugs and insoluble matter. US Patent 5,972,389, 1999 (DepoMed, Inc., Foster City, CA).

125. Shell JW, Louie-Helm J, Markey M. Extending the duration of drug release within the stomach during the fed mode. US Patent 6,340,475, 2002 (DepoMed, Inc., Menlo Park, CA).

126. Shell JW, Louie-Helm J, Markey M. Extending the duration of drug release within the stomach during the fed mode. US Patent 6,635,280, 2003 (DepoMed, Inc., Menlo Park, CA).

127. Gusler G, et al. Optimal polymer mixtures for gastric retentive tablets US Patent 6,723,340, 2004 (DepoMed, Inc., Menlo Park, CA).

128. Berner B, Louie-Helm J. Tablet shapes to enhance gastric retention of swellable controlled-release oral dosage forms. US Patent 6,488,962, 2002 (DepoMed, Inc., Menlo Park, CA).

129. Caldwell LJ, Gardner CR, Cargill RC. Drug delivery device which can be retained in the stomach for a controlled period of time. US Patent 4,767,627, 1988 (Merck & Co., Inc., Rahway, NJ).

130. Caldwell LJ, et al. Drug delivery device which can be retained in the stomach for a controlled period of time. US Patent 4,758,436, 1988 (Merck & Co., Inc., Rahway, NJ).

131. Pogany S, Zentner GM. Bioerodible thermoset elastomers. US Patent 5,047,464, 1991 (Merck & Co., Inc., Rahway, NJ).

132. Pogany S, Zentner GM. Bioerodible thermoset elastomers. US Patent 5,217,712, 1993 (Merck & Co., Inc., Rahway, NJ).

133. Edgren DE, Jao F, Wong PSL. Gastric retention dosage form having multiple layers. US Patent 6,797,283, 2004 (Alza Corporation, Mountain View, CA).

134. Edgren DE, et al. Method of fabricating a banded prolonged release active agent dosage form. US Patent 6,365,183, 2002 (ALZA Corporation, Mountain View, CA).

135. Wong PSL, et al. Prolonged release active agent dosage form adapted for gastric retention. US Patent 6,120,803, 2000 (ALZA Corporation, Mountain View, CA).

136. Wong PSL, et al. Prolonged release active agent dosage form adapted for gastric retention. US Patent 6,548,083, 2003 (Alza Corporation, Mountain View, CA).

137. Wong PSL, Edgren DE. Gastric retaining oral liquid dosage form. US Patent 6,635,281, 2003 (Alza Corporation, Palo Alto, CA).

138. Edgren DE, Magruder JA, Bhatti GK. Controlled release dosage form comprising different cellulose ethers. US Patent 4,871,548, 1989 (ALZA Corporation, Palo Alto, CA).

139. Curatolo WJ, Lo J. Gastric retention system for controlled drug release. US Patent, 5,002,772, 1991 (Pfizer Inc., New York, NY).

140. Curatolo WJ, Lo J. Gastric retention system for controlled drug release. US Patent, 5,443,843, 1995 (Pfizer Inc., New York, NY).

141. Flashner-Barak M, et al. Composition and dosage form for delayed gastric release of alendronate and/or other bis-phosphonates. US Patent 6,476,006, 2002 (Teva Pharmaceutical Industries, Ltd., Petah Tiqva, Israel).

142. Flashner-Barak M, Rosenberger V, Lerner EI. Compositions and dosage forms for gastric delivery of irinotecan and methods of treatment that use it to inhibit cancer cell proliferation. US Patent 6,881,420, 2005 (Teva Pharmaceutical Industries Ltd., Petah Tiqva, Israel).

143. Talwar N, Sen H, Staniforth JN. Orally administered drug delivery system providing temporal and spatial control. US Patent 6,960,356, 2005 (Ranbaxy Laboratories Limited, New Delhi, India).

144. Timmins P, Dennis AB, Vyas KA. Biphasic controlled release delivery system for high solubility pharmaceuticals and method. US Patent 6,475,521, 2002 (Bristol-Myers Squibb Co., Princeton, NJ).

145. Timmins P, Dennis AB, Vyas KA. Method of use of a biphasic controlled release delivery system for high solubility pharmaceuticals and method. US Patent 6,660,300, 2003 (Bristol-Myers Squibb Co., Princeton, NJ).

146. Lerner EI, Flashner M, Penhasi A. Gastrointestinal drug delivery system. US Patent 5,840,332, 1998 (Perio Products Ltd., Jerusalem, Israel).

147. Krumme M. Expandable gastroretentive therapeutical system with prolonged stomach retention time. US Patent 6,776,999, 2004 (LTS Lohmann Therapie-Systeme AG, Andernach, Germany).

148. Park K, Park H. Super absorbent hydrogel foams. US Patent, 5,750,585, 1998 (Purdue Research Foundation, West Lafayette, IN).

149. Park K, Chen J, Park H. Hydrogel composites and superporous hydrogel composites having fast swelling, high mechanical strength, and superabsorbent properties. US Patent 6,271,278, 2001 (Purdue Research Foundation, West Lafayette, IN).

13

ORAL TARGETED DRUG DELIVERY SYSTEMS: ENTERIC COATING

WENDY DULIN

Pharmaceutical Development, Wyeth Research, Pearl River, NY, USA

13.1 INTRODUCTION

Enteric coating polymers have been of significant importance and versatility as a drug release-controlling tool in targeted drug delivery. They are used in a wide variety of delivery systems to exert their pH-dependent effect to facilitate drug release according to specific mechanisms and/or locations.

13.1.1 History of Enteric Coating

The word *enteric* literally means "of, relating to, or affecting the intestines." However, it is commonly understood in pharmaceutical development as "being or having a coating designed to pass through the stomach unaltered and disintegrate in the intestines [1]." Thus, it is a coating or product with a delayed release, usually by means of a pH-dependent coating on a tablet, pellet, or granule. Other types of dosage forms may also be used. The terms "gastro resistant," "enterosoluble," "delayed release," and "pH sensitive" are some expressions used to refer to enteric dosage forms. Reasons for applying an enteric coat include

Reason	Example
Preventing irritation of the gastric mucosa or esophagus by the drug	Aspirin, bisphosphonates
Protecting the drug from destruction by gastric acid or enzymes	Duloxetine, proton pump inhibitors

(continued)

Reason	Example
Drug targeting to a specific region for a local effect, a high concentration, or systemic absorption	Colon-specific therapy such as 5-acetylsalicylic acid for colitis; colonic absorption of peptides and proteins
Delayed release for a delayed or double-pulse effect	Valsartan and hydrocortisone for chronotherapy

Any of the above reasons for applying an enteric coat can result in therapeutic benefits and cost savings due to improved efficacy, patient compliance, reduced dose, and reduced side effects (and thus reduced additional medical care). In addition, enteric coatings are used in studies to understand drug absorption and as a research tool.

Some enteric polymeric systems can be added to the matrix of the dosage form, rather than used as a coating to delay release. Other dosage forms may delay release by a mechanism other than pH control such as slowly swelling and then disintegrating, or digestion by certain intestinal enzymes. In these cases, the dosage form is usually referred to as "delayed action" rather than enteric.

Today, there are over 40 products listed in the PDR containing an enteric polymer [2] and many more products are found on pharmacy shelves in generic, nonprescription, and nutraceutical categories. The most common polymers used with these products today are methacrylic acid copolymer and cellulose esters. Enteric coating materials are polymers containing carboxylic acid groups, and the concentration and distribution of these acidic groups in the

Oral Controlled Release Formulation Design and Drug Delivery: Theory to Practice, Edited by Hong Wen and Kinam Park
Copyright © 2010 John Wiley & Sons, Inc.

polymer are the primary determinant of how it performs at gastrointestinal pH.

Shellac: Although the concept of enteric coating can be traced back to 1867 [3], the first enteric coating material to be widely used is shellac, which was introduced in 1930 [4] and is still used today in many products. It is listed in the *Handbook of Pharmaceutical Excipients* [5] as an enteric coating agent and is also used as a moisture barrier coating and an ingredient of pharmaceutical glaze and printing ink. It is a complex mixture of natural origin and therefore the exact composition may vary depending on the source or the manufacturing process. The NF monograph for shellac lists four types of shellac according to the method of preparation. The Ph.Eur. also lists four types of shellac and the JPE lists three. The shellac resin is composed primarily of esters of aleuritic and shellolic acids that impart the pH-sensitive character to the shellac films and its pK_a is reported to be between 6.9 and 7.5 [6]. Wax, gluten, and other substances may also be present. It has typically been used in pharmaceuticals by applying a 35% alcoholic solution, but success has been made in developing totally aqueous systems for shellac coating [7–9]. One of the disadvantages of shellac is the slowing down of its disintegration on storage [10]. Selection of the appropriate grade of shellac and modification by hydrolysis or with certain additives has been reported to improve the aging and enteric release properties [6, 11–16]. In addition to storage instability, shellac also was noted to have delayed intestinal release and even caused intestinal obstruction [4, 17].

Zein, a long-chain prolamine obtained as a by-product of corn processing, has been reported to be a replacement material for shellac in enteric coating [5]. However, little is reported on its use for this application. Some recent publications suggest that it may be useful as a sustained release material [18–20]. *In vitro* studies indicate that the release rate slows down as more zein coating is applied, but contrary to typical enteric coatings, the release actually gets slower as the pH of the dissolution media increases. The addition of pepsin to the dissolution medium degrades the zein proteins and the dissolution rate increases [18]. Unfortunately, there are no reported *in vivo* data to correlate to the *in vitro* dissolution.

Cellulose Acetate Phthalate: Following shellac, cellulose acetate phthalate (CAP) gained popularity by offering improvements over shellac. Cellulose acetate phthalate, at the time it was introduced in 1940, was regarded as the first true enteric polymer, with the film coating predictably disintegrating at a pH of ≥ 6.2 [4, 21]. It is listed in the Japanese Pharmacopoeia and Ph.Eur. and also under the name Cellacefate in the NF. It is produced by the reaction of the partial acetate ester of cellulose with phthalic anhydride. The phthalyl groups comprise 30.0–36.0% and the acetyl groups comprise 21.5–26.0% of the CAP. Approximately half of the hydroxyl groups are acetylated and approximately one-fourth

are esterified with one of the two phthalic acid groups. CAP film coats can be applied from organic solvent or aqueous systems with a plasticizer. An aqueous reconstituted colloidal dispersion is available for coating using 10–30% concentrations. The low-viscosity dispersion uses polymer particles having an average size of approximately 0.2 μm [5]. CAP is compatible with a wide range of plasticizers, and it was the first enteric coating reported to be useful for application to hard or soft shell capsules [22, 23].

Use of CAP has decreased due to the susceptibility of the phthalate ester to undergo hydrolysis upon storage with an accompanying loss of gastric resistance [24], but addition of a plasticizer or $MgCO_3$ can improve its moisture resistance [5, 21]. CAP has generated a great deal of interest recently since it was discovered to have anti-infective activity against sexually transmitted diseases, with especially potent inhibitory activity against the HIV-1 virus by its ability to block infection and transmission in both cell-free and cell-associated primary HIV-1 isolates [25].

Others: Following CAP, other cellulose ester polymers and enteric polymers were introduced: cellulose acetate trimellitate (CAT, not listed in NF or Ph.Eur. and rarely used today), polyvinyl acetate phthalate (PVAP), hypromellose phthalate (HPMCP), methacrylic acid polymers and copolymers, and, more recently, hypromellose acetate succinate (HPMCAS). Each of these offered certain advantages when introduced, such as dissolution at a particular pH between 5 and 7, lower viscosity, more resistance to moisture permeability, and better stability. Along with the introduction of new polymers came the development of aqueous-based coating systems. These materials give the formulator a wide array of individual polymers and coating systems that can be designed to fulfill the needs of a particular drug for optimized drug delivery, manufacturability, and stability. Properties of these coating polymers are summarized in Table 13.1. By combining with other polymers or excipients, and designing the dosage form (e.g., combining pellets with different release profiles or bioadhesive polymers), targeted oral drug delivery may be achieved.

Enteric polymer films are generally applied to a substrate, usually tablets or pellets, by traditional methods of pan or fluid bed coating. A solution or suspension of the polymer in water or organic solvent is applied to the moving substrate as a fine spray, and the applied droplets coalesce into a film as water or solvent is evaporated by warm air. Suppliers of enteric polymers and coating formulations usually provide helpful technical information and support for application of their products in manufacturing equipment.

In the past, release from enteric-coated dosage forms was tested only by a disintegration test to verify that the formulation did not disintegrate after 2 h in 0.1 N HCl. Today, a two-stage dissolution test is used. USP dissolution testing for enteric-coated formulations is listed under General Chapter <711> under the heading "Delayed-Release

TABLE 13.1 Summary of Enteric Coating Polymers and Properties

Polymer	Structure	Trade Names	How Supplied	Dissolution pH	Comments
Cellulose Acetate Phthalate (CAP)		Aquacoat® CPD (FMC) Powdered CAP (various)	30% Aqueous pseudo-latex dispersion For solvent coating	6.2	Store aq. dispersion refrigerated Requires plasticizer BP/NF/JP/PhEur
Hypromellose Phthalate (HPMCP; Hydroxypropyl Methylcellulose Phthalate)		HPMCP HP-50, HP-55 and HP-55S (Shin-Etsu)	Granular powder	5.0(HP-50) 5.5 (HP-55)	HP-55S has a higher MW and forms stronger films Usually requires plasticizer Only coated from organic solvent BP/NF/JP/PhEur
Polyvinyl Acetate Phthalate (PVAP)		Sureteric® (Colorcon) Opadry® Enteric "91 series" (Colorcon)	Powder for Aqueous Suspension For solvent coating	4.5–5.0	Sureteric usually requires anti-foam. Opadry Enteric usually requires plasticizer. NF

(Continued)

TABLE 13.1 (*Continued*)

Polymer	Structure	Trade Names	How Supplied	Dissolution pH	Comments
Hypromellose Acetate Succinate (HPMCAS)	R = H or –CH$_3$ –COCH$_3$ –COCH$_2$CH$_2$COOH –CH$_2$CHCH$_3$ –CH$_2$CHCH$_3$ OH OCOCH$_3$ –CH$_2$CHCH$_3$ OCOCH$_2$CH$_2$COOH	Aqoat® (Shin-Etsu)	AS-L AS-M AS-H For G also refer to fine and granular powders (for aqueous dispersions and organic solutions, respectively)	5.8[a] 6.3[a] 7.0[a]	L, M, and H designations depend on substitution and molecular weight. Aq. suspensions typically require plasticizer, glidant and surfactant. NF/JPE
Methacrylic Acid Polymers and Copolymers	for Eudragit L55 and Kollicoat MAE grades (NF type C): R^1, R^3 = H, CH$_3$ R^2 = H R^4 = CH$_3$, C$_2$H$_5$ for Eudragit L and S grades (NF types A and B): R^1,R^3 = CH$_3$ R^2 = H R^4 = CH$_3$	Eudragit® (Evonik)	L100-55 powder (C)	5.5	NF grades of Methacrylic Acid Copolymer – A, B & C – correspond to the Methacrylic Acid content and viscosity. Type C contains surfactant. Methacrylic Acid Copolymer Dispersion uses C grade. FS30D is for coating particles.
			L30D-55 30% aqueous dispersion(C) L100 powder(A) S100 powder(B)	 6.0 7.0	
			FS30D	7.0	
		Kollicoat® MAE (BASF)	MAE 100P powder MAE 30DP 30% aq. disp. (C)	5.5	Plasticizers and glidants are required for all but Acryl-Eze®. Acryl-Eze® MP is for coating particles. NF/Ph.Eur./JPE
		Eastacryl® (Eastman Chemical)	30D 30% aq. dispersion (C)	5.5	
		Acryl-Eze® (Colorcon)	Fully formulated powder for dispersion containing Eudragit L	5.5	

[a] Values are approximate and depend on ionic strength (dissolution occurs at lower pH in the presence of NaCl).

Dosage Forms" [26]. Six individual dosage units are placed in USP dissolution apparatus 1 (basket) or 2 (paddle) and tested for 2 h in 0.1 N HCl, and then the same dosage form continues to be tested for 45 min (or other specified time) in pH 6.8 phosphate buffer. Sinkers may be used for dosage units that would otherwise float in apparatus 2. Alternatively, apparatus 3 (reciprocating cylinder) or 4 (flow-through cell) may be used with the same acid and buffer stage media. The product meets USP acceptance criteria for stage 1 testing when no individual tablet exceeds 10% dissolved in the acid stage (level A_1) and releases at least $Q + 5\%$ in the buffer stage (level B_1) for the specified product. The length of time for testing in buffer and the Q value depend on the attributes of the product and is specified by the manufacturer.

13.1.2 Dosage Forms with Enteric Coating

Tablets, pellets, and granules are the most popular enteric-coated dosage forms. Pellets/granules offer the well-known benefits of more uniform release and reproducibility [27]. This is due to the *in vivo* dispersion of the multiparticulates as well as the physiology of the gastrointestinal tract. Transit of a single-unit dosage form through the stomach may take from minutes to over 10 h. Multiparticulates, on the other hand, being dispersed, will "trickle" from the stomach and have a more reproducible overall transit time. This transit of multiparticulates from the stomach also explains how formulations containing both immediate release and enteric-coated pellets give sustained release profiles rather than a two-pulse effect [28].

Coated pellets or granules are usually filled into capsule shells but also have been compressed into tablets. Another approach is to prepare coated minitablets that are then filled into capsules. Directly compressed minitablets have the benefit of avoiding solvent when incorporating the drug into the substrate for coating. Enteric coatings tend to be brittle and when compressing pellets into tablets, the formulator must be careful not to introduce breaks or cracks in the coating. This can be done by using appropriate (1) components in the tablet formulation, (2) compression force, and (3) amount and type of polymer(s) and plasticizer(s) in the enteric coat formulation [29, 30]. Ando et al. [31] have described the use of a novel double-structured punch with a center punch and an outer punch for forming dry-coated capsule-shaped tablets with a multiparticulate core.

Patient information for enteric-coated products will usually include a statement such as "swallow whole, do not crush or chew." For patients who have difficulty swallowing, the dose may be administered in a granule or pellet form, with instructions for how to sprinkle the granules onto, for example, applesauce prior to swallowing, or suspended in orange juice prior to drinking. Some drugs, such as acid-labile proton pump inhibitors, may also be prepared

extemporaneously by a pharmacist for oral or enteral administration to patients who have difficulty or cannot swallow. One method is to prepare a bicarbonate suspension of the drug with 8.4% sodium bicarbonate solution, using either the drug substance or the enteric-coated product (the coating will dissolve in the sodium bicarbonate solution). The resulting suspension is stable for 30 days in the refrigerator. There are questions, however, about the bioequivalence of this suspension to the enteric-coated product since the bicarbonate may not provide enough buffer capacity to protect from stomach acid, and the taste of the bicarbonate may lead to poor compliance [32, 33]. The coadministration of an enteric-coated product with a commercial antacid product, either for ease of swallowing a liquid or to achieve a more rapid effect, has produced varying reports regarding pharmacokinetics, which may depend on the particular antacid administered. In some cases, it is recommended for coadministration and in other cases to avoid it due to having no effect or lower bioavailability [34, 35]. Pharmacists must also exercise care when making product substitutions. Enteric-coated divalproex has been erroneously substituted for prescribed extended release divalproex and a suffix such as "ER" must not be confused for enteric release when it means extended release [36].

An enteric coating may not always be the best formulation approach in terms of both preventing degradation and drug delivery. The commercial product Zegerid® is a combination of the proton pump inhibitor omeprazole and sodium bicarbonate in either capsule form or as a powder for oral suspension. The sodium bicarbonate causes a prompt rise in gastric pH, preventing acid degradation of omeprazole. By avoiding a delayed release enteric coating, the product achieves relatively faster and higher peak plasma omeprazole levels, while maintaining equivalent exposure. Thus, it provides an immediate release alternative to an enteric-coated proton pump inhibitor for rapid relief of heartburn and acid reflux disease [35].

Tablets, pellets, and granules are the most popular enteric-coated dosage forms, but capsules may sometimes require coating, particularly in R&D studies. Gelatin capsules have tended to be difficult to coat due to poor adhesion to the smooth, shiny surface of the capsule shell and usually require a subcoat prior to aqueous enteric coating. HPMC capsules, on the other hand, have a somewhat rough surface. During the coating process, the HPMC shell itself will undergo partial dissolution that makes for better adhesion of a film coat. Cole et al. [37] describe the use of Eudragit® coatings on HPMC capsules with good performance as demonstrated by film integrity (microscopy), and *in vitro* dissolution and *in vivo* scintigraphic studies. Mercier et al. [38] also describe the use of a blend of Eudragit® L100 and S100 coating on HPMC capsules for the delivery of an oral vaccine to the intestine. The dissolution showed partial release at pH 6.0–6.5 in 60 min and complete release at pH 7.0–7.5. They also reported no change in the dissolution after 6 months storage of the coated capsules.

Enteric-coated peppermint oil softgel capsules are a popular nutraceutical for delayed release to the site of action for treatment of irritable bowel syndrome. Enteric-coated fish oil softgel capsules are also used for maintenance treatment of Crohn's disease and other ailments. For this product, the enteric coating has the added benefit of reducing the fish oil taste that patients may experience with noncoated capsules. Coating of softgel capsules is challenging due to the dynamic nature of the interplay between the gelatin shell, fill contents, film coating, and moisture. Prewarming the capsules prior to coating and the selection of appropriate plasticizer in the enteric film are two critical elements that require special attention [39]. Recently, the FDA approved a new delayed release softgel product by Banner Pharmacaps called Stavzor® (valproic acid). The patent-pending Enteri-Care® technology contains the enteric polymer in the gelatin shell itself rather than a coating, resulting in a small and clear dosage form [40].

In another method of incorporating the enteric substance in the capsule shell itself, Mehuys et al. [41] used hot melt extrusion to form hollow cylinders with PVAP or HPMCAS. The hollow cylinders are filled with drug followed by sealing the ends. These enteric capsules had good gastric resistance followed by fast drug release and can be used as an alternative to coating tablets or pellets.

Compression-coated tablets avoid the use of water or organic solvents for application of the enteric barrier layer. Using special tablet presses, powders are evenly compressed in a layer around a tablet core to produce a tablet-in-tablet dosage form. Incorporation of an enteric polymer into a thin outer layer can prevent gastric acid dissolution of an active substance from the tablet core. Law and Zhang [42] compression coated pure Eudragit® L100-55 powder with and without a controlled release polymer (HPC) onto tablets containing an active enzyme. The coating protected the enzyme from acid degradation, but, even without HPC, was slow to release the enzyme in the intestinal pH media.

A number of investigators have described a powder-layering process for coatings that avoid the use of water or organic solvents. Obara et al. [43] used a process of feeding enteric polymer powder (HPMCAS) and spraying plasticizer onto moving tablets or pellets followed by a curing step. Their studies showed that 30% and 8% polymers (weight gain) were required for pellets and tablets, respectively, to achieve gastric resistance. However, a small amount of water was needed to facilitate coalescence of the film. Kablitz et al. [44] took this a step further by improving the process of feeding the powder and plasticizer, adding 25% polymer to the pellets, and curing without the need for any water. The total processing time, with curing, was only 68 min. Sauer et al. [45] used hot melt extrusion to preplasticize polymers such as Eudragit® L100-55, which was then layered onto

tablets, followed by curing. This process provided gastric resistance but also used more polymer and plasticizer and less glidant than aqueous film coating.

13.1.2.1 Matrix Systems
At low levels, enteric polymers can be used as binders in granulations for tablets and capsules. At higher levels, these same polymers can be formulated into directly compressed or wet granulated matrix dosage forms. Adding an enteric polymer to a dry blend followed by direct compression seems a simple and attractive alternative to the coating process. However, the ability to do this is not straightforward, and once the enteric polymer is added to a matrix, it can have a sustained rather than a delayed release effect. In 1983, McGinity et al. [46] mixed Eudragit® polymer powders with theophylline to make tablets. The blends had good compressibility, flow, weight variation, and sustained release effect. The enteric L100 powder, however, compared to the sustained release Eudragit® polymers had little effect on dissolution. Oren and Seidler [47] in 1990 described a matrix combination of a drug with a hydrogel and an enteric polymer. Similar to the concept of blended polymers in coating, the combination of hydrogel/enteric polymer was proposed to facilitate steady release for a drug that has pH-dependent solubility. When they used an acrylic powder or dispersion (e.g., 0.5–25% Eudragit L100-55) and hypromellose in the tablet matrix, release in acid was generally 12–30% in 30 min and the profile was also sustained but faster at higher pH.

Estaban et al. [48] used Eudragit® FS30D and L30D-55 acrylic dispersions at 16.6% to granulate tablets containing diltiazem as a model drug. The dissolution profiles for the two matrix tablets are shown in Figures 13.1 and 13.2. The tablets prepared with L30D-55 show a more enteric release profile having delayed release in acid and faster release at pH greater than 6. The tablets prepared with FS30D, which is used in enteric films to release at a pH above 7.0, have a pH-independent profile. The authors explain this behavior as a result of salt formation and water uptake, the FS30D polymer taking up more water than L30D-55.

Tatavarti and Hoag [49] describe a sustained release effect produced in matrix tablets prepared with Eudragit® L30D-55 and a drug with pH-dependent solubility. They noted that the effect of the enteric polymer on drug release was dependent on the microenvironmental pH, the basicity of the drug, and the water uptake and matrix permeability.

Guo et al. [50] studied the compression properties of CAP and HPMCAS. HPMCAS produced tablets with a higher tensile strength, while CAP exhibited more elastic recovery. They described the CAP as having enteric behavior, inhibiting drug release in acid and dissolving and bringing about tablet erosion at intestinal pH.

The enteric polymer can be added to the tablet matrix by conventional techniques of direct compression, wet granulation, or roller compaction, but care must be taken in doing

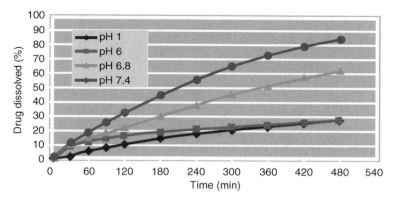

FIGURE 13.1   Release profile of diltiazem HCl with Eudragit® L30D-55 matrix tablet at different pH values. From Ref. 48.

this. Addition of a plasticizer when wet granulating with a polymethacrylate dispersion facilitates the coalescence of the polymer particles and increases the effect of the polymer in the matrix [51]. Mixing acid-labile drugs or amines with enteric polymers can cause degradation or interaction, leading to loss of active [52, 53] or loss of enteric effect or both.

Some matrix sustained release tablets will incorporate sodium alginate or alginic acid as release modifiers to impart a pH dependency to the drug release. The alginate polymer will swell at low pH but not dissolve. Its acidic functional groups will begin to ionize at a pH of approximately 3–4. Drug release will decrease at pH 1.2 and increase at pH 6.8. Thus, it behaves in a tablet matrix as a blended polymer with HPMC or other matrix-forming polymers and can achieve pH-independent release for weak bases with pH-dependent solubility [54]. Many grades of alginate are available and the molecular weight and the particle size can have a significant effect on drug dissolution. Other naturally derived polymers, such as gum cordia, have been evaluated for their ability to form matrix tablets that have both enteric resistance and sustained release properties [55].

Palmieri et al. [56] evaluated five enteric polymers by preparing directly compressed tablets and by making spray-dried microspheres that were then compressed into tablets. These products contained only drug and polymer to avoid any effects of other excipients. Of the polymers tested, only CAP provided a quasi-zero-order release, with all others providing a delayed or partial release in acid followed by faster release in buffer. Slower release was observed from the tablets compressed from microspheres, indicating that the spray-drying step facilitated the formation of a matrix, as well as better compressibility.

Hot melt extrusion can be used to prepare a polymer–drug extrudate containing amorphous drug that is released at intestinal pH. This is currently an area of much study for enteric and other forms of controlled release. When the formulation is adjusted by varying the level of drug, polymer, plasticizer, and other additives, the resulting product may be optimized to produce a physically stable product having a specific release profile. Methacrylic acid copolymer has been used to prepare hot melt extrudates containing guaifenesin that are physically stable after addition of hydrophilic

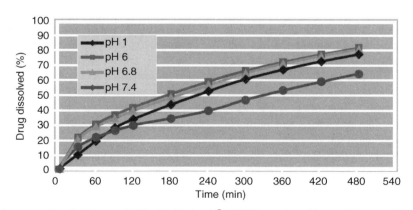

FIGURE 13.2   Release profile of diltiazem HCl with Eudragit® FS30D matrix tablet at different pH values. From Ref. 48.

polymer additives. The hot melt produced a delayed release, but not as complete as an enteric film, and release in pH 6.8 media was slow [57]. Follonier et al. [58] reported a pH-independent release of diltiazem rather than a delayed release effect after enteric polymers were added to a sustaining polymer in a hot melt system. They noted that the enteric polymers were progressively releasing from the pellets at high pH, but remained intact at pH 1. One explanation offered for this very slow release behavior was a pore constriction mechanism.

13.1.3 Effects of Gastrointestinal Physiology

The most important physiological consideration for enteric dosage forms is gastric emptying. Since an enteric coating disintegrates once it reaches a higher pH environment after leaving the stomach, the residence time in the stomach profoundly affects the drug's pharmacokinetics. Pharmacokinetic modeling must take into account the gastric emptying rate and the intestinal transit of the formulation to accurately reflect plasma concentrations [28]. As noted previously, the variability in gastric emptying is great, lasting from minutes to over 10 h. The formulator, when developing and testing an enteric-coated formulation, should consider this variability. A 2 h dissolution test in 0.1 N HCl may show an acceptably low level of drug release, but it would also be prudent to test at 4 or 6 h to verify the acid resistance for longer stomach residence times.

Food, size, and shape of the dosage form, coadministered drugs, gender, age, disease state, and exercise may influence the gastric fluid composition and residence time. Age may have an effect since geriatric patients have been reported to have a higher gastric pH and longer residence time. Gender plays a role since gastric transit is reported to be faster in males than females [59, 60]. Females also have a gastric emptying rate that is dependent on the menstrual cycle [61]. High variability can confound the results of a study if gender effects are not taken into account. No difference has been reported, however, in chemical composition and pH of the stomach and the jejunum for men and women [62]. Patients with achlorhydria or hypochlorhydria, having a higher than normal gastric pH, are likely to experience dose dumping in the stomach from an enteric-coated formulation. As suggested by Agyilirah and Banker [3], *in vitro* testing should include intermediate pHs such as 3, 4, and 5 when developing new products and in order to have meaningful tests that reflect *in vivo* performance. Haddish-Berhane et al. [63] have developed a stochastic model that provides a prediction of drug release using the pH gradient in the GI tract and can be used for formulation design and optimization.

The size of the dosage form, or the ability of the dosage form to break up into smaller pieces, will impact the gastric emptying rate for that product. The cutoff size threshold of an indigestible solid, at which it will remain in the stomach until phase 3 contractions ("housekeeper waves"), has been controversial. It has been claimed that if the solid is in a size range of approximately 2 mm or less, it behaves like a liquid and will readily pass through the pylorus at any time. Some reports indicate that 7 mm is the cutoff for solid materials to be retained in the stomach prior to being emptied with a phase 3 contraction. The ability of the GI tract to crush the dosage form and dump the drug was evaluated by Kamba et al. [64]. They determined that the ability of the formulation to withstand a crushing force of 1.2 N would enable it to withstand the forces in the small intestine of a dog or human.

13.1.3.1 Food Effects

The pH of the stomach may be raised with food from a median value of 1.7 to 6.7 [65]. Thus, enteric formulations, in order to maintain the enteric effect, should be administered on an empty stomach. Food also affects the gastric residence time. The higher the energy content of the meal, the longer the delay in gastric emptying, irrespective of the type of energy (fat, carbohydrate, or protein). The frequency of meals may also affect the residence time of a dosage form. Repetitive dosing of an enteric-coated tablet with meals can actually lead to gastric accumulation of several doses. The delay is more pronounced with a single-unit dosage form rather than multiparticulates. Meal effects on multiparticulates vary, sometimes slowing down the rate of absorption, but often having no effect. Additional food effects that should be considered include the effect of fruit juice, coffee, tea, and chewing gum—all of which affect the acidity of gastric contents and may change gastric emptying. Food effects can be very complex and depend on multiple factors such as type of food (i.e. high fiber), enterocyte and hepatocyte metabolism, biliary recycling, and diet-induced effects on elimination [66].

To evaluate the effect of food, timing of the dose relative to food intake is also critical based on the changes in peristaltic contractions with food. A delay of 15 to 30 min after food is ingested allows for adjustments in contraction patterns to an ordered fed state. If the dose is administered concomitant with food, gastric emptying may not be delayed and no food effect will be observed [67].

13.1.3.2 Disease Effects

Changes in GI physiology may occur in a number of disease states. For enteric formulations, the worst case would be a patient who has a high stomach pH and/or rapid stomach emptying, thereby defeating the built-in delay. Diseases that affect gastric emptying or pH include diabetes, anxiety disorders, peptic ulcers, migraines, pancreatitis, anorexia, certain malignancies, scleroderma, gastroenteritis, and pyloric stenosis. Since patients may be unaware or may not report some of these conditions, it can be a challenge to achieve optimal drug delivery, particularly for drugs that are given on a repetitive or chronic basis. As an example, diabetic patients experience gastric stasis and are often advised to have frequent small meals, which can lead to prolonged and irregular gastric emptying. Since diabetic patients are at higher risk of heart disease, physicians will often prescribe daily enteric-coated aspirin tablets. These competing situations lead to less than optimum therapy for patients.

13.2 APPLICATIONS OF ENTERIC COATING

13.2.1 Reduced GI Toxicity

A good understanding of the mechanism of GI side effects and the target population is necessary to know whether an enteric coating will lessen such effects in short- or long-term use. While endoscopic studies have shown that enteric-coated aspirin alleviated gastric irritation and ulceration in healthy young volunteers, there is still some debate about the utility of an enteric coating to alleviate GI toxicity for aspirin and other drugs. Studies in humans are usually reported for short-term effects in healthy volunteers. In one study, the effectiveness of the enteric coat to provide protection for older long-term aspirin users was not demonstrated [68]. This lack of protection may be explained at least in part by the known occurrence of achlorhydria or hypochlorhydria in the elderly. Enteric-coated erythromycin was reported to have the same or higher levels of GI side effects (abdominal pain and cramps, nausea, vomiting, diarrhea, and gas) compared to a formulation without an enteric coat [69, 70]. Potassium chloride formulations are now sustained release rather than enteric coated due to the risk of injury to the intestinal mucosa with enteric-coated products [71]. GI irritation by nonsteroidal anti-inflammatory drugs (NSAIDs) is now understood to be due to a combination of systemic inhibition of cyclooxygenase (COX) and a high local concentration effect [72]. Phenylbutazone, on the other hand, has shown good improvement in gastrointestinal side effects after long-term studies with enteric formulations [73].

Mycophenolic acid (MPA) was recently introduced as an immunosuppressive agent for preventing organ transplant rejection. It does not have the side effects associated with other immunosuppressives such as leukopenia and nephrotoxicity. A morpholinoethyl ester prodrug, mycophenolate mofetil (MMF; CellCept®) [2], was developed to have advantageous pharmacokinetic properties such as good solubility in the delivery environment (e.g., the stomach), fast absorption, and high peak plasma concentration compared to mycophenolic acid. However, the most common adverse events are GI side effects (nausea, vomiting, diarrhea, constipation, gastritis, and ulcers), cytopenias, and opportunistic infection. Patients with these GI or hematologic effects can experience some relief by dose division or reduction, but these changes are associated with higher incidences of graft loss [74–76].

To overcome the GI adverse events, an enteric-coated formulation of the sodium salt of MPA was developed (EC mycophenolate sodium or EC-MPS; Myfortic®) [77–79]. The commercial product is a delayed release tablet with a hypromellose phthalate coating. Pharmacokinetic studies in patients confirmed the delayed release, and the time of peak plasma concentration (T_{max}) of MPA was 1.5–2.75 h after oral administration of MPS—later than that of MMF (T_{max} 0.5–1 h) [80].

At standard doses, EC-MPS was able to consistently achieve $AUC_{(0-24h)}$ >30 μg h/mL, which is believed to be the level needed for efficacy [81]. Both efficacy *and* adverse events were reported to be similar for equivalent doses of MMF and EC-MPS (for active moiety mycophenolate, 1000 mg of MMF is equivalent to 720 mg of MPS) [82, 83]. Since pharmacokinetic analysis showed that EC-MPS reached statistically higher plasma concentrations at equimolar doses, and it was not associated with an increase in adverse events, EC-MPS may provide increased tolerability relative to systemic exposure, thus providing more patients with therapeutic concentrations [84]. It is difficult to quantify GI adverse events following transplantation due to many confounding factors such as the stress of surgery, data collection techniques, concomitant medications, and high prevalence of GI events even without MPA treatment. One study where patients were switched from MMF to EC-MPS did indicate a reduction in severity of GI events for EC-MPS [85]. Other reports showed similar outcomes after the switching from MMF to MPS [86–88]. In heart transplantation, there was significant difference in dose reduction of MMF 42.1% versus MPS 26.9% ($p < 0.05$) [89]. Conflicting reports have been published for liver transplants [90, 91]. These authors concluded that it is beneficial to convert MMF to MPS.

It is interesting to note that the original rationale for development of EC-MPS did not result in remarkable improvement in tolerability as listed in the PDR [92]. There have been a number of suggestions for this. One is the toxicity of MPA due to its pharmacological effect on purine synthesis, thus producing an antiproliferative effect on intestinal cells. This view is not favored since purines are abundant in the GI tract and this effect would not be rate limiting for nucleoside absorption in the gut wall. Another possibility is local, direct contact between MPA and luminal wall. But it is becoming more likely that the GI toxicity is related to the absorption phase, whereby a toxic active metabolite, acyl-MPAG, is formed in the enterocytes. Acyl-MPAG is also a metabolite of NSAIDs, which have well-known GI adverse events via the formation of protein adducts. Biliary recycling of MPA metabolites that are deconjugated back to MPA provides a second source for further local metabolism and toxicity. Additional studies have been suggested for deeper understanding of these events in the gut, with the ultimate goal of improved tolerability via a drug delivery strategy [93].

Didanosine is a drug that eventually used enteric coating to solve formulation-related side effects. It was first introduced as a buffered formulation to stabilize the drug against degradation in stomach acid. However, these formulations tended to produce side effects such as nausea, bloating, gas, diarrhea, GI upset and cramps, and the buffer altered the absorption of other drugs. Enteric coating was able to alleviate the side effects of the buffer along with providing bioequivalent exposure [94, 95].

13.2.2 A Tool for Understanding Drug Absorption and Toxicity

Evaluation of the regional absorption of a drug is crucial prior to initiating a modified release formulation development program. Basit et al. [96] describe the use of formulation technology to assess regional absorption as a way to avoid some of the difficulties and costs associated with intubation or remote control devices. Using ranitidine, they prepared formulations for immediate release and also for release in the small intestine and the colon. Their data showed a good general agreement with a previous intubation study.

Cora et al. [97] used enteric-coated magnetic HPMC capsules to study gastrointestinal motility and transit and residence times. *In vitro* tests could be modified to give a good correlation to disintegration times in the *in vivo* studies.

Bours et al. [98] used enteric-coated ATP to investigate its ability to protect the intestinal mucosa from indomethacin-induced permeability changes.

13.2.3 Absorption Enhancement

Sometimes additional ingredients such as surfactants, absorption enhancers, superdisintegrants, pH modifiers, and enzyme inhibitors may be added to the formulation in addition to the enteric polymer to enhance drug absorption once it is released from the dosage form. Some formulations of enteric-coated pancreatic enzymes also include a high pH buffer or are dosed with a proton pump inhibitor to compensate for an often too low duodenal pH. This approach may not provide a benefit over simple enteric coating [99]. Siepe et al. [100] have described strategies for incorporating pH modifiers into the tablet core to achieve a controlled microenvironmental pH.

13.2.4 Formulation Design and Drug Targeting

Some of the reasons for using enteric polymers have already been given (e.g., avoid gastric irritation) and simple formulations of an enteric coat on tablets or pellets may be used to achieve this effect. Targeting release of a drug to a specific site is desired to maximize the localized drug's concentration, hence requiring a dose lower than a systemic dose, which may also improve the drug's safety. Budesonide, a potent corticosteroid used in the treatment of Crohn's disease, can cause less adrenal suppression when given as

an enteric formulation targeted to the distal ileum, its site of action, compared to a systemic dose [101]. Higher levels have been measured in the small intestine and colon (69% of dose) from an enteric-coated/ethylcellulose SR formulation compared to a nonenteric formulation (30% of dose) [102]. Enteric coating is also often used in combination with a sustained release polymer to achieve a particular therapeutic or compliance goal by delaying and then slowly releasing the active substance.

The use of enteric coating to target drugs to the colon is covered in another chapter. Additional uses of enteric coating to design formulations for specific purposes are given in the following section.

13.2.4.1 Achieving Sustained Release of Drugs Having pH-Dependent Solubility
At gastric pH, a basic drug will have high solubility. At some point as the drug travels down the GI tract and the pH increases, the solubility decreases. A conventional insoluble polymer for sustained release of such a weakly basic drug may not produce the desired sustained effect due to the lack of a concentration gradient driving diffusion at high pH. By blending a pH-dependent enteric polymer with a pH-independent insoluble polymer, it may be possible to compensate for the changes in GI solubility of the drug with changes in permeability of the coating. At low pH where the drug solubility is high, the enteric and insoluble polymers will allow only a small amount of drug to be released. At high pH, the enteric polymer may, at least partially, leach out of the coating and allow more of the drug, which now has a lower solubility, to be released. The result is pH-independent release of drug from the formulation as illustrated in Figure 13.3. The formulator can manipulate the level of the two polymers (enteric and SR) as well as the possible addition of pore formers to the coating or acidifiers to the core to achieve the desired release. Blends prepared using the aqueous polymers of Eudragit® NE:Eudragit® L and of Kollicoat® SR:Kollicoat® MAE have provided this pH-independent release [103, 104]. Kim et al. [105] also describe the blending of an aqueous ethylcellulose coating (Surelease®) with HPMCP and water-soluble pore formers to achieve targeted drug release for the basic drug tamsulosin. Siepmann et al. [106] have recently published an excellent review paper on blended polymers in coatings, with detailed discussion of the use of insoluble and enteric polymer blends and underlying mechanisms for release. Siepmann et al. [107] have also described film coatings that allow the formulator to create pH-independent release for weakly basic drugs by blending an aqueous Aquacoat® ECD ethylcellulose dispersion with pH-dependent propylene glycol alginate.

Venkatesh [108] reports the use of blends of Eudragit® RS/RL polymers along with an acid component to facilitate dissolution of a pH-dependent drug with very low solubility in the intestine. The pellet formulation is designed with tartaric acid crystals as the core material, a low permeability

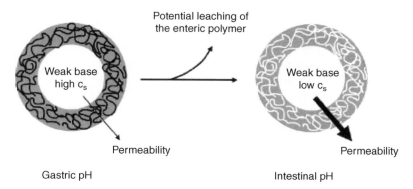

FIGURE 13.3   Schematic of a GIT-insoluble:enteric polymer blend as controlled release film coating. From Ref. 106.

coating, a drug/tartaric acid layer, and an outer layer having more permeability. An illustration of the buffered pellet is shown in Figure 13.4. The solubility of tartaric acid is higher than the drug, but the author suggests that even better results could be achieved with a less soluble acid such as succinic or fumaric to maintain the micro-pH needed for solubilizing the drug.

There has been much study of blended polymers to gain a better mechanistic understanding of the interactions between the polymers. These interactions may cause both precipitation and complex formation and only when the polymer blends are miscible can drug release be adjusted by varying composition. Knowledge of the underlying mechanisms may facilitate a rational choice of polymer blends for tailored drug release. Menjoge and Kulkarni [109, 110] reviewed interactions of polymers with a reverse enteric polymer and with Eudragit® E using thermal analysis and FTIR to quantify the degree of interaction and understand the functional group interactions. Familiarization with mechanisms allows choosing blends that will be miscible and give a known amount of swelling at given pHs.

13.2.4.2   Targeting to Specific Regions of the GI Tract

For some drugs, targeting the release of active to the small intestine can offer therapeutic benefits. Cysteamine is one drug that has been reported to give the highest exposure when targeted to release in the small intestine compared to the stomach or cecum [111]. However, site-specific delivery to the proximal small intestine may be difficult due to the perceived slow *in vivo* dissolution of dosage forms once the enteric coat disintegrates [112]. As early as the 1970s, it was suggested to use effervescent formulations with an enteric coat to get rapid disintegration in the proximal small intestine [113]. Indomethacin and erythromycin are two examples of the many drugs that have their principal absorption site in the proximal small intestine. Formulating these drugs in an enteric product must be done carefully to ensure that the dosage form is delivered intact from the stomach and then quickly made available for absorption. One technique that has been recommended in these instances for good reproducibility and low intersubject variability is to enteric coat formulations having diameters no greater than 5 mm [71].

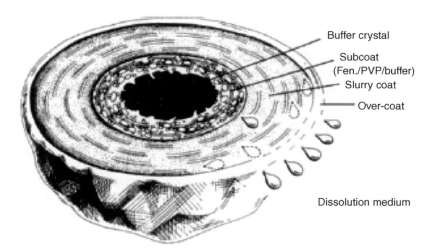

FIGURE 13.4   Controlled-release slurry-coated buffer crystals, showing water penetration and dissolved buffer and drug (fenoldopam) diffusing through the outer coat. From Ref. 108.

Huyghebaert et al. [114] sought to target interleuken-10 to the ileum, the major site of inflammation in Crohn's disease. To achieve this, they wanted an enteric coat that disintegrated at pH 6.8 and then released the IL-10 within 40 min. They evaluated several polymer systems and none were satisfactory, either releasing the active below pH 6.8 or not meeting the 40 min criteria.

Eiamtrakarn et al. [115] developed a mucoadhesive patch consisting of layered films for oral delivery to the small intestine. The patch was placed in a capsule and the capsules were coated with different enteric polymers that disintegrate at pH 5.5, 6.2, and 7.0. The results show that drug release was parallel to the polymer used, with 2.3, 3.3, and 5.0 h for the respective T_{max} values in beagle dogs. The mucoadhesive was included to keep the patch at the intestinal wall for site-specific drug delivery [116].

Ibekwe et al. [117] studied the *in vivo* performance of the Eudragit® polymers S (organic), S (aqueous), and FS (aqueous) that have a dissolution pH of approximately 7, with FS being more flexible for coating pellets. Gamma scintigraphy of coated tablets revealed that the S (aqueous) tablets all disintegrated in the proximal to mid small intestine. For S (organic) tablets, three of eight tablets did not disintegrate at all and the remaining five disintegrated at the ileocecal junction or ascending colon. The FS (aqueous) tablets likewise disintegrated at the ileocecal junction or ascending colon, with 2 out of 16 failing to disintegrate. The authors also reported a better correlation to dissolution testing in a physiological bicarbonate buffer rather than phosphate buffer.

Sangalli et al. [118] used varying thickness of HPMC tablet coatings to introduce a delay in the release of active. By adding an enteric coat, they introduce the HPMC-produced delay after the drug passes through the stomach. Since the small intestine transit time (3–5 h) is less variable than gastric transit, the HPMC coat can be used to introduce a lag until a specific intestinal site is reached. Their predictions were verified by *in vivo* tests (Figure 13.5) and built upon work previously done by Wilding et al. [119] and others. The enteric coating may also be used to delay release until the colon is reached where the formulation is disintegrated due to the presence of colonic bacteria to release the drug substance [120].

Miller et al. [121] discuss the potential therapeutic benefits of targeting poorly soluble drugs to specific regions of the GI tract where the released drug is at supersaturated levels. Their work showed that a solid dispersion of itraconazole with Eudragit® L100-55 and a polymeric stabilizer resulted in substantial improvements in levels of supersaturated itraconazole at the targeted pH for release *in vitro*, which was also reflected in animal studies.

An enteric coating has been utilized in an intestinal-retentive dosage form developed to allow the dosage form to pass through the stomach and then, once the coating

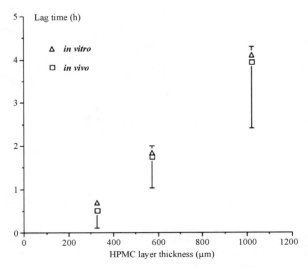

FIGURE 13.5 Effect of HPMC layer thickness on *in vitro* and *in vivo* lag times for a model tablet formulation. Bars represent standard deviation. From Ref. 119.

disintegrates in the intestine, the fast-swelling superporous hydrogel becomes fixed and is retained for at least 45–60 min. This system allows for a prolonged residence time and targeted drug delivery to intestinal sites [122].

Enteric coatings have been used to facilitate the oral delivery of vaccines. Safety and efficacy have been demonstrated in monkey models for HIV immunization. The adenoviral vector passes the harsh digestive environment of the stomach and elicits both a mucosal and a systemic immune response from the small intestine [38, 123].

Enteric polymers are used in experiments with sensor elements to trigger release of a drug substance at a particular site in the GI tract. The enteric membrane will disintegrate and the gastrointestinal contents will trigger an electric circuit, which causes a gas-producing cell to empty the drug from the capsule [124].

13.2.4.3 Pore Formers
A pore former is a substance that is added to a functional polymer film to create a channel (pore) or less permeable barrier to drug release through that film. The pore former may be a soluble material that dissolves and leaves a hole or a channel in the film, or it may hydrate and form a path for easier drug diffusion. Enteric polymers are used *as* pore formers in other polymers (usually sustained release polymers) and pore formers are *added to* enteric polymers for enhancing release in the low-pH environment. As noted earlier, blended polymers essentially use an enteric polymer as a pore former in the sustained polymer film to allow faster drug release at high pH (Figure 13.3). The addition of other more soluble pore formers, such as lactose, may be necessary for faster release in acid. Typical pore

formers used with enteric polymers include hypromellose, PEG, and lactose. Pore formers that have been reported with enteric and other coatings are mannitol, sorbitol, sucrose, hydroxypropylcellulose, hydroxyethylcellulose, sodium carboxymethyl cellulose, dicalcium phosphate, and PVP. For coatings that contain only an enteric polymer, it is sometimes desired to have partial release of the active in the gastric region and a pore former will fulfill this need. Alternatively, some of the active substances could be added to the coating to give an initial immediate release of drug and to form a pore in the film. Compatibility with the enteric formulation may not happen with some pore formers such as cationic polymers. However, pH adjustment of the coating suspension could provide a suitable coating formulation. The hygroscopicity of the pore former also needs to be considered for the long-term stability of the formulation. The ability of the pore former to rapidly dissolve or gel will also affect the lag time for drug release. Pore former levels range from 3% to over 100% of the polymer weight. Depending on the pore former, high levels may lead to weakened films that could prematurely burst.

13.3 POLYMER CONSIDERATIONS

Rafati et al. [125] describe an approach for aqueous enteric coating via salt formation to solubilize the enteric polymer with ammonium hydrogen carbonate that subsequently dissociates during coating to convert the polymer to its nonionized form and release ammonia and carbon dioxide.

Drug–polymer interactions are a possibility that may influence the stability or adhesive strength of the film coating on storage. Sarisuta et al. [126] reported a potential interaction between ranitidine and Eudragit® E100 in a tablet that had reduced film adhesive strength after storage for 4 weeks at 40°C/75% RH. A drug may also diffuse into the coating, during or after application, and this has been reported for potassium iodide and riboflavin with cellulose acetate phthalate coatings. Changes in the core composition or adding a subcoat can minimize premature release of the drug due to this migration [5, 127, 128]. A subcoat may also be necessary to prevent degradation of an acid-labile active via a solid-state interaction between the active and the enteric polymer. Addition of alkaline stabilizers to the core or even the subcoat has also been used to prevent degradation of the active [52, 129]. Typical subcoat materials are hypromellose, HPC, PVP, and polyvinyl alcohol.

13.4 SUBSTRATE CONSIDERATIONS

Choice of polymer in the tablet core may play a critical role in the performance of a sustained release matrix core that has been enteric coated. Miyazaki et al. [130] demonstrated that enteric-coated cores containing theophylline with HPC or a polyion complex (a hydrophilic dextran mixture) had very different AUCs because the matrix with HPC could not swell adequately in the upper GIT. Uncoated cores had similar AUCs since the enteric coating did not restrict swelling.

The solubility of the active will influence the enteric resistance of the coating, with highly soluble compounds known to require more coating than poorly soluble ones. This may also affect the type and amount of subcoat needed. Other ingredients in the core may affect the enteric resistance. A waxy substrate was found to produce less diffusion of riboflavin from the core into the coat than lactose [127]. The effectiveness of an Eudragit® L 30D-55 coating on pellets with the highly water-soluble drug chlorpheniramine maleate was affected by the type of subcoat and was found to be correlated with the water vapor transmission rates of these subcoatings [131].

13.5 PROCESS CONSIDERATIONS

It is critical for an enteric-coated product to have the proper film integrity and thickness to resist the gastric environment, yet not an excessive amount that may impede dissolution. When organic solvents are used, a 4–8% coating level (based on tablet weight) is usually sufficient to provide acid resistance. Aqueous coatings generally need higher levels of 10–15% to produce the same effect. The amount required depends on many factors and should be optimized for individual formulations. It is also better to express the coating level in terms of mg/cm^2 or the thickness in microns rather than wt%. Fourman et al. [132] discuss the importance of proper baffle design for good tablet mixing in the coating pan and found that for their aqueous enteric coating system, weight gain variability of ~8% relative standard deviation was an acceptable process variation. In addition to dissolution, other tools such as SEM can be used to evaluate the uniformity and integrity of the film coat. Newer process analytical technology (PAT) techniques with improved process monitoring and control produce the most uniform and reproducible coatings.

Siepmann et al. [133] have discussed the effect of polymer level, plasticizer content, and curing on drug release from HPMCAS-coated pellets. At low coating levels, the relatively large HPMCAS particles do not coalesce sufficiently to form a continuous film and drug release is fast in the acid stage. As the level of polymer increases, the particles coalesce into a coating with enteric resistance in the acid medium. Curing studies showed that high humidity and temperature slowed down release and altered the mechanism of release from the pellets.

A detackifier is needed with some enteric coatings to prevent stickiness during the coating process and to obtain good films. Talc has been traditionally used for this purpose, but an improvement in recent years is the use of glyceryl monostearate (GMS; mono- and diglycerides NF) as a detackifier. Talc requires high levels in the formulation, tends to settle out of suspension, and may also clog spray nozzles. GMS, on the other hand, overcomes all of these disadvantages, being incorporated as an emulsion and requiring low usage levels [134, 135].

13.5.1 Other Processes

Many processes other than traditional fluid bed or pan coating have been studied for enteric polymers. These include nanosuspensions, solid dispersions, emulsification–evaporation, coprecipitation, solvent evaporation, emulsification– diffusion, spray drying/congealing, deposition, salting out, and simple coacervation [136]. These are not popular methods for producing commercial products with enteric coatings due to many challenges such as use of organic solvents and difficulties in scale-up. However, it is an area of active research and these techniques may offer benefits such as high surface area for quick release at high pH, small size for better compressibility, and maintaining the drug in a high-energy state for enhanced solubility and bioavailability. Some of these processes are discussed below.

Nanosuspensions. Drugs with very low solubility can be prepared as nanoparticle or nanocrystal suspensions, usually by a high-pressure homogenization technique. This can then be layered onto nonpareil seeds in a fluidized bed, followed by the addition of an enteric polymer in the same equipment. Release at the desired site is faster for the nanosized drug compared to the same formulation using micronized drug. Moschwitzer and Muller [137] created nanocrystals of hydrocortisone acetate by high-pressure homogenization and included a mucoadhesive in the layering dispersion. Dissolution testing of the enteric-coated pellets from this approach gave faster dissolution and increased release compared to micronized drug.

Coacervation. Alginates have been used in microparticles to prepare beads by aqueous coacervation with calcium chloride. The beads will often be formulated with an enteric polymer or chitosan. Scocca et al. [138] describe the formation of alginate microspheres containing an enteric polymer by spraying into a calcium chloride solution and then freeze drying. The proteins within the resultant particles were protected from gastric acid during dissolution testing and dissolved in 3 min at intestinal pH.

Dong and Bodmeier [136] describe coacervation via dissolving the lipophilic drug and enteric polymer in an organic phase and a hydrophilic polymer in the aqueous phase. Organic phases investigated were acetone, ethanol, and isopropanol. Optimization of the encapsulation efficiency is achieved by controlling the composition, pH, and type of solvent. Drugs that were soluble in the enteric polymer were successfully encapsulated in a noncrystalline state with high encapsulation efficiency. A photo of the dried coacervates is shown in Figure 13.6.

Deposition Technique. An electrostatic coating process by Phoqus® Pharmaceuticals places the coating on specific areas of the tablet (such as coating only one side), which in combination with the core formulation can produce drug release that is delayed, pulsed, increased/delayed, or other profiles [139].

Emulsion Solvent Evaporation. Drug-loaded microspheres of pantoprazole have been successfully prepared using the emulsion solvent evaporation technique with enteric polymers Eudragit® S100 and HPMCP [140]. Theophylline and piroxicam microcapsules were also successfully prepared at 25% loading with Eudragit® S100 [141]. Yang et al. [142] used an o/o emulsion solvent evaporation technique for the highly water-soluble polypeptide thymosin to achieve high encapsulation efficiency with an acrylic acid resin.

Emulsion Diffusion. De Jaeghere et al. [143] used the emulsion–diffusion method to produce spheres in the

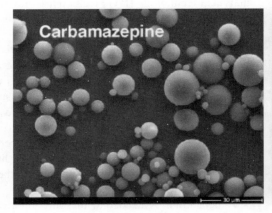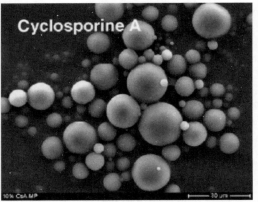

FIGURE 13.6 Scanning electron micrographs of carbemazepine- and cyclosporin A-loaded (10% w/w) Eudragit® L100-55 microparticles. From Ref. 137.

nanoparticle size range. Lipophilic drug and Eudragit® L100-55 were dissolved in benzyl alcohol and an aqueous phase containing polyvinyl alcohol as emulsifier was added. A 50% drug load was achieved with 98% encapsulation efficiency. Their formulation was tested in dogs and fourfold increase in exposure was observed compared to the drug as a gavage suspension (AUC of 27.1 and 6.5 µg h/mL, respectively).

Spray drying and Spray Congealing. The formulation and processing conditions need to be carefully evaluated for spray drying/congealing to produce an acceptable product in terms of size, density, uniformity of active and polymer(s), reproducibility, and acid resistance [144, 145]. Palmieri et al. prepared spray-dried microspheres of ketoprofen by preparing dilute aqueous solutions of the drug and enteric polymer in 1 N NH_4OH. They evaluated five enteric polymers and Eudragit® L and S were most effective for the encapsulation, and all microspheres were best at the lowest drug:polymer ratio (1 : 4) tested [146].

13.6 ADDITIONAL STUDIES WITH ENTERIC POLYMERS

Yuan and Clipse [147] reported an enteric formulation of ibuprofen in a liquid-filled softgel capsule containing cellulose acetate phthalate. The drug and CAP are dissolved in a PEG-based vehicle and filled into softgel capsules. Upon contact with water, the liquid becomes solid and dissolution in 0.1 N HCl after 2 h is ≤7.0%.

Lau and Gleason [148] describe a solventless coating method via initiated chemical vapor deposition (iCVD). The synthesis and coating of methacrylic acid and ethyl acrylate copolymers is confirmed by describing the kinetics of the copolymerization process, by shifts in the bond stretching modes, and by pH-dependent release of model drugs.

Bacteria have been encapsulated and enteric coated for delivery to the lower intestine for a therapeutic effect. Oxalobacter formigenes delivery to degrade endogenous oxalate leads to a significant reduction in urinary and plasma oxalate, which can be used to treat hyperoxaluria that can result in kidney stones and renal failure [149].

Reverse enteric polymers, that is, polymers that are soluble at low pH and insoluble at high pH, have found utility for taste masking applications. Additional applications for tailoring drug release profiles may be possible based on mechanistic investigations [110].

13.7 CONCLUSIONS

Enteric dosage forms, when used with an understanding of gastrointestinal physiology, therapeutic needs, and formula-

tion strategies, can be a potent tool in the formulator's arsenal for skillfully designing a product to target drugs for optimal patient outcomes. Even difficult drugs with solubility and compatibility challenges can be formulated using materials and technologies into stable and bioavailable products with delayed, targeted release.

REFERENCES

1. Merriam-Webster Online Dictionary: www.m-w.com/dictionary/enteric.
2. *Physician's Desk Reference Electronic Library*, Thomson's Healthcare Products, 2008.
3. Agyilirah GA, Banker GS. Polymers for enteric coating applications. In Tarcha PJ, editor. *Polymers for Controlled Drug Delivery*, CRC Press, Boca Raton, FL, 1990.
4. Chambliss WG. Forgotten dosage form: enteric coated tablets. *Pharm. Technol.* 1983;7(Sep):124, 126, 128, 130, 132, 138, 140.
5. *Handbook of Pharmaceutical Excipients*, Medicines Complete/American Pharmacists Association, 2008.
6. Limmatvapirat S, et al. Enhanced enteric properties and stability of shellac films through composite salts formation. *Eur. J. Pharm. Biopharm.* 2007;67(3):690–698.
7. Chang RK, Iturrioz G, Luo CW. Preparation and evaluation of shellac pseudolatex as an aqueous enteric coating system for pellets. *Int. J. Pharm.* 1990;60(Apr 30):171–173.
8. Krause KP, Muller RH. Production of aqueous shellac dispersions by high pressure homogenisation. *Int. J. Pharm.* 2001;223(1–2):89–92.
9. Zambito Y, DiColo G. Preparation and *in vitro* evaluation of chitosan matrices for colonic controlled drug delivery. *J. Pharm. Pharm. Sci.* 2003;6(2):274–281.
10. Luce GT. Disintegration of tablets enteric coated with CAP (cellulose acetate phthalate). *Manuf. Chem. Aerosol News* 1978;49(Jul):50, 52, 67.
11. Sheorey DS, Kshirsagar MD, Dorle AK. Study of some improved shellac derivatives as microencapsulating materials. *J. Microencapsul.* 1991;8(3):375–380.
12. Sheorey DS, Kshirsagar MD, Dorle AK. Evaluation of shellac derivatives as coating materials. *Indian J. Pharm. Sci.* 1992;54(5):169–173.
13. Limmatvapirat S, et al. Modification of physicochemical and mechanical properties of shellac by partial hydrolysis. *Int. J. Pharm.* 2004;278(1):41–49.
14. Pearnchob N, Dashevsky A, Bodmeier R. Improvement in the disintegration of shellac-coated soft gelatin capsules in simulated intestinal fluid. *J. Control. Release* 2004;94(2–3):313–321.
15. Limmatvapirat S, et al. Effect of alkali treatment on properties of native shellac and stability of hydrolyzed shellac. *Pharm. Dev. Technol.* 2005;10(1):41–46.
16. Qussi B, Suess WG. Investigation of the effect of various shellac coating compositions containing different water-soluble

polymers on *in vitro* drug release. *Drug Dev. Ind. Pharm.* 2005;31(1):99–108.

17. Andrus CH, Ponsky JL. Bezoars: classification, pathophysiology, and treatment. *Am. J. Gastroenterol.* 1988;83(5): 476–478.

18. O'Donnell PB, et al. Aqueous pseudolatex of zein for film coating of solid dosage forms. *Eur. J. Pharm. Biopharm.* 1997;43(Jan):83–89.

19. Liu X, et al. Microspheres of corn protein, zein, for an ivermectin drug delivery system. *Biomaterials* 2005;26 (1):109–115.

20. Georget DM, Barker SA, Belton PS. A study on maize proteins as a potential new tablet excipient. *Eur. J. Pharm. Biopharm.* 2008;69(2):718–726.

21. Park ES, et al. A new formulation of controlled release amitriptyline pellets and its *in vivo/in vitro* assessments. *Arch. Pharmacal Res.* 2003;26(7):569–574.

22. Jones BE. Production of enteric coated capsules. *Manuf. Chem. Aerosol News* 1970;41(May):53–57.

23. *Enteric Coating Soft-Shell Gelatin Capsules with Eastman(R) C-A-P(TM) Enteric Coating Material, Application Brochure,* Eastman Chemical Co., 1987.

24. Healey JNC, In Hardy JG, Davis SS, Wilson CG, editors. *Drug Delivery to the Gastrointestinal Tract,* Ellis Horwood, Chichester, UK, 1989, pp. 83–96.

25. Lu H, et al. Cellulose acetate 1,2-benzenedicarboxylate inhibits infection by cell-free and cell-associated primary HIV-1 isolates. *AIDS Res. Hum. Retroviruses* 2006;22(5): 411–418.

26. *United States Pharmacopeia/National Formulary Online,* United States Pharmacopeial Convention, 2008.

27. Bogentoft C, et al. Influence of food on the absorption of acetylsalicylic acid from enteric-coated dosage forms. *Eur. J. Clin. Pharmacol.* 1978;14(5):351–355.

28. Watanalumlerd P, Christensen JM, Ayres JW. Pharmacokinetic modeling and simulation of gastrointestinal transit effects on plasma concentrations of drugs from mixed immediate-release and enteric-coated pellet formulations. *Pharm. Dev. Technol.* 2007;12(2):193–202.

29. Beckert TE, Lehmann K, Schmidt PC. Compression of enteric coated pellets to disintegrating tablets. *Int. J. Pharm.* 1996;143(Oct 25):13–23.

30. Dashevsky A, Kolter K, Bodmeier R. Compression of pellets coated with various aqueous polymer dispersions. *Int. J. Pharm.* 2004;279(1–2):19–26.

31. Ando M, et al. Development and evaluation of a novel dry-coated tablet technology for pellets as a substitute for the conventional encapsulation technology. *Int. J. Pharm.* 2007;336(1):99–107.

32. Ferron GM, et al. Oral bioavailability of pantoprazole suspended in sodium bicarbonate solution. *Am. J. Health Syst. Pharm.* 2003;60(13):1324–1329.

33. Erickson MA. Compounding hotline, *Pharmacy Times, Feb 2005* www.pharmacytimes.com/issues/articles/2005-02 2015.asp.

34. Iwao K, et al. Decreased plasma levels of omeprazole after coadministration with magnesium-aluminium hydroxide dry suspension granules. *Yakugaku Zasshi* 1999;119(3): 221–228.

35. Howden CW. Review article: immediate-release proton-pump inhibitor therapy—potential advantages. *Aliment. Pharmacol. Ther.* 2005;22 Suppl 3:25–30.

36. Meinhold JM, et al. XR vs XL vs CR vs ER vs SA: What do these suffixes really mean as it pertains to modified-release mood stabilizers and antiepileptic drugs? *ASHP Midyear Clinical Meeting* 2007;42: December.

37. Cole ET, et al. Enteric coated HPMC capsules designed to achieve intestinal targeting. *Int. J. Pharm.* 2002;231(1): 83–95.

38. Mercier GT, et al. Oral immunization of rhesus macaques with adenoviral HIV vaccines using enteric-coated capsules. *Vaccine* 2007;25(52):8687–8701.

39. Felton LA, McGinity JW. Enteric film coating of soft gelatin capsules. *Drug Deliv. Technol.* 2003;3(6).

40. Chidambaram N, Fatmi A. *Enteric Valproic Acid,* US Pat. Appl. 2007/0098786A1, May 3, 2007 Banner Pharmacaps Inc.

41. Mehuys E, Remon JP, Vervaet C. Production of enteric capsules by means of hot-melt extrusion. *Eur. J. Pharm. Sci.* 2005;24(2–3):207–212.

42. Law D, Zhang Z. Stabilization and target delivery of nattokinase using compression coating. *Drug Dev. Ind. Pharm.* 2007;33(5):495–503.

43. Obara S, et al. Dry coating: an innovative enteric coating method using a cellulose derivative. *Eur. J. Pharm. Biopharm.* 1999;47(1):51–59.

44. Kablitz CD, Harder K, Urbanetz NA. Dry coating in a rotary fluid bed. *Eur. J. Pharm. Sci.* 2006;27(2–3):212–219.

45. Sauer D, et al. Influence of processing parameters and formulation factors on the drug release from tablets powder-coated with Eudragit L 100-55. *Eur. J. Pharm. Biopharm.* 2007;67(2):464–475.

46. McGinity JW, Cameron CG, Cuff GW. Controlled release theophylline tablet formulations containing acrylic resins. 1. Dissolution properties of tablets. *Drug Dev. Ind. Pharm.* 1983;9(1–2):57–68.

47. Oren PL, Seidler WMK. Sustained release matrix. US Patent 4,968,508, Nov. 6, 1990.

48. Estaban E., Gallardo D., et al. Comparison of Eudragit FS30D and L30D-55 as Matrix Formers in Sustained Release Tablets, Degussa technical information, Evonik Roehm GmbH.

49. Tatavarti AS, Hoag SW. Microenvironmental pH Modulation Based Release Enhancement of a Weakly Basic Drug from Hydrophilic Matrices. *J. Pharm. Sci.* 2006;95:1459–1468.

50. Guo HX, et al. Direct compression of sustained-release matrix tablets of enteric cellulose esters. *S. T. P. Pharma Sci.* 2002;12 (5):287–292.

51. *Eudragit Application Guidelines,* 10th edition, Evonik Industries, 06/ 2008.

52. Stroyer A, McGinity JW, Leopold CS. Solid state interactions between the proton pump inhibitor omeprazole and various

enteric coating polymers. *J. Pharm. Sci.* 2006;95(6): 1342–1353.

53. Jansen PJ, et al. Characterization of impurities formed by interaction of duloxetine HCl with enteric polymers hydroxypropyl methylcellulose acetate succinate and hydroxypropyl methylcellulose phthalate. *J. Pharm. Sci.* 1998;87 (1):81–85.

54. Huang YB, et al. Optimization of pH-independent release of nicardipine hydrochloride extended-release matrix tablets using response surface methodology. *Int. J. Pharm.* 2005; 289(1–2):87–95.

55. Mukherjee B, Dinda SC, Barik BB. Gum cordia: a novel matrix forming material for enteric resistant and sustained drug delivery—a technical note. *AAPS PharmSciTech* 2008;9 (1):330–333.

56. Palmieri GF, et al. Polymers with pH dependent solubility: possibility of use in the formulation of gastroresistant and controlled release matrix tablets. *Drug Dev. Ind. Pharm.* 2000;26(8):837–845.

57. Bruce C, et al. Crystal growth formation in melt extrudates. *Int. J. Pharm.* 2007;341(1–2):162–172.

58. Follonier N, Doelker E, Cole ET. Various ways of modulating the release of diltiazem hydrochloride from hot-melt extruded sustained release pellets prepared using polymeric materials. *J. Control. Release* 1995;36:243–250.

59. Mojaverian P, et al. Effect of food on the absorption of enteric-coated aspirin: correlation with gastric residence time. *Clin. Pharmacol. Ther.* 1987;41(1):11–17.

60. Hermansson G, Sivertsson R. Gender-related differences in gastric emptying rate of solid meals. *Dig. Dis. Sci.* 1996;41 (10):1994–1998.

61. Gill RC, et al. Effect of the menstrual cycle on gastric emptying. *Digestion* 1987;36(3):168–174.

62. Lindahl A, et al. Characterization of fluids from the stomach and proximal jejunum in men and women. *Pharm. Res.* 1997;14(Apr):497–502.

63. Haddish-Berhane N, et al. A multi-scale stochastic drug release model for polymer-coated targeted drug delivery systems. *J. Control. Release* 2006;110(2):314–322.

64. Kamba M, et al. Comparison of the mechanical destructive force in the small intestine of dog and human. *Int. J. Pharm.* 2002;237(1–2):139–149.

65. Dressman JB, et al. Upper gastrointestinal (GI) pH in young, healthy men and women. *Pharm. Res.* 1990;7(Jul):756–761.

66. Walter-Sack IE. The significance of nutrition for the biopharmaceutical characteristics of oral controlled release products and enteric coated dosage forms. In Gundert-Remy U, Moller H, editors. *Oral Controlled Release Products*, Wissenschaftliche Verlagsgesselschaft, Stuttgart, 1990.

67. Rhie JK, et al. Drug marker absorption in relation to pellet size, gastric motility and viscous meals in humans. *Pharm. Res.* 1998;15(2):233–238.

68. Walker J, et al. Does enteric-coated aspirin result in a lower incidence of gastrointestinal complications compared to normal aspirin? *Interact. Cardiovasc. Thorac. Surg.* 2007;6 (4):519–522.

69. Ellsworth AJ, Christensen DB, Volpone-McMahon MT. Prospective comparison of patient tolerance to enteric-coated vs nonenteric-coated erythromycin. *J. Fam. Pract.* 1990;31 (3):265–270 (see comment).

70. Guthrie R. Erythromycin in acute pharyngitis: a comparison of efficacy and patient tolerance of two twice-daily preparations. *Clin. Ther.* 1988;10(5):530–535.

71. Takada K. Oral drug delivery, traditional. In Mathiowitz E, editor. *Controlled Drug Delivery*, Vol. 2, John Wiley & Sons, Inc., New York, 1999.

72. Liversidge GG, Conzentino P. Drug particle size reduction for decreasing gastric irritancy and enhancing absorption of naproxen in rats. *Int. J. Pharm.* 1995;125(2):309–313.

73. Cardoe N, Fowler PD. Butacote: a six-year follow-up of patients with gastric intolerance to other medications. *J. Int. Med. Res.* 1977;5 Suppl 2:59–66.

74. Pelletier RP, et al. The impact of mycophenolate mofetil dosing patterns on clinical outcome after renal transplantation. *Clin. Transplant.* 2003;17(3):200–205.

75. Knoll GA, et al. Mycophenolate mofetil dose reduction and the risk of acute rejection after renal transplantation. *J. Am. Soc. Nephrol.* 2003;14(9):2381–2386 (see comment).

76. Boots JM, Christiaans MH, van Hooff JP. Effect of immunosuppressive agents on long-term survival of renal transplant recipients: focus on the cardiovascular risk. *Drugs* 2004;64 (18):2047–2073.

77. Haeberlin B, et al. Enteric coated pharmaceutical compositions. US Patent 6306900, October 23, 2001, Novartis AG, Basel, Switzerland.

78. Haeberlin B, et al. Enteric-coated pharmaceutical compositions of mycophenolate, US Patent 6025391, February 15, 2000, Novartis AG, Basel, Switzerland.

79. Haeberlin B, et al. Enteric-coated pharmaceutical compositions, US Patent 6172107, January 9, 2001, Novartis AG, Basel, Switzerland.

80. Mycophenolate Sodium (Myfortic), National PBM Drug Monograph, March 2007.

81. Shaw LM, et al. Current issues in therapeutic drug monitoring of mycophenolic acid: report of a roundtable discussion. *Ther. Drug Monit.* 2001;23:305–315.

82. *Physician's Desk Reference: Myfortic Tablets (Novartis)*, 2006.

83. Salvadori M, et al. Long-term administration of enteric-coated mycophenolate sodium (EC-MPS; myfortic) is safe in kidney transplant patients. *Clin. Nephrol.* 2006;66(2):112–119.

84. Gabardi S, Tran JL, Clarkson MR. Enteric-coated mycophenolate sodium. *Ann. Pharmacother.* 2003;37(11):1685–1693.

85. Budde K, et al. Enteric-coated mycophenolate sodium can be safely administered in maintenance renal transplant patients: results of a 1-year study. *Am. J. Transplant.* 2004;4(2): 237–243 (see comment).

86. Tomlanovich S, et al. A three-month, prospective, open-label, two cohort study to investigate the efficacy, safety and tolerability of EC-MPS in combination with cyclosporin or tacrolimus in renal transplant recipients with GI intolerance. *Transplantation* 2006;82(1S2):487.

87. Chan L, et al. Patient-reported gastrointestinal symptom burden and health-related quality of life following conversion from mycophenolate mofetil to enteric-coated mycophenolate sodium. *Transplantation* 2006;81(9):1290–1297 (see comment).

88. Walker R, et al. Analysis of conversion from mycophenolate mofetil to enteric-coated mycophenolate sodium in renal transplant patient subpopulations using patient-reported outcomes. *Transplantation* 2006;82(1S2):514.

89. Kobashigawa JA, et al. Similar efficacy and safety of enteric-coated mycophenolate sodium (EC-MPS, myfortic) compared with mycophenolate mofetil (MMF) in de novo heart transplant recipients: results of a 12-month, single-blind, randomized, parallel-group, multicenter study. *J. Heart Lung Transplant.* 2006;25(8):935–941 (see comment).

90. Dumortier J, et al. Conversion from mycophenolate mofetil to enteric-coated mycophenolate sodium in liver transplant patients presenting gastrointestinal disorders: a pilot study. *Liver Transplant.* 2006;12(9):1342–1346 (see comment).

91. Cantisani GP, et al. Enteric-coated mycophenolate sodium experience in liver transplant patients. *Transplant. Proc.* 2006;38(3):932–933.

92. Filler G, Buffo I. Safety considerations with mycophenolate sodium. *Expert Opin. Drug Saf.* 2007;6(4):445–449.

93. Arns W. Noninfectious gastrointestinal (GI) complications of mycophenolic acid therapy: a consequence of local GI toxicity? *Transplant. Proc.* 2007;39(1):88–93.

94. Kunches LM, et al. Tolerability of enteric-coated didanosine capsules compared with didanosine tablets in adults with HIV infection. *J. Acquir. Immune Defic. Syndr.* 2001;28(2):150–153.

95. Damle B, et al. Pharmacokinetics and gamma scintigraphy evaluation of two enteric coated formulations of didanosine in healthy volunteers. *Br. J. Clin. Pharmacol.* 2002;54(3): 255–261.

96. Basit AW, et al. The use of formulation technology to assess regional gastrointestinal drug absorption in humans. *Eur. J. Pharm. Sci.* 2004;21(2–3):179–189.

97. Cora LA, et al. Enteric coated magnetic HPMC capsules evaluated in human gastrointestinal tract by AC biosusceptometry. *Pharm. Res.* 2006;23(8):1809–1816.

98. Bours MJ, et al. Effects of oral adenosine 5'-triphosphate and adenosine in enteric-coated capsules on indomethacin-induced permeability changes in the human small intestine: a randomized cross-over study. *BMC Gastroenterol.* 2007; 7: 23.

99. Kalnins D, et al. Enteric-coated pancreatic enzyme with bicarbonate is equal to standard enteric-coated enzyme in treating malabsorption in cystic fibrosis. *J. Pediatr. Gastroenterol. Nutr.* 2006;42(3):256–261.

100. Siepe S, et al. Strategies for the design of hydrophilic matrix tablets with controlled microenvironmental pH. *Int. J. Pharm.* 2006;316(1–2):14–20.

101. Fedorak RN, Bistritz L. Targeted delivery, safety, and efficacy of oral enteric-coated formulations of budesonide. *Adv. Drug Deliv. Rev.* 2005;57(2):303–316.

102. Edsbacker S, et al. A pharmacoscintigraphic evaluation of oral budesonide given as controlled-release (Entocort) capsules. *Aliment. Pharmacol. Ther.* 2003;17(4):525–536.

103. Lecomte F, et al. pH-Sensitive polymer blends used as coating materials to control drug release from spherical beads: elucidation of the underlying mass transport mechanisms. *Pharm. Res.* 2005;22(7):1129–1141.

104. Dashevsky A, Kolter K, Bodmeier R. pH-independent release of a basic drug from pellets coated with the extended release polymer dispersion Kollicoat SR 30 D and the enteric polymer dispersion Kollicoat MAE 30 DP. *Eur. J. Pharm. Biopharm.* 2004;58(1):45–49.

105. Kim MS, et al. Influence of water soluble additives and HPMCP on drug release from Surelease-coated pellets containing tamsulosin hydrochloride. *Arch. Pharmacal Res.* 2007;30(8):1008–13.

106. Siepmann F, et al. Polymer blends for controlled release coatings. *J. Control. Release* 2008;125(1):1–15.

107. Siepmann F, Wahle C, et al. Ethylcellulose-based controlled release film coatings with pH-triggered permeability, Poster presentation, 2006 AAPS annual meeting, San Antonio, TX, www.aapspharmsci.org/abstracts/AM_2006/AAPS2006-000743.pdf.

108. Venkatesh GM. Development of controlled-release SK&F 82526-J buffer bead formulations with tartaric acid as the buffer. *Pharm. Dev. Technol.* 1998;3(4):477–485.

109. Menjoge AR, Kulkarni MG. Mechanistic investigations of phase behavior in Eudragit E blends. *Int. J. Pharm.* 2007;343 (1–2):106–121.

110. Menjoge AR, Kulkarni MG. Blends of reverse enteric polymer with Enteric and pH-independent polymers: mechanistic investigations for tailoring drug release. *Biomacromolecules* 2007;8(1):240–251.

111. Dohil R, et al. Understanding intestinal cysteamine bitartrate absorption. *J. Pediatr.* 2006;148(6):764–769 (see comment).

112. Degenis G, Sandefer E. Gamma scintigraphy and neutron activation techniques in the *in vivo* assessment of orally administered dosage forms. *Adv. Drug Deliv. Rev.* 7 1991 309–345.

113. Nishimura K, et al. Effervescent enteric coated L-dopa formulation and method of using the same. U.S. Patent 3,961,041, June 1, 1976.

114. Huyghebaert N, Vermeire A, Remon JP. *In vitro* evaluation of coating polymers for enteric coating and human ileal targeting. *Int. J. Pharm.* 2005;298(1):26–37.

115. Eiamtrakarn S, et al. Gastrointestinal mucoadhesive patch system (GI-MAPS) for oral administration of G-CSF, a model protein. *Biomaterials* 2002;23(1):145–152.

116. Eaimtrakarn S, et al. Retention and transit of intestinal mucoadhesive films in rat small intestine. *Int. J. Pharm.* 2001;224 (1–2):61–67.

117. Ibekwe VC, et al. An investigation into the *in vivo* performance variability of pH responsive polymers for ileo-colonic drug delivery using gamma scintigraphy in humans. *J. Pharm. Sci.* 2006;95(12):2760–2766.

118. Sangalli ME, et al. *In vitro* and *in vivo* evaluation of an oral system for time and/or site-specific drug delivery. *J. Control. Release* 2001;73(1):103–110.

119. Wilding IR, et al. Enteric coated timed release systems for colonic targeting. *Int. J. Pharm.* 1994;111(Oct 6):99–102.

120. Omar S, et al. Colon-specific drug delivery for mebeverine hydrochloride. *J. Drug Target.* 2007;15(10):691–700.

121. Miller DA, et al. Targeted intestinal delivery of supersaturated itraconazole for improved oral absorption. *Pharm. Res.* 2008;25(6):1450–1459.

122. Dorkoosh FA, et al. Feasibility study on the retention of superporous hydrogel composite polymer in the intestinal tract of man using scintigraphy. *J. Control. Release* 2004;99(2):199–206.

123. Gomez-Roman VR, et al. Oral delivery of replication-competent adenovirus vectors is well tolerated by SIV- and SHIV-infected rhesus macaques. *Vaccine* 2006;24(23):5064–5072.

124. Groning R, Danco I, Muller RS. Development of sensor elements to control drug release from capsular drug delivery systems. *Int. J. Pharm.* 2007;340(1–2):61–64.

125. Rafati H, Ghassempour A, Barzegar-Jalali M. A new solution for a chronic problem; aqueous enteric coating. *J. Pharm. Sci.* 2006;95(11):2432–2437.

126. Sarisuta N, et al. The influence of drug-excipient and drug-polymer interactions on butt adhesive strength of ranitidine hydrochloride film-coated tablets. *Drug Dev. Ind. Pharm.* 2006;32(4):463–471.

127. Guo HX, Heinamaki J, Yliruusi J. Diffusion of a freely water-soluble drug in aqueous enteric-coated pellets. *AAPS PharmSciTech* 2002;3(2):E16.

128. Guo HX, Heinamaki J, Yliruusi J. Amylopectin as a subcoating material improves the acidic resistance of enteric-coated pellets containing a freely soluble drug. *Int. J. Pharm.* 2002;235(1–2):79–86.

129. Lovgren KI, et al. Pharmaceutical preparation for oral use. US Patent 4,786,505 1988.

130. Miyazaki Y, et al. *In vivo* drug release from hydrophilic dextran tablets capable of forming polyion complex. *J. Control. Release* 2006;114(1):47–52.

131. Bruce LD, Koleng JJ, McGinity JW. The influence of polymeric subcoats and pellet formulation on the release of chlorpheniramine maleate from enteric coated pellets. *Drug Dev. Ind. Pharm.* 2003;29(8):909–924.

132. Fourman GL, Hines CW, Hritsko RS. Assessing the uniformity of aqueous film coatings applied to compressed tablets. *Pharm. Technol.* 1995;19(Mar):70, 72, 74, 76.

133. Siepmann F, et al. Aqueous HPMCAS coatings: effects of formulation and processing parameters on drug release and mass transport mechanisms. *Eur. J. Pharm. Biopharm.* 2006;63(3):262–269.

134. Petereit HU, Assmus M, Lehmann K. Glyceryl monostearate as a glidant in aqueous film-coating formulations. *Eur. J. Pharm. Biopharm.* 1995;41(4):219–228.

135. Lin YA, Dulin WA, Ku S. Enteric-coated pellet formulation and process scale-up improvement using mono- and di-glycerides as a glidant. Poster Presentation, 2007 AAPS Annual Meeting, San Diego, CA, www.aapsj.org/abstracts/AM_2007/AAPS2007-001493.PDF.

136. Dong W, Bodmeier R. Encapsulation of lipophilic drugs within enteric microparticles by a novel coacervation method. *Int. J. Pharm.* 326(1–2): 2006 128–138.

137. Moschwitzer J, Muller RH. Spray coated pellets as carrier system for mucoadhesive drug nanocrystals. *Eur. J. Pharm. Biopharm.* 2006;62(3):282–287.

138. Scocca S, et al. Alginate/polymethacrylate copolymer microparticles for the intestinal delivery of enzymes. *Curr. Drug Deliv.* 2007;4(2):103–108.

139. Cardinal Health/Phoqus Drug Delivery Systems Brochure 2005.

140. Comoglu T, et al. Development and in vitro evaluation of pantoprazole-loaded microspheres. *Drug Deliv.* 2008;15 (5):295–302.

141. Obeidat WM, Price JC. Preparation and evaluation of Eudragit S 100 microspheres as pH-sensitive release preparations for piroxicam and theophylline using the emulsion-solvent evaporation method. *J. Microencapsul.* 2006;23 (2):195–202.

142. Yang JF, et al. Thymosin-loaded enteric microspheres for oral administration: preparation and *in vitro* release studies. *Int. J. Pharm.* 301 2005; (1–2):41–47.

143. De Jaeghere F, Allémann E, Cerny R, Galli B, Steulet AF, Müller I, Schütz H, Doelker E, Gurny R. pH-dependent dissolving nano- and microparticles for improved peroral deliver of a highly lipophilic compound in dogs. *AAPS Pharmsci.* 2001;3(1):E8.

144. Raffin RP, et al. Sodium pantoprazole-loaded enteric microparticles prepared by spray drying: effect of the scale of production and process validation. *Int. J. Pharm.* 2006;324 (1):10–18.

145. Giunchedi P, Conte U. Spray-drying as a preparation method of microparticulate drug delivery systems: overview. *S. T. P. Pharma Sci.* 1995;5(4):276–290.

146. Palmieri GF, et al. Gastro-resistant microspheres containing ketoprofen. *J. Microencapsul.* 2002;19(1):111–119.

147. Yuan J, Clipse N, In situ enteric coating protection: new application of C-A-P, AAPS annual meeting abstract, 2005. www.aapspharmsci.org/abstracts/AM_2005/Staged/aaps2005-002032.pdf.

148. Lau KK, Gleason KK. All-dry synthesis and coating of methacrylic acid copolymers for controlled release. *Macromol. Biosci.* 2007;7(4):429–434.

149. Hoppe B, et al. Oxalobacter formigenes: a potential tool for the treatment of primary hyperoxaluria type 1. *Kidney Int.* 2006;70(7):1305–1311 (see comment).

14

ORALLY ADMINISTERED DRUG DELIVERY SYSTEMS TO THE COLON

Mirela Nadler Milabuer, Yossi Kam, and Abraham Rubinstein
Faculty of Medicine, School of Pharmacy, The Hebrew University of Jerusalem, Jerusalem, Israel

14.1 INTRODUCTION

Spatial placement of drugs in the alimentary canal, following their oral ingestion, is an attractive yet complicated challenge for drug delivery designers. Apparently, localizing drugs in specific targets along the small bowel is beyond attainment. Nonetheless, a variety of interesting pharmaceutical techniques for the prolongation of residence time of drugs in the stomach [1–4] and targeting the colon with minimal drug loss during small bowel passage have been described in the past quarter of century.

There are a number of medical grounds for shuttling drugs to the colon. These include

a. local treatment of those inflammatory bowel diseases (IBD) that are confined to the colonic epithelium (e.g., ulcerative colitis) [5];

b. local treatment of colon motility disorders, such as irritable bowel syndrome (IBS), which respond to topical application of sedative or antispasmodic drugs [6–8];

c. the possibility to interfere with the proliferation of colon polyps (first stage in colon carcinoma) with some nonsteroidal anti-inflammatory drugs (NSAIDs) [9] such as sulindac (metabolized in the colon to the active moiety, sulindac sulfide) [10, 11] or NSAIDs that may act directly on the colon epithelium, such as celecoxib [12–14] in mono or combinatorial therapy [15];

d. accumulated evidence that drug absorption enhancement works better in the colon than in the small intestine [16–18];

e. the anticipation that protein drugs can be absorbed better from the large bowel due to hypothetic reduced proteolytic activity in this organ [19];

f. the unique metabolic activity of the colon, which makes it an attractive organ for drug delivery system designers [20, 21].

Essentially, orally administered colonic drug delivery systems are designed to minimize the interaction between their drug loads and the fluids of the stomach and the small intestine, to avoid premature release of the drug. Commonly, the drug is anchored to its platform either chemically (as a prodrug or polymeric prodrug), or physically (by its entrapment with polymeric coatings). It is therefore obvious that the successful functioning of such systems depends on longitudinal physiological restrictions of the gastrointestinal (GI) tract. Parameters such as residence time, gastric emptying, regional pH, chyme composition, and viscosity and enzymatic activity are crucial in the engineering of colonic delivery systems, due to the fact that these systems have to pass the entire length of the digestive tube prior to their arrival at the colon. Accordingly, they need to cope with complex, altering biological environments. A simple solution for colonic arrival would be the use of rectal administration of drugs by means of enemas or suppositories. However, this mode of administration suffers from compliance problems and will not be discussed in this chapter. The purpose of this chapter is to present the major design strategies that have been described in the literature to ferry drugs to the large intestine after oral administration and their susceptibility to longitudinal GI parameters. It should also be taken into

Oral Controlled Release Formulation Design and Drug Delivery: Theory to Practice, Edited by Hong Wen and Kinam Park
Copyright © 2010 John Wiley & Sons, Inc.

consideration that common medical use of colonic delivery is commercially employed in the context of local treatment of mucosal inflammation associated with IBD. All other medical opportunities are wishful thinking and await their exploration by pharmaceutical scientists.

14.2 ANATOMICAL AND PHYSIOLOGICAL CONSIDERATION RELEVANT TO COLONIC DELIVERY OF DRUGS

The colon (about 1.5 m long), which arches around the small intestine, is divided into the ascending segment (right colon), transverse segment, descending segment (left colon), and the sigmoid region bordering the rectum. Transversely, the layers of the colonic wall are the mucosa, consisting of the epithelium, the lamina propria, and the muscularis mucosa; the submucosa; the muscularis propria, consisting of the inner circular muscle layer, the intermuscular space, and the outer longitudinal muscle layer; and the serosa (Figure 14.1) [22].

14.2.1 Mucosal Barriers of the Colon

Similar to the small intestine, a single cell layer of epithelium forms the interface of the colonic wall with the lumen. The apical surface is dominated by columnar epithelial cells (colonocytes) and M-cells. The colonocytes are constantly being replaced with a half-life of about 6 days [23]. The epithelial cells are bound together by tight junction, creating a physical barrier to the uptake of molecules from the normal healthy intestinal lumen into the mucosa, with pore size openings in the human colon, estimated to be 2.3 Å, in contrast to the large pores of the ileum (4 Å) and the jejunum

(8 Å) [24]. It has been suggested that peptides with flexible conformation may be able to transverse these pores [25].

It should be recognized, however, that the opening of the tight junction may be altered by disease states [26]. Such is the case with IBD, where tight junction opening is associated with inflammation of the colonic epithelium. A break in the continuity of the epithelial surface may offer the opportunity for direct drug uptake into the affected tissue of not only chemotherapies but also pathological agents. The permeability of the intestinal mucosa is increased in most patients with Crohn's disease and 10–20% of their clinically healthy relatives. Permeability is also increased in individuals with celiac disease, trauma, burns, and administered nonsteroidal anti-inflammatory drugs. [27].

14.2.2 Motility of the Colon

The motility of the large bowel is different from that of the small intestine aiming to fulfill three functions: water reabsorption, residual carbohydrates fermentation, and storage and propulsion of fecal material in its distal portion. These functions are time consuming and require a slow exposure of the colonic mucosa to the luminal content. The contractions in the large intestine are less organized than in the small intestine. Thus, it is believed that the typical colon motility is less propulsive, and results in movements that mix, knead, and churn the luminal content [28, 29]. Colonic transit time is influenced by various factors, such as the size of the dosage forms, the presence of food, gender, diseased state, and stress [22]. Overall, the prolonged colonic transit time (up to an average of 45 h and more) increases the residence time of drug delivery systems in the colon, which in turn increases contact time with the mucosa, thus enhancing drug absorption.

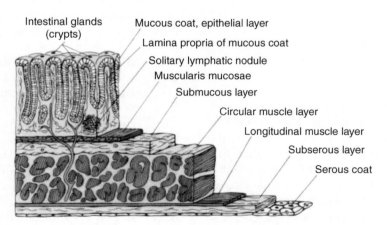

FIGURE 14.1 Schematic cross section of the large bowel wall (source: http://images.google.co.il/imgres?imgurl=http://www.cancerline. com/gUserFiles/structure_of_the_colon.gif&imgrefurl=http://www.cancerline.com/CancerlineHCP/9898_22878_0_0_0.aspx&h=309&w= 516&sz=89&hl=en&start=17&um=1&usg=__-Guqxzke_xnC4W_8w8_Z_M0Cuxg=&tbnid=2_cQLVlNuvQpYM:&tbnh=78&tbnw= 131&prev=/images%3Fq%3Dcolon%2Banatomy%26um%3D1%26hl%3Den%26client%3Dfirefox-a%26rls%3Dorg.mozilla:en-US:official%26sa%3DN).

14.2.3 Colonic Mucus Lining

Mucus is synthesized and released slowly and continuously to function as the major housekeeper for preventing mucosal injury or contamination by the rich microflora of the colon lumen. The mucus gel of the colon differs from that of the small intestine. Its mucins (the proteoglycan skeleton of mucus) are larger in size and have a higher protein content than proximal intestinal mucin. Sulfate and sialic acid are present in greater proportions in colonic mucus [30]. In the context of colonic drug delivery, sialic acid was found to overexpress in malignant human cell lines and rat colonic tissues. Its expression correlated to the metastatic stage *in vitro* [31]. The binding of the cationic copolymers to the cell lines and tissues correlated with the charge density of the polymer and with the metastatic stage of the cell lines. It was concluded that cationic polymers could be used as a targeting tool to colonic malignant epithelium [31].

14.2.4 Colonic Microflora and Metabolism

One attractive feature of the colon is its metabolic activity resulting due to its typical microflora. While the gastric microflora is predominantly aerobic and its bacterial concentration is approximately 10^3 colony-forming units (CFU)/mL, the large bowel ecosystem is anaerobic in nature with a typical bacterial concentration of 10^{11} CFU/mL [20, 32]. The huge amount of microorganisms in the lumen of the colon merits considering the colon as the body's "microbial organ" [33]. The colonic microflora is highly diverse [34]. Out of the 400 distinct bacterial species in the colon, 30% of fecal isolates are of the genus *Bacteroides*, a group which is characterized by a relatively high growth yield even at low rates, which enables it to compete successfully for energy sources in the lumen of the colon [35]. For that purpose, the colonic bacterial flora ferments various substrates, primarily polysaccharides that, under normal conditions, produce typical end products such as short-chain fatty acids (acetic, propionic, and butyric acids) and gases, such as methane, carbon dioxide, and hydrogen [36]. For drug delivery design purposes, this means that the pH of the ascending colon is mildly acidic (around 5.5) [37]. In addition to the typical saccharolytic activity of the colon, it is capable of reducing azo bonds, S–S bonds, and hydrolyzing esters [21], all being exploited for drug delivery purposes [38]. Some examples are shown in Table 14.1.

Microbially activated drug delivery systems [39, 40] had been suggested almost 25 years ago in the context of targeting the inflamed colon with steroid prodrugs and oral administration of peptide drugs (insulin and lysine vasopressin) [41–43]. As will be described later, the design was based on the ability of the colonic microbial flora to hydrolyze glycosidic bonds in the former and to reduce azo bonds in the latter case.

TABLE 14.1 Drug-Metabolizing Enzymes in the Human Intestinal Flora

Enzyme	Activity Determined in	Metabolic Reaction Catalyzed
Reducing enzymes		
Nitroreductase	Lumen	Reduction of aromatic and heterocyclic nitro compounds
Azoreductase	Feces	Cleavage of azo compounds
N-oxide reductase, sulfoxide reductase hydrogenase	Feces	Reduction of cabonyl groups and aliphatic double bonds
Hydrolytic enzymes		
Esterases and amidases	Feces	Cleavage of esters or amides of carboxylic acids
Glycosidases	Lumen	Hydrolysis of saccharidic β-glycosidic bonds
Glucuronidase	Feces	Hydrolysis of β-glucuronic acid conjugates
Sulfatase	Lumen	Hydrolysis of *O*-sulfate and sulfamate

From Ref. 21 with permission.

14.2.5 Luminal pH in the GI Tract

The pH of the resting stomach is acidic, ranging between 1.7 in young people to 1.3 in elderly subjects. Immediately after meal ingestion, the pH in the stomach increases sharply (to values of 5–6) and decreases again slowly when gastric acid secretion overrides the buffering capacity of the food [44, 45]. In fasting humans, the duodenal pH is around 6.4. After a meal, the pH of the proximal duodenum drops because of the influx of acidic chyme. The chyme is then rapidly buffered by the bicarbonate secretion as it moves distally. The intestinal postprandial pH can be as high as 7.75 [46]. The pH of the colon varies depending on the type of food ingested. In general, a drop in pH to 6 and below is monitored in the ascending colon due to the fermentation processes in that region. Also, the pH changes as a result of disease states. For example, inflammation causes the colonic pH to decrease [47, 48].

The shallow pH gradient between the small intestine and the colon is a serious obstacle in the design of colonic delivery drug carriers based on pH-sensitive polymers. For years, it was assumed that the pH in the digestive tube elevates gradually. Accordingly, colonic delivery systems of salicylates and steroids were developed, assuming a specific biodegradation of the vehicle coat upon colon arrival [49–52]. However, it has been shown that specificity is very low with pH-dependent colonic platforms and, depending on small bowel transit time and coat thickness, they may release their drug load in a premature manner in the ileum or in an undesired slow manner in the descending colon [53–55].

14.2.6 Colonic Drainage

Blood flow to the colon, which is likely to determine the rate of drug absorption from the lumen, is less than that of the small intestine, with the proximal colon receiving a more prolific supply than distal regions of the organ [56]. Drainage from the upper colon appears distinct from that of the more distal regions of the large intestine, where the former is drained by both the hepatic portal veins and the lymphatics and the latter is drained predominantly by the lymphatics [57].

The blood flow rate in humans is 1000–2000 times greater than the rate of lymphatic flow [58]. Although blood circulation is considered more significant for drug distribution, drugs entering the blood through the hepatic portal vein circulate directly to the liver, where they may undergo rapid biotransformation (hepatic first-pass metabolism) before they are able to act. However, introduction of drugs into the lymphatics has been considered as a means of improving oral bioavailability, because drugs could potentially pass from the lymphatics into the blood circulation before reaching the liver [59]. The mode of uptake into the lymphatics could occur through the M-cells or the epithelial cells [60, 61].

14.3 THE EVOLUTION OF COLON-SPECIFIC DRUG DELIVERY: FROM LAXATIVE DRUGS TO 5-ASA [62]

It has long been recognized that some drugs are activated in the colon after oral administration. For example, senna extract contains anthracene derivatives that exist in the form of glycoside (sennosides) and can be hydrolyzed to anthraquinones, anthranols, and oxanthrones. The sennosides are much more effective as laxatives when administered intact than are the sugar-free aglycones [63, 64]. Hattori et al. suggested that the metabolism involves both β-glucosidase hydrolysis and a reduction reaction via the electron-carrier cofactors FAD, FMN of both the sennoside and the sennidine (Figure 14.2). Similar cofactors are involved in the reduction of azo compounds in the colon [65].

The laxative, sodium sulisatin (Figure 14.2), acts as a prodrug. Being an anionic compound, it is poorly absorbed, and therefore arrives almost intact at the colon after oral ingestion. In the colon, it is hydrolyzed by bacterial

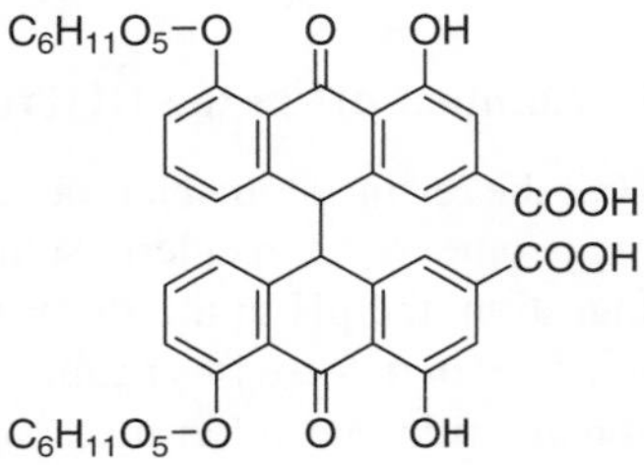

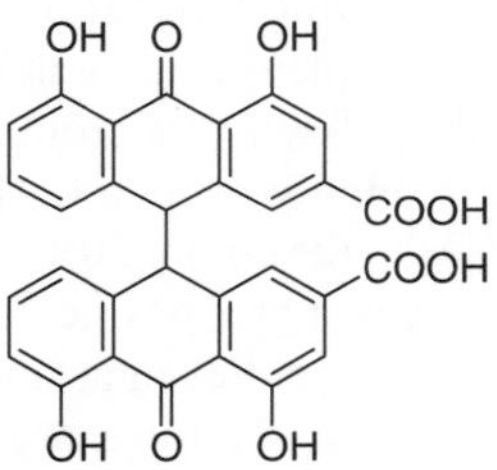

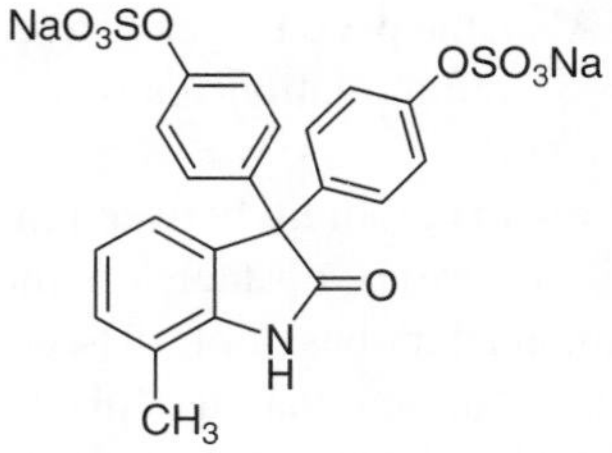

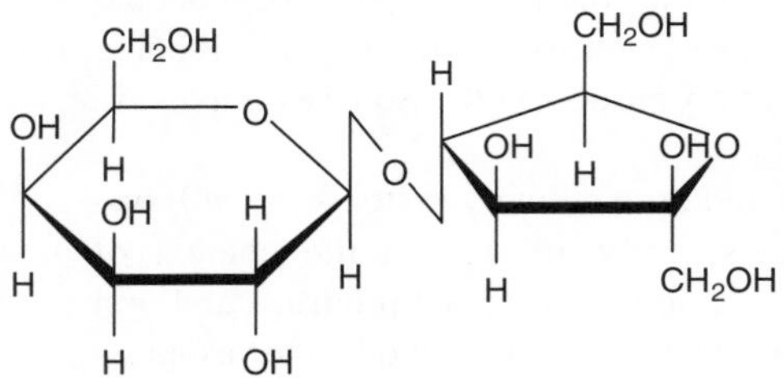

FIGURE 14.2 Colon-specific laxative drugs: sennoside (I) and its aglycone sennidine (II), sodium sulisatin (III), and lactulose (IV).

arylsulfate sulfohydrolase to a diphenolic derivative, which presumably is the active laxative [66]. The osmotic laxation effect of lactulose, semisynthetic disaccharide (Figure 14.2), is also colon specific. After oral administration, it is neither degraded by the disaccharidases of the small intestine nor is it absorbed. In the colon, it is fermented rapidly, leading to a transitory pH decrease, rapid gas production, and volume increase, as well as increase in osmolarity and the local production of short-chain fatty acids [67, 68].

The ability of naloxone and the narcotic antagonist nalmefene to induce diarrhea in morphine-dependent rats was compared to the glucuronide conjugates of the two drugs [69]. Their oral administration caused diarrhea, while in the glucuronide conjugates of either of the narcotic antagonists diarrhea was delayed for 1–3 h, reflecting the transit time to the distal ileum. A direct colonic administration of the naloxone and nalmefene glucuronides caused diarrhea within 5–8 min, indicating that the pharmacological response to the glucuronide conjugates of naloxone and nalmefene was initiated by bacterial β-glucuronidases in the rat colon. This hydrolysis of the drug glucuronides was specific to the colonic bacterial activity, since they were found to be inactive when administered subcutaneously.

The most illustrious molecule that demonstrates the concept of colon-specific drug delivery is sulfasalazine, the sulfonamide prodrug of 5-aminosalicylic acid (5-ASA). This azo compound was not the first to be used in GI disorders. A former compound, Prontosil (Figure 14.3), a sulfanilamide prodrug was used in the 1930s for the treatment of intestinal streptococcal infections [70]. Sulfasalazine was developed with the intention of causing a specific affinity between the salicylate moiety and the inflamed connective tissues. The distribution hypothesis was found to be true [71, 72]. However, it was also found that, as a result of a local azo reduction, sulfasalazine was split in the human colon [73] and that, in fact, the active moiety was 5-ASA. [74].

Based on the understanding of the mode of action of sulfasalazine, several other colon-specific azo compounds were synthesized: benzalazine (Intestinol®), balsalazide [75], and the most unique olsalazine (Dipentum®) (Figure 14.3). All compounds were designed to be reduced by azo reductases in the colon and to locally release 5-ASA. Olsalazine is composed of two 5-ASA molecules azo linked in their amine tail. When cleaved by colonic bacteria, the azodisalicylate delivers twice the amount of 5-ASA as compared to the other prodrugs and avoids the undesired action of the nontherapeutic moiety such as the sulfapyridine of the sulfasalazine [76, 77]. Indeed, the optimal dose of olsalazine for maintaining remission in ulcerative colitis was found to be 1 g/day [78] as compared to 2–4 g/day of sulfasalazine or 5-ASA in enteric coating formulations.

14.4 DESIGN CONSIDERATIONS IN COLONIC DRUG DELIVERY SYSTEMS

The medical insight on the mode of action of colon mucosal-active drugs and the recognition that some molecules function as prodrugs of the active moieties triggered rigorous R&D activity during the past 20 years of the last century. A variety of tactics were developed and tested in preclinical and clinical studies. For example, 5-ASA was formulated into numerous drug platforms, starting with prodrugs, polymeric prodrugs, and biodegradable polymers, ending with enteric-coated tablets or capsules. The relative success accomplished with the specific degradation of azo compounds in the human colon prompted the initiation of studies in which colon-specific platforms were suggested and tested in the context of shuttling proteinecious drugs to the large intestine [19].

In the following sections of this chapter, some most significant strategies for colonic delivery will be discussed. However, one should realize that not a lack in technologies

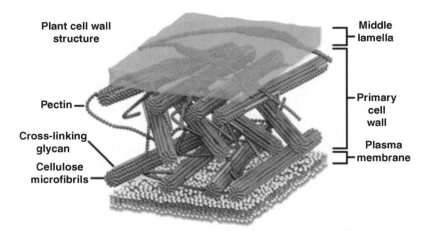

FIGURE 14.3 A schematic presentation of the plant cell wall (source: http://www.molecularexpressions.com/cells/plants/images/cell-wallfigure1.jpg).

restricted the development of feasible colonic drug delivery. It is the dearth in medical rationale and the complexity of clinical trials that hampered the progress of this existing area in drug delivery.

The technologies that will be reviewed in this chapter are slow release polymer coats, prodrugs, polymeric prodrugs, biodegradable matrices and hydrogels, and particulate carriers.

14.5 SLOW RELEASE AND ENTERIC-COATED BARRIERS

The most common approach to achieve colonic arrival of drugs is to use pharmaceutical techniques that minimize drug absorption along the small intestine, allowing drug unload in the large bowel only. A variety of techniques are being used here, the majority of which exploit enteric coating polymers (low pH-resistant polymers that swell and dissolve in moderate acidic to neutral pH values) [79]. Clinical studies have demonstrated that an effective enteric coating that is designed to disintegrate after a finite lag time could be used as a drug platform to the distal ileum and colon. An extended lag time of 3–4 h is considered to be sufficient for colonic delivery [80]. The lag time can be achieved either by using a thick enough coating layer or by using polymers that disintegrate at pH values larger than 7. The most popular materials utilized in colonic delivery of 5-ASA are the methacrylic acid–methylmethacrylate copolymers, commercially distributed as Eudragit® products. The methacrylic acid content in Eudragit L is 46–50% and it dissolves at a pH above 6. The methacrylic acid content in Eudragit S is 28–31% and it dissolves at a pH above 7 [79]. If applied with the same coating thickness, tablets coated with Eudragit L would release their 5-ASA load earlier than tablets coated with Eudragit S [81]. Therefore, the use of Eudragit L requires thicker coats to get as close as possible to the colon [50].

The delivery of 5-ASA to the colon is effective when local treatment of ulcerative colitis is needed. In the case of Crohn's disease, in which inflammation is spread all over the intestine, a slow release of 5-ASA along the entire gut mucosa is favorable and requires a different design [82]. Pentasa® is made of drug containing pellets coated with ethyl cellulose, a nonionic polymer, commonly used for sustained release products. The product design (pellets) allows maximal drug distribution as well as a constant release rate of 5-ASA over the full length of the intestinal tube (including the colon) after oral ingestion.

The enteric coating technique has also been utilized to deliver steroids to the large intestine. For example, when prednisolone in Eudragit-S coated capsules was delivered orally to ulcerative colitis and Crohn's disease patients, a delayed absorption was observed indicating colon arrival [83]. Wilkinson and coworkers have demonstrated

that although drug blood levels of prednisolone metasulfabenzoate were lower than prednisolone acetate after oral administration of both drugs in Eudragit-S coated tablets, a better clinical remission was accomplished after treatment with the former steroid, suggesting a pronounced local activity [84]. Better steroid candidates for topical action in the colon are beclomethasone dipropionate, which has been reported to be 500 times more potent topically than dexamethasone and 5000 times more potent topically than hydrocortisone [85] and budesonide [86, 87]. Cellulose acetate phthalate (CAP) enteric coating formulation of beclomethasone dipropionate was found to be effective in the delivery of the drug to the distal ileum as evaluated in proctocolectomy and ileostomy patients [51]. Apart from their poor absorption properties, budesonide and beclomethasone are unique because of the rapid first-pass metabolism they undergo after being absorbed (hepatic as well as in circulating erythrocytes). This feature contributes significantly to their reduced side effects and justifies their delivery in colonic delivery systems [88].

Enteric coating has also been used for the colonic delivery of peppermint oil, an antispasmodic agent when administered orally. Colpermine® capsules containing peppermint oil are now prescribed for the specific therapy of IBS. It was found that after colon arrival of the drug, constant drug blood levels were maintained over 8 h [89].

As has been mentioned, although the enteric coating approach provides an easy solution for colonic delivery of drugs, some major drawbacks are involved with this technique. In contrast to what was believed in the past [52], the pH of the ascending colon is slightly acidic and elevates gradually in the caudal direction [46]. Moreover, inflammation may reduce the luminal pH [48]. Therefore, if an enteric-coated dosage form, especially when coated with Eudragit-S type polymer, does not open in the distal ileum, it will not be able to dissolve in the ascending colon. The successful functioning of enteric coating formulations in colonic delivery of drugs depends primarily on the kinetics of the coat dissolution. This, in turn, depends on the transit time in the small bowel. The shorter the transit time, the more chances are that the dosage form will malfunction. Using gamma scintigraphy, Ashford, Fell, and coworkers have found that in seven volunteers, the transit time of enteric coating tablets varied from 5 to 15 h and the site of tablet disintegration varied from the ileum to the splenic flexus in the colon [54]. In a set of *in vitro* studies performed by the same group, the pH was altered within the limits of physiological values. It was concluded that enteric-coated tablets may be able to release their content as early as the duodenum or may not release the drug at all [53]. Nonetheless, despite these serious drawbacks, enteric-coated formulations are the only commercialized products in the market. Typical colonic delivery systems of the enteric coating type are summarized in Table 14.2.

TABLE 14.2   Some Typical Colonic Delivery System Designs of the Enteric Coating and Sustained Release (Mainly Ethylcellulose Film Coating) Types

Polymer	Coated Drug	Brand Name	Comments	References
CAP	Beclomethasone dipropionate		Gelatin capsule coated with the CAP film	[51]
Eudragit S, Eudragit RL30D + RS30D, Eudragit L100-55 + Eudragit S100, Eudragit-L, Eudragit FS 30D, Eudragit S; Eudragit FS 30D, Eudragit S100	5-ASA, 4-ASA, budesonide		Tablets coated with the various Eudragit products or their mixtures	[151–156]
Eudragit L-100 + Eudragit S-100	Diclofenac		*Multireservoir system*: chitosan microcores entrapped within acrylic microspheres, all coated with the Eudragit blends	[157]
Calcium pectinate and Eudragit S-100	5-Fluorouracil		Calcium pectinate beads coated with the Eudragit film	[158]
Pectin, chitosan, and Eudragit RS	Theophylline		A film made of the three polymers, loaded with a priming dose. A burst release occurs due to gradual increase in the film porosity	[159]
Eudragit L100, Eudragit S100	Prednisolone		Chitosan-succinyl-prednisolone conjugate coated with the Eudragit enteric coating films	[160]
Eudragit RS, Eudragit RL, Eudragit FS, or Eudragit S	Indomethacin or theophylline		Inulin-Eudragit films with varying swelling and permeation properties	[161]
Eudragit S	5-ASA	Oros-CT®	Capsule containing miniosmotic pumps or coated osmotic pump device with Eudragit S	[162, 163]
Eudragit L-100	Budesonide		Osmotic pump	[164]
Eudragit L	5-ASA	Salofalk® Mesasal® Claversal®	5-ASA tablets coated with Eudragit	
Ethylcellulose (95%) and diethylphthalate (5%)	Dofetilide	Pulsincap®	A capsule fitted with a hydrogel plug (cross-linked PEG), which swells upon exposure to the fluids of the gut causing drug ejection in a pulsatile manner. Colonic arrival depends on film layer	[165, 166]

The polymers are commonly employed as protective coats on top of solid dosage forms, or cocasted into films with biodegradable polymers (e.g., polysaccharides), or coformulated with biodegradable polymers into particulate carriers (e.g., microspheres) to increase their colon specificity.

A diffusional barrier to slow down drug release while traveling toward the colon can be achieved by a thick, biodegradable hydrogel coat such as hydroxypropyl methylcellulose (HPMC) [90]. With this drug carrier, the lag time to disintegration is determined by the thickness of the polymeric coat covering the drug containing core [91]. Similarly, the Geomatrix® system, a multilayered hydrophilic matrix tablet has been designed to achieve two goals: a delayed drug release and maintenance of constant release rates [92, 93].

Ueda et al. developed a dosage form that absorbs fluid through an outer membrane into a swellable core after oral administration at a predetermined rate. As the core swells, it exerts pressure on the coat wall and eventually breaks, resulting in a burst release of an entrapped drug [94].

14.6   POLYMERS WITH ENZYME-DRIVEN DEGRADATION PROPERTIES

The unique, enzyme-rich biological milieu of the large intestine lumen can be exploited to increase the specificity of colonic drug delivery. Instead of, or in addition to, relying on pH changes as a discharge mechanism, a polymeric component is added, designed to cleave by enzymes that reside only in the colon [21]. These polymers contain functional groups that undergo site-specific changes (most commonly hydrolysis or reduction) that cause the polymeric backbone to collapse. Since residence time in the colon is restricted to several [8–72] hours [95, 96], hydrophilic polymers are commonly used to allow their swelling so that upon colon arrival their breakable moieties reexposed to enzymatic cleavage [39].

TABLE 14.3 Some Typical Examples of Colon-Specific Drug Delivery Systems, Based on azo-Polymers

Polymer and Mode of Application	Carrier Type	Drug Load	Tested in	References
Polystyrene-HEMA cross-linked with DVAB	Film coating	Insulin, vasopressin	Rats, dogs	[43, 97]
Acrylic polymer cross-linked with DVAB	Film coating		Colitis-induced rats, *in vitro*	[167, 168]
HEMA, MMA, and MA, containing *N,N'*-bis(methacryloyloxyethyl)oxy (carbonylamino) azobenzene	Film coating	Theophylline	Rats	[169]
Polyurethane containing azo aromatic groups	Film coating	Experimental drugs	Culture of intestinal flora	[170]
N,N-Dimethylacrylamide, *N-t*-butylacrylamide, and acrylic acid cross-linked with azoaromatic compounds	Hydrogel		Rats	[171]
HEMA and methacryloyloxy azobenzene	Hydrogel	5-Flurouracil	Simulated gastric and intestinal fluids	[172]
Methacrylated inulin, copolymerized with bis(methacryloylamino)azobenzene and 2-hydroxyethyl methacrylate or methacrylic acid	Hydrogel	Prednisolone	–	[173]

DVAB: divinilazobenzene; HEMA: 2-hydroxyethyl methacrylate; MMA: methyl methacrylate; MA: methacrylic acid.

14.7 AZO POLYMERS

Saffran et al. coated insulin and lysine-vasopressin solid dosage forms (pellets, gelatin minicapsules, or simply, paper strips) with copolymers of styrene and hydroxyethylmethacrylate cross-linked with divinilazobenzene. It was postulated that this polymer would be able to protect the entrapped protein drugs in the stomach and the upper portion of the small intestine and degrade in the colon [43]. A sustained pharmacological response to the protein drugs, antidiuresis for lysine-vasopressin and hypoglycemia for insulin, was observed when the coated delivery systems were orally administered to rats and later to dogs [97]. This lag time was related to a delayed release caused by a specific degradation of the polymer coat as a result of azo reduction in the cecum of the rat. These revolutionary studies later appeared to lack crucial control studies employing non-azo-cross-linked acrylic polymers. In the absence of such studies, it is unavoidable to deduce that the delayed pharmacological response could be also a result of the polymer swelling, similarly to the behavior of pH-dependent acrylic copolymers such as the Eudragit® products. Alternatively, germ-free rats could be used as control experiments.

Brondsted and Kopecek [98] described the synthesis and biological evaluation of a series of hydrogels made by cross-linking copolymerization of *N,N*-dimethylacrylamide, *N-tert*-butylacrylamide, acrylic acid, and 4,4'-di(methacryloylamino)-azobenzene [99]. The various polymers differed from each other by the pH-dependent equilibrium degree of swelling, modulus of elasticity, and permeability (as determined by measuring insulin diffusion at pH 2 and 7.4). In addition to their ability to degrade due to the azo cross-linker used, their specificity was increased because of their degree of swelling that was lower at pH 2 than at pH 7.4. The swelling properties were inversely proportional to the modulus of elasticity and directly proportional to the permeability of the entrapped drug used (insulin). The polymers' degradability *in vitro* and *in vivo* was found to be dependent upon the degree of swelling of the hydrogels. The higher the degree of swelling, the faster degradation was observed [98]. It appears then that the nature of the azo cross-linker is less important than the swelling characteristics of the azo polymer. Indeed, the role of the azo aromatic chain length of the cross-linker in the degradation of copolymers of 2-hydroxyethyl methacrylate and methyl methacrylate was investigated by Van den Mooter et al. and was found to be of minor importance when different polymer films were tested for intrinsic viscosity and permeability after incubation in human feces [100–102].

Colonic azo reduction occurs because electron carriers (redox mediators), such as benzyl viologen and flavin mononucleotide, act as electron shuttles between the intracellular enzyme and the substrate [103]. The initial substrate thought to be involved in cellular electron transport requires the presence of NADPH as an electron source [104]. The activity of the latter depends on flavoproteins that act as electron donors. Thus, the rate-determining steps in anaerobic azo reduction are the ratio between the redox potential of the mediator and the substrate, the permeability of the mediator to the intracellular compartment, the specific affinity between the flavoprotein reducing enzymes and the mediator, and steric and electrostatic factors [103]. It is noteworthy that, similarly to azo reduction, the final reduction cascade of the laxative sennoside to its active moiety rhein anthrone (Figure 14.2) requires the involvement of NADH and FAD, or benzyl viologen as electron carriers [65].

Lloyd et al. have suggested that colon-specific drug delivery is a valid approach not because a particular organism possessing a specific azo reductase exists in the colon, but because low molecular weight electron mediators, such as NADPH, are present and able to diffuse throughout a swollen polymeric matrix [105].

14.8 NATURAL POLYSACCHARIDES AND RELATED DESIGN OUTLINES

The most distinctive enzymatic activity of the colon is the ability to hydrolyze glycosidic bonds. Most of plant polysaccharides that escape small intestine digestion are consumed efficiently in the proximal large bowel, which can be regarded as a small fermentor. Saccharolytic bacteria hydrolyze efficiently plant- as well as mucopolysaccharides, converting them to oligo- and monosaccharides, short-chain fatty acids, and CO_2 [106]. Moreover, although the small intestine epithelium exerts some glycosidase activity [107], the major hydrolysis of glycosidic bonds occurs in the colon. Apart from the β-D-glucosidase and β-D-galactosidase that have been mentioned earlier, amylase, pectinase, xylanase, β-D-xylosidase, and dextranases are also active in the human colon [108]. In accord with this physiological feature, a variety of plant cell wall polysaccharides (Figure 14.3) have been tested for their ability to serve as colon-specific drug shuttles. Solid dosage forms of fermentable polysaccharides could be used as specific drug carriers to the colon if formulated properly, that is, if their water solubility was reduced in a manner that did affect their specific degradation in the human colon. Appropriate candidates for that purpose are polysaccharides that specifically ferment in the human colon such as pectin [109, 110], guar gum [32, 111], dextrans, amylose, and the mucopolysaccharide chondroitin sulfate [112]. They can be used as such [113–115], or after chemical modification in order to reduce their water solubility, thus enabling them to avoid drug leakage in the small bowel. In our group, we showed that such adaptations, mostly cross-linking, did not alter the polysaccharides' capability of being degraded by colonic microflora. Cross-linked chondroitin sulfate with diaminododecane [116, 117], calcium pectinate [118, 119], and cross-linked guar [120–123] were tested *in vitro* (rat cecal contents, bench chemostates), *in vivo* (dog studies), and in human studies [124], employing a variety of anti-inflammatory drugs. Instead of chemical modification, physical pressure could be employed in order to attenuate drug release rate prior to colon arrival. For example, instead of using

calcium pectinate, plain pectin tablets can be used as was demonstrated by Ashford, Fell, and coworkers [125, 126]. Table 14.4 summarized a decade of studies that employed natural polysaccharide matrices for colon-specific drug delivery.

In the design of polysaccharide matrices, the enzymatic activity of the colon microflora in the lumen should be taken into consideration for efficient, specific biodegradation. If the enzymes are periplasmic [127, 128] or are not released freely by the bacteria, a direct contact between the microflora and the surface of the solid polysaccharide should be taken into consideration. In some cases, this could lead to a potential functional risk due to a possible formation of a bacterial biofilm [129, 130]. It has been shown that a biofilm formation can interfere with the enzymatic biodegradation of pectin films in the rat cecum. In the absence of sufficient agitation, the attraction of pectinolytic bacteria attenuates or even prevents drug release from the surface of the eroded polysaccharide [131].

Another important feature of polysaccharide matrices is their swelling properties. This point was demonstrated by Brondsted and Kopecek in their work with cross-linked copolymers of *N,N*-dimethylacrylamide, *N-tert*-butylacrylamide, acrylic acid, and 4,4'-di(methacryloylamino)-azobenzene [98, 99]. Swelling creates a diffusional barrier at the surface of the solid dosage form as it starts to travel through the GI tract. At the same time, the hydrated polymer allows the penetration of colonic enzymes or bacteria that will cause its degradation at a later stage [121]. This essential swelling may cause a premature release of hydrophilic and water-soluble drug molecules and could limit the use of polysaccharide matrices to low water-soluble drugs (e.g., steroid drugs).

14.9 SYNTHETIC SACCHARIDIC POLYMERS

A more sophisticated design of colonic delivery systems would be the combination of a film-forming polymer together with a biodegradable component, such as a polysaccharide. It is thought that a specific increase in the porosity of such films may cause a spatial drug release in the colon from solid dosage forms. For example, galactomannans were (physically) incorporated into Eudragit-RL films to increase the specificity of enteric coating films [132]. Similarly, physical mixtures made of methacrylic acid copolymers (Eudragit-RS) and β-cyclodextrins were used to prepare biodegradable films. The porosity of the films increased significantly when analyzed *in vitro* in a simulated colonic environment [133]. In another study, ethylated oligogalactomannan was cross-linked with diisocyanate and further incorporated into polyurethanes [134]. The products were found to possess good film-forming properties and degraded

TABLE 14.4 Typical Examples of Colon-Specific Drug Delivery System, Based on Natural Polysaccharides

Polysaccharide	Platform	Drug Load	Tested in	References
Cross-linked chondroitin sulfate	Tablets	Indomethacin	Rat cecal content	[116]
Chondroitin sulfate complexed with chitosan	Films	–	–	[174]
Cross-linked dextran	Capsule, hydrogel	Hydrocortisone, bovine serum albumin, salmon calcitonin	Simulated gastric and intestinal fluids	[175–177]
Carboxymethyl high amylose starch and chitosan	Tablet	–	–	[178]
Lauroyldextran cross-linked with galactomannan	Tablets	Theophylline	Simulated gastric and intestinal fluids	[179]
Pectin	Tablets	^{99}Technetium-EDTA	Humans	[125]
Calcium pectinate	Tablets	Indomethacin, insulin	Rat cecal content, dogs	[118, 119]
Amidated pectins	Matrix beads	Acetaminophen, chloroquine	Rat	[180, 181]
Chitosan	Capsules	5-ASA	Rat	[182]
Cross-linked chitosan	Microspheres	5-FU	Rat	[183]
Chitosan and guar gum	Films	Celecoxib	Rats	[14]
Guar gum	Tablets	Dexamethasone, indomethacin	Humans, rat cecal content	[184–186]
Cross-linked guar gum	Tablets	Hydrocortisone	Rat	[187]
Alginate beads coated with chitosan	Microspheres	Saccharomyces boulardii	Rats	[188]
Alginate-chitosan cross-linked blends	Beads	Bovine serum albumin	Simulated gastric and intestinal fluids	[189]

The polysaccharide matrices, in either natural form or chemically modified (cross-linked), are commonly used as wet or dry gels, compressed (tablets) or casted into films.

in vitro in the above simulated colonic fermentation system and in the presence of hemicellulase. In both cases, a significant reduction in the tensile strength of the polymers, followed by a reduction in their molecular weight, was observed within few hours. The same concept was exploited to fabricate new biodegradable films made of glassy amylose and ethyl cellulose [195]. The colon arrival and specific degradation of pellets coated with the new film were verified by *g*-scintigraphy and $^{13}CO_2$ breath test in volunteers.

A series of water-insoluble acrylic polymers containing disaccharide side groups were synthesized and evaluated *in vitro* [135]. A cellobiose-derived monomer of methacrylic acid, 4-*O*-β-D-glucopyranosyl-1-methacrylamido-1-deoxy-D-glucitol, was prepared. The saccharide-methacrylic acid monomers were then polymerized with additional molecules of methacrylic acid or with themselves to produce a series of copolymers or homopolymers, respectively. The degradation of the disaccharide side groups in the homo- or copolymers was evaluated by measuring the amount of glucose cleaved upon incubation with β-glucosidase. It was found that although the homopolymers contained relatively more saccharidic groups, the copolymers degraded faster. This observation was related to the better hydration properties of the copolymers due to the presence of more methacrylic acid in their polymeric backbones. Table 14.5 summarizes some colon-specific drug delivery system based on polysaccharide mixtures with synthetic polymers.

14.10 COLON-SPECIFIC PRODRUGS: GLYCOSIDIC AND GLUCURONIC ACID PRODRUGS OF STEROIDS

5-ASA prodrugs were mentioned earlier in this chapter. One of the pioneering works in the delivery of prodrugs to the human colon was done by Friend and Chang, who engineered glycosidic prodrugs of steroids for the topical treatment of ulcerative colitis [41, 42]. Dexamethasone-β-D-glucoside and later budesonide-β-D-glucuronide prodrugs were tested *in vivo* [136, 137]. It was concluded that those glycosilated steroids possessed colon selectivity because they could hydrolyze easier in the animals' cecum, due to their increased hydrophilicity (which reduced their absorption in the proximal GI tract) and because of the expected shorter transit time in the small intestine. In some of the tests (scoring evaluation), the budesonide prodrug was found to be superior to the parent drug. In addition, adrenal suppression, which is the major side effect of steroid drugs, was not caused by the prodrug of budesonide whereas free budesonide treatment did cause adrenal suppression.

14.11 POLYMERIC PRODRUGS

The term polymeric prodrug refers to drug carriers in which the drug molecule(s) is linked directly to a high molecular

TABLE 14.5   Some Typical Examples of Colon-Specific Drug Delivery System Based on Polysaccharide Mixtures with Synthetic Polymers

Polysaccharide	Platform	Drug Load	Tested in	References
Inulin reacted with glycidyl methacrylate	Hydrogel	–	–	[190]
Methacrylated dextran and methacrylated α,β-poly(N-2-hydroxyethyl)-DL-aspartamide	Hydrogels	Beclomethasone	Caco-2 cells	[191]
Amylose-ethylcellulose	Film	5-ASA	Simulated gastric and intestinal fluids	[192]
Pectin and ethylcellulose	Tablet pellets	Paracetamol 5-FU	Dogs	[193, 194]
Poly(acrylic acid-co-acrylamide)/O-carboxy-methyl chitosan	Superporous hydrogels	Insulin	Rats	[195]
Methacrylate-grafted chondroitin sulfate copolymerized with acrylic acid	Hydrogel	BSA-FITC	Simulated gastric and intestinal fluids	[196]

weight backbone. The linkage between the drug and the polymer is designed so that it will be susceptible to specific enzymatic cleavage, which is expected to cause a preferential release of the drug only after large intestine arrival. In the past 15 years, many efforts were invested in improving 5-ASA specificity, not only by synthesizing novel prodrugs but also by the fabrication of polymeric prodrugs of the drug. The advantage of this design is the ability to encompass additional features into the carrier, for example, mucoadhesion properties that would increase specificity by prolongation of the residence time in the vicinity of the mucosal injury [138]. At Dynapol, sulfasalazine was linked to a high molecular polymeric backbone [139]. Upon azo reduction, the 5-ASA molecule was released from the polymer (named poly-ASA) while the sulfapyridine moiety remained attached to the polymer to be excreted in the feces. Burroughs Wellcome has brought the poly-ASA to clinical phase and has reported that it was well tolerated for as long as 18 weeks by patients previously intolerant to sulfasalazine.

Kopecek and coworkers have suggested the use of the water-soluble N-(2-hydroxypropyl)methacrylamide copolymers (HPMA copolymers) azo linked to 5-ASA, in which specificity is increased by providing mucoadhesion properties to the polymers. This work was a continuation of their previous studies, in which lysosomotropic HPMA carriers for anticancer drugs were developed. The specificity of the new anticancer carriers was increased by incorporating targeting moieties that could "sense" cell surface receptors or antigens. For example, galactosamine containing HPMA was shown in vitro and in vivo to interact specifically with leukemia L1210 cells [140]. The rationale for the development of mucoadhesive HPMA colon-specific polymers was the finding that some bacteria adhere to the guinea pig colonic epithelium, probably through specific interactions between glucose and fucose moieties to mucosal lectins [141]. Since, in many cases, the reduction of polymeric azo bonds is relatively slow, it would be reasonable to increase the delivery system's colonic residence time as long as possible to increase the efficiency of 5-ASA hydrolysis.

The polymeric systems synthesized were fabricated from the water-soluble HPMA copolymers with different contents of pendent amino saccharide moieties, such as fucosylamine. The fucosylamine was supposed to anchor the polymer to mucosal lectins of the colon [142]. In addition, 5-ASA molecules were azo linked to the backbone of the polymer. Various polymers of the same type, containing increasing amounts of fucosylamine, were tested in vitro in everted sacks from guinea pigs [99].

14.12   DEXTRAN PRODRUGS

Because of its documented safety and the linearity of its polymeric chains (enables reproducibility), it is tempting to use dextran as a polymeric backbone to design polymeric prodrugs. Some examples are nicotinic acid, benzoic acid, chromoglygate (for the slow release in the lung), procainamide, acyclovir, and methotrexate [143]. Most of these polymeric prodrugs were developed for intravenous injection. Recognizing the dextranase activity in the human large intestine [144], Larsen and coworkers suggested the use of dextran prodrugs for colonic delivery. They synthesized a series of dextran prodrugs such as naproxen-dextran, ketoprofen-dextran, and ibuprofen-dextran and tested them in vitro and in vivo (pigs). Using pharmacokinetic and enzyme kinetic considerations, they have demonstrated that dextran prodrugs may be efficient colon-specific drug carriers [145–147]. They are activated in the colon by hydrolysis of the glucosidic bonds to give shorter prodrug oligomers after which colonic esterases [148] split the link between the drug and the chopped oligomers.

Dextran prodrugs of steroid drugs were prepared and tested recently by McLeod and coworkers. In this case, dicarboxylic (succinic or glutaric) acids were used as spacers to allow linkage between the steroid drug and the dextran. In this manner methylprednisolone and dexamethasone were attached to dextran with an average molecular weight of 72,600. The products were mostly stable to hydrolysis

TABLE 14.6 Typical Examples of Colon-Specific Drug Delivery System Based on Polymeric Prodrugs

Polymer	Drug	Cleavable Bond	References
Dextran	Naproxen, nalidixic acid, dexamethasone, methylprednisolone 5-ASA	Ester	[143, 145, 146, 149, 197–199]
		Amine	
α-, β-, and γ-Cyclodextrins	Biphenylylacetic acid, prednisolone 21-succinate 5-ASA	Ester or amide	[200–202]
Polyanhydride	5-ASA	Amide	[203]
HEMA	5-ASA	Azo	[138]
HPMA	9-Aminocamptothecin	Aromatic azo	[204]
HPMA	Hyaluronan-doxorubicin	Hydrazide	[205]

(probably caused by mucosal type A carboxylesterase) in the proximal small intestine contents of the rat and degraded rapidly in its cecal contents. However, the steroid–dextran conjugates released steroid hemiesters (i.e., the steroids linked to the dicarboxylic acid spacers) in the cecal contents. These hemiesters needed to pass further hydrolysis prior to releasing the free drug. Nonetheless, the polymeric prodrugs of dexamethasone and methylprednisolone were much more effective than the parent drugs when tested in colitis-induced rats [149]. Table 14.6 presents some polymeric prodrugs that were tested in the context of colon-specific drug delivery systems.

14.13 CONCLUSION AND PROSPECTS

A huge number of techniques for targeting the large bowel via the oral route have been reported during the past quarter of a century, most of which did not advance to clinical phases. The main reasons for the failure to move from promise to reality are (a) the medical rationale for employing colon-specific drug carriers remains questionable (excluding topical treatment of IBD, which employs, in most cases, unsophisticated technologies and large doses of old drugs), (b) nonrealistic assumptions regarding the therapeutic advantage of targeting the colon with molecules that probably would have worked after systemic administration (in other words, the lack of accompanying in-depth pharmacological research), and (c) the ongoing difficulty to launch oral protein products ("biologics"), attenuating the development of colonic platforms for this complicated group of molecules. However, there are some intriguing reports that inspire further efforts in the research of colonic drug delivery [150]. Colonocyte targeting is one example. Existing methods of colon drug delivery are appropriate for ferrying large drug doses, but their ability to "target" is certainly poorly developed. It is unlikely that a new biologic aimed at topical treatment of colon inflammation at the colonocyte level, or a new anticancer drug, designed to anchor the cell membrane, can arrive intact to its site of action if left in the large bowel

lumen. Thus, it is about time to explore realistic, efficient, and safe modes of targeted drug delivery to colonocytes.

REFERENCES

1. Davis SS. Formulation strategies for absorption windows. *Drug Discov. Today* 2005;10:249–257.
2. Hwang SJ, Park H, Park K. Gastric retentive drug-delivery systems. *Crit. Rev. Ther. Drug Carrier Syst.* 1998; 15:243–284.
3. Streubel A, Siepmann J, Bodmeier R. Gastroretentive drug delivery systems. *Expert Opin. Drug Deliv.* 2006;3:217–233.
4. Waterman KC. A critical review of gastric retentive controlled drug delivery. *Pharm. Dev. Technol.* 2007;12:1–10.
5. Kesisoglou F, Zimmermann EM. Novel drug delivery strategies for the treatment of inflammatory bowel disease. *Expert Opin. Drug Deliv.* 2005;2:451–463.
6. Somerville KW, Richmond CR, Bell GD. Delayed release peppermint oil capsules (Colpermin) for the spastic colon syndrome: a pharmacokinetic study. *Br. J. Pharmacol.* 1984;18:638–640.
7. Passaretti S, Sorghi M, Colombo E, Mazzotti G, Tittobello A, Guslandi M. Motor effects of locally administered pinaverium bromide in the sigmoid tract of patients with irritable bowel syndrome. *Int. J. Clin. Pharmacol. Ther. Toxicol.* 1989;27:47–50.
8. Nolen HW. III Friend DR. Menthol-β-D-glucuronide: a potential prodrug for treatment of irritable bowel syndrome. *Pharm. Res.* 1994;11:1707–1711.
9. Arber N. Do NSAIDs prevent colorectal cancer?. *Can. J. Gastroenterol.* 2000;14:299–307.
10. Ciolino HP, Bass SE, MacDonald CJ, Cheng RY, Yeh GC. Sulindac and its metabolites induce carcinogen metabolizing enzymes in human colon cancer cells. *Int. J. Cancer* 2008;122:990–998.
11. Williams CS, Goldman AP, Sheng H, Morrow JD, DuBois RN. Sulindac sulfide, but not sulindac sulfone, inhibits colorectal cancer growth. *Neoplasia* 1999;1:170–176.
12. Glebov OK, Rodriguez LM, Lynch P, Patterson S, Lynch H, Nakahara K, Jenkins J, Cliatt J, Humbyrd CJ, Denobile J,

Soballe P, Gallinger S, Buchbinder A, Gordon G, Hawk E, Kirsch IR. Celecoxib treatment alters the gene expression profile of normal colonic mucosa. *Cancer Epidemiol. Biomarkers Prev.* 2006;15:1382–1391.

13. Arber N, Eagle CJ, Spicak J, Racz I, Dite P, Hajer J, Zavoral M, Lechuga MJ, Gerletti P, Tang J, Rosenstein RB, Macdonald K, Bhadra P, Fowler R, Wittes J, Zauber AG, Solomon SD, Levin B. Celecoxib for the prevention of colorectal adenomatous polyps. *N. Engl. J. Med.* 2006; 355:885–895.

14. Haupt S, Zioni T, Gati I, Kleinstern J, Rubinstein A. Luminal delivery and dosing considerations of local celecoxib administration to colorectal cancer. *Eur. J. Pharm. Sci.* 2006; 28:204–211.

15. Torrance CJ, Jackson PE, Montgomery E, Kinzler KW, Vogelstein B, Wissner A, Nunes M, Frost P, Discafani CM. Combinatorial chemoprevention of intestinal neoplasia. *Nat. Med.* 2000;6:1024–1028.

16. Ishizawa T, Hayashi M, Awazu S. Enhancement of jejunal and colonic absorption of fosfomycin by promoters in the rat. *J. Pharm. Pharmacol.* 1987;39:892–895.

17. Tomita M, Shiga M, Hayashi M, Awazu S. Enhancement of colonic drug absorption by the paracellular permeation route. *Pharm. Res.* 1988;5:341–346.

18. Sintov A, Simberg M, Rubinstein A. Absorption enhancement of captopril in the rat colon as a putative method for captopril delivery by extended release formulations. *Int. J. Pharm.* 1996;143:101–106.

19. Haupt S, Rubinstein A. The colon as a possible target for orally administered peptides and proteins. *Crit. Rev. Ther. Drug Carrier Syst.* 2002;19:499–545.

20. Simon GL, Gorbach SL. Intestinal flora in health and disease. *Gastroenterology* 1984;86:174–193.

21. Scheline RR. Metabolism of foreign compounds by gastrointestinal microorganisms. *Pharmacol. Rev.* 1973;25: 451–523.

22. Phillips SF, Pemberton JH, Shorter RG, editors. *The Large Intestine: Physiology, Pathophysiology, and Disease*, Raven Press, New York, 1991.

23. Barkla DH, Gibson PR. The fate of epithelial cells in the human large intestine. *Pathology* 1999;31:230–238.

24. Fordtran JS, Rector FC Jr, Ewton MF, Soter N, Kinney J. Permeability characteristics of the human small intestine. *J. Clin. Invest.* 1965;44:1935–1944.

25. Pauletti GM, Gangwar S, Knipp GT, Nerurkar MM, Okumn FW, Tamura K, Siahaan TJ, Borchardt RT. Structural requirements for intestinal absorption of peptide drugs. *J. Control. Release* 1996;41:3–17.

26. Cereijido M, Contreras RG, Flores-Benitez D, Flores-Maldonado C, Larre I, Ruiz A, Shoshani L. New diseases derived or associated with the tight junction. *Arch. Med. Res.* 2007;38:465–478.

27. Hollander D. Intestinal permeability, leaky gut, and intestinal disorders. *Curr. Gastroenterol. Rep.* 1999;1:410–416.

28. Sarna SK. Physiology and pathophysiology of colonic motor activity. Part 1. *Dig. Dis. Sci.* 1991;36:827–862.

29. Sarna SK. Physiology and pathophysiology of colonic motor activity. Part 2. *Dig. Dis. Sci.* 1991;36:998–1018.

30. Podolsky DK. Oligosaccharide structures of human colonic mucin. *J. Biol. Chem.* 1985;260:8262–8271.

31. Azab AK, Kleinstern J, Srebnik M, Rubinstein A. The metastatic stage-dependent mucosal expression of sialic acid is a potential marker for targeting colon cancer with cationic polymers. *Pharm. Res.* 2008;25:379–386.

32. Salyers AA. Bacteroides of the human lower intestinal tract. *Annu. Rev. Microbiol.* 1984;38:293–313.

33. Jia W, Li H, Zhao L, Nicholson JK. Gut microbiota: a potential new territory for drug targeting. *Nat. Rev. Drug Discov.* 2008;7:123–129.

34. Eckburg PB, Bik EM, Bernstein CN, Purdom E, Dethlefsen L, Sargent M, Gill SR, Nelson KE, Relman DA. Diversity of the human intestinal microbial flora. *Science* 2005;308: 1635–1638.

35. Macfarlane GT, Macfarlane S. Human colonic microbiota: ecology, physiology and metabolic potential of intestinal bacteria. *Scand. J. Gastroenterol. Suppl.* 1997;222:3–9.

36. Cummings JH, Englyst HN. Fermentation in the human large intestine and the available substrates. *Am. J. Clin. Nutr.* 1987;45:1243–1255.

37. Fallingborg J. Intraluminal pH of the human gastrointestinal tract. *Dan. Med. Bull.* 1999;46:183–196.

38. Rowland IR. Factors affecting metabolic activity of the intestinal microflora. *Drug Metab. Rev.* 1988;19:243–261.

39. Rubinstein A. Microbially controlled drug delivery to the colon. *Biopharm. Drug Dispos.* 1990;11:465–475.

40. Sinha VR, Kumria R. Microbially triggered drug delivery to the colon. *Eur. J. Pharm. Sci.* 2003;18:3–18.

41. Friend DR, Chang GW. A colon-specific drug-delivery system based on drug glycosides and the glycosidases of colonic bacteria. *J. Med. Chem.* 1984;27:261–266.

42. Friend DR, Chang GW. Drug glycosides: potential prodrugs for colon-specific drug delivery. *J. Med. Chem.* 1985; 28:51–57.

43. Saffran M, Kumar GS, Savariar C, Burnham JC, Williams F, Neckers DC. A new approach to the oral administration of insulin and other peptide drugs. *Science* 1986;233: 1081–1084.

44. Russell TL, Berardi RR, Barnett JL, Dermentzoglou LC, Jarvenpaa KM, Schmaltz SP, Dressman JB. Upper gastrointestinal pH in seventy-nine healthy, elderly, North American men and women. *Pharm. Res.* 1993;10:187–196.

45. Dressman JB, Berardi RR, Dermentzoglou LC, Russell TL, Schmaltz SP, Barnett JL, Jarvenpaa KM. Upper gastrointestinal (GI) pH in young, healthy men and women. *Pharm. Res.* 1990;7:756–761.

46. Evans DF, Pye G, Bramley R, Clark AG, Dyson TJ, Hardcastle JD. Measurement of gastrointestinal pH profiles in normal ambulant human subjects. *Gut* 1988;29:1035–1041.

47. Charalambides D, Segal I. Colonic pH: a comparison between patients with colostomies due to trauma and colorectal cancer. *Am. J. Gastroenterol.* 1992;87:74–78.

48. Fallingborg J, Christensen LA, Jacobsen BA, Rasmussen SN. Very low intraluminal colonic pH in patients with active ulcerative colitis. *Dig. Dis. Sci.* 1993;38:1989–1993.

49. Dew MJ, Hughes PJ, Lee MG, Evans BK, Rhodes J. An oral preparation to release drugs in the human colon. *Br. J. Clin. Pharmacol.* 1982;14:405–408.

50. Klotz U, Maier KE, Fischer C, Bauer KH. A new slow-release form of 5-aminosalicylic acid for the oral treatment of inflammatory bowel disease. Biopharmaceutic and clinical pharmacokinetic characteristics. *Arzneimittelforschung* 1985;33:636–639.

51. Levine DS, Raisys VA, Ainardi V. Coating of oral beclomethasone dipropionate capsules with cellulose acetate phthalate enhances delivery of topically active antiinflammatory drug to the terminal ileum. *Gastroenterology* 1987;92:1037–1044.

52. Ritschel WA. Targeting in the gastrointestinal tract: new approaches. *Methods Find. Exp. Clin. Pharmacol.* 1991; 13:313–336.

53. Ashford M, Fell JT, Attwood D, Woodhead PJ. An *in vitro* investigation into the suitability of pH dependent polymers for colonic targeting. *Int. J. Pharm.* 1993;91:241–245.

54. Ashford M, Fell JT, Attwood D, Sharma H, Woodhead PJ. An *in vivo* investigation into the suitability of pH dependent polymers for colonic targeting. *Int. J. Pharm.* 1993;95:193–199.

55. Hanauer SB. Medical therapy of ulcerative colitis. *Lancet* 1993;342:412–417.

56. Edwards CA. Anatomical and physiological basis: physiological factors influencing drug absorption. In Bieck PR, editor. *Colonic Drug Absorption and Metabolism*, Marcel Dekker, New York, 1993, pp. 1–28.

57. Caldwell L, Nishihata T, Rytting JH, Higuchi T. Lymphatic uptake of water-soluble drugs after rectal administration. *J. Pharm. Pharmacol.* 1982;34:520–522.

58. Kararli TT. Gastrointestinal absorption of drugs. *Crit. Rev. Ther. Drug Carrier Syst.* 1989;6:39–86.

59. Ichihashi T, Kinoshita H, Yamada H. Absorption and disposition of epithiosteroids in rats. 2. Avoidance of first-pass metabolism of mepitiostane by lymphatic absorption. *Xenobiotica* 1991;21:873–880.

60. Florence AT. The oral absorption of micro- and nanoparticulates: neither exceptional nor unusual. *Pharm. Res.* 1997;14:259–266.

61. Gershkovich P, Hoffman A. Uptake of lipophilic drugs by plasma derived isolated chylomicrons: linear correlation with intestinal lymphatic bioavailability. *Eur. J. Pharm. Sci.* 2005;26:394–404.

62. Rubinstein A. Approaches and opportunities in colon-specific drug delivery. *Crit. Rev. Ther. Drug Carrier Syst.* 1995;12:101–149.

63. Hardcastle JD, Wilkins JL. The action of sennoside and related compounds on human colon and rectum. *Gut* 1970; 11:1038–1042.

64. Staumont G, Frexinos J, Fioramonti J, Bueno L. Sennoside and human colonic motility. *Pharmacology* 1988;36(Suppl. 1):49–56.

65. Hattori M, Namba T, Akao T, Kobashi K. Metabolism of sennosides by human intestinal bacteria. *Pharmacology* 1988;36(Suppl. 1):172–179.

66. Moreto M, Gonalons E, Mylonakis N, Giraldez A, Torralba A. 3,3-bis-(4-hydroxyphenyl)-7-methyl-2-indolinone (BHMI), the active metabolite of the laxative sulisatin. *Arzneim Forsch/Drug Res.* 1979;29:1561–1564.

67. Gullikson GW, Bass P. Mechanisms of action of laxative drugs. In Csaky TZ, editor. *Pharmacology of Intestinal Permeation II*, Springer-Verlag, Heidelberg, 1984, pp. 419–459.

68. Bennett A, Eley KG. Intestinal pH and propulsion: an explanation of diarrhoea in lactase deficiency and laxation by lactulose. *J. Pharm. Pharmacol.* 1976;28:192–195.

69. Simpkins JW, Smulkowski M, Dixon R, Tuttle R. Evidence for the delivery of narcotic antagonists to the colon as their glucuronide conjugates. *J. Pharmacol. Exp. Ther.* 1988; 244:195–205.

70. Ryde ME. Low-molecular-weight azo compounds. In Friend DR, editor. *Oral Colon-Specific Drug Delivery*, CRC Press, Boca Raton, 1992, pp. 143–152.

71. Hanngren A, Hansson E, Svartz N, Ullberg S. Distribution and metabolism of salicyl-azo-sulfapyridine. I. A study with C-14-5-amino-salicylic acid. *Acta Med. Scand.* 1963; 173:61–72.

72. Hanngren A, Hansson E, Svartz N, Ullberg S. Distribution and metabolism of salicyl-azo-sulfapyridine. II. A study with S35-salicyl-azo-sulfapyridine and S35-sulfapyridine. *Acta Med. Scand.* 1963;173:391–399.

73. Klotz U. Clinical pharmacokinetics of sulphasalazine, its metabolites and other prodrugs of 5-aminosalicylic acid. *Clin. Pharmacokinet.* 1985;10:285–302.

74. Peppercorn MA, Goldman P. The role of intestinal bacteria in the metabolism of salicylazosulfapyridine. *J. Pharmacol. Exp. Ther.* 1972;181:555–562.

75. McIntyre PB, Rodrigues CA, Lennard-Jones JE, Barrison IG, Walker JG, Baron JH, Thornton PC. Balsalazide in the maintenance treatment of patients with ulcerative colitis, a double-blind comparison with sulphasalazine. *Aliment. Pharmacol. Ther.* 1988;2:237–243.

76. Willoughby CP, Aronson JK, Agback H, Bodin NO, Truelove SC. Distribution and metabolism in healthy volunteers of disodium azodisalicylate, a potential therapeutic agent for ulcerative colitis. *Gut* 1982;23:1081–1087.

77. Bartalsky A. Salicylazobenzoic acid in ulcerative colitis. *Lancet* 1982;1:960.

78. Travis SPL, Tysk C, de Silva HJ, Sandberg Gertzen H, Jewell DP, Jarnerot G. Optimum dose of olsalazine for maintaining remission in ulcerative colitis. *Gut* 1994;35:1282–1286.

79. Agyilirah GA, Banker GS. Polymers for enteric coating applications. In Tarcha PJ, editor. *Polymers for Controlled Drug Delivery*, CRC Press, Boca Raton, 1991, pp. 39–66.

80. Hardy JG, Healy JNC, Lee SW, Reynolds JR. Gastrointestinal transit of an enteric-coated delayed release 5-aminosalicylic acid tablet. *Aliment. Pharmacol. Ther.* 1987;1:209–216.

81. Myers B, Evans DN, Rhodes J, Evans BK, Hughes BR, Lee MG, Richens A, Richards D. Metabolism and urinary

excretion of 5-amino salicylic acid in healthy volunteers when given intravenously or released for absorption at different sites in the gastrointestinal tract. *Gut* 1987;28:196–200.

82. Rasmussen SN, Rondesen S, Hvidberg EF, Hansen S, Binder V, Halskov S, Flachs H. 5-Aminosalicylic acid in a slow-release preparation: bioavailability, plasma level, and excretion in humans. *Gastroenterology* 1982;83:1062–1070.

83. Thomas P, Richards D, Richards A, Rogers L, Evans BK, Dew MJ, Rhodes J. Absorption of delayed-release prednisolone in ulcerative colitis and Crohn's disease. *J. Pharm. Pharmacol.* 1985;37:757–758.

84. Ford GA, Oliver PS, Shepherd NA, Wilkinson SP. An Eudragit-coated prednisolone preparation for ulcerative colitis: pharmacokinetics and preliminary therapeutic use. *Aliment. Pharmacol. Ther.* 1992;6:31–40.

85. Harris DM. Properties and therapeutic uses of some corticosteroids with enhanced topical potency. *J. Steroid Biochem.* 1975;6:711–716.

86. Fedorak RN, Bistritz L. Targeted delivery, safety, and efficacy of oral enteric-coated formulations of budesonide. *Adv. Drug Deliv. Rev.* 2005;57:303–316.

87. Marin-Jimenez I, Pena AS. Budesonide for ulcerative colitis. *Rev. Esp. Enferm. Dig.* 2006;98:362–373.

88. Hanauer SB. Review: evolving concepts in treatment and disease modification in ulcerative colitis. *Aliment. Pharmacol. Ther.* 2008;27(Suppl. 1):15–21.

89. White DA, Thompson SP, Wilson CG, Bell GD. A pharmacokinetic comparison of two delayed-release peppermint oil preparations, Colpermin and Mintec, for treatment of the irritable bowel syndrome. *Int. J. Pharm.* 1987;40:151–155.

90. Wilding IR, Davis SS, Pozzi F, Furlani P, Gazzaniga A. Enteric coating timed release system for colonic targeting. *Int. J. Pharm.* 1994;111:99–102.

91. Gazzaniga A, Maroni A, Foppoli A, Palugan L. Oral colon delivery: rationale and time-based drug design strategy. *Discov. Med.* 2006;6:223–228.

92. Maggi L, Bruni R, Conte U. High molecular weight polyethylene oxides (PEOs) as an alternative to HPMC in controlled release dosage forms. *Int. J. Pharm.* 2000;195:229–238.

93. Conte U, Maggi L, Colombo P, La Manna A. Multi-layered hydrophilic matrices as constant release devices (Geomatrix systems). *J. Control. Release* 1993;26:39–47.

94. Ueda Y, Hata T, Yamaguchi H, Ueda S, Kodani M. Time-controlled explosion systems and processes for preparing the same. US Patent 4,871,549, 1989.

95. Proano M, Camilleri M, Phillips SF, Brown ML, Thomforde GM. Transit of solids through the human colon: regional quantification in the unprepared bowel. *Am. J. Physiol.* 1990;258:G856–G862.

96. McLean RG, Smart RC, Gaston-Parry D, Barbagallo S, Baker J, Lyons NR, Bruck CE, King DW, Lubowski DZ, Talley NA. Colon transit scintigraphy in health and constipation using oral iodine-131-cellulose. *J. Nucl. Med.* 1990; 31:985–989.

97. Saffran M, Field JB, Pena J, Jones RH, Okuda Y. Oral insulin in diabetic dogs. *J. Endocrinol.* 1991;131:267–278.

98. Brondsted H, Kopecek J. Hydrogels for site-specific oral drug delivery: synthesis and characterization. *Biomaterials* 1991; 12:584–592.

99. Kopecek J, Kopeckova P, Brondsted H, Rathi R, Rihova R, Yeh P-Y, Ikesue K. Polymers for colon-specific drug delivery. *J. Control. Release* 1992;19:121–130.

100. Van den Mooter G, Samyn C, Kinget R. Azo polymers for colon-specific drug delivery. *Int. J. Pharm.* 1992;87:37–46.

101. Van den Mooter G, Samyn C, Kinget R. Azo polymers for colon-specific drug delivery. II. Influence of the type of azo polymer on the degradation by intestinal microflora. *Int. J. Pharm.* 1993;97:133–139.

102. Van den Mooter G, Samyn C, Kinget R. Characterization of colon-specific polymers: a study of the swelling properties and the permeability of isolated polymer films. *Int. J. Pharm.* 1994;111:127–136.

103. Kopecek J, Kopeckova P. *N*-(2-hydroxypropyl)methacrylamide copolymers for colon-specific drug delivery. In Friend DR, editor. *Oral Colon-Specific Drug Delivery*, CRC Press, Boca Raton, 1992, pp. 189–212.

104. Gingell R, Walker R. Mechanisms of azo reduction by *Streptococcus faecalis*. II. The role of soluble flavins. *Xenobiotica* 1971;1:231–239.

105. Lloyd AW, Martin GP, Soozandehfar SH. Azopolymers: a means of colon specific drug delivery?. *Int. J. Pharm.* 1994;106:255–260.

106. Hawksworth G, Drasar BS, Hill MJ. Intestinal bacteria and the hydrolysis of glycosidic bonds. *J. Med. Microbiol.* 1971;4:451–459.

107. Conchie J, Macdonald DC. Glycosidase in the mammalian alimentary tract. *Nature* 1959;184(Suppl. 16):1233.

108. Englyst HN, Hay S, Macfarlane GT. Polysaccharide breakdown by mixed populations of human fecal bacteria. *FEMS Microbiol. Ecol.* 1987;95:163–171.

109. Cummings JH, Southgate DAT, Branch WJ, Wiggins HS. The digestion of pectin in human gut and its effect on calcium absorption and large bowel function. *Br. J. Nutr.* 1979;41:477–485.

110. Werch SC, Ivy AC. On the fate of ingested pectin. *Am. J. Dig. Dis.* 1941;8:101–105.

111. Macfarlane GT, Hay S, Macfarlane S, Gibson GR. Effect of different carbohydrates on growth, polysaccharidase and glycosidase production by *Bacteroides ovatus*, in batch and continuous culture. *J. Appl. Bacteriol.* 1990;68:179–187.

112. Macfarlane GT, Cummings JH. The colonic flora, and large bowel digestive function. In Phillips SF, Pemberton JH, Shorter RG, editor. *The Large Intestine: Physiology, Pathophysiology and Disease*, Raven Press, New York, 1991, pp. 51–92.

113. Sinha VR, Kumria R. Polysaccharides in colon-specific drug delivery. *Int. J. Pharm.* 2001;224:19–38.

114. Rubinstein A. Natural polysaccharides as targeting tools of drugs to the human colon. *Drug Dev. Res.* 2000;50:435–439.

115. Hovgaard L, Brondsted H. Current applications of polysaccharides in colon targeting. *Crit. Rev. Ther. Drug Carrier Syst.* 1996;13:185–223.

116. Rubinstein A, Nakar D, Sintov A. Colonic drug delivery: enhanced release of indomethacin from cross-linked chondroitin matrix in rat cecal content. *Pharm. Res.* 1992;9:276–278.

117. Sintov A, Di Capua N, Rubinstein A. Crosslinked chondroitin sulfate: characterization for drug delivery purposes. *Biomaterials* 1995;16:473–478.

118. Rubinstein A, Radai R, Ezra M, Pathak S, Rokem JS. *In vitro* evaluation of calcium pectinate: a potential colon-specific drug delivery carrier. *Pharm. Res.* 1993;10:258–263.

119. Rubinstein A, Radai R. *In vitro* and *in vivo* analysis of colon-specificity of calcium pectinate formulations. *Eur. J. Pharm. Biopharm.* 1995;41:291–295.

120. Gliko-Kabir I, Yagen B, Penhasi A, Rubinstein A. Cross-linking guar with glutaraldehyde: a new method for preparing hydrogels for colon-specific drug delivery. In Ottenbrite RM, editor. *Frontiers in Biomedical Polymer Applications*, Technomic, Basel, 1998, pp. 83–93.

121. Gliko-Kabir I, Penhasi A, Rubinstein A. Characterization of crosslinked guar by thermal analysis. *Cabohydr. Res.* 1999;316:1–8.

122. Gliko-Kabir I, Yagen B, Penhasi A, Rubinstein A. Phosphated crosslinked guar for colon-specific drug delivery. I. Preparation and physical and chemical characterization. *J. Control. Release* 2000;63:121–127.

123. Gliko-Kabir I, Yagen B, Baluom M, Rubinstein A. Phosphated crosslinked guar for colon-specific drug delivery. II. *In vitro* and *in vivo* evaluation. *J. Control. Release* 2000;63:129–134.

124. Adkin DA, Kenyon CJ, Lerner EI, Landau I, Strauss E, Caron D, Penhasi A, Rubinstein A, Wilding IR. The use of scintigraphy to provide "proof of concept" for novel polysaccharide preparations designed for colonic drug delivery. *Pharm. Res.* 1997;14:103–107.

125. Ashford M, Fell JT, Attwood D, Sharma H, Woodhead P. An evaluation of pectin as a carrier for drug targeting to the colon. *J. Control. Release* 1993;26:213–220.

126. Ashford M, Fell J, Attwood D, Sharma H, Woodhead P. Studies on pectin formulations for colonic drug delivery. *J. Control. Release* 1994;30:225–232.

127. Salyers AA, O'brien M. Cellular location of enzymes involved in chondroitin sulfate breakdown by *Bacteroides thetaiotaomicron. J. Bacteriol.* 1980;143:772–780.

128. Kuritza AP, Salyers AA. Digestion of proteoglycan by *Bacteroides thetaiotaomicron. J. Bacteriol.* 1983;153:1180–1186.

129. Macfarlane S, Woodmansey EJ, Macfarlane GT. Colonization of mucin by human intestinal bacteria and establishment of biofilm communities in a two-stage continuous culture system. *Appl. Environ. Microbiol.* 2005;71:7483–7492.

130. Rubinstein A, Ezra M, Rokem JS. Adhesion of bacteria on pectin casted films. *Microbiosis* 1992;73:163–170.

131. Rubinstein A, Radai R, Friedman M, Fischer P, Rokem JS. The effect of intestinal bacteria adherence on drug diffusion through solid films under stationary condition. *Pharm. Res.* 1997;14:503–507.

132. Lehmann KOR, Dreher KD. Methacrylate-galactomannan coating for colon-specific drug delivery. In *Proceedings of the International Symposium on Controlled Release of Bioactive Materials* 1991;331–332.

133. Siefke V, Weckenmann HP, Bauer KH. β-Cyclodextrin matrix films for colon-specific drug delivery. In *Proceedings of the International Symposium on Controlled Release of Bioactive Materials* 1993;182–183.

134. Sarlikiotis AW, Bauer KH. Synthese und untersuchung von polyurethanen mit galactomannan-segmenten als hilfsstoffe zur freisetzung von peptid-arzneistoffen im dickdarm. *Pharm. Ind.* 1992;54:873–880.

135. Sintov A, Ankol S, Levy DP, Rubinstein A. Enzymatic cleavage of disaccharide side groups in insoluble synthetic polymers: a new method for specific delivery of drugs to the colon. *Biomaterials* 1993;14:483–490.

136. Fedorak RN, Haeberlin B, Empey LR, Cui N, Nolen H III, Jewell LD, Friend DR. Colonic delivery of dexamethasone from a prodrug accelerates healing of colitis in rats without adrenal suppression. *Gastroenterology* 1995;108:1688–1699.

137. Cui N, Friend DR, Fedorak RN. A budesonide prodrug accelerates treatment of colitis in rats. *Gut* 1994;35:1439–1446.

138. Mahkam M, Doostie L, Siadat SO. Synthesis and characterization of acrylic type hydrogels containing azo derivatives of 5-amino salicylic acid for colon-specific drug delivery. *Inflammopharmacology* 2006;14:72–75.

139. Garretto M, Riddel RH, Winans CS. Treatment of chronic ulcerative colitis with poly-ASA: a new nonabsorbable carrier for release of 5-aminosalicylate in the colon. *Gastroenterology* 1983;84:1162.

140. Duncan R, Hume IC, Kopeckova P, Uldrich K, Strohalm J, Kopecek J. Anticancer agents coupled to *N*-(2-hydroxypropyl)methacrylamide copolymers. 3. Evaluation of adriamycin conjugates against mouse leukaemia L1210 *in vivo. J. Control. Release* 1989;10:51–63.

141. Ashkenazi S. Adherence of non-fimbriate entero-invasive *Escherichia coli* O124 to guinea pig intestinal tract *in vitro* and *in vivo. J. Med. Microbiol.* 1986;21:117–123.

142. Rihova B, Rathi RC, Kopeckova P, Kopecek J. *In vitro* bioadhesion of carbohydrate-containing *N*-(2-hydroxypropyl) methacrylamide copolymers to the GI tract of guinea pigs. *Int. J. Pharm.* 1992;87:105–116.

143. McLeod AD, Friend DR, Tozer TN. Glucocorticoid-dextran conjugates as potential prodrugs for colon-specific delivery: hydrolysis in rat gastrointestinal tract contents. *J. Pharm. Sci.* 1994;83:1284–1288.

144. Hehre EJ, Servy TW. Dextran-splitting anaerobic bacteria from the human intestine. *J. Bacteriol.* 1952;63:424.

145. Harboe E, Larsen C, Johansen M, Olsen HP. Macromolecular prodrugs. XV. Colon-targeted delivery—bioavailability of naproxen from orally administered dextran-naproxen ester prodrugs varying in molecular size in the pig. *Pharm. Res.* 1989;6:919–923.

146. Larsen C, Harboe E, Johansen M, Olsen HP. Macromolecular prodrugs. XVI. Colon-targeted delivery—comparison of the rate of release of naproxen from dextran ester prodrugs in

homogenates of various segments of the pig gastrointestinal tract. *Pharm. Res.* 1989;6:995–999.

147. Nielsen LS, Weibel H, Johnsen M, Larsen C. Macromolecular prodrugs. XX. Factors influencing model dextranase-mediated depolymerization of dextran derivatives *in vitro*. *Acta Pharm. Nord* 1992;4:23–30.

148. Inoue M, Morikawa M, Tsuboi M, Yamada T, Sugiura M. Hydrolysis of ester-type drugs by the purified esterase from human intestinal mucosa. *Jpn. J. Pharmacol.* 1979;29:17–25.

149. McLeod AD, Fedorak RN, Friend DR, Tozer TN, Cui N. A glucocorticoid prodrug facilitates normal mucosal function in rat colitis without adrenal suppression. *Gastroenterology* 1994;106:405–413.

150. Steidler L, Hans W, Schotte L, Neirynck S, Obermeier F, Falk W, Fiers W, Remaut E. Treatment of murine colitis by *Lactococcus lactis* secreting interleukin-10. *Science* 2000;289:1352–1355.

151. Mardini HA, Lindsay DC, Deighton CM, Record CO. Effect of polymer coating on fecal recovery of ingested 5-amino salicylic acid in patients with ulcerative colitis. *Gut* 1987;28:1084–1089.

152. Li Y, Li HJ, Yang GR, Gu WP, Ma YK, Zhang MH, Sun J, Sun SJ. Colon-specific delivery tablets of sodium 4-aminosalicylic acid. *Yao Xue Xue Bao* 2006;41:927–932.

153. Khan MZ, Prebeg Z, Kurjakovic N. A pH-dependent colon targeted oral drug delivery system using methacrylic acid copolymers. I. Manipulation of drug release using Eudragit L100-55 and Eudragit S100 combinations. *J. Control. Release* 1999;58:215–222.

154. Norlander B, Gotthard R, Strom M. Pharmacokinetics of a 5-aminosalicylic acid enteric-coated tablet in patients with Crohn's disease or ulcerative colitis and in healthy volunteers. *Aliment. Pharmacol. Ther.* 1990;4:497–505.

155. Rudolph MW, Klein S, Beckert TE, Petereit H, Dressman JB. A new 5-aminosalicylic acid multi-unit dosage form for the therapy of ulcerative colitis. *Eur. J. Pharm. Biopharm.* 2001;51:183–190.

156. Iruin A, Fernandez-Arevalo M, Alvarez-Fuentes J, Fini A, Holgado MA. Elaboration and "*in vitro*" characterization of 5-ASA beads. *Drug Dev. Ind. Pharm.* 2005;31:231–239.

157. Lorenzo-Lamosa ML, Remunan-Lopez C, Vila-Jato JL, Alonso MJ. Design of microencapsulated chitosan microspheres for colonic drug delivery. *J. Control. Release* 1998; 52:109–118.

158. Jain A, Gupta Y, Jain SK. Potential of calcium pectinate beads for target specific drug release to colon. *J. Drug Target.* 2007;15:285–294.

159. Ghaffari A, Navaee K, Oskoui M, Bayati K, Rafiee-Tehrani M. Preparation and characterization of free mixed-film of pectin/chitosan/Eudragit RS intended for sigmoidal drug delivery. *Eur. J. Pharm. Biopharm.* 2007;67:175–186.

160. Oosegi T, Onishi H, Machida Y. Novel preparation of enteric-coated chitosan-prednisolone conjugate microspheres and *in vitro* evaluation of their potential as a colonic delivery system. *Eur. J. Pharm. Biopharm.* 2008;68:260–266.

161. Akhgari A, Farahmand F, Afrasiabi Garekani H, Sadeghi F, Vandamme TF. Permeability and swelling studies on free films containing inulin in combination with different polymethacrylates aimed for colonic drug delivery. *Eur. J. Pharm. Sci.* 2006;28:307–314.

162. Theeuwes F, Wong PL, Burkoth TL, Fox DA. Osmotic systems for colon-targeted drug delivery. In Bieck PR, editor. *Colonic Drug Absorption and Metabolism*, Marcel Dekker, New York, 1993, p 137.

163. Chacko A, Szaz KF, Howard J, Cummings JH. Non-invasive method for delivery of tracer substances or small quantities of other materials to the colon. *Gut* 1990;31:106–110.

164. Liu H, Yang XG, Nie SF, Wei LL, Zhou LL, Liu H, Tang R, Pan WS. Chitosan-based controlled porosity osmotic pump for colon-specific delivery system: screening of formulation variables and *in vitro* investigation. *Int. J. Pharm.* 2007; 332:115–124.

165. Wilding IR, Davis SS, Bakhshaee M, Stevens HNE, Sparrow RA, Brennan JB. Gastrointestinal transit and systemic absorption of captopril from a pulsed-release formulation. *Pharm. Res.* 1992;9:654–657.

166. Stevens HN, Wilson CG, Welling PG, Bakhshaee M, Binns JS, Perkins AC, Frier M, Blackshaw EP, Frame MW, Nichols DJ, Humphrey MJ, Wicks SR. Evaluation of Pulsincap to provide regional delivery of dofetilide to the human GI tract. *Int. J. Pharm.* 2002;236:27–34.

167. Roldo M, Barbu E, Brown JF, Laight DW, Smart JD, Tsibouklis J. Orally administered, colon-specific mucoadhesive azopolymer particles for the treatment of inflammatory bowel disease: An *in vivo* study. *J. Biomed. Mater. Res. A* 2006;79:706–715.

168. Kakoulides EP, Smart JD, Tsibouklis J. Azocrosslinked poly (acrylic acid) for colonic delivery and adhesion specificity: *in vitro* degradation and preliminary *ex vivo* bioadhesion studies. *J. Control. Release* 1998;54:95–109.

169. Van den Mooter G, Samyn C, Kinget R. *In vitro* evaluation of a colon-specific drug delivery systems: an absorption study of theophylline from capsules coated with azo polymers in rats. *Pharm. Res.* 1995;12:244–247.

170. Yamaoka T, Makita Y, Sasatani H, Kim SI, Kimura Y. Linear type azo-containing polyurethane as drug-coating material for colon-specific delivery: its properties, degradation behavior, and utilization for drug formulation. *J. Control. Release* 2000;66:187–197.

171. Brondsted H, Kopecek J. Hydrogels for site-specific drug delivery to the colon: *in vitro* and *in vivo* degradation. *Pharm. Res.* 1992;12:1540–1545.

172. Shantha KL, Ravichandran P, Rao KP. Azo polymeric hydrogels for colon targeted drug delivery. *Biomaterials* 1995;16:1313–1318.

173. Maris B, Verheyden L, Van Reeth K, Samyn C, Augustijns P, Kinget R, Van den Mooter G. Synthesis and characterisation of inulin-azo hydrogels designed for colon targeting. *Int. J. Pharm.* 2001;213:143–152.

174. Chen WB, Wang LF, Chen JS, Fan SY. Characterization of polyelectrolyte complexes between chondroitin sulfate and

chitosan in the solid state. *J. Biomed. Mater. Res. A* 2005;75:128–137.

175. Brondsted H, Andersen C, Hovgaard L. Crosslinked dextran—a new capsule material for colon targeting of drugs. *J. Control. Release* 1998;53:7–13.

176. Chiu HC, Hsiue GH, Lee YP, Huang LW. Synthesis and characterization of pH-sensitive dextran hydrogels as a potential colon-specific drug delivery system. *J. Biomater. Sci. Polym. Ed.* 1999;10:591–608.

177. Basan H, Gumusderelioglu M, Orbey Tevfik M. Release characteristics of salmon calcitonin from dextran hydrogels for colon-specific delivery. *Eur. J. Pharm. Biopharm.* 2007; 65:39–46.

178. Calinescu C, Mateescu MA. Carboxymethyl high amylose starch: chitosan self-stabilized matrix for probiotic colon delivery. *Eur. J. Pharm. Biopharm.* 2008;70(2):582–589.

179. Hirsch S, Binder V, Schehlmann V, Kolter K, Bauer KH. Lauroyldextran and crosslinked galactomannan as coating materials for site-specific drug delivery to the colon. *Eur. J. Pharm. Biopharm.* 1999;47:61–71.

180. Wakerly Z, Fell J, Attwood D, Parkins D. Studies on amidated pectins as potential carriers in colonic drug delivery. *J. Pharm. Pharmacol.* 1997;49:622–625.

181. Munjeri O, Fell JT, Collett JH. Hydrogel beads based on amidated pectins for colon-specific drug delivery: the role of chitosan in modifying drug release. *J. Control. Release* 1997;46:273–278.

182. Tozaki H, Komoike J, Tada C, Maruyama T, Terabe A, Suzuki T, Yamamoto A, Muranishi S. Chitosan capsules for colon-specific drug delivery: improvement of insulin absorption from the rat colon. *J. Pharm. Sci.* 1997;86:1016–1021.

183. Zhao XL, Li KX, Zhao XF, Pang DH, Chen DW. Study on colon-specific 5-Fu pH-enzyme Di-dependent chitosan microspheres. *Chem. Pharm. Bull. (Tokyo)* 2008;56: 963–968.

184. Kenyon CJ, Nardi RV, Wong D, Hooper G, Wilding IR, Friend DR. Colonic delivery of dexamethasone: a pharmacoscintigraphic evaluation. *Aliment. Pharmacol. Ther.* 1997; 11:205–213.

185. Prasad YV, Krishnaiah YS, Satyanarayana S. *In vitro* evaluation of guar gum as a carrier for colon-specific drug delivery. *J. Control. Release* 1998;51:281–287.

186. Krishnaiah YSR, Satyanarayana S, Rama Prasad YV, Narashima Rao S. Evaluation of guar as a compression coat for drug targeting to the colon. *Int. J. Pharm.* 1998;171:137–146.

187. Gliko-Kabir I, Yagen B, Penhasi A, Rubinstein A. Low swelling, crosslinked guar and its potential use as colon-specific drug carrier. *Pharm. Res.* 1998;15:1019–1025.

188. Graff S, Hussain S, Chaumeil JC, Charrueau C. Increased intestinal delivery of viable *Saccharomyces boulardii* by encapsulation in microspheres. *Pharm. Res.* 2008;25: 1290–1296.

189. Xu Y, Zhan C, Fan L, Wang L, Zheng H. Preparation of dual crosslinked alginate-chitosan blend gel beads and *in vitro* controlled release in oral site-specific drug delivery system. *Int. J. Pharm.* 2007;336:329–337.

190. Vervoort L, Van den Mooter G, Augustijns P, Busson R, Toppet S, Kinget R. Inulin hydrogels as carriers for colonic drug targeting. I. Synthesis and characterization of methacrylated inulin and hydrogel formation. *Pharm. Res.* 1997; 14:1730–1737.

191. Pitarresi G, Casadei MA, Mandracchia D, Paolicelli P, Palumbo FS, Giammona G. Photocrosslinking of dextran and polyaspartamide derivatives: a combination suitable for colon-specific drug delivery. *J. Control. Release* 2007;119: 328–338.

192. Siew LF, Basit AW, Newton JM. The potential of organic-based amylose-ethylcellulose film coatings as oral colon-specific drug delivery systems. *AAPS PharmSciTech* 2000;1:E22.

193. Wakerly Z, Fell JT, Attwood D, Parkins D. Pectin/ethylcellulose film coating formulations for colonic drug delivery. *Pharm. Res.* 1996;13:1210–1212.

194. Fan LF, He W, Bai M, Du Q, Xiang B, Chang YZ, Cao DY. Biphasic drug release: permeability and swelling of pectin/ethylcellulose films, and *in vitro* and *in vivo* correlation of film-coated pellets in dogs. *Chem. Pharm. Bull. (Tokyo)* 2008;56:1118–1125.

195. Yin L, Fei L, Cui F, Tang C, Yin C. Superporous hydrogels containing poly(acrylic acid-*co*-acrylamide)/*O*-carboxymethyl chitosan interpenetrating polymer networks. *Biomaterials* 2007;28:1258–1266.

196. Tsai MF, Tsai HY, Peng YS, Wang LF, Chen JS, Lu SC. Characterization of hydrogels prepared from copolymerization of the different degrees of methacrylate-grafted chondroitin sulfate macromers and acrylic acid. *J. Biomed. Mater. Res. A* 2008;84:727–739.

197. Lee JS, Jung YJ, Doh MJ, Kim YM. Synthesis and properties of dextran-nalidixic acid ester as a colon-specific prodrug of nalidixic acid. *Drug Dev. Ind. Pharm.* 2001; 27:331–336.

198. Pang YN, Zhang Y, Zhang ZR. Synthesis of an enzyme-dependent prodrug and evaluation of its potential for colon targeting. *World J. Gastroenterol.* 2002;8:913–917.

199. Ahmad S, Tester RF, Corbett A, Karkalas J. Dextran and 5-aminosalicylic acid (5-ASA) conjugates: synthesis, characterisation and enzymic hydrolysis. *Carbohydr. Res.* 2006;341:2694–2701.

200. Minami K, Hirayama F, Uekama K. Colon-specific drug delivery based on a cyclodextrin prodrug: release behavior of biphenylylacetic acid from its cyclodextrin conjugates in rat intestinal tracts after oral administration. *J. Pharm. Sci.* 1998;87:715–720.

201. Yano H, Hirayama F, Arima H, Uekama K. Prednisolone-appended alpha-cyclodextrin: alleviation of systemic adverse effect of prednisolone after intracolonic administration in 2,4,6-trinitrobenzenesulfonic acid-induced colitis rats. *J. Pharm. Sci.* 2001;90:2103–2112.

202. Zou M, Okamoto H, Cheng G, Hao X, Sun J, Cui F, Danjo K. Synthesis and properties of polysaccharide prodrugs of 5-aminosalicylic acid as potential colon-specific delivery systems. *Eur. J. Pharm. Biopharm.* 2005;59:155–160.

203. Cai QX, Zhu KJ, Chen D, Gao LP. Synthesis, characterization and *in vitro* release of 5-aminosalicylic acid and 5-acetyl aminosalicylic acid of polyanhydride—P(CBFAS). *Eur. J. Pharm. Biopharm.* 2003;55:203–208.

204. Gao SQ, Sun Y, Kopeckova P, Peterson CM, Kopecek J. Pharmacokinetic modeling of absorption behavior of 9-aminocamptothecin (9-AC) released from colon-specific HPMA copolymer-9-AC conjugate in rats. *Pharm. Res.* 2008;25:218–226.

205. Luo Y, Bernshaw NJ, Lu ZR, Kopecek J, Prestwich GD. Targeted delivery of doxorubicin by HPMA copolymer-hyaluronan bioconjugates. *Pharm. Res.* 2002;19:396–402.

15

DISSOLUTION TESTING: *IN VITRO* CHARACTERIZATION OF ORAL CONTROLLED RELEASE DOSAGE FORMS

MICHELE XUEMEI GUO

Analytical and Quality Sciences, Wyeth Research, Pearl River, NY, USA

15.1 INTRODUCTION

Dissolution testing is an integral part of pharmaceutical development and batch quality assurance for product safety and efficacy. During formulation development, dissolution serves as a guide for formulation design, optimization, and selection of formulation for further clinical studies. It is one of the stability indicating tests for product shelf-life evaluation during product development and a quality control tool to ensure product batch-to-batch consistency during manufacturing. It is also a tool for the evaluation of product performance upon changes in formulation, manufacture site, and facility, in conjunction with bioequivalence testing in humans if required [1].

Drug needs to be appropriately dissolved within the gastrointestinal (GI) tract prior to permeation and absorption. Drug absorption after oral administration follows three steps: release of the drug from the dosage form, solubilization, and permeation of the drug. Dissolution is closely related to release and solubilization of the absorption process. Thus, the *in vitro* testing may be relevant to the *in vivo* performance. When *in vitro–in vivo* correlation (IVIVC) is established, dissolution testing can be used as a substitute for expensive bioequivalent clinical studies for product *in vivo* performance evaluation [2]. The importance of dissolution and its close relationship with bioavailability and batch *in vivo* performance are well recognized by regulatory authorities.

For controlled release formulations, drug release is modified by controlling excipients that extend the release profile and thus the absorption process. Matrix tablets with hydroxypropyl methylcellulose (HPMC), polyethylene oxide (PEO), or other water-soluble excipients, as well as polymer-coated pellets are typical sustained release formulations. For matrix formulation, upon contacting with dissolution medium the polymer gradually hydrates, swells, and forms a layer of hydrogel. This layer regulates further liquid penetration into the hydrogel matrix. Controlled release is achieved by drug diffusion through the hydrogel layer as well as by erosion and dissolution of the polymer [3–7]. The release process can be monitored through *in vitro* dissolution testing.

Dissolution testing depicts formulation characteristics. Properly developed dissolution methods provide valuable information to guide the selection of excipients, and grades and levels of excipients during formulation development. A dissolution method should be discriminative against formulation variables and process changes with the goal of predicting *in vivo* performance.

15.2 PARAMETERS AFFECTING DISSOLUTION

Dissolution is a combined result of intrinsic factors of the drug and extrinsic factors of the dissolution method. Intrinsic factors refer to the drug's physical properties such as crystallinity, wettability, solvation, solubility, amorphism, polymorphism, particle size, and so on, which are important considerations for formulation development. Extrinsic factors refer to the dissolution conditions such as hydrodynamic conditions, temperature, medium pH, volume, buffer concentrations, and so on.

Oral Controlled Release Formulation Design and Drug Delivery: Theory to Practice, Edited by Hong Wen and Kinam Park
Copyright © 2010 John Wiley & Sons, Inc.

Based on the Nerst–Brunner and Levich modification of the Noyes–Whitney model, dissolution rate can be described by the following equation [8]:

$$dX_d/dt = AD/\delta(C_s - X_d/V) \qquad (15.1)$$

where dX_d/dt is the dissolution rate, expressed as the change in the amount of drug dissolved (X) per unit of time (t), D is the diffusivity coefficient, A is the surface area, δ is the thickness of the diffusion film adjacent to the dissolving surface, C_s is the saturation solubility of the drug, X_d is the amount of drug dissolved, and V is the dissolution medium volume. Drug properties, dissolution conditions, and formulation compositions affect the equation parameters. For example, diffusivity is related to molecular size and the temperature and viscosity of the dissolution medium. When surfactant is used in the dissolution media, the micelles formed have a lower diffusivity due to their bulky size.

15.2.1 Intrinsic Factors

15.2.1.1 Solubility Solubility is the main intrinsic factor controlling dissolution rate, hence bioavailability. Solubility is related to the compound's hydrophilicity and crystalline structure.

According to the Biopharmaceutics Classification System (BCS), drugs are classified into the following four classes [9, 10]:

- Class 1: high solubility and high permeability
- Class 2: low solubility and high permeability
- Class 3: high solubility and low permeability
- Class 4: low solubility and low permeability

From a pharmaceutical point of view, the solubility classification of a drug is determined by dissolving the highest unit dose of the drug into 250 mL of buffer with a pH range from 1.0 to 7.5 [11]. The drug substance is considered highly soluble when the dose/solubility volumes are less than or equal to 250 mL.

For class 1 and 3 compounds, which have high solubility, the drug is expected to dissolve in the gastrointestinal tract quickly. For these types of compounds, BCS suggests that a criterion of 85% dissolved in 0.1 N HCl in 15 min can ensure that bioavailability is not limited by dissolution [10]. These types of compounds behave like a solution and do not present bioavailability problems. Dissolution profiles are generally not expected. Dissolution testing is done to verify that the drug can release from the formulation in a short time frame under mild aqueous conditions.

For class 2 compounds, which have low solubility and high permeability, dissolution is most likely the rate-limiting step for drug absorption. Dissolution profiles are generally

expected for this type of compounds, and an IVIVC may also be expected. Surfactants are often used in formulation to improve drug solubility in the gastrointestinal tract, and thus absorption. Other solubility enhancing excipients may be used as well. For example, citric acid is often used in controlled release formulation for poorly soluble basic compounds to create an acidic microenvironment and thus improve solubility.

When the dosage unit is fully dissolved in the medium, the solution concentration should be at least three times less than the drug solubility in this medium. This is known as sink condition, and should always be sought when developing a dissolution method. Absence of sink conditions may generate unpredictable kinetics and suppress drug release [12]. When sink conditions are met, it is more likely that the dissolution results will reflect the properties of the dosage form. For class 2 compounds, adjusting medium pH for ionizable compounds and/or adding a small amount of surfactant in media to improve solubility are common practices and will be discussed in further detail in Section 15.2.2.

15.2.1.2 Particle Size Particle size and wettability are important factors affecting drug dissolution and bioavailability. Decreasing particle size, thus increasing surface area, is an effective way to speed up dissolution of poorly soluble compounds [13–15]. This process is often achieved by milling or recrystallization of the drug substance. Studies also show that milling may be associated with changing surface properties that may affect the blending characteristic and flow properties during batch preparation [16, 17]. To increase solubility and dissolution rate, nanomilling poorly soluble drug substance using homogenizer has shown promise [18].

Large particles present in drug substance also impact product content uniformity and dissolution precision of replicate testing. Dosage form characteristics, such as dissolution rate and precision, and content uniformity, should be considered when establishing particle size specification.

15.2.2 Dissolution Method or Extrinsic Factors

Dissolution method development is an evolving process. Suitable *in vitro* testing methods should be developed to guide pharmaceutical development and support stability study. Adjustment of the method might be warranted to establish *in vitro–in vivo* relationships when pharmacokinetic data are available.

15.2.2.1 Apparatus Choosing appropriate apparatus is usually the first step of dissolution method development. Apparatuses I (rotating basket), II (rotating paddle), III (reciprocating cylinder), and IV (flow-through cell) are recommended for controlled release formulation dissolution testing. The configurations and system components of the

TABLE 15.1   Dissolution Apparatuses Commonly Used for Controlled Release Dosage Form Dissolution Testing

Apparatus	Configuration	Equipment Components
Apparatus I	Rotation basket	A set of assembly. Each assembly consists of a vessel, a cover, a motor, a metallic drive shaft, a cylindrical basket, and a temperature-controlling bath
Apparatus II	Rotation paddle	A set of assembly. Each assembly consists of a vessel, a cover, a motor, a metallic drive shaft with a paddle, and a temperature-controlling bath
Apparatus III	Reciprocating cylinder	A set of cylindrical, flat-bottomed glass vessels, a temperature-controlling bath, a set of glass reciprocating cylinders, inert fittings, screens, a motor, and a drive assembly
Apparatus IV	Flow-through cell	A reservoir, a pump, a flow-through cell, and a temperature-controlling bath

commonly used apparatuses are listed in Table 15.1. Each apparatus should be used at its compendially recognized rotation speed (e.g., 100 rpm for baskets and 50–75 rpm for paddles). Other speeds can be used if justification is provided, such as IVIVC data. Donato et al. studied dissolution methods for a softgel capsule of lopinavir, a poorly soluble drug [19]. A level A IVIVC was established based on a dissolution method using apparatus II with a paddle rotation speed of 25 rpm and a dissolution medium containing sodium lauryl sulfate (SLS) solution.

Apparatuses I and II are most frequently used. The configurations of the two systems are simple and both are easy to operate. Apparatus II is recommended as the starting apparatus for method development. When using apparatus II, the dosage form is dropped and must sink to the center of the vessel bottom in order to reduce variability. A sinker may be used to prevent the tablet from floating during dissolution.

Apparatuses III and IV have relatively short histories. They were added in the fourth supplement of USP 22 in 1991. Both systems allow changes of media and hydrodynamic conditions to provide more biorelevant *in vitro* conditions.

Apparatus III, a reciprocating cylinder system, consists of multiple rows of vessels that can be used to contain different biorelevant media. Dipping rate and residence time in each vessel can be programmed as desired. The system has the advantages of mimicking the changes in pH gradient, buffer concentrations, ionic strengths, and mechanical forces of the GI tract, and the capability of full automation of the operation process. Apparatus III offers sound hydrodynamic conditions. The dissolution medium continuously moves around the product being tested. No coning (poorly stirred) area exists. Deaeration or nondeaeration shows negligible impact on dissolution [20]. Recent studies have shown that apparatus III is an attractive system for dissolution testing of modified release formulations [21–25], especially for colonic delivery formulations, as it allows the dissolution tubes to move between successive rows of vessels containing different pH media [23].

Apparatus IV is a flow-through cell dissolution system, which can be used in "open-loop" or "closed-loop" mode. Dissolution medium continuously flows through the system so that poorly stirred spots are eliminated. The system allows a large quantity of medium to be used, which is an advantage

for low-solubility drug requiring a large amount of medium to reach sink condition. Similar to a reciprocating cylinder system, it has the advantage of easily changing medium to compensate GI tract pH gradient, buffer concentration, and so on, and thus offers more biorelevant conditions. Perng et al. reported reasonable correlation of *in vitro–in vivo* data using four sequential GI tract simulation media on flow-through cell apparatus for a poorly soluble drug substance [26]. Jinno et al. reported that an IVIVC was established on cilostazol tablets of submicron drug substance crystal by using flow-through cell in "closed-loop" mode, while dissolution data generated by using apparatus II did not achieve IVIVC [27].

Prediction dissolution for water-soluble compounds in a flow-through cell system was made using hydrodynamics [28]. It was reported that a flow-though system was used to combine with Caco-2 permeation cell for permeability measurement [29].

15.2.2.2 Hydrodynamics
Hydrodynamics is an important factor of dissolution testing. Agitation reduces the boundary layer thickness (δ in Equation 15.1) at the surface of the product unit in the medium, hence increasing dissolution rate. Study also showed that fast agitation weakens the structure of hydrogel and increases drug release for controlled release formulations [30].

Hydrodynamic conditions are related to both dissolution apparatus and the rotating/dipping speed or medium flow rate used. Studies were performed to evaluate the hydrodynamic conditions in dissolution vessels [31–35]. Bai and Armenante studied the velocity distribution and shear rate variability resulting from changes in the impeller location in apparatus II [31]. They found that the axial and radial components of the velocity are symmetrically distributed with respect to the vessel centerline and are very weak in the region below the paddle, especially under the shaft. The shear rate reached maximum at about 15–20° away from the centerline. They also found that the paddle system is extremely sensitive to the asymmetric position of the rotating paddle. This study is consistent with the coning phenomenon (a poorly stirred zone) often observed in apparatus II [25, 34, 35]. Kamba et al. reported similar observation from a study on rotation paddle apparatus [32]. In this study, dissolution rate of planar constant release (PCR) tablets was found to increase as the distance between the vessel center and the tablet position increased from 0 to 28 mm. These studies showed that there is a considerable position-dependent difference in the hydrodynamics at the bottom of the vessel in apparatus II, which may contribute to intratablet dissolution variability.

Hydrodynamic condition comparison among apparatuses I, II, and III was studied [23–25, 36]. A study on a colon-specific drug delivery system in multi-pH media showed a close similarity (f_2 of 80.6) between apparatus III at 5 dpm and apparatus II at 100 rpm [23]. Rohrs et al. reported that the estimated equivalent agitation rates for apparatuses I and II were 100 rpm and for apparatus III were about 10 dpm or less [36]. Missaghi and Fassihi conducted a dissolution study for dimenhydrinate from an eroding matrix system using apparatuses I, II, and III [25]. Their work showed the following release order: apparatus III at 8 dpm > apparatus II at 100 rpm > apparatus III at 5 rpm > apparatus II at 50 rpm > apparatus I at 100 rpm > apparatus I at 50 rpm.

15.2.2.3 Dissolution Media

Selection of dissolution medium is at the core of dissolution method development. Different media should be screened to find those suitable for the dissolution testing of a particular product. Commonly used media include diluted HCl solution, buffer with different pH, water, water with salts, and water with surfactant. Other media can be used if justified.

The choice of dissolution medium should be based on physical properties of the compounds, types of formulation, and the *in vitro* medium screening data of the formulation. The most suitable medium is the one providing the most discriminative and sensitive dissolution data with robust results and a minimum of 75% release at the last time point.

Dissolution should be performed at 37°C for oral dosage forms. Medium should be degassed for the dissolutions sensitive to bubble presence. Medium volume is usually 500–900 mL for apparatuses I and II. Other volumes can be used if justified. The main considerations for choosing medium volume generally are to meet sink conditions and provide adequate detection sensitivity.

Medium pH Dissolution characteristics of dosage form should be evaluated in the physiologic pH range. For controlled release formulation, pH 1.2–7.5 should be evaluated.

Solubility of the compound and/or solubility–pH profile is an important parameter to be considered when developing a dissolution method. Solubility–pH profile for ionizable compounds should be studied prior to formulation development. Sink condition should always be used, unless otherwise justified.

For basic or acidic compounds, solubility in dissolution media varies with medium pH. The solubility of a weak base is significantly higher in the medium with pH < pK_a. Similarly, for a weak acid the solubility is enhanced in medium with pH > pK_a. The effects of pH on drug solubility were extensively studied [37, 38]. Sheng et al. studied the effects of pH combined with surfactant on solubilization and dissolution for an insoluble weak acid model compound, ketoprofen [37]. The study showed that both pH and surfactant have significant impact on ketoprofen solubility. They reported that the solubility of ketoprofen, with pK_a of 4.76, increased 160-fold from pH 4 to 6.8 when no surfactant was used. The solubility increased 50-fold in medium with 2% SLS at pH 4. pH was found to have more dominant effect than SLS concentration on dissolution rate.

McNamara and Amidon studied the intrinsic dissolution of dipyridamole, a low-solubility weak base with pK_a of 6.05 [38]. They reported that the compound dissolution in pH 5 medium was increased to twice that of pH 6.8 medium when no buffer was used in either medium. In medium with pH 5, which is below its pK_a of 6.05, dipyridamole protonation is significantly greater than that in pH 6.8. As a result, solubility and intrinsic dissolution are increased.

Ionizable compound dissolution in different medium conditions is dependent on the compound's characteristics relating to ionization and solubility, which is an important consideration for both formulation development and dissolution method development. Figure 15.1 shows dissolution profiles of model compound A in capsules, a poorly soluble basic compound with estimated pK_a of 4.2, 4.4, and 7.9, in media with different pH and surfactant. The profiles show that the compound has fast release in media with pH 3 and below, which is expected for a basic compound with pK_a of 4.2. As pH increases, the dissolution rate decreases significantly. At pH 6.8, the final release is about 20% compared to 100% in 0.1 and 0.01 N HCl. The dissolution profiles suggest that without surfactant this compound formulation would have very limited release in weak acidic or neutral media.

Due to its short half-life (1.0 h in dog and 1.2 h in rat), compound A's formulation was prepared as a controlled release form. A considerable amount of absorption will occur in the lower GI tract where the pH is higher than in the upper GI tract. The dissolution profiles in Figure 15.1 indicate many challenges for both dissolution method development and formulation development to achieve reasonable bioavailability in the lower GI tract. Simultaneously, the profiles provide valuable information for both types development, and will be discussed later in a case study.

Figure 15.2 shows the release profile in different pH media of a sustained release matrix tablet of model compound B, a basic compound with pK_a of 6.8. The slow to fast release ranking order from high pH to low pH media shows the pH effect on the dissolution of an ionizable basic compound formulation.

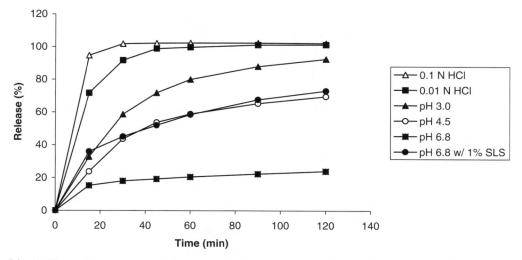

FIGURE 15.1   Dissolution medium screening for compound A (5.6 mg compound A mixed with capsule fill in each capsule). Dissolution conditions: apparatus II (paddle), rotation speed 50 rpm, medium volume 900 mL.

Medium pH has also been found to affect dissolution rate by interaction with controlling excipients. Zhang and McGinity studied chlorpheniramine maleate (CPM) sustained release tablets using PEO as the drug carrier [5]. They reported that CPM, a drug with solubility in excess of 1 mg/mL in all three media, had increased release in 0.1 N HCl compared to that in purified water and in pH 7.4 medium due to the formation of hydrogen bonds between PEO and acidic medium. The interaction expedites the dissolution of the PEO in the medium, resulting in faster dissolution in 0.1 N HCl.

Surfactants   For poorly soluble compounds, where sink conditions cannot be reached in media with various pH values, an appropriate amount of surfactant could be added into dissolution media to improve release. It is believed that small amount of surfactant can mimic the endogenous surfactant in the gastrointestinal tract. Surfactants can be used either as wetting agents or to solubilize the drugs. Different concentrations of surfactant should be studied to provide the appropriate dissolution profiles. When used below the critical micelle concentration (CMC) surfactants improve drug solubility by improving wettability, while when used above CMC surfactants improve solubility by forming micelles. The amount of surfactant should be as small as possible while ensuring that all strengths can have minimal 75% release at the last time point.

Surfactants were found to have significant impact on solubility of poorly soluble drugs. As cited earlier,

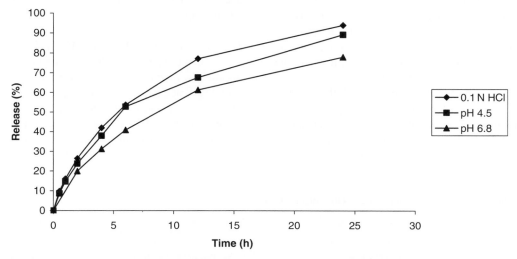

FIGURE 15.2   Dissolution profiles of 5 mg compound B sustained release tablets in different pH media. Dissolution conditions: apparatus II (paddle), rotation speed 50 rpm, medium volume 900 mL.

Sheng et al. reported that the solubility of a weak acid, ketoprofen, increased 50-fold in the media with 2% SLS at pH 4 [37]. However, despite this increase in solubility, the intrinsic dissolution rate of ketoprofen increased much more moderately due to the much smaller diffusivity of micelle species.

Three classes of surfactants are commonly used: anion surfactants, neutral surfactants, and cation surfactants. Table 15.2 lists the commonly used surfactants.

Many scientists choose SLS as the starting point of surfactant selection due to its excellent solubilization capacity and low cost, and often find success. Considerations of surfactant choice should include compound solubility in the media with the surfactant, dissolution characteristics, and compatibility of the surfactant with the compound, formulation, and type of capsule shells. SLS has been shown to significantly slow down the dissolution of gelatin shells at pH <5 [39]. This is believed to be caused by salt formulation of amino acid in the gelatin shell with SLS.

Medium Buffer Buffered media provide the drug with buffer capacity to maintain the medium pH throughout the dissolution process for a consistent environmental pH. In addition to maintaining the medium pH, buffer species and concentration were found to have a significant impact on the dissolution for low-solubility ionizable drugs when the drug ionization is increased in a medium with suitable pH value [38, 40–43]. Aunins et al. studied dissolution of indomethacin, a low-solubility weak acid (carboxylic acid) with pK_a of 4.17, and reported that dissolution in pH 7 phosphate buffer is significantly faster than that in pH 7 medium without buffer [38]. McNamara et al. reported that the dissolution of indomethacin was increased with greater bicarbonate concentration in the medium at pH 6.8 [43]. In medium with pH 6.8 (greater than indomethacin's pK_a), carboxylic acid ionization is increased as a result of the acid (carboxylic acid)–base (buffer) equilibrium. Dipyridamole, a low-solubility weak base with pK_a of 6.1, is not ionized and the dissolution was found to be independent of buffer species and concentrations in the medium with pH 6.8 (above its pK_a) [43].

High concentration buffer also interacts with the controlling excipients. A study of buffer molarity on verapamil release from modified release minitablets using apparatus III was reported [30]. Eudragit® and Carbopol® were used as the rate-retarding polymers in the modified release tablets. It was found that the dissolution rate of verapamil tablets increased significantly in media with 0.1 and 0.2 M phosphate buffer compared to that in medium with 0.05 M buffer. The facilitated dissolution is believed to be caused by a weakened hydrogel layer, in turn due to the interaction between buffer and the repulsive forces that exist between charged carbomer molecules in the system.

Biorelevant Media Simulated gastrointestinal fluids were introduced by Dressman et al. in 1998 [8, 44], with the goal of reconstructing GI tract physiological conditions in dissolution test media. Fasting-state simulated intestinal fluid (FaSSIF) and fed-state simulated intestinal fluid (FeSSIF) were proposed with the consideration of different bile salts, lecithin concentration, pH, buffer capacity, and osmolality in the GI tract. Since then, the biorelevant media were used extensively in dissolution to study IVIVC and food effect [8, 44–48]. Recently, Jantratid et al. studied the modified versions of biorelevant media FeSSIF-V2 and FaSSIF-V2 [49]. The newer media were reported to be more physiologically relevant.

A biorelevant dissolution medium containing lipid digestion products in the form of fatty acids/monoglycerides in addition to the bile salt/lecithin has been developed and used to successfully establish IVIVC for solid dosage forms of a poorly soluble compound [50]. Simulated gastric fluid (SGF) at pH 1.2 and simulated intestine fluid (SIF) at pH 6.8, with enzymes in both media, are included in USP as dissolution biorelevant media that provide dissolution scientists with additional choices for dissolution method exploration. Compared to literature reported biorelevant media, the compositions of USP simulated media are relatively simple.

Due to the complexity and high cost of medium preparation and their instability, the biorelevant media are mainly used for exploring biorelevance and not commonly used for QC testing.

15.2.2.4 Solution Detection Method

Although solution detection is not part of the drug release process, it is part of the dissolution method. A suitable detection of the dissolution samples is needed for dissolution testing. Dissolution samples can be quantitatively analyzed with online UV (fiberoptic) or offline UV. For both UV detections, UV absorbing excipients and capsule shells interfere with the assay.

TABLE 15.2 Commonly Used Surfactants for Dissolution Testing

Surfactant Type	Anionic Surfactants	Neutral Surfactants	Cationic Surfactants
Commonly used surfactants	Sodium lauryl sulfate, sodium deoxycholate	Tween 20, Tween 80, Triton X-100, Myri-52, Brij-35	Cetyl trimethylammonium bromide (CTAB), cetylpyridinium chloride (CPC)

Suitable correction solution, such as capsule shell solution, and/or baseline correction need to be established to ensure the method specificity.

Since commercial fiber-optic dissolution systems became available in late 1990s, they have gained increased acceptance in research and development. Recognition of fiber-optic dissolution systems in quality control environments is still in the early stage. Fiber-optic detection offers the advantage of eliminating the sampling step and consequently the offline detection. It provides the data in real time and is able to significantly increase working efficiency. The increased productivity is particularly important for both dissolution method and product development, where fast turnaround is necessary. In addition, dissolution profile monitoring provides dissolution data in a continuous fashion throughout the dissolution period and thus compensates for the disadvantage of discrete sampling where a critical part of the dissolution profile might have been missed due to insufficient sampling points.

While fiber-optic detection is gaining popularity in the pharmaceutical industry, special cautions should be exercised when using it for dissolution testing. Soluble UV absorbing excipients, insoluble excipients, and bubbles in media interfere with UV absorption. Insoluble excipients that flow around in the media may scatter or block light transmission, impacting the drug's UV absorbance. Two types of interference were reported [51, 52]: wavelength-independent scattering and wavelength-dependent scattering. Wavelength-independent interference typically produces a raised baseline over the entire UV range. A simple wavelength baseline correction is generally able to remove the interference. Wavelength-dependent interference tends to have greater effect in the low-wavelength region. In this case, a more complicated mathematical approach is needed to correct the interference, which is less practical in routine dissolution testing.

Liu et al. studied over 30 commonly used pharmaceutical excipients using Delphian fiber-optic dissolution systems and their scattering interference on accuracy, linearity, and precision of a model compound [53]. The study showed that light scattering by excipients was mostly wavelength independent. Simple baseline correction worked well to minimize the spectral interference from the insoluble excipients. Due to the presence of insoluble excipients, large variability is often observed for low UV absorbance level solution detection, which often leads to a high quantitation limit. When using fiber-optic dissolution for very low strength formulation, special caution should be taken to ensure data accuracy.

HPLC is a traditional and common method for dissolution sample analysis. HPLC analysis is more specific as excipients and other interference are separated from the drug substance. It is more suitable during formulation development when excipients still may change. The disadvantage is that offline analysis takes additional testing time.

15.3 DISSOLUTION'S ROLE IN CONTROLLED RELEASE DOSAGE FORM DEVELOPMENT

15.3.1 Dissolution Testing During Formulation Development

Based on the drug's physical properties and delivery objective, different formulation compositions are studied during formulation development. The types, grades, and levels of controlling excipients determine the release profile. For matrix formulation, the hydrogel formation upon contacting with medium is crucial in maintaining controlled release [3]. The time taken to form the hydrogel layer and the thickness of the layer are related to the proportion of the polymer and the manufacturing process. Viscosity of the water-soluble polymer plays an important role in the release mechanism [5]. In a formulation with high-solubility drug and high-viscosity polymer, the drug release will most likely be controlled by diffusion. A release from a formulation with low-viscosity polymer is often controlled by erosion of the hydrogel layer in addition to diffusion [5].

In vitro release kinetics can be described by the following empirical equation [3]:

$$Mt/M_\infty = kt^n$$

where M_t is the drug released at time t, M_∞ is the release at infinity, k is the kinetic constant, and n is the diffusional exponent. The value of exponent n ranges from 0.43 to 1.00. A value of n close to 0.5 suggests diffusion-controlled release while close to 1 implies erosion-controlled release [3].

Normally, multiple batches with different excipients, grades, or levels are prepared during formulation development. Controlled release prototype formulations should be screened in various dissolution media, hydrodynamics, and/or apparatus for an extended period of time. Dissolution provides the *in vitro* release profile and ranking order and release rate comparison, and characterizes the *in vitro* performance of a drug's delivery. It helps detect and understand how process variables, such as polymer grade, level, compress force, and so forth, affect drug release, and helps formulation scientists to select excipients and formulation for *in vivo* studies. Proper use of dissolution can facilitate formulation development and quickly identify potential issues during drug release in prototype formulations.

Often, the dissolution method is developed along with the formulation optimization. The method should be appropriately discriminating and capable of distinguishing significant changes in composition or manufacturing process that might be expected to affect *in vivo* performance. It should also be rugged and reproducible for day-to-day operation and technology transfer between laboratories.

The *in vitro* testing for controlled release formulation is usually carried out for 8–24 h depending on the drug dosing frequency. For offline analysis, multiple sampling points are needed to monitor the release. Early time points are used to evaluate whether there is a dose dumping. Middle time points are needed to evaluate the release profile and the final time point is to assess the release. An extended release profile with a minimum of 75% release at the last time point is generally desired.

An appropriately developed dissolution method should not have large variability. The relative standard deviation (RSD) should not be greater than 20% for early time points and 10% for later time points. Efforts should be made to reduce artifacts associated with the testing procedure, such as coning, dosage content not dispersed freely, and so on. Dosage forms often cause dissolution data variability, such as poor content uniformity, process inconsistencies, excipient interactions or interference, capsule shell aging, and so on. A suitable dissolution method is able to depict the dosage form characteristics and changes caused by stability conditions.

15.3.1.1 A Case Study

Dissolution method development is a dynamic process. It is typically being done at the same time as formulation development to provide critical drug release information of each particular formulation. When the major excipient changes in the formulation, the dissolution method often needs to change in order to closely depict drug release in the new formulation.

When pH-sensitive excipients are used, multistage dissolution testing with medium changing among different pH media is necessary to better relate the formulation *in vivo* release. Delayed release is a typical formulation using enteric coating that does not dissolve in upper gastrointestinal tract or low pH media.

Compound physical properties, especially solubility and/or solubility–pH profile, are usually the starting point when choosing dissolution medium. Figure 15.3 illustrates the solubility–pH profile of compound A, whose solubility is pH dependent and is less than 1 μg/mL in media with pH greater than 3. Dissolution of compound A in different media, as shown in Figure 15.1, suggests that the use of surfactant for higher pH media is necessary.

Due to the short half-life in dog and rat, the dosage form of compound A was developed as a sustained release formulation. The formulation consists of multilayer coatings with a layer of enteric coating. The dosage form includes two portions of drug formulated in the immediate release layer on top of the sustained release layer. To suit the multilayer formulation characteristics, a two-stage dissolution method was developed based on the solubility profile shown in Figure 15.3 and compound dissolution profiles shown in Figure 15.1. Stage 1 dissolution is carried out in 0.01 N HCl for 2 h for the immediate release portion of drug. Stage 2 was developed to depict the drug release of the sustained release portion. A pH 6 medium was used to suit the higher pH in the lower GI tract. 2% SLS was added to the medium to achieve a minimum of 75% release at the end point. Figure 15.4 shows the average release profile of six capsules of compound A in this multilayer coating formulation using the two-stage dissolution method. As the dissolution profile shows, the acid stage has a consistent release of about 30% corresponding to the immediate release portion in the formulation. The following 10 h dissolution shows a slow release profile corresponding to the sustained release portion.

15.3.2 In Vivo–In Vitro Correlation

A biorelevant dissolution test can be used to predict *in vivo* performance when an *in vitro–in vivo* correlation can be established. This practice has been accepted and encouraged by the FDA. IVIVC is defined as a predictive mathematical model describing the relationship between an *in vitro* prop-

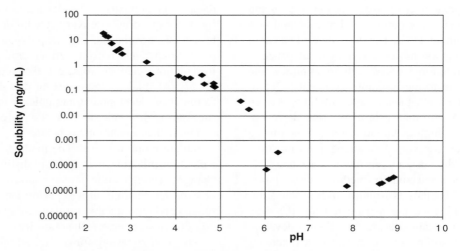

FIGURE 15.3 Solubility–pH profile of compound A.

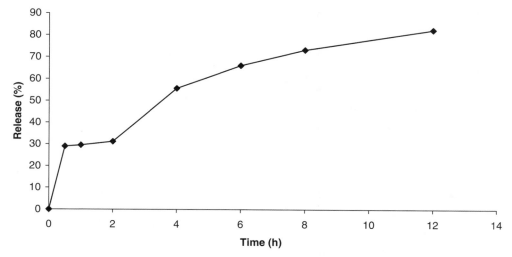

FIGURE 15.4 Two-stage dissolution profiles of compound A sustained release capsules Dissolution conditions: first stage, 900 mL of 0.01 N HCl for 2 h; second stage, 900 mL of phosphate buffer at pH 6.0 with 2% SLS for 10 h. Apparatus I (basket), rotation speed 100 rpm.

erty of an oral dosage form (usually the rate or extent of drug dissolution or release) and a relevant *in vivo* response (e.g., plasma drug concentration or amount of drug absorbed) [2]. With IVIVC established, expensive bioequivalence study may be avoided when formulation or other changes occur postapproval [2].

A common practice to establish IVIVC is to develop two or three significantly different *in vitro* release rate dosage forms. The batches are subjected to different dissolution condition tests to investigate the effect of dissolution variables. Various dissolution conditions should be explored at this stage to find the condition that generates *in vitro* profile simulating the *in vivo* data, thus establishing *in vitro–in vivo* correlation. Dissolution variables, such as apparatus, different medium (pH, enzyme, surfactants, osmotic pressure, ionic strength, etc.), and hydrodynamic conditions should be investigated to determine whether the *in vitro* release is independent of dissolution conditions and which conditions provide the most robust IVIVC. Among the variables, official apparatus and most commonly used conditions should be the starting point. Often biorelevant media, such as FaSSIF and FeSSIF, are used to establish IVIVC. Methods using biorelevant media have a good chance of relating *in vivo* performance, but are not convenient for QC testing.

Pharmacokinetic study is performed on the different *in vitro* release dosage forms. The plasma concentration profiles can be deconvoluted into absorption profiles. Drug absorbed and dissolved is then compared and possible IVIVC can be established under the FDA Guidance for Industry [2]. The IVIVC demonstrated is then validated via internal prediction when two formulations were used, or both internal and external predictions when three release profile formulations were used. Internal validation measures how well the IVIVC model describes the data used for the IVIVC study while external validation measures the accuracy of the

prediction of *in vivo* data from additional *in vitro* data that was not used for the IVIVC study.

Figure 15.5 shows the dissolution profiles of three release rate formulations of model compound B. Nine dissolution conditions covering various pH media and rotation speeds were used to generate *in vitro* releases for the three formulations. Among the dissolution conditions used, a level A IVIVC (a point to point correlation) was established on dissolution condition of using paddle at 50 rpm with phosphate buffer in pH 6.8 and verified though internal and external validation. As detailed IVIVC discussion is beyond the scope of this chapter, the *in vivo* data and correlation between *in vitro* and *in vivo* data are not being discussed here.

15.3.3 Product Manufacturing and Scale-Up Postapproval Changes

Dissolution serves as a test for manufacturing process and quality control. It detects tablet-to-tablet and batch-to-batch variability. A good dissolution method should be discriminating and able to detect batch quality change for the same formulation, yet sufficiently robust for day-to-day operation. Dissolution data with an appropriately established specification should fail nonbioequivalent batches.

An appropriate specification should be established to set the limits of acceptable batches. The product's dissolution characteristics should be understood by a well-developed dissolution method. The dissolution specification should be established based on batch historical data and in consultation with biopharmaceutical scientists and regulatory agencies. The acceptance criteria should be representative of multiple batches with the same nominal composition and manufacturing process, and representative of performance in stability studies [54]. For New Drug Applications (NDAs), the specification should be based on the dissolution characteristics of

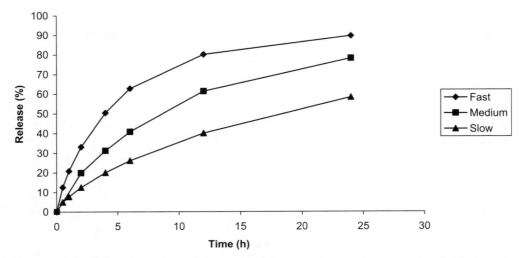

FIGURE 15.5 Dissolution profiles of three *in vitro* release rate tablets of compound B. Dissolution conditions: apparatus II, rotation speed 50 rpm, 900 mL phosphate buffer at pH 6.8.

batches used in the pivotal clinical studies and/or in confirmatory bioavailability studies [11]. When IVIVC is demonstrated, a meaningful dissolution specification range can be established for batches that provide bioequivalence to the biobatches based on the IVIVC study.

Dissolution is part of product stability testing. The dissolution test should appropriately reflect relevant changes in the drug product over time that are caused by storage conditions, such as temperature, humidity, photo exposure, and other stresses.

Often, changes of excipient component or composition, process, equipment, site, and/or scale-up/scale-down occur after drug approval. Depending on the level of changes defined in FDA's Guidance for Industry, various documents are needed to support product quality and performance [1]. For certain postapproval changes, appropriate additional dissolution tests on the new batches are needed and compared to reference batches [1, 11], Bioequivalence documentation may also be required. Dissolution profiles of a new batch and a reference batch are compared by similarity factor F_2 and difference factor F_1.

$$F_1 = \left\{ \left[\sum_{t=1}^{n} |R_t - T_t| \right] / \left[\sum_{t=1}^{n} R_t \right] \right\} \times 100$$

$$F_2 = 50 \log \left[1 + 1/n \sum_{t=1}^{n} (R_t - T_t)^2 \right]^{-0.5} \times 100$$

where log is logarithm to base 10, n is the number of sampling points, $\sum$ represents summation over all time points, R_t is dissolution at time point of the reference (unchanged drug product), and T_t is dissolution at time point of the test (changed product).

An F_2 value above 50 with a similarity factor F_1 typically not greater than 15% suggests that the two dissolution profiles are similar.

15.4 CONCLUSION

Dissolution is a vital part of pharmaceutical development, particularly for controlled release formulation. It depicts formulation characteristics and indicates drug release behavior. Dissolution data are a main parameter of product quality and performance. The *in vitro* test provides valuable information about batch-to-batch consistency, or bioequivalence if IVIVC has been established. The use of a biorelevant dissolution method can have a significant impact on new drug development resulting in a faster approval by regulatory authorities.

REFERENCES

1. Guidance for Industry. *SUPAC-MR: Modified Release Solid Oral Dosage Forms*, U.S. Department of Health and Human Services, Food and Drug Administration, Center for Drug Evaluation and Research (CDER), Rockville, MD, September 1997.

2. Guidance for Industry. *Extended Release Oral Dosage Forms: Development, Evaluation and Application of In-Vitro/In-Vivo correlations*, U.S. Department of Health and Human Services, Food and Drug Administration, Center for Drug Evaluation and Research (CDER), Rockville, MD, September 1997.

3. Colombo P, Bettini R, Santi P, Peppas NA. Swellable matrices for controlled drug delivery: gel-layer behavior, mechanisms

and optimal performance. *Pharm. Sci. Technol. Today* 2000; 3(6):198–204.

4. Li H, Hardy RJ, Gu X. Effect of drug solubility on polymer hydration and drug dissolution from polyethylene oxide (PEO) matrix tablets. *AAPS PharmSciTech* 2008;9(2): 437–443.

5. Zhang F, McGinity JW. Properties of sustained-release tablets prepared by hot-melt extrusion. *Pharm. Dev. Technol.* 1999; 4(2):241–250.

6. Shah N, Zhang GH. Prediction of drug release from hydroxypropyl methylcellulose (HPMC) matrices: effect of polymer concentration. *Pharm. Res.* 1993;10(12):1693–1695.

7. Mockeland JE, Lippold BC. Zero-order drug release from hydrocolloid matrices. *J. Pharm. Sci.* 1993;10(7):1066–1070.

8. Dressman JB, Amidon G, Reppas C, Shah VP. Dissolution testing as a prognostic tool for oral drug absorption: immediate release dosage forms. *Pharm. Res.* 1998;15(1):11–22.

9. Amiden GL, Lennernäs H, Shah VP, Crison JR. A theoretical basis for a biopharmaceutics drug classification: the correlation of *in vitro* drug product dissolution and *in vivo* bioavailability. *Pharm. Res.* 1995;12(3):413–420.

10. Guidance for Industry. *Dissolution Testing of Immediate Release Solid Oral Dosage Forms*, U.S. Department of Health and Human Services, Food and Drug Administration, Center for Drug Evaluation and Research (CDER), Rockville, MD, August 1997.

11. Guidance for Industry. *Waiver for In Vivo Bioavailability and Bioequivalence Studies for Immediate Release Solid Oral Dosage Form Based on a Biopharmaceutics Classification System*, U.S. Department of Health and Human Services, Food and Drug Administration, Center for Drug Evaluation and Research (CDER), Rockville, MD, August 2000.

12. Jamzad S, Fassihi R. Role of surfactant and pH on dissolution properties of fenofibrate and glipizide—a technical note. *AAPS PharmSciTech* 2006;7(2):E1–E6.

13. Battchar SN Wesley JA Fioritto A, Martin PJ Babu SR. Dissolution testing of a poorly soluble compound using the flow-through cell apparatus. *Int. J. Pharm.* 2002; 236:135–143.

14. Papp R, Luk P, Mullett WM Kwong E, Debnath S, Thibert R. Development of a discriminating *in vitro* dissolution method for a poorly soluble NO-donating selective cyclooxygenase-2 inhibitor. *J. Pharm. Biomed. Anal.* 2008;47:16–22.

15. Rasenack N, Muller BW. Dissolution rate enhancement by *in situ* micronization of poorly water-soluble drugs. *Pharm. Res.* 2002;19(12):1894–1900.

16. Machin L, Sartnurak S, Thomas I, Moore S. The impact of low levels of amorphous material (<5%) on the blending characteristics of a direct compression formulation. *Int. J. Pharm.* 2002;231:213–226.

17. Feeley JC, York P, Sumby BS, Dicks H. Determination of surface properties and flow characteristics of salbutamol sulphate, before and after micronisation. *Int. J. Pharm.* 1998; 172:89–96.

18. Kaqashima Y. Nanoparticulate systems for improved drug delivery. *Adv. Drug Deliv. Rev.* 2001;47(1):1–2.

19. Donato EM, Martins LA, Froehlich PE, Bergold AM. Development and validation of dissolution test for lopinavir, a poorly water-soluble drug, in soft gel capsules, based on *in vivo* data. *J. Pharm. Biomed. Anal.* 2008;47:547–552.

20. Borst I, Ugwu S, Beckett AH.New and extended applications for USP drug release apparatus 3, Dissolution Technologies, February 1997, pp. 11–18.

21. Joshi A, Pund S, Nivsarkar M, Vasu K, Shishoo C. Dissolution test for site-specific release ionized pellets in USP apparatus 3 (reciprocating cylinder): optimization using response surface methodology. *Eur. J. Pharm. Biopharm.* 2008;69: 769–775.

22. Wong D, Larrabee S, Clifford K, Tremblay J, Friend DR. USP dissolution apparatus 3 (reciprocating cylinder) for screening of guar-based colonic delivery formulations. *J. Control. Release* 1997;47:173–179.

23. Li J, Yang L, Ferguson SM, Hudson TJ, Watanabe S, Katsuma M, Fix JA. *In vitro* evaluation of dissolution behavior for a colon-specific drug delivery system (CODES™) in multi-pH media using United States pharmacopoeia apparatus II and III. *AAPS PharmSciTech* 2002;3(4): article 33.

24. Cacace J, Reilly EE, Amann A. Comparison of the dissolution of metaxalone tablets (Skelaxin) using USP apparatus 2 and 3. *AAPS PharmSciTech* 2004;5(1): article 6.

25. Missaghi S, Fassihi R. Release characteristic of dimenhydrinate from an eroding and swelling matrix: selection of appropriate dissolution apparatus. *Int. J. Pharm.* 2005;293:35–42.

26. Perng CY, Kearney AS, Palepu NR, Smith BR, Azzarano LM. Assessment of oral availability enhancing approaches of SB-248083 using flow through cell dissolution testing as one of the screens. *Int. J. Pharm.* 2003;250:147–156.

27. Jinno J, et al. *In-vitro–in-vivo* correlation of wet milled tablet of poorly soluble cilostazol. *J. Control. Release* 2008;130(1): 29–37.

28. Cammarn SR, Sakr A. Predicting of dissolution though hydrodynamics: salicylic acid tablets in flow through cell dissolution. *Int. J. Pharm.* 2000;201:199–209.

29. Motz SA, Schaefer UF, Balbach S, Eichinger T, Lehr C-M. Permeability assessment for solid oral drug formulations based on Caco-2 monolayer in combination with a flow through dissolution cell. *Eur. J. Pharm. Biopharm.* 2007;66:286–295.

30. Khamanga SMM, Walker RB. The effects of buffer molarity, agitation rate, and mesh size on verapamil release from modified-release mini-tablets using USP apparatus 3, Dissolution Technologies, May 2007, pp. 19–23.

31. Bai G, Armenante PM. Velocity distribution and shear rate variability resulting from changes in the impeller location in the USP dissolution testing apparatus II. *Pharm. Res.* 2008; 25(2):320–336.

32. Kamba M, Seta Y, Takeda N, Hamaura T, Kusai A, Nakane H, Nishimura K. Measurement of agitation force in dissolution test and mechanical destructive force in disintegration test. *Int. J. Pharm.* 2003;250:99–109.

33. Underwood FL, Cadwallader DE. Effects of various hydrodynamic conditions on dissolution rate determinations. *J. Pharm. Sci.* 1976;65:697–700.

34. Mirza T, Joshi Y, Liu Q, Vivilecchia R. Evaluation of dissolution hydrodynamics in the USP, Peak™ and flat-bottom vessels using different solubility drugs. *Dissolution Technologies* 2005, 11–16.

35. Collins CC, Nair RR. Comparative evaluation of mixing dynamics in USP apparatus 2 using standard USP vessels and Peak™ vessels, Dissolution Technologies, May 1998, pp. 17–21.

36. Rohrs BR, Bruch-Clark DL Witt MJ, Stelzer DJ. USP dissolution apparatus 3 (reciprocating cylinder): instrument parameter effects on drug release from sustained release formulations. *J. Pharm. Sci.* 1995;84:922–926.

37. Sheng JJ, Kasim NA, Chandrasekharan R, Amidon GL. Solubilization and dissolution of insoluble weak acid, ketoprofen: effects of pH combined with surfactant. *Eur. J. Pharm. Sci.* 2006;29:306–314.

38. McNamara DP, Amidon GL. Reaction plane approach for estimating the effects of buffers on the dissolution rate of acidic drugs. *J. Pharm. Sci.* 1988;77:511–517.

39. Zhao F, Malayev V, Rao V, Hussain M. Effect of sodium lauryl sulfate in dissolution media on dissolution of hard gelatin capsule shells. *Pharm. Res.* 2004;21:144–148.

40. Mooney KG, Mintun MA, Himmelstein KJ, Stella VJ. Dissolution kinetics of carboxylic acids II: effect of buffers. *J. Pharm. Sci.* 1981;70:22–32.

41. Aunins JG, Southard MZ, Meyers RA, Himmelstein KJ, Stella VJ. Dissolution of carboxylic acids III: effect of polyionizable buffers. *J. Pharm. Sci.* 1985;74:1305–1316.

42. Ramtoola Z, Corrigan OI. Influence of he buffering capacity of the medium on the dissolution of he drug excipient mixtures. *Drug Dev. Ind. Pharm.* 1989;15:2359–2374.

43. McNamara DP, Whitney KM, Goss SL. Use of a physiological bicarbonate buffer system for dissolution characterization of ionizable drugs. *Pharm. Res.* 2003;20:1641–1646.

44. Galia E, Nicolaides E, Horter D, Lobenberg R, Reppas C, Dressman JB. Evaluation of various dissolution media for predicting *in vivo* performance of class I and II drugs. *Pharm. Res.* 1998;15:698–705.

45. Lobenberg R, Kramer J, Shah VP, Amidon GL, Dressman JB. Dissolution testing as a prognostic tool for oral drug absorption: dissolution behavior of glibenclamide. *Pharm. Res.* 2000; 17(4):439–444.

46. Nicolaides E, Galia E, Efthymiopoulos C, Dressman JB, Reppas C. Forecasting *in vivo* performance of four low solubility drugs from their *in vitro* dissolution data. *Pharm. Res.* 1999; 16(12):1876–1882.

47. Dressman JB, Reppas C. *In vitro–in vivo* corrections for lipophilic, poorly water-soluble drugs. *Eur. J. Pharm. Sci.* 2000;11(2):S73–S80.

48. Nicolaides E, Symillides M, Dressman JB, Reppas C. Biorelevant dissolution testing to predict the plasma profile of lipophilic drugs after oral administration. *Pharm. Res.* 2001;18:380–388.

49. Jantratid E, Janssen N, Reppas C, Dressman JB. Dissolution media simulating conditions in the proximal human gastrointestinal tract: an update. *Pharm. Res.* 2008;25(7):1663–1676.

50. Lue BM, Nielsen FS, Magnussen T, Schou HM, Krisensen K, Jacobsen LO, Mullertz A. Using biorelevant dissolution to obtain IVIVC of solid dosage forms containing a poorly-soluble model compound. *Eur. Pharm. Biopharm.* 2008; 69:648–657.

51. Lu X, Lozano R, Shah P. *In-situ* dissolution testing using different UV fiber optic probes and instruments, Dissolution Technologies. *November* 2003 6–15.

52. Bynum K, Roinestad K, Kassis A, Procreva J, Gehriein L, Cheng F, Palermo P. Analytical performance of a fiber optic probes dissolution system, Dissolution Technologies, November 2001, pp. 13–22.

53. Liu L, Fitzgerald G, Embry M, Cantu R, Pack B. Technical evaluation of a fiber-optic probe dissolution system, Dissolution Technologies, 2008, pp. 10–20.

54. U.S. Pharmacopoeia 31, Rockville, MD.

16

CHALLENGES AND NEW TECHNOLOGIES OF ORAL CONTROLLED RELEASE

XIAOMING CHEN

Pharmaceutical Operations, OSI Pharmaceuticals, Cedar Knolls, NJ, USA

HONG WEN

Pharmaceutical Development, Novartis Pharmaceuticals Corporation, East Hanover, NJ, USA

KINAM PARK

Departments of Pharmaceutics and Biomedical Engineering, Purdue University, West Lafayette, IN, USA

16.1 INTRODUCTION

As discussed in previous chapters, significant advances have been attained in developing and commercializing oral controlled release products. Many platforms are available for delivering small molecule drugs with good aqueous solubility in a prolonged release or a delayed release. However, there are significant challenges in developing controlled release formulations for drugs with poor aqueous solubility, which require both solubilization and engineering of release profile. To deliver drugs at zero-order release rate, preferably independent of the gastrointestinal (GI) tract environment, besides osmotic pump drug delivery systems, many efforts and achievements have been made.

Moreover, many of new therapeutics under development are large molecules such as peptides, proteins, oligonucleotides, and vaccines. Their physical, chemical, and biopharmaceutical attributes distinct from small molecule drugs demand novel controlled release technologies to diminish barriers for oral delivery, such as instability in GI tract and poor absorption. Those unmet technology needs create great opportunities for research, development, and innovation. It is optimistic that breakthroughs in controlled oral delivery for water-insoluble drugs and biopharmaceuticals will have a significant impact on pharmaceutical and biotechnology industry. On the other hand, the continuous improvement in current delivery technologies is also important regarding the decrease of cost and the increase of efficiency. Those advancements include novel excipients, processes, and equipments as new tools for formulation scientists to develop oral controlled release formulations.

16.2 ORAL CONTROLLED DELIVERY FOR WATER-INSOLUBLE DRUGS

With few exceptions, drug products have to be dissolved in GI fluids to get absorbed for small molecule drugs. For a water-insoluble drug, the absorption and bioavailability could be restricted by dissolution rate and solubility in the GI tract. There are many established approaches to formulating water-insoluble drugs as oral dosage forms. Strategies include salt formation, microenvironmental pH control, solubilization by surfactants, complexation with cyclodextrins, solid dispersion, lipid-based formulation, and nanoparticles formulation (Table 16.1). Which strategy to choose is based on molecular and physical properties of a drug. It demands substantial formulation and process development to have a drug product with enhanced bioavailability for a water-insoluble drug.

Oral Controlled Release Formulation Design and Drug Delivery: Theory to Practice, Edited by Hong Wen and Kinam Park
Copyright © 2010 John Wiley & Sons, Inc.

TABLE 16.1 Formulation Approaches for Water-Insoluble Drugs

Strategy	Mechanism	Practical Approach
Salt formulation	Ionization at solid state for ionizable compounds	HCl, citrate, maleate salt et al. for weak bases; sodium, potassium, lysine salt et al. for weak acids
pH control	Promote or maintain ionization during dissolution for ionizable compounds	Alkalizing agents such as sodium bicarbonate, calcium carbonate for weak bases; acidifying agents such as citric acid, tartaric acid for weak acids
Surfactant	Improve wettability and solubility	Sodium lauryl sulfate, Poloxamer
Complexation	Form complex with cyclodextrin to improve aqueous solubility	Sulfobutylether beta cyclodextrin, hydroxypropyl beta cyclodextrin
Solid dispersion	Formation and maintain amorphous state with enhanced solubility	*Process*: spray drying, melt extrusion. *Polymers*: povidone, polyethyleneglycol, polymethacrylate, hydroxypropyl methylcellulose, ethylcellulose, hydroxypropyl cellulose
Lipid-based formulation	Formation of emulsion, microemulsion to enhance solubility and dissolution	*Liquid or semisolid excipients*: propylene glycol monocaprylate, macrogol glycerides, glyceryl monooleate, medium-chain triglycerides, diethylene glycol monoethylether
Nanoparticles	Reduce particle size and enhance dissolution	*Process*: milling, controlled crystallization. *Excipients*: Tween, phospholipid, Poloxamer

Developing a controlled release formulation for a water-insoluble drug is very challenging. A controlled oral delivery may be needed to achieve prolonged exposure or time-based release for a water-insoluble drug under certain circumstances. It could offer advantages in imporving efficacy, reducing side effect, or achieving a more desirable dose regimen. However, many platforms for controlled release have been established for drugs with acceptable aqueous solubility, which have been thoroughly reviewed in previous chapters. The release rate of a soluble drug in a solid dosage form is slowed down in a certain mode to achieve controlled release. It is clear that direct plug-in of current matrix-based or coating-based delivery systems without technology fabrication will fail to achieve acceptable controlled release of a water-insoluble drug. On the other hand, *in vitro* dissolution methods based on a sink condition generated by surfactants may provide misleading correlation for *in vivo* behaviors.

A combination of solubilization and modulating the release is needed to achieve controlled release for a water-insoluble drug. If a drug could be solubilized by a surfactant or a complex agent, inclusion of a solubilizing agent in polymer-based matrix tablets may provide a solution. Rao et al. studied the matrix tablet formulation of prednisolone, a sparingly water-soluble drug, using sulfobutylether-β-cyclodextrin (SBEBCD) as a solubilizing agent. SBEBCD promotes a sustained and complete release in a hydroxypropyl methylcellulose-based tablet formulation [1]. Another study also demonstrated that SBEBCD could also work as a solubilizing agent and an osmotic agent for controlled porosity osmotic pump pellets of prednisolone [2]. A complete and sustained release of prednisolone has been observed. It has been reported that controlled release felodipine tablets have been effectively prepared using Poloxamer as

a solubilizing agent and Carbopol as a controlled release matrix [3].

Many drugs need more complicated formulation approaches to enhance the dissolution, such as amorphous solid dispersion, emulsion, microemulsion, self-emulsifying, and nanoparticles. Among them, amorphous solid dispersion is the most popular approach to enhancing solubility and dissolution.

There are numerous publications about the development of controlled release formulations of water-insoluble compounds using solid dispersion as a solubilization approach. For a solid dispersion, drug molecules are stabilized in a high-energy state with hydrophilic polymers such as polyethylene glycol, polyvinyl povidone, and polyvinyl alcohol. Solid dispersions could be prepared by spray drying or melting extrusion. Mehramizi has reported that an osmotic pump tablet of glipizide has been developed using glipizide/polyvinylpyrrolidone dispersion as the core [4], where the solid dispersion has enhanced the solubility and ensured the complete release. Hong and Oh have studied the dissolution kinetics and physical characterization of three-layered tablets of nifedipine solid dispersion with poly(ethylene oxide) matrix capped by Carbopol [5]. They have discovered that the swelling and morphological change of Carbopol layers have minimized the release of rapidly erodible PEO200K (MW 200,000) and changed the nifedipine release to a diffusion-controlled process.

However, the physical stability of solid dipersions has to be monitored for polymer-based matrix systems, membrane coatings, or osmotic systems during prolonged release. All three approaches need water to diffuse inside formulation to solubilize drugs and advance the release. Drugs may crystallize out during the prolonged exposure to water due to supersaturation inside dosage form or change of glass transition temperature because of interaction with water. New formulation approaches have to be undertaken to ensure the physical stability of solid dispersions and achieve sustained release, which will be discussed in Section 16.3.

Nanoparticle formulation can be used to formulate poorly soluble drugs to enhance bioavailability. The drug dissolution rate is increased due to the increase of surface area. However, there are barely any available literatures about the controlled release of poorly soluble drugs with nanoparticle as the carrier. It is projected that an erosion-based system may be more suitable because the drug solubility is not changed. Diffusion-controlled matrix or membrane coating system is challenging to achieve the goal.

16.3 NEW FORMULATION DESIGNS IN ACHIEVING DESIRED RELEASE PROFILES

Many formulation designs have been pursued to achieve controlled release and minimize the impact of the GI environment. Osmotic pump drug delivery systems pioneered by Alza have many proved successes regarding those two aspects, as covered in previous chapters. Disintegration-controlled matrix tablet (DCMT) and erodible molded multilayer tablet by Egalet take an erosion approach and show some promises. On the other hand, bioadhesive polymers have advantages in improving gastroretentive delivery and enhancing localized therapy in GI tract. Moreover, significant progress has also been made in using computer modeling to design controlled release formulations.

16.3.1 Disintegration-Controlled Matrix Tablet

DCMT is an erosion-based controlled release platform. It has been developed for the sustained release of solid dispersions

by Tanaka et al. [6, 7]. Disintegration-controlled matrix tablets contain hydrogenated soybean oil as the wax matrix with solid dispersion granules uniformly distributed in the wax. The solid dispersion granules are formulated with low-substituted hydroxypropylcellulose as a disintegrant. Drug release is controlled by the process of tablet erosion. The wax only allows the penetration of water to the surface layer of tablet, and water triggers the swelling of the disintegrant on the surface, and subsequently tablet erosion results in the separation of solid dispersion granules from the tablet. A constant rate of tablet disintegration/erosion can be achieved by repeating the processes of water penetration and swelling/separating of solid dispersion granules (Figure 16.1).

DCMT has been successfully applied to the sustained release formulation of nilvadipine [6], a poorly soluble drug with an aqueous solubility of 1 µg/mL. The release profile of nilvadipine from DCMT has been modified by balancing the amount of wax and the amount of disintegrant. The wax matrix prevented water penetration into the tablet and ensuresd the amorphous state of solid dispersion during the dissolution process. Sustained release profiles of nilvadipine from DCMT were nearly identical in several dissolution mediums with varying pH and agitation speed [6]. An *in vivo* study in dogs has revealed that DCMTs successfully sustained the absorption of nilvadipine without reducing the bioavailability compared with IR coprecipitate tablets [7] (Figure 16.2). It suggests that DCMT is able to achieve the complete dissolution and absorption of a poorly water-soluble drug by maintaining the physical stability of solid dispersion in the GI tract.

16.3.2 Erodible Molded Multilayer Tablet (Egalet®)

Similar to DCMT, Egalet erodible molded tablets is an erosion-based platform. It has the advantage of delivering

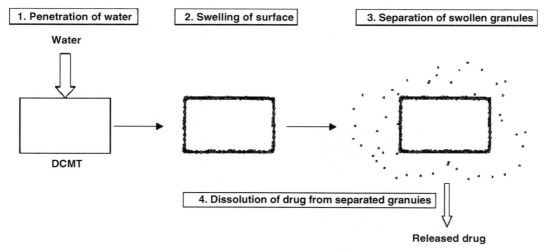

FIGURE 16.1 Proposed mechanism of drug release from DCMT [6].

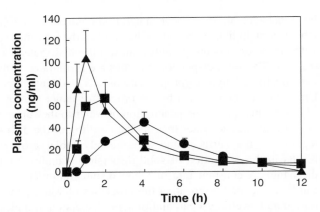

FIGURE 16.2 Plasma concentration profiles of nilvadipine after oral administration of DCMTs to beagle dogs under fasting condition [7]. key; (■) IR tablets, (■) DCMT-1, (●) DCMT-2.

zero-order or delayed release with minimal impact from the gastrointestinal conditions. Nevertheless, Egalet is a more sophisticated engineered delivery system with erosion occurred in one dimensional, whereas DCMT erosion is taken place in all three dimensions. It is obvious that Egalet could achieve a better zero-order release. Drug is dispersed in the matrix and the release is controlled by the rate of erosion in the two ends of tablets. The surface area for erosion is constant.

Egalet erodible molded multilayered tablets are prepared by injection molding (IM) [8]. As shown in Figure 16.3, a tablet produced via Egalet technology contains a coat and a matrix. Drug release is controlled through the gradual erosion of the matrix part. The mode and rate of release are designed and engineered by altering the matrix, the coat, and the geometry to achieve either a zero-order release or a delayed release.

For a zero-order release, a drug is dispersed through the matrix. The coat is biodegradable but has poor water permeability to prevent its penetration. The matrix tends to erode when in contact with available water. The erosion of the matrix is caused by GI fluids and promoted by gut movements in the GI tract. The drug release is mediated almost wholly by erosion because the dosage form is designed to slow down the water diffusion into the matrix. It is definitely more desirable for drugs with chemical and physical stability issues after contacting with water. For example, if the drug is solubilized as a solid dispersion and the solid dispersion tends to crystallize after interacting with water, the erosion-based delivery will ensure the stability. It is clear that Eaglet-based delivery system is suitable for the controlled delivery of water-insoluble compounds. The unique delivery system will also prevent hydrolysis and reduce luminal enzymatic activity.

As illustrated in Figure 16.3, the release rate of Egalet prolonged release is dependent on the erosion rate and drug concentration. It is clear that a zero-order release can be easily achieved if a uniform drug concentration in the matrix and a constant erosion rate are present. The erosion rate could be tailored through altering the composition of the matrix. For example, addition of polyethylene glycol could speed up the erosion [9]. However, the *in vivo* erosion rate may be affected by GI mobility. Due to the erosion controlled delivery, it is projected that the burst release effect can be minimized in the Egalet system.

On the other hand, the Egalet delivery system is easily fabricated for delayed release. Delayed release is gaining popularity for the enhancement of local effect or chronotherapy. The release of drug is delayed in a certain period of a time in GI tract and release in a bolus dose or a designated modified release. One area of delayed release applicable is to achieve colonic delivery for some therapeutic agents. On the other hand, the delayed release may provide advantages of a time release. The release of the drug can be timed to match the natural rhythms of a disease such as the morning stiffness and pain experienced by arthritis patients on waking.

A delayed release can be accomplished through three-compartment tablets including a coat, a drug release matrix, and a lag component. The lag component provides a predetermined delay for the drug release. After the lag component is eroded, the release drug is initiated in a designed mode as depicted in Figure 16.4.

Egalet delivery technology is developed based on standard plastic injection molding to ensure accuracy, reproducibility, and low production cost. It is being actively evaluated for development of numerous controlled release formulations by various companies.

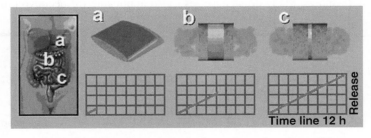

FIGURE 16.3 Egalet delivery for a zero-order release [9].

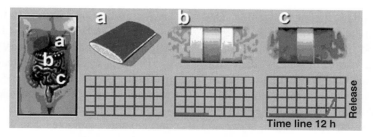

FIGURE 16.4 Egalet delivery for a delayed release [9].

16.3.3 Bioadhesive Oral Delivery

Bioadhesive delivery could be applied to oral controlled release. Bioadhesive polymers tend to adhere to the mucin/epithelial surface and find applications in buccal, ocular, nasal, and vaginal drug delivery. Those polymers could also help increase the residential time of solid dosage forms in the GI tract to improve gastroretentive delivery. On the other hand, bioadhesive polymers enable oral dosage forms stay close to epithelial layer and benefit the quick flux of drugs after dissolution. It could enhance oral absorption or improve localized therapy if the disease occurred in the GI tract.

Bioadhesion is an interesting phenomenon of the attachment of a synthetic or biological polymer to a biological tissue [10]. Adhesion could occur either with the epithelial cell layer or with the mucus layer. Adhesion to the mucus layer, namely, mucoadhesion, is more related for oral delivery. The GI tract is covered by a layer of mucus. Polymers containing hydrogen bonding groups tend to bind to the mucus layer. The mechanism of mucoadhesion is not fully understood. It is proposed that the attraction forces such as hydrogen bonding, van der Waals, and charges make polymers to have a close contact with the mucus. The close contact further promotes the penetration of polymers and formulation of entanglements with the mucin [10]. Many natural polymers and pharmaceutical ingredients show bioadhesive properties. Those polymers are carbomers, chitosan, starch, polymethacrylic acid, hydroxypropylcellulose, hydroxypropyl methylcellulose, and sodium carboxymethylcellulose. Bioadhesive polymers could be formulated with drugs in monolithic or multiparticulate forms to achieve controlled release.

Bioadhesive delivery could benefit the controlled release of drugs with narrow absorption window. Many drugs have a narrow absorption window from the proximal part of the GI tract due to transporter-mediated absorption. Increasing the residence time in the upper GI tract could extend and enhance the absorption. It has been reported that mucoadhesive microspheres of acyclovir made from ethylcellulose and Carbopol achieved a better bioavailability than a suspension formulation [11]. The mucoadhesive microspheres has an AUC_{0-t} of 6055.7 ng h/mL and a mean residence time of 7.2 h whereas the suspension has an AUC_{0-t} of 2335.6 ng h/mL

and an MRT of 3.7 h. Combination of bioadhesive polymers with another mechanism possibly will improve the degree of success of gastroretention, which is very challenging to achieve. Chavanpatil et al. have discussed the development of a novel sustained release, swellable, and gastroretentive drug delivery system for ofloxacin with additional bioadhesive properties [12]. Varshosaz et al. have reported the design and *in vitro* test of a bioadhesive and floating drug delivery system of ciprofloxacin [13].

Bioadhesive delivery is advantageous in providing sustained release for localized therapy. Deshpande et al. have published a study about the design and evaluation of oral bioadhesive-controlled release formulations of miglitol, intended for the prolonged inhibition of intestinal α-glucosidases and the enhancement of plasma glucagon like peptide-1 levels [14]. Pectin-based microspheres for the colon-specific delivery of vancomycin have been developed by Bigucci et al. [15]. The microspheres made of pectin and chitosan show desirable mucoadhesive properties.

16.3.4 Computer Modeling

The biggest challenge for developing of controlled release products is *in vitro–in vivo* correlation. Fabricating a drug formulation with a zero-order release *in vitro* is less demanding than predicting the *in vivo* performance. The GI tract is much more complicated than a dissolution unit. Influences from pH, food, GI mobility, regional absorption, bile salts, and gut metabolism are too complex to formulate as several simple mathematical equations. Computer simulation software are useful tools for formulators to study the drug molecules and propose target release profiles. Softwares are also applicable for deconvoluting animal and human pharmacokinetic (PK) data and provide feedback for further formulation optimization.

Advanced software packages for modeling oral absorption are physiology-based simulation softwares including iDEA, GastroPlus, and SimCYP.

iDEA, a software for predicting oral drug absorption, has been developed according to a physiologically multiple-compartment model invented by Grass [16]. The Grass' model divides the GI tract to five compartments including the stomach, duodenum, jejunum, ileum, and colon. Oral adsorption of a drug is simulated on three parameters—solubility, permeability, and tissue surface area in all compartments. iDEA could use permeability data from Caco-2 cell and have the capability to calculate solubility and permeability from a simple structure input. iDEA is convenient to use for predicting oral absorption of new chemical entities in early discovery. Several successful cases have been published about using iDEA to predict oral absorption, including ganciclovir, ketorolac, naproxen, and atenolol [17]. However, iDEA does not consider important factors such as first-pass metabolism, transport process, dissolution, and precipitation. It could not be used for evaluating formulations. iDEA is gradually phasing out while GastroPlus, a more sophisticated software, is gaining popularity.

GastroPlus is one of widely used software packages for modeling oral absorption of solid dosage forms. GastroPlus is an advanced software to simulate oral absorption, pharmacokinetics, and pharmacodynamics in animals and human based on molecular, physical, biopharmaceutical, and pharmacological attributes of a drug. The oral absorption module is based on an absorption model named advanced compartmental and transit model (ACAT) [18]. In GastroPlus, there are nine physiological compartments for human in line with different segments of the GI tract [19] (Table 16.2). A set of differential equations are used to simulate the kinetics of drug release, precipitation, and absorption in the nine compartments. It takes into consideration pH-dependent solubility, gut, and liver metabolism, and regional absorption. GastroPlus is very useful in modeling the effect of solubility, particle size, and dose on the extent of oral absorption for a specific drug (Table 16.3). GastroPlus also contains sophisticated pharmacokinetics (PK) simulations. The PK simulations could calculate PK parameters from intravenous and oral PK data, fit complex nonlinear metabolism and transport, and predict PK profiles based on PK parameters and kinetics of oral absorption. GastroPlus has been used as a tool in preclinical compound selection to assess absorption liability [20], establish *in vitro–in vivo* correlation (IVIVC) to justify biowaiver [21], and study food effect on pharmacokinetics [19].

There is a controlled release module in the software package. The oral absorption could be simulated with an input of *in vitro* controlled release profile. The data input could be one of three ways, tabulated data with linear interpolation, tabulated data with cubic spline interpolation, or Weibull function parameters. The absorption–time profile and bioavailability can be simulated with various release rates and profiles. The impact of physicochemical properties such as pH-dependent solubility and physiological parameters can be inspected at the same time. With the help of the PK module, animal and human intravenous (IV) and immediate release (IR) PK data could be deconvoluted to help the simulation of controlled release formulations. Various controlled release formulations could be studied in GastroPlus, including gastric retention, multiparticulate, integral tablet, and enteric-coated tablet and capsule. A study by Lukacova et al. demonstrated that GastroPlus is very useful in modified release formulation development. GastroPlus has been applied to obtain simulated models for adinazolam and metoprolol about the absorption, PK, and pharmacodynamics (PD) after IV and immediate release oral administration [19]. The fitted model for adinazolam was then utilized to study the PD profile for a modified release (MR) formulation and to propose a new formulation with desired onset and duration of action. The obtained metoprolol model is useful for understanding the *in vivo* profiles of MR formulations.

Similar to GastroPlus, SimCYP ADME simulator is a powerful platform for the prediction of drug absorption, PK, and drug–drug interaction [23]. The simulator consists of a huge database of human physiological, genetic, and epidemiological information. By incorporating *in vitro* data, it simulates a virtual clinical trial to predict pharmacokinetic profiles in a "real-world" population. The advanced dissolution, absorption, and metabolism (ADAM) model in the SimCYP simulator is a compartmental transit model having

TABLE 16.2 Nine Compartments and Physiological Parameter for a Fasting Adult Human (85 kg) Proposed in GastroPlus [19]

Compartment	Volume (mL)	Transit Time (h)	pH
Stomach	50	0.25	1.3
Duodenum	48	0.26	6.0
Jejunum 1	175	0.95	6.2
Jejunum 2	140	0.76	6.4
Ileum 1	109	0.59	6.6
Ileum 2	79	0.43	6.9
Ileum 3	56	0.31	7.4
Cecum	53	4.50	6.4
Ascending colon	57	13.5	6.8

TABLE 16.3 Input Variables for GastroPlus [22]

Parameter	Input
Chemical structure	ISIS structure
Dose information	Dose strength,
	subsequent dose,
	dose interval,
	dose volume,
	drug particle density,
	effect particle radius,
	dose form
Solubility	Solubility at different pH (1.5–7.5)
Permeability	CaCO2 permeability or calculate Peff
Molecular properties	pK_a, $\log P$, $\log D$

seven compartments for the small intestine. It takes into account of physiological variables such as gastric emptying time, small intestinal transit time, the radius, and length of the small intestine. It considers the dynamics of drug dissolution, precipitation, degradation, intestinal metabolism, and active transport. It calculates the effect of bile salts on dissolution and the effects of pH and transporter on permeability. It was reported that the plasma concentration profiles for three modified release formulations (fast, moderate, and slow) of metoprolol have be successfully predicted [24].

It is clear that both GastroPlus and SimCYP will play a more important role in future controlled release formulation development.

16.4 ORAL DELIVERY OF BIOPHARMACEUTICALS

Development of biopharmaceutical agents is thriving for past two decades owing to many breakthroughs in genomics and biotechnology. Hormones, growth factors, and cytokines are important therapeutic agents for managing and curing many diseases including diabetes, anemia, and hepatitis. Monoclonal antibodies with specific molecular targets are blockbuster drugs for cancer and rheumatoid arthritis. Many new vaccines have been developed to cope with some challenging infection diseases such as HIV, HCV, and HBV. Oligonucleotides such as small interferon RNA appear to be very promising as specific therapeutic agents for many unmet medical needs.

16.4.1 Challenges in Oral Delivery of Biopharmaceuticals

Though many bioactive molecules have either become successful commercial products or reached clinical trials, most of them are delivered by intravenous or subcutaneous injection. The development of noninvasive administration of peptide drugs especially through oral delivery route still remains not only a great challenge but also an exciting mission.

Oral delivery of biopharmaceuticals is challenged by instability in gastrointestinal fluids and poor permeability. First, biopharmaceuticals have poor stability in the GI tract [25–27]. Proteins and nucleotides would be denatured at acidic pH in the stomach, which results in the loss of biological activities. Proteases and nucleases in GI fluids also degrade biopharmaceuticals. Pepsin in the stomach and proteases secreted from the pancreas cleave peptide bonds and break down active therapeutic proteins and peptides. Moreover, proteins and nucleotides have poor intestinal permeability and are difficult to get absorbed due to their large molecular weight and hydrophilic nature [28, 29]. It has been reported that the mucus layer in the intestinal lumen

tends to bind charged molecules such as proteins and nucleotides, which also prevent oral absorption [30, 31].

16.4.2 Approaches in Overcoming Challenges in Oral Delivery of Biopharmaceuticals

Progress has been made for oral delivery of peptides or small proteins because of their relative small molecular weight [32–34]. Several approaches are being taken to improve the GI stability and enhance absorption, like enteric coating, protease inhibitors, and permeation enhancers.

Enteric coating of oral dosage forms has been utilized to avoid the release of therapeutic agents in the stomach, which mitigate the denature caused by acidic pH and the degradation catalyzed by pepsin. Enteric polymers like Eudragit S100 can ensure that oral dosage forms such as tablets, capsules, or pellets pass through the stomach without release and deliver therapeutic peptides to the small intestine or the colon [32]. Moreover, specific colonic delivery has been proposed to be advantageous to the oral delivery of peptides or proteins due to their lower proteolysis activity and greater responsiveness to absorption enhancers [33].

On the other hand, the enzymatic cleavage by pancreatic proteases in the intestine might be reduced by the inclusion of protease inhibitors in formulations. Protease inhibitors applicable are either specific enzyme inhibitors such as aprotin or trypsin inhibitors originated from animals or plants [34–36]. It has been reported that organic acids are also effective in inhibiting pancreatic proteases through controlling local pH, which are more active in neutral and alkaline pH than acidic pH [37].

In addition, permeation enhancers are needed to reduce the barrier for the oral absorption of peptides. Many permeation enhancers are surfactants or detergents such as sodium cholate, fatty acids, bile salts, and phospholipids [38–40]. Aside from solubilizing poorly soluble peptides, absorption enhancers wield their effects by increasing the permeability of biological membrane and fluidizing the lipid membrane. Consequently, the tight junction between mucosal cells are loosened and become more permeable to peptides. Calcium chelating agents enhance the permeation by disrupting the tight junction [41]. Another efficient agent to loosen tight junction is zonula occludens toxin or Zot [42–44]. Zot is a single polypeptide chain of 399 amino acids present in toxigenic strains of *V. cholera*. It has the ability to reversibly alter intestinal epithelial tight junction, which allows the passage of macromolecules through mucosal barriers. An encouraging 10-fold increase in insulin absorption has been observed in both rabbit jejunum and ileum *in vivo* with Zot [44]. However, there is a concern that the loosening of tight junction may have long-term toxic effects and pose undesirable consequences for the absorption of other drugs, nutrients, or pathogens. Another strategy to improve absorption of peptide is to use bioadhesive

TABLE 16.4 Ingredients Used for Oral Delivery of Peptides

Ingredients	Function	Examples
Enteric coating polymer	Prevention of degradation in the stomach; targeted delivery to the colon	Eudragit L30D-55, Eudragit FS30D
Protease inhibitor	Inhibition of pancreatic enzymes	Aprotin, trypsin inhibitor, chymotrypsin inhibitor, Bowman-Birk inhibitor, organic acids
Permeability enhancer	Decrease of the interaction with mucus layer; improvement of paracellular and transcellular permeability	Acylcarnitine, salicylates, sodium cholate, long-chain fatty acids, bile salts, surfactants

polymers. Bioadhesive polymers could increase the residence time and provide a prolonged delivery. Moreover, bioadhesive carriers localize the associated drugs to absorption sites, which also promotes oral adsorption [45].

It is possible that combination of above strategies could result in the improvement of bioavailability for therapeutic peptides (Table 16.4).

16.4.3 Several Oral Delivery Systems for Biopharmaceuticals

Unigene (Fairfield, NJ) has developed a proprietary technology to deliver various therapeutic peptides orally [46]. The formulation is an enteric-coated capsule or tablet with organic acids and permeability enhancers mixed with a peptide (Figure 16.5).

It has been claimed that the oral delivery system by Unigene could achieve 1–10% bioavailability for various peptides and small proteins, depending on size, charge, and structure [46]. Capsules and tablets of salmon calcitonin, a 32-amino acid peptide for the treatment of postmenopausal osteoporosis, have been studied in rats, dogs, and humans. Calcitonin is a polypeptide hormone that regulates the calcium and phosphorous metabolism. A significant amount of intact salmon calcitonin has been detected after oral dosing in rats, dogs, and humans [46]. The C_{max} (250–3500 pg/mL) in dog and human was linear with dose from 0.33 to 4.58 mg of salmon calcitonin. The phase I study in human has

demonstrated that an oral dose of 0.5 mg was able to achieve a mean C_{max} of 300 pg/mL with T_{max} ranging from 90 to 180 min. Unigene has entered a phase II clinical trial for oral salmon calcitonin. Oral formulations of therapeutic peptides, such as human parathyroid hormone, glucagon-like peptide-1, and leuprolide, are being developed using the same platform and are in preclinical or early clinical test.

There are remarkable interests in developing oral formulation for insulin. The progress of oral delivery of insulin has gained significant media attention. Oral delivery of insulin has been demonstrated in diabetic dog using enteric-coated microcapsule with sodium cholate as an absorption enhancer and the soybean trypsin inhibitor as a blocker to proteolysis degradation [47]. A clinical study of oral insulin by Emisphere did not generate promising results for further development, revealing the significant hurdle for oral delivery of therapeutic peptides and protein. Nanoparticle formulation is gaining attention right now, which may be able to enhance the overall bioavailability of oral doses of proteins and peptides [48].

It is clear that the route to reach approval and market for oral delivery of peptide is still formidable. Tremendous effort in research and development is being made due to the significant benefit of oral dosage form regarding patient compliance and acceptance. Besides formulation approach, chemical modifications have been utilized to improve the stability or absorption of peptides [49, 50].

Complexation hydrogel has been reported to be a promising carrier for the intestinal delivery of peptides such as interferon beta and calcitonin [51]. Therapeutic peptides could form complex with hydrogels such as poly(methacrylic acid) grafted with poly(ethylene glycol). The hydrogels have multiple functions to ensure the stability of peptide. Peptides are protected in the acidic environment in the stomach because the hydrogels do not swell in acidic conditions. The hydrogels are designed to swell in neutral or alkaline pH to release the peptides in the intestinal lumen. In addition, the advantages of those hydrogels are their mucoadhesive characteristic to enhance adsorption and potentials to inhibit proteolysis degradation by chelating calcium. It was demonstrated that the absorption of interferon beta and calcitonin has been significantly improved through complexing with hydrogels compared with a solution formulation. The area

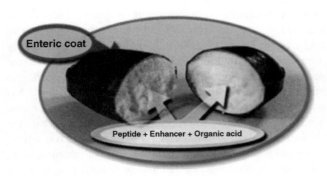

FIGURE 16.5 Oral delivery system developed by Unigene [46].

TABLE 16.5   AUC Values Following *In Situ* Administration of Interferon *β*-Loaded P(MAA-*g*-EG) Microparticles into the Ileal Segments [51]

Dose ($\times 10^6$ IU/kg)	AUC	
	Interferon β Solution	Interferon β-loaded P(MAA-*g*-EG)
0.75	6.7 ± 6.0 $\Big]^a$	48.4 ± 6.7^b $\Big]^a$
2.0	74.6 ± 14.2	154.4 ± 29.0
6.0	84.8 ± 32.1	257.9 ± 80.7

AUC: area under the curve.

Each value represents the mean ± S.E. from $n = 3–8$.

$^a p < 0.05$, significant difference among doses of "interferon β solution" or "interferon β-loaded P(MAA-*g*-EG)."

$^b p < 0.01$, significant difference compared with the corresponding "interferon β solution."

under the curve (AUC) of hydrogel formulation was 2–6-fold higher than the solution formulation after being delivered specifically to the jejunum area. It is clear that this approach is a significant breakthrough to pursue oral delivery of peptides (Table 16.5).

16.4.4   Oral Delivery for Vaccines

Vaccination is a long established therapy against various infectious diseases caused by bacteria and viruses. Vaccine is a biopharmaceutical preparation that establishes or strengthens the immunity of animals or humans to a particular infectious disease. There are many types of vaccines available, which include killed microorganisms, live and attenuated microorganisms, inactive toxoids, protein subunits, polysaccharide–protein conjugates, and DNA vaccines. The immunization of vaccines is mainly relied on parental route. However, there are substantial interests in developing oral vaccines for humans and animals due to their more desirable acceptance, avoidance of needles, easy administration, and low cost.

Challenges are expected to be encountered for the oral immunotherapy of vaccines, such as instability and poor absorption in gastrointestinal tract. Various ingredients have been explored to improve oral bioavailability, which include hexose monosaccharide, ethyl alcohol and water vehicles, oxygen-containing metal salts, enteric coating, and poly(lactic-*co*-glycolic) acid microspheres [52]. It has been reported that lipid formulations were capable of enhancing the oral vaccination in animal studies [53]. In a study reported by Vipond et al., the oral vaccination of Bacille Calmette–Guérin (BCG) against tuberculosis has showed promising results in animal studies [53]. Lipid formulations containing BCG strains induced significant γ-interferon response in mice and provided proved protection against TB aerosol challenge in guinea pigs. The effectiveness of vaccination of lipid formulation was superior to unformulated BCG and equivalent to

subcutaneous immunization. Nayak et al. have studied the formulation of rotavirus with polylactide (PLA) and polylactide-*co*-glycolide (PLGA) for vaccination [54]. Rotavirus is a virus that causes severe diarrhea, mostly in babies and young children. Rotavirus vaccine is an oral vaccine delivered as a solution. In this study, rotavirus was trapped in PLA and PLGA to form microparticles, which offered enhanced stability in the stomach and a sustained exposure. A single-dose oral immunization with 20 mg of antigen entrapped in PLA and PLGA particles has exhibited improved and long-lasting immunization than the soluble antigen.

16.4.5   Oral Delivery for Nucleotides

In contrast to protein-based drugs, nucleotide-based therapeutics have not attained major breakthroughs of reaching patients despite several decades of effort in research and development. Biopharmaceuticals based on antisense RNA showed early promises in animal studies but were barely able to demonstrate reliable and sufficient efficacy data in clinical trials. The main hurdle is effective delivery. Nucleotide-based drugs have to enter inside cells to be effective whereas many protein therapeutics act on receptors on cell surfaces. In recent years, development of nucleotide-based biopharmaceuticals, specific on RNA interference (RNAi), has gained new waves of interest. Major pharmaceutical companies such as Merck, Novartis, and Pfizer are rushing internal resources and external investment to develop RNAi therapeutics.

RNAi is a RNA-dependent gene silencing process in life cells found in eukaryotes including animals [55]. RNAi is an important cell defence against parasitic genes such as viruses and transposons. On the other hand, RNAi also directs gene expression and regulates development of eukaryotes. Small interference RNA (SiRNA), a class of 20–25 nucleotide-long double-stranded RNA molecules is central to RNA interference. The siRNA has been successfully used for genomic studies and drug target validation recently. It is anticipated that siRNA based on therapeutics will play a significant role in fighting against infectious diseases such as HIV, HCV, and HBV in the future. By the way, progression of many diseases such as cancers, Alzheimer's diseases, and diabetes is related to the activity of multiple genes. It is expected that turning off of a gene with a siRNA may provide a therapeutic effect to cure or modulating various diseases. The siRNA has the potential to become next generation of new therapies for many unmet medical needs with even greater impact than the introduction of monoclonal antibodies.

Some promising results have been obtained in cell-based *in vitro* models and *in vivo* animal models through localized delivery. A study carried out by Crowe has demonstrated that siRNAs targeted for HIV chemokine receptors, CXCR4, and CCR5, are effective in blocking the entry of HIV virus *in vitro* [56]. The silencing of hepatitis A and hepatitis B, influenza, and measles genes using siRNA is being actively pursued for potential therapy [57–59]. However, effective systemic administration of siRNA for therapeutic use faces mammoth challenges such as poor stability and limited cellular uptake. siRNAs could not freely pass cell membrane due to its strong anionic charge of its backbone and large molecular weight (about 13 kDa). On the other hand, unmodified, naked siRNAs have poor stability in blood and serum, as they are rapidly degraded by endo- and exonucleases.

Chemical modifications in the backbone, base, or sugar have been engaged to enhance stability without adversely affecting gene silencing activity [60]. Delivery through viral vectors has been reported to be successful in many cases of animal studies. There are some safety concerns about the clinical use, such as potential mutagenicity or oncogenesis. Development of nonviral delivery system has been actively advanced including chemical modification, linking to cell penetrating peptide, and lipid-based delivery [60]. A major breakthrough of nonviral delivery is to utilize antibodies to direct the delivery of siRNA. Liposomes or protein complexes carried with siRNA can be coated with antibodies. Antibodies will trigger the cell membrane internalization after binding specific cell surface receptors to achieve the delivery of siRNA. It has been published that HER-2 siRNA could be delivered to tumor cells both *in vitro* and *in vivo* with liposome coated with antitransferrin scFv [57]. Kumar et al. have demonstrated that a complex with antiviral siRNA and a CD7-specific single-chain antibody conjugated to oligo-9-arginine peptide was able to achieve significant anti-HIV activity in an HIV-infected mice model [58].

Many ongoing trials with siRNA mainly relied on localized delivery or parental route. Oral delivery of siRNA is clear to face even bigger hurdle. It is optimistic that the technologies under development for oral delivery of proteins and knowledge gained for siRNA will promote a new generation of formulation scientists to innovate and bring this very promising category of drugs to patients.

16.5 NEW PLATFORM TECHNOLOGIES FOR ORAL CONTROLLED RELEASE

As discussed above, oral controlled release has been advanced to new frontiers such as controlled delivery of poorly soluble compounds and biopharmaceutical oral formulation. Moreover, new technologies and novel processes are being innovated to achieve lower cost, higher efficiency, more robustness, and better quality. These new technologies include hot melt extrusion (HME), injection molding, printing techniques, and dry coating. It is obvious that these new technologies will have a huge impact on formulation development of sustained release, modified release, and targeted release oral delivery systems.

16.5.1 Hot Melt Extrusion for Controlled Release

Hot melt extrusion (HME) is a widely used process in plastic, rubber, and food industry. In recent decades, HME is emerging as a powerful process technology for drug delivery, HME is applicable to the manufacture of different variety of solid dosage forms including granules, pellets, tablets, rods, and films for oral, transdermal, and implant delivery. Oral dosage forms from HME can be immediate or controlled released. The primary application is to make solid dispersion formulations to enhance solubility for poorly soluble compounds. However, its value in developing controlled release formulations has gained extensive attention.

There are many advantages of HME for controlled release formulation (Table 16.6). HME process is solvent free and anhydrous, which is more environment friendly and is more desirable for drugs with physical and chemical stability sensitive to water. It only involves few stages in a single continuous unit operation that simplifies process development and scale-up. Very high load could be achieved through HME. Due to the intensive mixing and agitation during the process, drug particles are uniformly suspended or melted in a polymer matrix to achieve superior content uniformity. Many kinds of dose forms such as tablets, pellets, rods, and films and different geometries can be produced by HME. For drugs with poor solubility, amorphous form can be generated to achieve sufficient solubility for controlled release. However, HME does demand drug molecules to have acceptable thermal stability to withstand the thermal process. Many commercial products have been successfully launched via HME process, such as Kaletra (lopinavir/ritonavir/copovidone) tablets, Certican tablets (everolimus/HPMC), Rezulin tablets (troglitazone/PVP), and Sporanox capsules (itraconazole/HPMC).

TABLE 16.6 Advantages and Disadvantages of HME

Advantages	Disadvantages
• Continuous process	• Thermal process (drug/polymer stability)
• Few unit operations	
• Solvent free and anhydrous process	• Flow properties of polymer are essential
• High drug load	• Limited number of polymer
• Better content uniformity	• High density and low compressibility
• Formation of amorphous	
• Variety of dosage forms such as tablets, pellets, implants, films	
• Wide range of geometry	

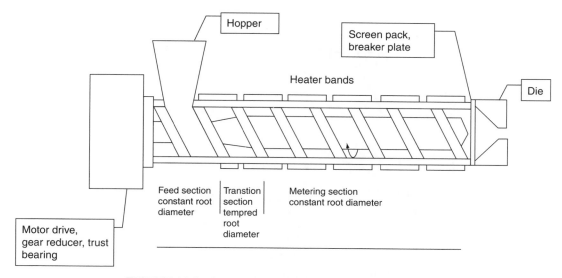

FIGURE 16.6 Component parts of single-screw extruder [61].

HME is a single-unit operation with four stages:

1. melting/softening of drug/polymer mix in a heated chamber;
2. mass transport of melted or softened drug/polymer mix through the barrel using a screw system;
3. extrusion of the drug/polymer mixture through a die;
4. cooling and fabricating the mixture to a designed shape.

The structure of a typical single-screw extruder is shown in Figure 16.6 [61].

Formulation development of a controlled release formulation using HME involves selection of polymers, plasticizers, and releasing modulators that are compatible with a targeted drug. The drug release rate is modified through the optimization of drug load, type and level of polymers, amount of plasticizers, and quantity of releasing modulators. Polymers used for melt extrusion need to have good thermal stability. Some commonly used polymers for controlled release, such as hydroxypropyl methylcellulose, ethylcellulose, polyethylene oxide (PEO), are suitable for HME for-

mulation (Table 16.7). Plasticizers are commonly included to HME formulations to lower the glass transition temperature and reduce the viscosity of melt mixture. The presence of plasticizers increases the flexibility of the extrudate and enables the process to take place in a much lower temperature. Lower process temperature helps mitigate the degradation problems. On the other hand, amount of plasticizers could influence the drug release rate. The drug release rate could also be modified by the addition of releasing modulators such as lactose, povidone, or low molecular HPMC. Process optimization for HME involves the study of process temperature, holding time, and feeding rate.

Formulation development of a controlled release system using HME is a combination of science and art. The design space for a target dissolution profile is substantially broad for formulation scientists to explore, which includes the selection and optimization of drug load, carrier polymer, plasticizer, and releasing agent. In certain circumstances, combination of two polymer carriers will help to develop a formulation with a custom dissolution profile. Coppens et al. have demonstrated that a HME formulation of acetaminophen (APAP) could be prepared using both hypromellose (HPMC) and polyethylene oxide to attain different mode of

TABLE 16.7   Polymer Carriers Applicable for Controlled Release Formulation via HME Process

Polymer	Trade Name	Physical Properties
Polyethylene oxide	Polyox WSR	Highly crystalline powder, hydrophilic, melting point 70°C, glass transition -40 to -60°C, thermal stability up to 350°C
Hydroxypropyl methylcellulose	Methocel	Hydrophilic, glass transition 160–210°C, thermal stability up to 250°C
Hydroxypropylcellulose	Klucel	Hydrophilic, thermal stability up to 260°C, glass transition 130°C
Ethylcellulose	Ethocel, Aqualon EC	Hydrophobic, amorphous glass transition 129–133°C, crystalline melting point 180°C, thermal stability up to 250°C

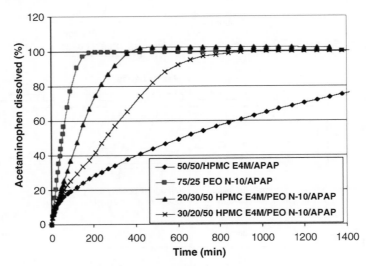

FIGURE 16.7 HME formulations of acetaminophen [62].

release profiles [62]. As shown in Figure 16.7, combination of HPMC and PEO helps the development of formulations with a 6–10 h sustained delivery with a complete release, which is more versatile for further modification than single polymer carrier.

A more efficient way to adjust release profiles is to select a suitable plasticizer or release modifier and vary its amount.

The most striking feature of HME process is its ability to prepare a controlled release formulation for poorly soluble compounds. Molecular dispersion could be prepared to enhance the solubility at the same time the release rate is customized to achieve immediate release, modified release, or sustained release depending on the targeted profile. It has been demonstrated in a case study by Zhu et al. that solubilization and controlled release can be accomplished at same time [63]. Tablets of indomethacin, a poorly soluble drug, were prepared by HME with Eudragit RD 100, a copolymer of acrylate, and methacrylate. It has been found that indomethacin formed a solid solution with Eudragit RD 100 and the release of indomethacin was controlled by the matrix of Eudragit RD 100. The inclusion of Pluronic® F68, Eudragit® L100, or Eudragit® S100 in the HME formulations facilitated the release rate [63].

16.5.2 Injection Molding for Controlled Release

Like hot melt extrusion, injection molding (IM) is a thermal process widely used in plastic industry. A polymer or a polymer mixture is melted and injected to a mold to form a finished product. Injection molding shares some common features with melt extrusion, such as the mass transport of polymer using a screw and the melting process. However, HME is more tailored for production of bulk products such as sheets, rods, and tubes. IM is designed to manufacture finished products in a single step, including bottles, caps, discs, or any designed articles. As shown in Figure 16.8, polymer resin is supplied to a machine through a hopper. After entrance into the barrel, the resin is heated to a proper melting temperature. The melted resin is injected into a mold by a reciprocating screw or a ram injector. The mold is cooled constantly to a temperature that lets the resin to solidify.

IM has been utilized for producing numerous medical designs in plastic parts, such as drug implants, delivery systems, and product containers. For example, polymeric microneedle for transdermal delivery can be fabricated using microinjection molding techniques [64].

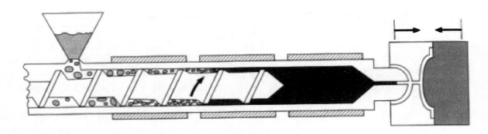

FIGURE 16.8 Drawing of a basic injection molding.

Similar to melt extrusion, IM can be applied in developing controlled release formulations to achieve sustained release, modified release, or delayed release. A major advantage of IM is that finish products are produced in a single step, whereas materials from melt extrusion may need further formulation effort to become a dosage form. A study has been reported by Quinten et al. about developing sustained release matrix tablets of metoprolol tartrate using IM. Matrix tablets of metoprolol were formulated with ethylcellulose as the matrix polymer, dibutyl sebacate as the plasticizer, and hypromellose as the release modifier. It has been found that metoprolol tartrate existed as a solid dispersion in the matrix tablets prepared by IM. Combination of 50% hypromellose with ethylcellulose resulted in a complete and first-order drug release profile with drug release controlled via the combination of diffusion and swelling/erosion [65].

Moreover, IM is more sophisticated than HME in the manufacture process so that it could be used for design and production of more complicated delivery system. It is projected that IM will advance significantly due to its ability to fabricate complex delivery system. The prominent case of IM for controlled release is the invention of Eaglet delivery system, an erosion-based delivery system feasible for prolonged release and delayed release, described in the previous section.

16.5.3 Printing Techniques for Controlled Release

While melt extrusion has become a practical approach for controlled release and IM is advancing the frontier, printing techniques for controlled release is more a technology for tomorrow. It is a fabulous platform but it may take long way to be applicable for market products.

Inkjet printer is a cheap and common consumer goods. However, the process of printing using inkjet is a complicated technical process. The printing process engages the rapid creation and release of liquid droplets through hundreds of delivery nozzles with a predetermined mode. The deposition position on a substrate is precisely controlled and can be preprogrammed.

It is very interesting to apply such a sophisticated technology to fabricate delivery systems. Delivery systems can be precisely designed and manufactured in three dimensions. The concept of printing itself can be prototyped to be a novel delivery system. Based on its own inkjet technology, Hewlett-Packard is developing a smart patch for transdermal delivery [66]. The patch consists of numerous microneedles linked to respective drug reservoirs. Four hundred cylindrical reservoirs, each one attached to its own microneedle, can be present in one in.^{2} of the patch. Drug administration is governed by an embedded microchip and powered by a low-power battery. This novel transdermal delivery system is applicable to numerous drugs that are not feasible for traditional delivery. It also has the flexibility to deliver various drugs and the potential to program according to biological signals.

It has been reported that printing technology can be utilized to fabricate artificial bones. The shape was designed based on the image data of bone deformity. And internal structure was built layer by layer through printing a water-based polymer adhesive onto thin layers of powdered α-tricalcium phosphate (TCP). The polymer hardened the TCP. An artificial bone has been reproduced precisely through repeating powder laying and polymer printing [67].

Inkjet printing also offers significant advantages for the coating of small medical devices including coronary artery stents [68]. Coronary artery stents coated with a variety of pharmacological agents such as sirolimus and paclitaxel are used to prevent restenosis. Traditional processes for the manufacture of drug-coated stents are dipping, ultrasonic spray coating, airbrush painting, and deposition along the struts using syringes. These processes lack tight manufacture control, retain high variability in drug concentration, and challenge the delivery in a more controlled manner. It is clear that inkjet printing is going to transform this field due to its preciseness and sophistication. The drug and polymer solutions can be put very precisely in both location and amount onto a stent. Drug gradient is flexible to engineer and thickness of polymer at different location is feasible to fabricate to gain desirable release kinetics. Coating could be very complicated, including multiple layers of polymers or several drugs. Moreover, the development process is more efficient than traditional processes. Drug waste is much less.

Inkjet printing also finds its edge in biomedical fields such as targeted gene delivery [69], tissue engineering [70–72], and the development of biodegradable implants [73].

Application of inkjet printing in controlled oral delivery has been advanced steadily since the invention of three-dimensional printing techniques by MIT in 1990s. Three-dimensional printing is a novel prototype technique with wide applications in the rapid and flexible production of prototype parts, end-use parts, and tools directly from a CAD model [74]. It has exceptional flexibility in building parts with complicated internal and external geometry and unprecedented control of location, material composition, microstructure, and surface texture, using various materials including ceramics, metals, polymers, and composites.

The process of three-dimensional printing is illustrated in Figure 16.9. A three-dimensional product is built layer by layer by powder delivery and printing of binder solution. Each run starts with a thin layer of powder delivered over the surface of a powder bed. A binder material is sprayed by inkjet head to join particles at defined positions based on design from CAD. The powder bed supported by a piston is lowered to allow next layer to be printed. This layer-by-layer process recurs until the product is finished. Following

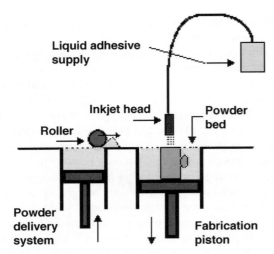

FIGURE 16.9 Diagram of three-dimensional printing [74].

a thermal treatment or other curing methods, loose powder is removed, leaving the fabricated product.

It is obvious that three-dimensional printing is applicable to developing sophisticated oral controlled release formulations due to its superior spatial and composition control. It has been reported that tablets with zero-order release characteristics could be attained through manipulating drug gradient, polymer gradient, geometry, and internal structure using three-dimensional printing [75]. Yu et al. have found out that a gradient distribution of release retardant polymer is functional in developing an erosion-based tablet with zero-order release [75]. The tablet consisted of a sandwiched structure with top and bottom layers impermeable to water penetration and drug diffusion. The drug release was controlled by the rate of erosion in the middle layer. A near-linear erosion rate has been achieved through printing release retardant polymers such as ethylcellulose in a decrease gradient from the periphery to the center. Another approach is to increase the drug concentration along the central axis to fulfill zero-order kinetics. It is clear that only printing techniques could deliver such precise concentration and texture control. It does provide another alternative to formulate zero-order release if not superior to traditional technologies.

Printing technology is also applicable in developing delayed release formulations such as time release or site-specific release. Current approaches for such formulations mainly rely on coating processes that involve multiple polymers or several layers of coating. The batch process approach is difficult to ensure the consistent coating of different single units. It is obvious that the performance of those dosage forms is variable and may lack reproducibility. It could be foreseen that printing technology is capable of advancing this field greatly due to its ability to be preprogrammed and to be precisely controlled to fabricate a sophisticated delivery system. Thickness of delayed layer, amount of permeability enhancer, geometry, and locations of drugs can be easily

engineered and precisely controlled. Katstra et al. have developed a pulsatory tablet of sodium diclofenac with two pulses of release using three-dimensional printing technology [76]. Sodium diclofenac was printed to two separate areas in a continuous enteric excipients phase in the tablet. Three-dimensional printing has endless flexibility to accommodate many ways of design and substantial engineering capability. It could be used for delivering multiple drugs in various modes. However, significant effort is needed to make this technology feasible for commercial manufacture.

16.5.4 Dry Coating

Coating is one of the major technologies to develop controlled release formulations including sustained release, modified release, and delayed release oral dosage forms. Coating is applicable to powder, granules, pellets, minitablets, tablets, and capsules to achieve desirable delivery profiles. Pan coating and fluid bed coating using solvent or latex are well established for many decades. For liquid coating, polymers, pigments, and excipients are mixed in an organic solvent or water to form a solution or dispersion. The coating solution or dispersion is sprayed into solid dosage forms in a pan coater or a fluid bed dryer and dried by hot air. Liquid coating has the disadvantages of significant solvent consumption, long process, and considerable energy use. On the other hand, there is a considerable challenge to develop very thick coating using liquid coating for delayed release or erosion-based controlled delivery. Dry coating, which is more environment friendly, is perceived to have the potential to eliminate some of the drawbacks of wet coating.

Two approaches could be used for dry coating, which are powder coating and compression coating. Powder coating is stemmed from metal coating. Powder materials are directly applied to a solid surface. Polymer particles are adhesive to the surface physically by electrostatic forces. The formation of coating is finished through a curing step by heat. Compression coating is more an extension of tableting technology. Coating material is compressed to surround an inner tablet through a multistep tableting process. Both techniques will be discussed in more detail in following paragraphs.

Powder coating is attained by applying fine particles to solid dosage forms and forming film by heat. Thermoplastic polymers are used for pharmaceutical coating. Plasticizers are often included to reduce the glass transition temperature, which allows the formation of film at a reduced temperature and with an improved flexibility. Depending on the way to promote the adhesion of particles onto the surface, powder coating is classified as plasticizer-dry-coating, electrostatic-dry-coating, heat-dry-coating, and plasticizer-electrostatic-heat-dry-coating.

For plasticizer-dry-coating, powder and a liquid plasticizer are sprayed using separate nozzles onto the dosage surface at the same time. The dosage surface and powder particles are wetted with the plasticizer to promote the adhesion between them. A continuous film is formed after curing above the glass transition temperature of the polymer for a predetermined time. Heat curing helps the particle deformation and coalescing to form a film. It has been reported that the spraying of a small amount of water or hypromellose solution during curing was beneficial to the film quality of hypromellose succinate acetate-coated spheres and tablets [77]. Dry powder coating can be accomplished in a centrifugal granulator, a fluid bed dryer, or a pan coater. As illustrated in Figure 16.10, powder mixture is carried by a stream of compressed air and sprayed onto tablets while a plasticizer solution is sprayed at the same time. It has been demonstrated that powder coating was able to achieve similar coating efficiency and performance to liquid coating for spheres and tablets with the enteric coating polymer, hypromellose succinate acetate [77].

Another way to promote the adhesion of particles to solid dosage is to use heat. The technique is called heat-dry-coating, which was invented by Cerea et al. [78]. Coating is performed in a spheronizer. Polymer particles are continuously spread onto tablets while heated by an infrared lamp to promote the binding and film forming. However, this technique is only applicable to polymers with low glass transition temperature, such as Eudragit E PO, a copolymer of dimethylaminoethyl methacrylate and methacrylate. For a polymer with high glass transition temperature, a preplasticization with plasticizer is needed. Both plasticizer-dry-coating and heat-dry-coating have difficulties in preparing uniform and smooth coating with controllable coating thickness.

However, electrostatic-dry-coating is capable of achieving uniform coating with controlled coating thickness. Electrostatic coating is widely applied in coating of metal surface. Dry powder is propelled by compressed air and forced through a spray gun. During the process, particles become electrically charged and tend to adhere to metal surfaces via electrostatic binding. However, it is not straightforward to do electrostatic-dry-coating for pharmaceutical dosage forms due to the weak electrical conductivity. One of the approaches to improve electrostatic-dry-coating is to ground the tablet more effectively and direct charged particles onto the tablet surface more specifically through a unique instrument design. Based on its proprietary instrumentation of electrostatic-dry-coating, Phoqus Co. has developed Chronocort, a once-daily modified release formulation for the treatment of adrenal insufficiency. In another report, the electrostatic coating has been accomplished through an electrically grounded pan coater [79]. Coating materials with plasticizer were sprayed to tablets by an electrostatic spray gun for a predetermined length to achieve certain thickness of coating. The curing step by heat helped the formation of continuous and uniform coating.

Powder coating is an improved coating technique, which utilizes equipments similar to liquid coating. On the other hand, compression coating is just a compaction technique. Coating material is directly compressed to surround a core by tablet compression. The main advantage of compression coating is its capability to allow much greater weight gain than liquid coating or powder coating. This aspect makes it very desirable to develop time-based release formulations for targeted delivery or chronotherapy. Targeted delivery is to delay the release of a dosage form until reaching a specific gastrointestinal location such as the colon to treat localized diseases. Chronotherapy is to deliver drugs to match the circadian rhythm of some diseases such as asthma, arthritis, epilepsy, migraine, and allergic rhinitis. Drugs will be released after a preprogrammed lag time. Tablet formulations for time-based delivery are typically made of a rapid release core tablet covered by a barrier layer to delay the release.

It has been demonstrated in numerous studies that compression coating is an exceptional choice for developing time release delivery. Ghimirea et al. have developed a time release tablet of theophylline using compression coating [80]. A barrier layer consisting of glycerol behenate and low-substituted hydroxypropylcellulose was pressed around an immediate release core of theophylline. The coated tablets showed pulsatile release with a lag time dependent on the

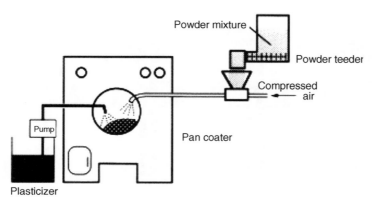

FIGURE 16.10 Illustration of dry coating with a tablet coating machine [77].

relative amount of glycerol behenate and low-substituted hydroxypropylcellulose. A tablet formulation of mesalamine has been developed by coating with layers of hypromellose Methocel E15 and Eudragit S100 to achieve a site-specific colonic delivery [81]. Lin et al. have reported that the lag time could be modulated from 1 to 16 h by modifying the type and amount of excipients of outer layer for a compressed coated tablet of sodium diclofenac [82].

Those cases were performed in lab scale and tablets were manufactured by manual processes. To achieve a reliable production, sophisticated instrumentation is needed to ensure the reproducible central position of core tablet, uniform thickness of coating layer, and consistent porosity of coating layer. One-step dry coating (OSDrC) technology invented by SKK in Japan is greatly advancing

the possibility of manufacturing dry coating products in production scale.

OSDrC is accomplished through a uniquely designed rotary tableting machine [83]. This machine has a variable double punch configuration to support single-step production of dry-coated tablets. The position of core tablets is precisely controlled and the coating layer is very flexible for fabrication in geometry and thickness.

The tableting process is depicted in Figure 16.11. The rotary-type tableting machine uses only a single set of punches and dies. All punches have a double structure consisting of a center punch and an outer punch. The manufacture process involves three compressions. At the first compression, the lower-outer layer is formed. The core is compressed with the lower layer at the second compression.

FIGURE 16.11   Procedure of OSDrC manufacturing process [83].

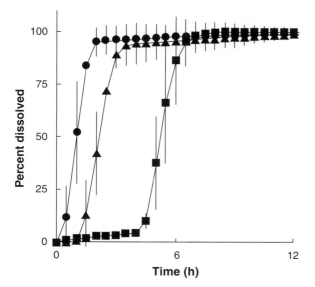

FIGURE 16.12 Effect of the outer layer thickness on the release of acetaminophen from OSDrC. The thickness of outer layer was (●) 0.5 mm, (▲) 1.0 mm, and (■) 2.0 mm [83].

The upper and side layers are added to form a whole tablet after the third compression.

OSDrC is applicable in controlled release formulation. It has the flexibility of including multiple cores with any shapes and varying the position of those cores. The inner core and outer layer can be formulated in a specific way to achieve a desired release profile such as sustained release, delayed and pulsatile release, or delayed and sustained release. Ozeki et al. have demonstrated that the delay time of a dry-coated acetaminophen tablet is dependent on the thickness of the outer hypromellose layer [83]. As shown in Figure 16.12, the thickest coating, 2.0 mm, has a lag time of 4.5 h whereas the 0.5 mm thick coating shows a nearly immediate release.

One-step dry-coated tablets have been applied in preparing a colonic delivery system for 5-fluorouracil (5-FU) with the mixture of Eudragit L100-55 and chitosan as the outer layer. It was shown by *in vitro* that the outer layer provided a sufficient delay for the colonic delivery of 5-FU [84].

16.6 NEW MATERIALS FOR ORAL CONTROLLED RELEASE

The advancement of oral controlled release relies not only on the invention of novel processes but also on the development of new materials. New materials with superior thermal stability are more desirable for formulation development using melt extrusion and injection molding. Novel excipients having unique hydration or erosion behavior may provide flexibility for crafting various release modes. However, the progress of new material invention is less significant than that of new processes innovation. The introduction of a material with new chemical entity is as challenging as the development of a new drug. It needs substantial safety data to pass the huge regulatory hurdle. On the other hand, pharmaceutical industry is reluctant to accept new materials to their commercial products if the performance is not superior and the usage is not necessary. Improvement of existing materials through physical changes is a more popular option.

Major excipient vendors like Dow, BASF, ISP, and FMC biopolymer are key players in innovating new material for oral controlled release. Their main strategy is to improve their flagship products in the aspect of particle size, compressibility, flow properties, and easiness of use. Ethylcellulose is widely used as a film coating agent for controlled release. Its regular grade is usually granular with the average particle size of 250 μm. Dow has developed a micronized version of ethylcellulose, called Ethocel FP. It can be used for the formulation of matrix tablets by direct compression. A direct compression grade of Methocel has also been developed by Dow to improve flow and eliminate the wet granulation step. A type of very low viscosity hypromellose, Methocel VLV from Dow, has been introduced for high productivity tablet coating applications. Kollidon SR, a combination of polyvinyl acetate and polyvinyl pyrrolidone, is desirable for direct compressed and nonerodible tablets. Kollidon SR has high dry binding capacity, which is suitable for developing porous floating systems.

There are interests in using modified starches as controlled release agents. Starch is a polysaccharide carbohydrate consisting of a large number of glucose monosaccharide. Starch is a main nutrient for human and a common pharmaceutical excipient. It is a multifunctional excipient in tablet and capsule formulations. Starch can work as a binder, disintegrant, lubricant, or flow aid. Many researches have showed that modified starch can be used as a hydrophilic polymer for matrix tablets. High-amylase carboxymethyl starch is formed by chemically modifying hydroxyl group of amylase by an etherification process. Both high-amylase carboxymethyl starch and high-amylase sodium carboxymethyl starch are suitable for developing matrix tablets [85–87]. High-amylose sodium carboxymethyl starch, produced by spray drying, was shown to have desirable properties as sustained drug release tablet excipient for direct compression. A cross-linked high amylase, Contramid, developed by Labopharm, is claimed to be desirable for developing controlled release formulations with high drug loading. Contramid is a free-flowing, highly compressible powder. A long-lasting, uniform surface membrane is formed after contramid is wetted, which helps control the release of orally administered drugs under a broad range of in-body conditions. Highly substituted starch acetate has been also introduced as a matrix-forming excipient for oral controlled delivery [88]. Release profile from the tablets could be easily adjusted over a very wide range for various drugs.

On the other hand, there are ongoing efforts in exploring some natural food ingredients as controlled release agents. Mucilage from *Hibiscus rosasinensis* Linn, a hydrophilic excipient, has been demonstrated to be applicable for the development of sustained release tablets [89]. Pectin and alginate are natural polysaccharides and have been used in food and beverage industries for many years. Pectin and alginate tend to form a complex with metal ions. The complex can be used as a matrix or membrane for the controlled delivery of drugs [90]. It has been reported that wax materials such as Dynasan 118 can be used a matrix-forming excipient. Cellulose acetate, an acetate ester of cellulose fiber, has been introduced as a pharmaceutical ingredient by Eastman Kodak. Cellulose acetate used to be a key ingredient for film base and magnetic tape. Data from Eastman Kodak suggest that cellulose acetate can be used as a controlled release agent for matrix tablets.

16.7   SUMMARY

As discussed in above paragraphs, oral controlled release is being advanced in many frontiers. New technologies such as melt extrusion, injection molding, printing technologies, and dry coating provide a great opportunity and flexibility for formulation scientists to design and develop oral controlled release formulations. Controlled release delivery of poorly soluble compounds could be pursued through several approaches with combination of solubilization and release modification. New formulation designs together with computer modeling help to achieve desired release profiles for drugs with different properties. Early promises have been shown for oral delivery of proteins and vaccines. It is optimistic that oral dosage form of biopharmaceuticals including proteins, peptides, vaccines, and nucleotides will become a reality eventually with the advancement of new technologies and new materials.

REFERENCES

1. Rao VM, Haslam JL, Stella VJ. Controlled and complete release of a model poorly water-soluble drug, prednisolone, from hydroxypropyl methylcellulose matrix tablets using (SBE)7m-β-cyclodextrin as a solubilizing agent. *J. Pharm. Sci.* 2001;90:807–816.

2. Sotthivirat S, Haslam JL, Stella VJ. Controlled porosity-osmotic pump pellets of a poorly water-soluble drug using sulfobutylether-beta-cyclodextrin, (SBE) 7M-beta-CD, as a solubilizing and osmotic agent. *J. Pharm. Sci.* 2007; 96:2364–2367.

3. Lee KR, Kim EJ, Seo SW, Choi HK. Effect of poloxamer on the dissolution of felodipine and preparation of controlled release matrix tablets containing felodipine. *Arch. Pharm. Res.* 2008;31:1023–1028.

4. Mehramizi A, Alijani B, Pourfarzib M, Dorkoosh FA, Rafiee-Tehrani M. Solid carriers for improved solubility of glipizide in osmotically controlled oral drug delivery system. *Drug Dev. Ind. Pharm.* 2007;33:812–823.

5. Hong SI, Oh SY. Dissolution kinetics and physical characterization of three-layered tablet with poly(ethylene oxide) core matrix capped by Carbopol. *Int. J. Pharm.* 2008;356: 121–129.

6. Tanaka N, Imaia K, Okimotob K, Uedaa S, Tokunagab Y, Ohikea A, Ibukic R, Higakid K, Kimura T. Development of novel sustained-release system, disintegration-controlled matrix tablet (DCMT) with solid dispersion granules of nilvadipine (I). *J. Control. Release* 2005;108:386–395.

7. Tanaka N, Imaia K, Okimotob K, Uedaa S, Tokunagab Y, Ohikea A, Ibukic R, Higakid K, Kimura T. Development of novel sustained-release system, disintegration-controlled matrix tablet (DCMT) with solid dispersion granules of nilvadipine (II): *in vivo* evaluation. *J. Control. Release* 2006; 112:51–56.

8. Pederson AV. Erosion-based drug delivery. *Manuf. Chem.* 2006; 11:1–6.

9. Washington N, Wilson CG. Erosion-based delivery. *Drug Deliv. Technol.* 2006;10:71–74.

10. Andrews GP, Laverty TP, Jones DS. Mucoadhesive polymeric platforms for controlled drug delivery. *Eur. J. Pharm. Biopharm.* 2009;71:505–518.

11. Tao Y, Lu Y, Sun Y, Gu B, Lu W, Pan J. Development of mucoadhesive microspheres of acyclovir with enhanced bioavailability. *Int. J. Pharm.* 2009;378:30–36.

12. Chavanpatil MD, Jain P, Chaudhari S, Shear R, Vavia PR. Novel sustained release, swellable and bioadhesive gastroretentive drug delivery system for ofloxacin. *Int. J. Pharm.* 2006; 316:86–92.

13. Varshosaz J, Tavakoli N, Roozbahani F. Formulation and *in vitro* characterization of ciprofloxacin floating and bioadhesive extended-release tablets. *Drug Deliv.* 2006;13:277–285.

14. Deshpande MC, Venkateswarlu V, Babu RK, Trivedi RK. Design and evaluation of oral bioadhesive controlled release formulations of miglitol, intended for prolonged inhibition of intestinal alpha-glucosidases and enhancement of plasma glucagon like peptide-1 levels. *Int. J. Pharm.* 2009;380:16–24.

15. Bigucci F, Luppi B, Monaco L, Cerchiara T, Zecchi V. Pectin-based microspheres for colon-specific delivery of vancomycin. *J. Pharm. Pharmacol.* 2009;61:41–46.

16. Grass GM. Simulation models to predict oral drug absorption from *in vitro* data. *Adv. Drug Deliv. Rev.* 1997;23:199–219.

17. Norris DA, Leesman GD, Sinko PJ, Grass GM. Development of predictive pharmacokinetic simulation models for drug discovery. *J. Control. Release* 2000;65:55–62.

18. Agoram B, Woltosz WS, Bolger MB. Predicting the impact of physiological and biochemical processes on oral drug bioavailability. *Adv. Drug Deliv. Rev.* 2001;50:S41–S67.

19. Lukacova V, Woltosz WS, Bolger MB. Prediction of modified release pharmacokinetics and pharmacodynamics from *in vitro*, immediate release, and intravenous data. *AAPS J.* 2009; 11:323–334.

20. Kuentz M, Nicka S, Parrotta N, Röthlisbergera D. A strategy for preclinical formulation development using GastroPlus™ as pharmacokinetic simulation tool and a statistical screening design applied to a dog study. *Eur. J. Pharm. Sci.* 2006;27:91–99.

21. Okumua A, DiMasob M, Löbenberg R. Computer simulations using GastroPlus™ to justify a biowaiver for etoricoxib solid oral drug products. *Eur. J. Pharm. Biopharm.* 2009;72:91–98.

22. Parrott N, Lave T. Prediction of intestinal absorption: comparative assessment of GASTROPLUS and IDEA. *Eur. J. Pharm. Sci.* 2002;17:51–61.

23. Jamei M, Turner D, Yang J, Neuhoff S, Polak S, Rostami-Hodjegan A, Tucker G. Population-based mechanistic prediction of oral drug absorption. *AAPS J.* 2009;11:225–237.

24. Polak S, Jamei M, Turner D. Prediction of the *in vivo* behaviour of modified release formulations of metoprolol from *in vitro* dissolution profiles using the ADAM model (SimCYP®v8). *10th European ISSX Meeting*, May 2008, pp. 18–21.

25. Langer R, Peppas N. Advances in biomaterials, drug delivery, and bionanotechnology. *AIChE J.* 2003;49:2990–3006.

26. Malik DK, Baboota S, Ahuja A, Hasan S, Ali J. Recent Advances in protein and peptide drug delivery systems. *Curr. Drug Deliv.* 2007;4:141–151.

27. Sinha V, Singh A, Kumar RV, Singh S, Kumria R, Bhinge J. Oral colon-specific drug delivery of protein and peptide drugs. *Crit. Rev. Ther. Drug Carrier Syst.* 2007;24:63–92.

28. Morishita M, Aoki Y, Sakagami M, Nagai T, Takayama K. *In situ* ileal absorption of insulin in rats: effects of hyaluronidase pretreatment diminishing the mucous/glycocalyx layers. *Pharm. Res.* 2004;21:309–316.

29. Aoki Y, Morishita M, Asai K, Akikusa B, Hosoda S, Takayama K. Region-dependent role of the mucous/glycocalyx layers in insulin permeation across rat small intestinal membrane. *Pharm. Res.* 2005;22:1854–1862.

30. Aoki Y, Morishita M, Asai K, Akikusa B, Hosoda S, Takayama K. Region-dependent role of the mucous/glycocalyx layers in insulin permeation across rat small intestinal membrane. *Pharm. Res.* 2005;22:1854–1862.

31. Hamman JH, Enslin GM, Kotzé AF. Oral delivery of peptide drugs: barriers and developments. *BioDrugs* 2005;19:165–177.

32. Hosny E, Al-Shora H, Elmazar M. Oral delivery of insulin from enteric-coated capsules containing sodium salicylate: effect on relative hypoglycemia of diabetic beagle dogs. *Int. J. Pharm.* 2002;237:71–76.

33. Sinha VR, Singh A, Kuma RV, Singh S, Kumria R, Bhinge JR. Oral colon-specific drug delivery of protein and peptide drugs. *Crit. Rev. Ther. Drug Carrier Syst.* 2007;24:63–92.

34. Morishita M, Morishita I, Takayama K, Machida Y, Nagai T. Site dependent effect of aprotinin, sodium caprate, Na_2EDTA and sodium glycocholate on intestinal absorption of insulin. *Biol. Pharm. Bull.* 1993;16:68–72.

35. Lueben HL, Lehr CM, Rentel CO, Noach AB, Boer AG, Verhoel JC. Bioadhesive polymers for the peroral delivery of peptide drugs. *J. Control. Release* 1994;29:329–338.

36. Tozaki H, Nishioka J, Komoike J, Okada N, Fujita T, Muranishi S. Enhanced absorption of insulin and (Asu(1,7)) eel-calcitonin using novel azopolymer-coated pellets for colon-specific drug delivery. *J. Pharm. Sci.* 2001;90:89–97.

37. Rick S. Oral protein and peptide drug delivery. In Binghe W, Teruna S, Richard AS, editors. *Drug Delivery: Principles and Applications*, Wiley Interscience, New Jersey, 2005, p. 189.

38. Aungst B. Intestinal permeation enhancers. *J. Pharm. Sci.* 2000;89:429–442.

39. Lecluyse EL, Sutton SC. *In vitro* models for selection of development candidates. Permeability studies to define mechanisms of absorption enhancement. *Adv. Drug Deliv. Rev.* 1997;23:163–183.

40. Dong ZL, Lecluyse EL, Thakker DR. Dodecylphosphocholine-mediated enhancement of paracellular permeability and cytotoxicity in Caco-2 cell monolayers. *J. Pharm. Sci.* 1999;88:1161–1168.

41. Sood A, Panchagnula R. Peroral route: an opportunity for protein and peptide drug delivery. *Chem. Rev.* 2001;101:3275–3303.

42. Noha NS, Natalie DE, Alessio F. Tight junction modulation and its relationship to drug delivery. *Adv. Drug. Deliv. Rev.* 2006;58:15–28.

43. Karyekar CS, Fasano A, Raje S. Zonula Occludens toxin increases the permeability of molecular weight markers and chemotherapeutic agents across the bovine brain microvessel endothelial cells. *J. Pharm. Sci.* 2003;92:414–423.

44. Fasano A, Uzzau S. Modulation of intestinal tight junctions by zonula occludens toxin permits enteral administration of insulin and other macromolecules in an animal model. *J. Clin. Invest.* 1997;99:1158–1164.

45. Peppas NA, Thomas JB, McGinty J. Molecular aspects of mucoadhesive carrier development for drug delivery and improved absorption. *J. Biomater. Sci. Polym. Ed.* 2009;20:1–20.

46. Mehta NM. Oral delivery and recombinant production of peptide hormones. *Biopharm. Int.* 2004;6:1–6.

47. Ziv E, Kidron M, Raz I, Krausz M, Blatt Y, Rotman A, Bar-On H. Oral administration of insulin in solid form to nondiabetic and diabetic dogs. *J. Pharm. Sci.* 1994;83:792–794.

48. Damgé C, Reis CP, Maincent P. Nanoparticle strategies for the oral delivery of insulin. *Expert Opin. Drug Deliv.* 2008;5:45–68.

49. Kipnes M, Dandona P, Tripathy D, Still JG, Kosutic G. Control of postprandial plasma glucose by an oral insulin product (HIM2) in patients with type 2 diabetes. *Diabetes Care* 2003;26:421–426.

50. Hashimoto M, Takada K, Kiso Y, Muranishi S. Synthesis of palmitoyl derivatives of insulin and their biological activities. *Pharm. Res.* 1989;6:171–176.

51. Kamei N, Morishita M, Chiba H, Kavimanda NJ, Peppas NA, Takayama K. Complexation hydrogels for intestinal delivery of interferon β and calcitonin. *J. Control. Release*, 2009;134:98–102.

52. Polovic N, Velickovic TC. Novel formulations for oral allergen vaccination. *Recent Pat. Infalmm. Allergy Drug Discov.* 2008;2:215–221.

53. Vipond J, Cross ML, Lambeth MR, Clark S, Aldwell FE, Williams A. Immunogenicity of orally-delivered lipid-formulated BCG vaccines and protection against Mycobacterium tuberculosis infection. *Microbes Infect.* 2008;10:1577–1581.

54. Nayak B, Panda AK, Ray P, Ray AR. Formulation, characterization and evaluation of rotavirus encapsulated PLA and PLGA particles for oral vaccination. *J. Microencapsul.* 2008;6:1–12.

55. Saha DWY. Therapeutic potential of RNA interference for neurological disorders. *Life Sci.* 2006;79:1773–1780.

56. Crowe S. Suppression of chemokine receptor expression by RNA interference allows for inhibition of HIV-1 replication. *AIDS* 2003;17 (Suppl 4): 103–1055.

57. Hogrefe RI, Lebedev AV, Zon G, Pirollo KF, Rait A, Zhou Q, Yu W, Chang EH. Chemically modified short interfering hybrids (siHYBRIDS): nanoimmunoliposome delivery *in vitro* and *in vivo* for RNAi of HER-2. *Nucleosides Nucleotides Nucleic Acids* 2006;25:889–907.

58. Kumar P, Ban HS, Kim SS, Wu H, Pearson T, Greiner DL, Laouar A, Yao J, Haridas V, Habiro K, Yang YG, Jeong JH, Lee KY, Kim YH, Kim SW, Peipp M, Fey GH, Manjunath N, Shultz LD, Lee SK, Shankar P. T cell-specific siRNA delivery suppresses HIV-1 infection in humanized mice. *Cell* 2008;134:577–586.

59. Kim WJ, Kim SW. Efficient SiRNA delivery with non-viral polymeric vehicles. *Pharm. Res.* 2009;26:657–666.

60. Bumcrot D, Manoharan M, Koteliansky V, Sah DW. RNAi therapeutics: a potential new class of pharmaceutical drugs. *Nat. Chem. Biol.* 2006;2:711–719.

61. Chokshi R, Zia H. Hot-melt extrusion technique: a review. *Iranian J. Pharm. Res.* 2004;3:3–16.

62. Coppens KA, Hall MJ, He VY, Koblinski BD, Larsen PS, Read MD, Shrestha V. Hypromellose and polyethylene oxide blends in hot melt extrusion. *Dow Chemical Form* 2006, No#198-02152.

63. Zhu Y, Shah N, Waseem MA, Martin I, McGinity J. Controlled release of a poorly water-soluble drug from hot-melt extrudates containing acrylic polymers. *Drug Dev. Ind. Pharm.* 2006; 32:569–583.

64. Kang JJ, Lee SH, Jung TS, Heo YM. 200 μm-class polymeric microneedle fabricated by a micro injection molding technique. *Nanotechnology* 2007;3:371–374.

65. Quinten T, Beer TD, Vervaet C, Remon JP. Evaluation of injection moulding as a pharmaceutical technology to produce matrix tablets. *Eur. J. Pharm. Biopharm.* 2009; 71:145–154.

66. Anonymous. HP takes inkjet technology to medical device market. HP news room URL: http://h41131.www4.hp.com/vn/en/stories/hp-takes-inkjet-technology-to-medical-device-market22928.html.

67. Igawa K, Chung UI, Tei Y. Custom-made artificial bones fabricated by an inkjet printing technology. *Clin. Calcium* 2008;18:1737–1743.

68. Tarcha PJ, Verlee D, Hui HW, Setesak J, Antohe B, Radulescu D, Wallace D. The application of ink-jet technology for the coating and loading of drug-eluting stents. *Ann. Biomed. Eng.* 2007;35:1791–1799.

69. Xu T, Rohozinski J, Zhao W, Moorefield EC, Atala A, Yoo JJ. Inkjet-mediated gene transfection into living cells combined with targeted delivery. *Tissue Eng. A* 2009;15:95–101.

70. Mironov V, Kasyanov V, Drake C, Markwald RR. Organ printing: promises and challenges. *Regen. Med.* 2008;3(1): 93–103.

71. Campbell PG, Weiss LE. Tissue engineering with the aid of inkjet printers. *Expert Opin. Biol. Ther.* 2007;7:1123–1127.

72. Abukawa H, Papadaki M, Abulikemu M, Leaf J, Vacanti JP, Kaban LB, Troulis MJ. The engineering of craniofacial tissues in the laboratory: a review of biomaterials for scaffolds and implant coatings. *Dent. Clin. North Am.* 2006;50:205–216.

73. Lin S, Chao PY, Chien YW, Sayani S, Kuma S, Mason M, Wes T, Yang A, Monkhouse D. *In vitro* and *in vivo* evaluations of biodegradable implants for hormone replacement therapy: effect of system design and PK-PD relationship. *AAPS PharmSciTech* 2001;2:E16.

74. Sachs EM, Cima MJ, Williams P, Brancazio D, Cornie J. Three-dimensional printing: rapid tooling and prototypes directly from a CAD model. *J. Eng. Ind.* 1991;114:481–488.

75. Yu DG, Yang XL, Huang WD, Liu J, Wang YG, Xu H. Tablets with material gradients fabricated by three-dimensional printing. *J. Pharm. Sci.* 2007;96:2446–2456.

76. Katstra WE, Palazzolo RD, Rowe CW, Giritlioglu B, Teung P, Cima MJ. Oral dosage forms fabricated by three dimensional printing. *J. Control. Release* 2000;66:1–9.

77. Obara S, Maruyama N, Nishivama Y, Kokubo H. Dry coating: an innovative enteric coating method using a cellulose derivative. *Eur. J. Pharm. Biopharm.* 1999;47:51–59.

78. Cerea M, Zheng W, Young C, McGinity JW. A novel powder coating process for attaining taste masking and moisture protective films applied to tablets. *Int. J. Pharm.* 2004; 279:127–139.

79. Zhu J, Luo YF, Ma Y, Zhang H. Direct coating solid dosage forms using powdered materials. US Patent Application 20070128274, 2007.

80. Ghimirea M, McInnesa FJ, Watsona DG, Mullena AB, Stevens HNE. *In vitro/in vivo* correlation of pulsatile drug release from press-coated tablet formulations: a pharmacoscintigraphic study in the beagle dog. *Eur. J. Pharm. Biopharm.* 2007; 67:515–523.

81. Gohel MC, Parikh RK, Nagori SA, Dabhi MR. Design of a potential colonic drug delivery system of mesalamine. *Pharm. Dev. Technol.* 2008;13:447–456.

82. Lin SY, Li MJ, Lin KH. Hydrophilic excipients modulate the time lag of time-controlled disintegrating press-coated tablets. *AAPS PharmSciTech* 2004;5:e54.

83. Ozeki Y, Ando M, Watanabe Y, Danjo K. Evaluation of novel one-step dry-coated tablets as a platform for delayed-release tablets. *J. Control. Release* 2004;95:51–60.

84. Sakurai M, Ozeki Y, Ando M, Okamoto H, Danjo K. Drug delivery to the colon using novel one-step, dry-coated tablets (OSDrC). *Yakugaku Zasshi* 2008;128:951–957.

85. Moghadam SH, Wang HW, Saddar El-Leithy E, Chebli C, Cartilier L. Substituted amylose matrices for oral drug delivery. *Biomed. Mater.* 2007;2:S71–S77.

86. Nabaisa T, Brouilleta F, Kyriacosc S, Mrouehc M, Amores da Silvad P, Batailleb B, Cheblia C, Cartiliera L. High-amylose carboxymethyl starch matrices for oral sustained drug-release: *in vitro* and *in vivo* evaluation. *Eur. J. Pharm. Biopharm.* 2007;65:371–378.

87. Brouilleta F, Batailleb B, Cartiliera L. High-amylose sodium carboxymethyl starch matrices for oral, sustained drug-release: formulation aspects and *in vitro* drug-release evaluation. *Int. J. Pharm.* 2008;356:52–60.

88. Korhonen O, Raatikainen P, Harjunen P, Nakari J, Suihko E, Peltonen S, Vidgren M, Paronen P. Starch acetates—multifunctional direct compression excipients. *Pharm. Res.* 2000; 12:1138–1143.

89. Jani GK, Shah DP. Evaluation of mucilage of *Hibiscus rosasinensis* Linn as rate controlling matrix for sustained release of diclofenac. *Drug Dev. Ind. Pharm.* 2008; 34:807–816.

90. Novosel'skaya IL, Vorropaeva NL, Semenova LN, Rashidova SS. Trends in the science and applications of pectins. *Chem. Nat. Compd.* 2000;36:1–10.

17

ORAL CONTROLLED DRUG DELIVERY: QUALITY BY DESIGN (QbD) Approach to Drug Development

Shailesh K. Singh[1], Thirunellai G. Venkateshwaran[2], and Stephen P. Simmons[2]

[1]Wyeth Research, Pearl River, NJ, USA
[2]Wyeth Pharmaceuticals, Collegeville, PA, USA

17.1 INTRODUCTION

Over the past few years, pharmaceutical companies have been facing an increasingly difficult economic climate. An increase in the regulatory hurdles for the approval of new molecular entities, patent expirations and increased healthcare costs have resulted in more focus in the costs associated with the manufacturing and development of pharmaceuticals. It has been estimated that many pharmaceutical processes operate at 2.5–4.5 sigma quality levels, but resource intensive pharmaceutical company quality systems achieve 5 sigma quality levels by sorting, reworking, and so on to prevent defective product leaving the factory [1]. During the heydays of the pharmaceutical industry, there was lesser focus on the yields, number of defects, etc., and the quality organizations of the companies were more focused on compliance based on inspection of the final products. Traditional development focused on the formulation and the delivery of the product to the next phase of the clinical studies. Most of the formulation development tended to be iterative and empirically designed. Thus, changes were driven by the need to modify the process during scale-up or due to the formulation failing to meet the desired shelf life of the product. During phase 3, changes were kept to a minimum to avoid the need for expensive bioequivalence studies to bridge between the Clinical Trial Material (CTM) and the commercial product. Thus, manufacturing processes were fixed and the quality of the product was measured by end product testing (commonly referred to as quality by testing). In this case,

quality is not built in to the product and is achieved by end product testing. This approach is inefficient and does not facilitate continual improvement. In the past, there also existed a notion that the regulatory processes and requirements prohibited manufacturing enhancements, which in turn prevented the modernization of the pharmaceutical industry. The initiation of the cGMPs for the 21st Century Initiative [2] and the publication of the Process Analytical Technology guidance [3] in 2004 by the FDA paved the way for the modernization of the pharmaceutical industry.

In July 2003, the experts from the three regional grouping (USA, EU, and Japan) working on the Quality Topics within ICH (International Conference on Harmonization of Technical Requirements for Registration of Pharmaceuticals for Human Use) created a vision for the future pharmaceutical quality system (Figure 17.1). This vision recognizes that regulatory agencies will also benefit from this initiative as it will enable them to prioritize and allocate resources more efficiently, and patients will also benefit from improved access to medicines and an enhanced assurance of quality. Since the meeting in 2003, the key framework guidance documents ICH Q8(R1) Pharmaceutical Development (now incorporating the Annex finalized in November 2008), ICH Q9 Quality Risk Management, and ICH Q10 Pharmaceutical Quality System have been developed [4–6]. In the EU, the EMEA PAT team have published a reflection paper suggesting how PAT information might be included in a common technical document (CTD) regulatory submission [7]. If the principles described in the ICH Q8/Q9 and Q10 guidance

Oral Controlled Release Formulation Design and Drug Delivery: Theory to Practice, Edited by Hong Wen and Kinam Park
Copyright © 2010 John Wiley & Sons, Inc.

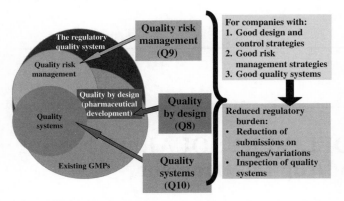

FIGURE 17.1 ICH vision for the future pharmaceutical quality system.

documents are implemented together in a holistic manner, then an effective system that emphasizes a harmonized science and risk-based approach to product development and maintenance is in place. This provides an even greater (quality) assurance that the patient will receive product that meets the critical quality attributes (CQA). (Figure 17.2).

"Quality by design (QbD)," although a new concept to the pharmaceutical industry, is a tried and tested concept that has been in existence for quite a few years and has been extensively applied in the automotive, the semiconductor, and the petrochemical industry. The concept of building quality into products has been extensively documented by Deming and Juran. The common theme of the various initiatives is "planning for quality," that is, building quality into the products compared to the traditional paradigm of testing the product to ensure quality. The Juran trilogy concept identifies quality planning, quality control, and quality improvement as three fundamental aspects of quality planning [8]. Quality planning is the process of identifying the needs of the customer and designing the product and the process to meet the needs of the customer. Figure 17.3 provides an example of a quality planning roadmap for a sustained release product. This map serves as the founda-

tion for the development of the product/process and is the primary requisite for QbD.

In the QbD paradigm, it is possible to use knowledge and data from product development studies to create a design space within which a continuous improvement can be implemented, but the management for such change control lies with the industry without need for regulatory approval. This will enable the use of the latest pharmaceutical science and engineering principles throughout the life cycle of the product. Prior knowledge and risk assessment are central to the understanding of the product and the process. In this paradigm, the processes can be adjusted to accommodate for the variability in the input material, thereby resulting in a product that is always of consistent quality. This can be accomplished only with enhanced process/product understanding. Tools such as process analytical technology (PAT) and design of experiments (DoE), in combination with enablers such as quality risk management (QRM), knowledge management, and so on, help in the enhanced understanding of the process/product. Alignment of quality systems to manage the concepts of QbD will enable the implementation of QbD in the manufacturing sites and thereby facilitate regulatory flexibility, an outcome of the process/product understanding.

The principles of QbD apply to all types of drugs and drug delivery systems. However, the complexity of the drug/ delivery platform and development model will influence the application of QbD [9]. The application of quality by design to various solid dosage forms is usually very similar as the manufacturing platforms and the, excipients used in the development of dosage forms are standard.

The objective of controlled delivery is to deliver a pharmacologically active agent in a predetermined, predictable, and reproducible manner. Controlled release (CR) denotes systems that are able to provide actual therapeutic control, whether this is of temporal nature, spatial nature, or both. These systems typically attempt to control drug concentration in the target tissue. In contrast, sustained release constitutes any dosage form that provides medication over an extended period of time. On the other hand, delayed release systems maintain the drug within the dosage form for a certain period of time before release starts. Commonly, the release rate of the drug is not altered once the release starts (e. g., enteric-coated beads).

Controlled release systems offer enhance patient compliance due to reduced dosing, a decreased incidence of side effects, and greater selectivity of pharmacological activity, improved bioavailability, *in vivo* stability for some drugs, and reduced cost of goods. At the same time, we also have to be careful with some of the disadvantages of controlled release dosage forms such as potential for dose dumping, less accurate dose adjustment, increased potential for first-pass metabolism, dependence on the residence time in the GI tract, and possible delay in onset of action.

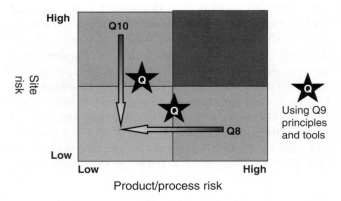

FIGURE 17.2 Synergistic impact of ICH Q8, Q9, and Q10 to consumer.

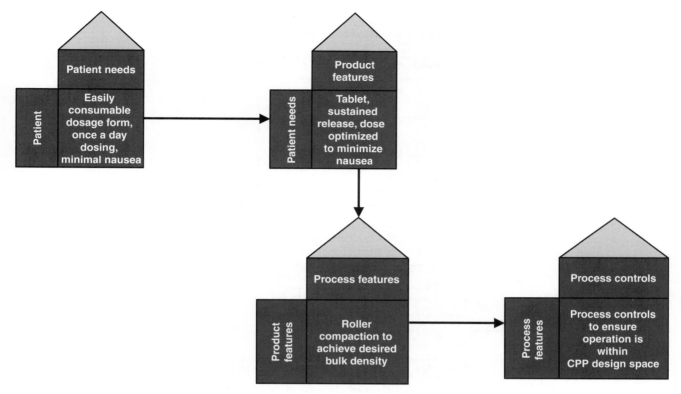

FIGURE 17.3 Quality planning roadmap for a sustained release tablet.

From a development perspective, oral controlled release products are no different from the traditional formulation and/or the IR products. However, more scrutiny needs to be given to the release mechanism and the associated complexities in the technology used therein. The choice of excipients, the type of delivery system, active pharmaceutical ingredient (API) properties, manufacturing platform, and so on will influence the quality attributes that are critical to safety, efficacy, and manufacturability (elements of the target product quality profile). Subsequent discussions will focus on some of the typical elements that will be important in the application of QbD to oral controlled release products. In particular, the factors influencing the performance of the drug product will be highlighted with examples where possible.

The next section elaborates what are the basic elements of the science and risk-based approach to pharmaceutical development. This enables us to gain a more systematic, enhanced understanding of the product and process under development.

17.2 ENABLERS OF QUALITY BY DESIGN

Knowledge management and quality risk management are two of the primary enablers of QbD. They play a critical role both in development and in the implementation of QbD. They are instrumental in achieving product realization, establish-

ing and maintaining a state of control, and lastly facilitating continual improvement [10]. A brief description of the two enablers and their utility is provided in the following sections.

17.2.1 Quality Risk Management

Quality risk management (QRM) is a key enabler for the development and application of QbD. During development, it enables resources to be focused on the perceived critical areas that affect product and process. It is one of the tools that provide a proactive approach to identifying, scientifically evaluating, and controlling potential risks to quality. It also facilitates continual improvement in the product and process performance throughout the product life cycle.

Quality risk assessment in combination with the prior scientific and engineering knowledge, helps in identifying material attributes and process parameters that can have a potential effect on the product critical quality attributes (CQAs) [5]. The tools used in risk assessment can help to identify and rank or prioritize parameters/attributes that have a potential impact on the product CQAs. As one advances through the various phases of development and gains process/ product knowledge, risk assessment enables the differentiation of true versus perceived risks. This in turn enables the efforts to be focused on the high-risk areas and develop strategies to mitigate them. A pictorial representation of the

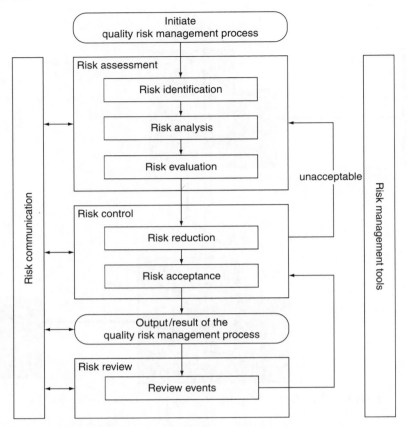

FIGURE 17.4 Risk management during drug development.

risk management process is provided in Figure 17.4. A number of tools are available for QRM. The choice of the tool is dictated by the situation at hand. Irrespective of the risk assessment tool used, the general outcome can be classified into acceptable risk, potential risk, significant risk, and unacceptable risk. A good review with examples of QRM in pharmaceutical development is provided [11].

17.2.2 Knowledge Management

Product and process knowledge management is an essential component of quality by design and must be managed from development through the commercial life of the product, including discontinuation. It is a systematic approach to acquiring, analyzing, storing, and disseminating information related to products, processes, and components. This also emphasizes on a transparency of information from development to commercial and vice versa. Prior knowledge comprises previous experience and understanding of what has been successful or unsuccessful, and recognition of issues, problems, or risks that may occur and need to be addressed.

Examples of prior knowledge include the following:

- Knowledge gained about the drug substance and/or drug product from early development work

- Knowledge of the properties of materials and components used in other products and the variability of associated physicochemical and functional properties
- Knowledge from related products, manufacturing processes, test methods, equipment, systems, and so on
- Knowledge from previous product and process development projects, both successful and unsuccessful
- Knowledge from the published scientific literature
- Experience from the manufacture and testing of related dosage forms and products, including deviations, customer complaints, etc.

Prior knowledge, be it from the literature, experience with prior compounds/processes that are similar provides the basis for the initial risk assessments and influences a number of decisions that are made. Therefore, a good understanding of the documentation relating to prior knowledge referenced in risk assessments and DoEs is a must for the success of QbD.

17.3 ELEMENTS OF QUALITY BY DESIGN

ICH Q8(R2): Pharmaceutical Development Annex discusses the various elements of quality by design. These in

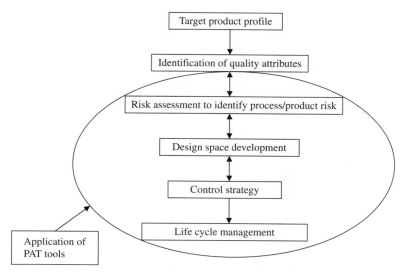

FIGURE 17.5 Elements of quality by design.

combination with the enablers form the fundamental basis for the QbD approach to development. Figure 17.5 provides a pictorial representation of the typical elements of QbD. This section describes the various elements in detail and provides examples of the elements for controlled release (CR) products.

17.3.1 Identifying a Quality Target Product Profile(QTPP)

The quality target product profile (QTPP) as defined in ICH Q8(R1) [12, 13] is a summary of the quality characteristics or attributes of a drug product that ideally will be achieved and thereby ensure the safety and efficacy of a drug product. The QTPP forms the basis of design for the development of the product and is developed with the end in mind. It is both prospective, that is, it describes the goals for the development team, and dynamic, that is, the QTPP may be updated or revised at various stages of development as new information is obtained during the development process. The FDA has published a guidance defining the Target Product Profile (TPP) [14], that focuses on the consumer (patient) and the desired product label. The QTPP is a subset of the TPP and is more oriented towards the chemistry, manufacturing and controls (CMC) aspects of development.

For CR formulations, there may be more than one method to formulate the product to meet the same QTPP. Thus, between brand and generic the formulation design may be different but the QTPP could be similar. Similarly, reformulation of the product does not mandate a change in the QTPP. The quality attributes that measure the performance for each formulation can be different. QTPP is a quantitative surrogate for aspects of clinical safety and efficacy and the QTPP can be used to design and optimize both the formulation and the process [15–17]. A typical QTPP includes quantitative targets

for attributes such as dissolution, potency, impurities, stability, release, and other product-specific requirements. Depending on the type of controlled release, certain functionality related test(s) could be incorporated if they are important, for example, polymer viscosity, tortuosity, glass transition of composite, and so on. For a generic CR product, bioequivalence is a quality attribute that is of extra significance.

The QTPP should typically include the following:

- Type of CR dosage form and route of administration, dosage form strength(s)
- Appearance, identity, size, ease of administration, patient population
- Container closure
- Therapeutic moiety release or delivery and pharmacokinetic characteristics (e.g., dissolution and aerodynamic performance) appropriate to the drug product dosage form being developed
- Drug product quality criteria (e.g., sterility, purity) appropriate for the intended marketed product, that is, manufacturing considerations

Consideration should also be given to the following when developing the QTPP:

- The clinical usage setting, for example, how and when is the product used? Is the product self-administered or administered by a health care professional? Does the patient have any special needs that may influence the product design?
- Special requirements for the supply chain, for example, cold chain distribution and multi-country packaging.
- Special requirements for customers in particular markets, for example, drug product, raw material, and

component cosmetic appearance requirements for Japan; pack sizes for developing markets; and so on.

Several examples of QTPP can be found in the mock quality overall summary (QOS) presented in the Office of Generic Drugs (OGD) website [18, 19]. Similarly, the European Federation of Pharmaceutical Industries and Associations (EFPIA) Mock P2 was developed to facilitate scientific and regulatory discussion between the industry and the European Federation and the regulatory bodies [20]. This document not only highlights the QTPP but also addresses which of these quality targets are critical based on prior knowledge.

17.3.2 Identification of Critical Quality Attributes

Pharmaceutical development consists of product and process design and development. The TPP provides the basis for the ideal dosage form. While designing a product and process, it is may be important to focus on the clinical performance, manufacturability, and global acceptability of the drug product. In the QbD paradigm, it is imperative that the manufacturing process is capable of accommodating typical variability in the inputs, resulting in a product that always meets the requirements of the QTPP. Figure 17.6 shows the typical flow of activities during the various phases of development. It is important to understand that the activities (process and product development) demonstrated in the figure are not independent and the interrelationship of the material attributes and the process plays a key role in the variability of the process. An understanding of the interrelationships and a prediction of the variability such that the process can be adjusted to keep the output a constant is the ultimate goal of QbD.

17.3.2.1 Critical Quality Attributes (CQA) A critical quality attribute as defined by ICH Q8(R2) is a physical, chemical, biological, or microbiological property or characteristic that should be within an appropriate limit, range, or distribution to ensure the desired product quality.

CQAs are generally associated with raw materials (drug substance, excipients), intermediates (in-process materials), and drug product. Drug product CQAs derived from the QTPP are used to guide the product and process development.

Drug product CQAs are the properties that are important for product performance, that is, the desired quality, safety, and efficacy. Depending on the CR dosage form, these may include the aspects affecting the purity, potency, stability, drug release, microbiological quality, and so on.

CQAs can also include those properties of a raw material that may affect drug product performance or manufacturability. An example of this would be drug substance particle size distribution (PSD) or bulk density that may influence the flow of a granulation and therefore the manufacturability of the drug product. Similarly, the dissolution from a controlled release dosage form is dependent on the particle size of the polymer and the hardness of tablet. In this example, PSD and hardness can be designated as CQA's. They are also commonly referred to as critical material attributes (CMA). (Table 17.1).

17.3.2.2 Quality Attributes Important to the Performance of the Drug Product From a clinical perspective, safety and efficacy (product performance) is of prime importance. Thus, for an oral CR product, it is important to consider attributes that are potential surrogate(s) for performance. This may be drug dissolution/release, potency, polymer concentration, polymer viscosity, glass transition temperature (T_g) of composite, etc., or any other attribute that can either be substituted for drug release or clinical design space.

Importance of Dissolution/Release Dissolution is typically used as a measure of the *in vivo* performance of oral controlled release formulations. During the course of development, product formulation and manufacturing processes are designed to achieve the desired dissolution characteristics for a product. Dissolution may relate product attributes to clinical performance. Discriminatory dissolution methods [21] are highly desirable and serve as

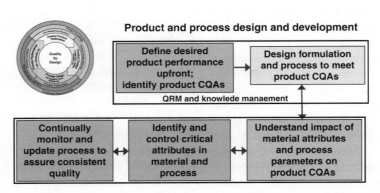

FIGURE 17.6 Relationship between material attributes, process parameters, and critical quality attributes.

TABLE 17.1   Example of a QTPP for a Typical Oral Controlled Release

Summary Quality Target Product Profile and Identification of Critical Quality Attributes for a Typical Oral Controlled Release Product

Quality Attribute	Target	Criticality
Dosage form	Dosage form could be matrix tablet, maximum weight *XX* mg	
Potency	Dosage form label claim	
Dosing	One tablet per dose, once daily	
Pharmacokinetics	For example, controlled release over a period of 12 or 24 h	Related to dissolution
Appearance	Dosage form description	Critical
Identity	Positive for drug name	Critical
Assay	95.0–105.0%	Critical
Impurities	List specified impurities with appropriate limit; unspecified impurities with limit; total impurities with limit	Critical
Water	Current limit (e.g., NMT 1.0%)	Critical/not critical: depending on API sensitivity to moisture
Content uniformity	Meets USP/EP/other pharmacopoeia	Critical
Hardness	NLT × SCU (preferred for film coating) for a tablet	For example, can be critical if related to dissolution
Friability	Current limit (e.g., NMT 1.0%)	
Dissolution	Conforms to USP (e.g., use a 5-point profile or NLT 10% in 0.1 N HCl for enteric coated)	Typically critical
Microbiology	If testing required, meets harmonized ICH criteria	Critical only if drug product supports microbial growth

a good *in vitro* tool to differentiate products that potentially have differences in pharmaceutical attributes (formulation and/or manufacturing processes differences) and therefore may have a different *in vivo* performance. Based on an understanding of how variability in formulations and manufacturing processes impact product performance, dissolution methods and acceptance criteria are established to provide continued assurance of clinical performance. Sometimes, different methods are used for product development and quality control purposes. Dissolution is a powerful tool in evaluating multivariate processes and factors that can affect drug product performance. To understand and identify critical quality attributes of the product and the processing parameters that impact dissolution performance, design of experiments (DoE) should be utilized by taking dissolution failures into consideration. When a reliable prediction of dissolution from another CQA is established and demonstrated in a design space, dissolution testing may not be required as part of specifications if the alternative CQA can be controlled by a process parameter or as an input function. If, for example, in CR drug products, if polymer concentration in the matrix tablet or as part of a functional film coating on beads or tablets can be correlated to the dissolution rate, then, polymer concentration measurements (% w/w solids, polymer concentration, T_g of polymer) using a process analyzer may serve as a control to dissolution. However, there needs to be a mechanistic understanding

and development of risk-based acceptance criteria for each of these situations.

Dissolution has been used as a key test for quality control test and specifications at batch release and shelf life serve as important measures of product quality. A good review of the present challenges and relevance of dissolution testing is provided in the attached reference [22]. The role of dissolution testing in determining the bioavailability/bioequivalence especially for CR products is well established [23]. The Biopharmaceutics Classification System (BCS) assumes that the rate and the mass absorbed are fundamentally related to the concentration (or solubility) and the permeability of the drug. It provides scientific rationale to lower regulatory burden and justify a biowaiver under certain circumstances.

BCS classification has also been used in the approval process of oral controlled release dosage forms. For IR products, most of the absorption window is in the upper small gastro-intestinal tract (GIT), that is, duodenum and jejunum. However, in case of controlled/modified release dosage forms, this may be quite varied especially after administration of dosage in the fasted state. The dosage forms may be exposed to different parts i.e. from the stomach to the colon. The physiology, surface area, pH, and residence in each of these regions varies with the dosage form. The presence of food can also impact the above conditions. Thus, the use of simulated media is important. If the mechanism for release is fairly robust to the changing physiological GI environment, then absorption might not be affected. This may be true for B Class I compounds (high solubility and permeability). In such cases, an IVIVC is possible using a single media approach. As an example, mesalazine is a MR product used for Crohn's disease and ulcerative colitis. To release the drug at the local site of inflammation, it is important to design dosage forms with different release patterns depending on the site of inflammation. A personalized dosage selection can be made based on each patients profile. Testing of the pH of the segment targeted for release may not be sufficient. If the drug is released at sites proximal to the target segment, than it gets prematurely absorbed to the systemic circulation and is no longer locally available to exert an anti-inflammatory effect. In such cases, the use of biorelevant dissolution media is very important. The Bio-Dis (USP apparatus 3) is typically preferred for such media change methods. Thus, depending on the site of inflammation and the combination of release patterns of various dosage forms, a selection of the appropriate dosage form can be made for each specific patient [24]. However, in a large number of studies it may be observed that the release is sensitive to the physiology. In such cases, the absorption is slowed down at the distal end of the small intestine. Also, depending on the nature of the API, for example, a weak base or acid that shows pH dependent solubility, there may be variability in the absorption and slowing down of the absorption process. There are also instances were the gut wall metabolism may also impact the absorption process—particularly the cytochrome P450 3A4 isozyme activity has shown a lot of such potential. Therefore, when extending the BCS model to oral CR products, it is important to understand some of the above issues including gut wall metabolism in the absorption process [25].

The use of sequential media for enteric-coated products is well known when performing simulations. Another key issue with such products is the robust performance, irrespective of whether they are administered in the fasted or fed conditions. Factors such as interaction with meal components; increases in gastric, bile, and pancreatic secretions; and changes in motility pattern can impact the absorption process. In fed state, the residence time is longer in the upper GI for nondisintegrating formulations due to switch in gastric and intestine motility pattern from fasted to fed state. For example, in the case of salbutamol MR, slower release in the fed state may be due to slower hydration of the film coating or a matrix or inhibition of erosion, to name a few possibilities. A greater concern would be if there is a faster release of drug when given with food causing "dose dumping". In such instances, the use of biorelevant dissolution media with Bio-Dis (USP Type 3) would help in better prediction.

There are several approaches using QbD to develop dissolution test, for example, identifying the desired *in vivo* performance, estimating desired therapeutic range and maximum variability to develop a dissolution test, and continue to optimize the test during clinical studies. QbD approaches enable movement within the design space and provide a greater potential for post-approval changes with minima regulatory burden. A validated IVIVC model allows us to predict the impact of changes to a formulation on the *in vivo* performance without having to conduct a bioeqivalence (BE) study [26]. The SUPAC MR guidance discusses in details the application of an IVIVC for various levels of scale-up and postapproval changes [27]. Thus, a level A IVIVC can help in formulation development, establishing biorelevant dissolution specifications and for supporting pre- and postapproval changes that would otherwise require a BE study.

If an IVIVC is established, then it provides an opportunity to utilize dissolution can be used as a surrogate for clinical efficacy. The risk assessment enables the identification of process and product parameters/attributes that impact the clinical safety and efficacy. A key factor in the development of an IVIVC is the development of a bio-relevant method that is discriminatory.

If IVIVC cannot be established, then it become rather challenging to identify the test or attribute that is correlated to clinical safety and efficacy. In such circumstances, a potential approach is to perform a risk assessment that identifies and creates a rank order of the key variables that impact dissolution [28]. Using some of these as possible variants in human study would help understand the impact of such variables *in vivo*. If the results suggest that there is an impact, than it would help define the boundaries or specifications for dissolution. If, however, there is no impact of such changes, a wider dissolution specification can be justified on the basis of the results of the study that demonstrated that the variables have no impact on the *in vivo* behavior [29]. Other risk-based approaches to clinical performance have also been discussed [30, 31].

Impact of Drug Substance on Dissolution/Release   It is also important to understand the impact of drug substance and excipients that can influence the CQAs of the drug product. It is important to understand the influence of physico-chemical properties of the drug substance on the drug product [32].

Crystal form of the drug substance is an important property. It is highly imperative that a comprehensive understanding of crystalline transformations and amorphous to crystalline transitions is undertaken. This will enable the development of a process that is not susceptible to processing challenges as a result of crystalline transformations, and also results in the design of a product that is stable and bioavailable. Several studies discuss the impact of crystalline transformations on the product quality and performance [33]. Biopharmaceutical properties of the active should be assessed for every form for which there is interest in development. Identification of appropriate salt form (salt selection), the suitable polymorphic form etc., is performed during pre-formulation stages of development. This assessment provides the information required to select a solid form, evaluate its developability, processability, and determine its classification based on BCS [34–36]. The BCS guidance helps in identifying class boundaries for solubility, permeability, and dissolution criteria.

Similarly, if PSD of drug substance has an influence on drug product dissolution, then its impact of this variability on the drug product CQA needs to be understood. Other properties such as intrinsic solubility, wettability, and partition coefficient may also influence dissolution or release. The effect of API properties on bioavailability can also be better understood from absorption modeling [37, 38]. A thorough understanding of the stability under various processing conditions as well as the *in vivo* activity of the drug substance including identification of all known impurities is desired. ICH Q6A describes the circumstances when drug product studies are required [39]. These studies help in investigating the potential impact of the drug substance on the drug product performance and help justify the development of suitable specifications.

Since the drug substance and excipients are in intimate contact, it is important to understand the compatibility between them. There are no universal methods to assess drug-excipient compatibility studies. Typically thermal analysis, isothermal calorimetry and HPLC analysis are performed to understand compatibility of various drug product components. The goal is to ensure a stable formulation and get an enhanced understanding of the drug-excipient interactions.

Impact of Excipients on Dissolution/Release For most oral controlled release solid dosage forms, the design and development depends on the material attributes of the active ingredient and the excipient components. The objective is to deliver drug to the patient in the required amount, the required rate, consistently within a batch, from batch to batch, and throughout the shelf life of the product. Typically, it consists of the API, the controlled release polymer (natural or synthetic, sometimes more than one or a combination), and other excipients (filler, binder, release modifier, plasticizer, glidant, lubricant, etc.), each with a specific function depending on the type of dosage form. Most controlled release excipients are based on physical and/or physicochemical characteristics/properties that are critical to the typical use of these polymers, for example, viscosity of polymer, percentage of chemical substituent, T_g of composite, and so on. These physical tests are referred to as functionality related characteristics (FRCs). These FRCs seem to impact the drug product CQA like dissolution. In addition to polymers, certain excipients depending on the type of formulation also serve as release modifiers. This is dependent on the mechanism used for development of the controlled release formula, for example, the use of weak acid or bases as microenvironment pH modifiers for drugs that show pH-dependent solubility, the effect of plasticizer to

modify release from plasticized films, the ratio of certain ingredients, the influence of binders and disintegrants, and/or processing effects on release profiles. The above are examples where interaction of more than one excipient is important to get the desired release profile. Guidance is available that helps to standardize the identification of such raw materials are available [40].

17.3.2.3 Challenges with Controlled Release Excipients

There are several issues and challenges related to availability and characteristics of CR excipients. Recent issues related to contaminated excipients, bioterrorism, counterfeiting, allergens, BSE/TSE, and other cost reduction goals have made it imperative to maintain a control on the supply chain and traceability of these ingredients to improve quality [41]. Ideally, a robust formulation should be able to accommodate all the variabilities associated with API, excipient, and the manufacturing process without compromising the performance. Most oral dosage form designs depend on the material attributes of the active and the excipient components. These may be linked to the drug product CQAs or specifications such as purity, uniformity, dissolution, appearance, and so on. Typically, excipients are selected with the goal to complement the properties of the drug substance. Sometimes, they may have more than one function. These variables can be studied during development, while others manifest themselves during the manufacturing process. Often during manufacturing process, problems emerge unexpectedly during the life cycle of the product. These occurrences are a result of the lack of understanding of the interactive effect of excipient mixtures. The non-availability of methods to understand and quantify the interactive effects is a limiting factor. Typically, pharmaceutical substances are organic solids and are difficult to characterize. Also, in a multicomponent system, properties may be specific for each formulation. Properties such as PSD can vary between lots. Behavior during processing is a complex combination of particulate and powder properties that can get amplified by equipment design and input energy [42].

Some of the sources of variability may be as follows:

a. Source dependent such as lot-to-lot variability from the same manufacturer; different manufacturing sites of the same manufacturer; different manufacturer of the same material using different processes; effects due to aging and degradation; effects due to transportation, handling, and storage;

b. Raw material dependent, such as conditions during growing season; conditions during harvest; variation in growing season year to year; environmental conditions, for example, storm and floods.

It may also be worth noting that most excipient suppliers are chemical industries for which pharma is only a small part of their business. It makes it difficult for them to make a product to a specific need, especially if it is a small part of the business. There also may be lack of understanding of needs.

The regulators are working with the global pharma industry to engage in the continuous quality improvement concepts in manufacturing. Thus, in a QbD/PAT realm, it is desirable that most formulators select well characterized excipients that are manufactured using well-defined controls and are specific to the needs of the pharmaceutical community. It is imperative that the pharmaceutical manu-facturers work in collaboration with excipient manufacturers to define the requirements for various excipients typically used in product design and development. In addition, it is important to communicate the typical requirements for materials that may be required in small scales for development studies. Examples of these include: Materials with a wide range of certain functionalities so that the impact of the variability on COA's can be investigated.

For a mechanistic understanding there may be a need to study the variability of a polymer at the edges of the functionality specification. For e.g. a grade of polymer (hypromellose) has a viscosity range of 100,000 +/- 30,000 cps for the functionality. For example, a grade of polymer (hypromellose) has a viscosity range 100,000 ± 30,000 cps for the functionality. It may be important to communicate with the supplier to ensure that such lots are available for proof of principle [43–47].

The road to this should start with identifying and establishing industrially acceptable methods for measurements and estimation of properties of pharmaceutical excipients (ASTM). Special initiatives with National Institute of Pharmaceutical Education and Research (NIPTE) and Three regions of International Pharmaceutical Excipient Council (TRIPEC) to institute these methods in a harmonized approach are in progress. This will help further to our enhanced understanding of the process and help develop robust formulations.

17.3.2.4 Quality Attributes Impacting the Manufacturability of the Drug Product

Physical and mechanical properties of either the drug substance or the excipient can influence the CQAs of drug product (hardness, appearance, moisture, friability, etc.) and its manufacturability [48]. The knowledge of mechanical properties of the drug substance and excipients plays a significant role in dosage form development and processing [42].

A good understanding of the drug substance and excipient properties can help in rational selection of excipients such that the properties are complementary (balance of ductile and brittle material) to the drug substance. This enables the development of a robust process and this rationale decision making enables targeted trouble shooting during investigation of challenges during scale-up or manufacturing. Most pharmaceutical materials can be elastic, plastic, viscoelastic, hard,

brittle, and soft. Properties such as tensile strength [47], brittle fracture index (BFI), elasticity (Young's modulus), plastic deformation pressure (hardness), degree of viscoelasticity, and so on are commonly evaluated during process/product development. The PSD, hardness, and brittleness of the API and excipients may influence the porosity of the product. Similarly, the bonding index, brittle fracture index, and compression profiles will contribute to the understanding of the hardness of a tablet. The particle shape, size, and distribution of the drug substance and the other excipients can significantly affect their flow behavior. This is particularly important for products that are manufactured using direct compression or dry granulation. Spherical particles are ideal for dry mixing, whereas flakes, rods, or needle shaped particles are difficult to process during for dry mixing. Most pharmaceutical components are somewhere between these two extremes.

Bulk and tap densities of the drug substance and excipients provide valuable information on the flow property of the blends and influence the choice of the manufacturing equipment and process. Typically, low bulk density materials show poor flow and require additional processing to avoid manufacturing problems. Materials with high angle of repose, low floodability, and bulk density are candidates for wet granulation, slugging, or roller compaction. Hygroscopic properties of the excipients/API may also influence the choice of the process as it may influence the equilibrium moisture content of the product. For example, certain excipients like PEG are also hygroscopic and if present is large amounts, may be pose difficulty in manufacturing.

The preliminary list of quality attributes are identified, need to be constantly reassessed during the various development stages as product and process knowledge and understanding is increased. Relevant CQAs can be identified by using an iterative process of quality risk management and experimentation that assesses the extent to which the variation can impact quality of a product.

17.3.2.5 Risk of Dose Dumping for CR Products

Dose dumping is associated with unintended, rapid drug release (entire or significant fraction of the drug contained in an oral controlled/modified release dosage form), in a short period of time. Depending on the therapeutic indication and the therapeutic index of the drug, dose dumping can pose a significant risk to patients, due to safety issues, diminished efficacy (if local effect is desired), or both. This happens when the release controlling mechanism is compromised. The likelihood of dose dumping for some controlled release products when administered with food has been recognized and a regulatory process has been established to address it [49, 50].

In the case of some modified release dosage forms, the drug as well as the excipients may exhibit higher solubility in ethanolic solutions compared to water. In such cases, there is rapid release of drug in the presence of alcohol. Thus, in theory, concomitant use of alcohol or alcoholic beverages along with these products may lead to a potential for dose dumping. This potential risk has gotten a lot of attention from the regulatory authorities recently. In 2005, the opiate drug, hydromorphone, formulated as a controlled delivery over 24 h (Palladone™) was withdrawn from the U.S. market by its manufacturer due to alcohol-induced dose dumping [51]. This drug was formulated into MR pellets using ethyl cellulose, Eudragit RS, and stearyl alcohol. Coingestion with alcohol resulted in 4–16-fold increase in C_{max} in different volunteers. Most opiates are narrow therapeutic index drugs, and the main complications of dose dumping are respiratory depression and hypotension.

Therefore, the FDA recognized the need to develop a surrogate like an *in vitro* test to study the (a) vulnerability, (b) ruggedness, and (c) uncertainty of the formulation so that any such risks can be mitigated at early stages of development and design phase. Preventing dose dumping thus became a part of the desired product design. The FDA issued draft guidance for industry on the dissolution of various extended release products in presence of different concentrations of ethanol in HCL equivalent to 5, 20, and 40% for 2 h [52]. Several products to date have been included as part of this guidance, for example, tramadol, morphine sulfate, bupropion, and metoprolol succinate. Thus, it may be important during the conceptual design phase to understand the influence of the retardant polymer during dissolution. Some contributing factors to dumping are product composition and design, release mechanism, retardant type, solubility of components, and the interaction of the above factors. Understanding the above factors will lead to the design of products that minimizes the alcohol induced dose dumping phenomenon. Ensure the design of a product that is risk averse to such influences. Although the 2 h exposure to ethanol is conceived as the worst case scenarios, several new studies have shown that in some cases this may be a little more complex [53]. Alternatively, in some cases the *in vitro* test provided information regarding the vulnerability of the product to alcohol. However, *in vivo* studies performed suggest that alcohol coadministration did not change the C_{max} of the drug (carvedilol) when taken after a meal [54]. There seems to be several possible reasons for this. Extended exposure to alcohol does not occur physiologically. Since carvedilol is recommended to be taken with food, there seems to be an interaction effect of food, alcohol, and gastric pH on the absorption. Thus, a careful understanding of the impact of the formulation, gastric emptying, release medium, and duration of exposure on the form and extent of impairment induced by alcohol needs to be screened before any conclusion can be made.

Dose dumping continues to emerge in one form or another. In the absence of the *in vivo* knowledge, it is important to perform the test to ensure this aspect is covered in the designing phase. This phenomenon needs to be continually monitored during the life cycle of the product and adjustments/ improvements if any may need to be made as a result

of the data obtained during this phase. It is recommended that information regarding dose dumping be discussed with the regulators. It is also important to note that the *in vitro* test should be applicable for generic oral controlled release that typically has a different release mechanism than the branded product.

17.3.3 Quality Risk Assessment

A key objective of risk assessment in pharmaceutical development is to identify which material attributes and process parameters affect the drug product CQAs, that is, to understand and predict sources of variability in the manufacturing process so that an appropriate control strategy can be implemented to ensure that the CQAs are within the desired requirements.

The identification of critical process parameters (CPP) and critical material attributes is an iterative process and occurs throughout development. During the initial phases of development, prior knowledge serves as the primary basis for the designation as there is not sufficient process/product understanding on the product under development. Therefore, the risks identified at the initial phases are perceived risks and as further process/product understanding is gained, the actual risks become clearer and a control strategy can be better defined. The risk assessment tools used in earlier phases of development therefore tend to be more qualitative and serve as a means to prioritize the experimentation. Typical tools used include risk ranking and filtering, input–process–output diagrams, Ishikawa diagram, and so on. Risk filtering and ranking is a tool for comparing and ranking risks. Risk ranking of complex systems typically requires evaluation of multiple diverse quantitative and qualitative factors for each risk. The tool involves breaking down a basic risk question into as many components as needed to capture factors involved in the risk. These factors are combined into a single relative risk score that can then be used for ranking risks. Table 17.2 is a typical example of risk filter that is used in early development to prioritize parameters/attributes with higher risk. This is typically qualitative in nature.

An Ishikawa diagram, also known as a fishbone or cause-and-effect diagram, is a graphic tool used to explore and display opinion about sources of variation in a process. The purpose of the Ishikawa diagram is to map the key sources that contribute to the problem being examined and their interrelationships. Figure 17.7 is an example of Ishikawa diagram.

The more advanced tools take considerably more effort to set up and maintain but give a comprehensive understanding of the risks involved. Thus, tools such as risk prioritization matrix and failure modes and effect analysis (FMEA) are used at later stages of development when more understanding and knowledge is available to formalize the risk.

A risk prioritization matrix is a simple tool to visualize a process and identify the risk associated with each of the steps

TABLE 17.2 Example of Risk Filter During Initial Drug Development

Initial risk assessment DP QRA, showing the impact of critical parameters/attributes/process and its impact on the CQA

Critical parameters/ factors / DP CQA	Polymer	Roll gap/roll force (ribbon porosity)	Comp. force	Amount of fines after RC	Lubricant distribution	Press speed	API PSD	Precompression force
Appearance	Low	Low	High	Low	High	Low	Low	Low
Identity	Low	Low	Low	Low	Low	Low	Low	Low
Assay	Low	Low	Low	Low	Low	High	Low	Low
CU	Low	Medium	Low	Low	Low	Low	Low	Low
Impurity	Low	Low	Low	Low	Low	Low	Low	Low
Dissolution	High	Low	Low	Low	Low	Low	Low	High
Tablet hardness	High	High	High	High	High	High	Low	High
Friability	High	High	High	High	Low	Low	Low	Low
Yield	Low	Low	Low	Low	Low	Low	Low	Low

through different review cycles. A risk prioritization matrix is category based and therefore semiquantitative. "Filters," in the form of weighting factors or cutoffs for risk scores, can be used to scale or fit the risk ranking to management or policy objectives. Figure 17.8 shows a typical example of risk prioritization that is used to rank the most important parameters influencing the drug product CQAs.

FMEA is a systematic technique by which the consequences of an individual fault mode are identified and evaluated. FMEA can be extended to include the degree of severity of the consequences, their respective probabilities of occurrence, and their detectability and become the so-called FMECA (failure mode effect and criticality analysis)

$$RPN = (S) \times (O) \times (D)$$

Risk priority number (RPN) also has a useful role to play as a KPI. Throughout the development, risk will become lower (due to increased knowledge, increased controls, better robustness, or improved design) and this will be demonstrated with RPN that will decrease.

Several other tools are also available that help to prioritize the attributes/variables [5]. Some of these include Preliminary Hazard Analysis (PHA), Fault Tree Analysis (FTA), Hazard and Operability Analysis (HAZOP), Hazard Analysis and Critical Control Points (HACCP), Root cause Analysis (RCA), Decision Trees (DT), Probabilistic Risk Analysis (PRA), and so on.

17.3.3.1 Risk Assessment Process A simple risk assessment process starts with the identification of the problem statement. Once a problem statement is determined, a team

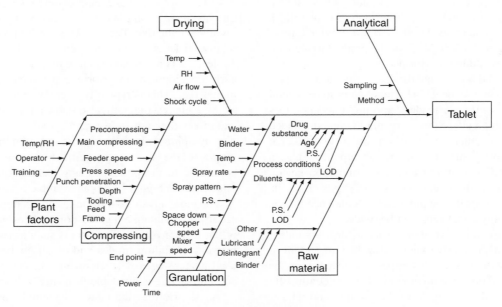

FIGURE 17.7 An example of Ishikawa diagram.

Risk prioritization matrix												
Quality attributes	Polymer concentration	Porosity/roll force/roll gap	Compression machine speed	Roller compactor feeder speed	Compressing force	Final blend lubrication time	API particle size	PSD of intragranular blend	Primary blend time	Excipient particle size	Precompression force	Quality compliance priority	Manufacturing priority
Dissolution	10	1	1	1	1	2	1	1	1	2	1	10	3
Assay/potency	5	7	1	1	1	5	5	1	3	2	1	10	2
Uniformity	2	7	5	7	1	2	5	5	8	4	1	7	2
Hardness	6	9	7	7	10	10	6	5	6	5	8	3	10
Thickness	3	2	3	1	10	1	1	1	1	2	7	2	5
Flow	7	5	4	7	2	2	5	8	5	5	2	1	4
Appearance	5	5	3	5	5	3	3	5	1	5	3	6	4
Stability	1	1	1	1	1	1	1	1	1	1	1	6	1
Yield	2	4	10	4	1	1	2	3	1	1	1	6	10
Totals (quality)	243	225	170	159	121	148	156	134	139	134	97		
Percent importance	14.08%	13.04%	9.85%	9.21%	7.01%	8.57%	9.04%	7.76%	8.05%	7.76%	5.62%		
Rank	1	2	3	4	10	6	5	8	7	8	11		
Totals (manufacturing)	188	212	229	183	196	156	141	153	125	129	153		
Percent importance	10.08%	11.37%	12.28%	9.81%	10.51%	8.36%	7.56%	8.20%	6.70%	6.92%	8.20%		
Rank	4	2	1	5	3	6	9	7	11	10	7		

Where priority = importance of the process parameter (column) to the quality attribute (row) as determined by the fmea team

Totals = the sum of each priority in the specific column × weighted average for the quality attribute

Percent importance = the total result for the column/sum of all totals
Rank = the rank of each parameter in respect to other parameters

FIGURE 17.8   A typical example of risk prioritization.

leader and a team comprising of multidisciplinary and cross-functional members are identified. The team leader could vary but in most cases the product or process development lead serves as the quality risk assessment (QRA) team lead. A QRA team usually comprises of representatives from regulatory, quality, manufacturing, analytical development, process development, product development, raw material characterization, and so on. The team selects the tool based on the problem statement. To ensure that all the members of the team have the same level of understanding, a brainstorming session is recommended. In this case, brainstorming techniques assist in developing a list of potential quality/material attributes that could impact performance and/or the manufacturability of the product. An assessment of the product, raw materials, process using a tool helps rank the risks. A tool such as the IPO diagram for each of the unit operations helps identify the process parameters that could impact the CQAs/CMAs. These can then be prioritized using the risk ranking/filtering approach and screening DOEs designed to enhance the process understanding.

17.4   PROCESS

17.4.1   Criticality

The assignment of attribute or parameter as critical or noncritical is an important outcome of the development process that provides the framework for the design space development/ designation. The primary assessment and designation of criticality should be based on the impact that quality attributes or parameters have on the safety, efficacy, and the quality (manufacturability) of the product.

Currently, there seems to be a lack of consensus in the approach to criticality within the pharmaceutical industry. Some groups classify criticality into two distinctive classes, that is, critical or noncritical, while other classify criticality as a continuum on the basis of a risk assessment. The risk associated with the process is a function of severity, probability, and detectability. Thus, if the severity is high, the chance is that the risk will be high. A well designed control strategy will decrease the occurence and thereby mitigate the overall risk. As additional process information is obtained over the life cycle of the product, it is possible that the risk associated with the attribute or parameter may change. This may make the parameter or attribute a low-risk category. Attributes and parameters in this form may still warrant some attention and their importance should not be overlooked. ISPE PQLI task force is currently collaborating with the various parties (regulators, ICH, and the industry) to define one approach to the criticality process and develop a decision tree [55].

Some of the important consequence of this classification is as follows:

- Risk assessment is used to prioritize experimental work and the allocation of resources based on which variables or areas are determined to be critical
- Attributes or parameters may be designated as critical because of their impact on quality, and by extension, safety and efficacy of the product for the patient
- Attributes and parameters determined to be critical may, as a consequence, be subject to specification in the compliance sections of a regulatory dossier. As such, modification of their designation, acceptance criteria, or associated measurement methods could necessitate some kind of regulatory approval. The use of process capability [56] and other methods for determination of criticality has previously been discussed and is an alternative approach to designating criticality [30].

17.4.2 Critical Process Parameter

A parameter is a measurable or quantifiable characteristic of a system or process [57], while a process parameter is an attribute of the manufacturing system [57]. Parameters are usually thought of as characteristics of equipment, processes, or manufacturing systems (e.g., temperature, mixing speed, airflow), whereas attributes can be considered as characteristics or properties of materials (e.g., melting point, viscosity, and sterility). However, there is not a rigid distinction and some characteristics or properties may be described as attributes or parameters. In some cases, an attribute of an output unit operation may be the parameter for the subsequent unit operation.

CPP is a process parameter whose variability has an impact on a critical quality attribute and therefore should be monitored or controlled to ensure that the process produces material of the desired quality [13].

CPPs may change during the product life cycle as new knowledge is gained. For example, new CPPs may be identified in a unit operation in the manufacturing process of the drug substance or drug product, or the acceptable range of values of a CPP may change based on improved product and process knowledge. These changes may impact the registered manufacturing process or design space.

It may be worth noting that for the same process you may have different set of CPPs. This is dependent on the type of product, the properties of the materials in the product, and the desired characteristics of the product. The operating parameter depends on the type of engineering controls on given equipment. For example, the drying process in a fluid bed dryer can be based on product temperature or the inlet air temperature. In some cases, the control could be equipment based or based on an attribute. If we develop controls based on an attribute (e.g., moisture of granulation), than it can be scale independent and thus easy to measure and controlled during scale-up. These also include CMAs. However, if the control is based on an equipment-dependent parameter, then the controls on the unit operation may be scale and equipment dependent, and scale-up/process transfer becomes more challenging.

17.4.3 Process Understanding and Knowledge

A structured approach to development is fundamental to enhanced process understanding. However, due to finite amount of resources, at the time of commercialization, the process understanding will be limited to a combination of primarily qualitative knowledge and limited quantitative knowledge. This learning will continue over the life cycle of product to enhance the understanding. A process is considered well understood [58] when

1. All critical sources of variability are identified and explained
2. Variability is managed by the process
3. Product quality attributes can be accurately and reliably predicted over the design space established for materials used, process parameters, manufacturing, environmental, and other conditions

As process understanding increases, the risk of producing poor quality product should decrease because the process robustness increases as a consequence of increased process knowledge. This in turn may enable risk-based regulatory approaches to managing change. Figure 17.9 illustrates this relationship of process understanding.

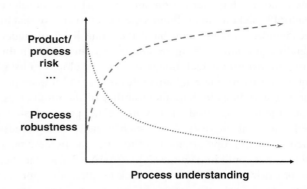

FIGURE 17.9 The relationship between process understanding and robustness.

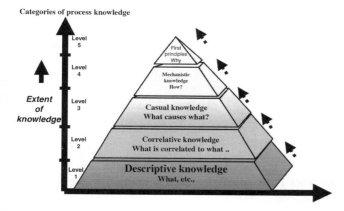

Categories of process knowledge

FIGURE 17.10 The different categories of process knowledge.

As described in ASTM E55 Standard on Process Understanding [58], describes process knowledge as the cornerstone of process understanding and that it exists in various categories. Figure 17.10 shows the various levels of process knowledge. These categories of process knowledge include (lowest to highest state of understanding) the following:

1. *Descriptive Knowledge*: It is derived solely from observation and reflects basic facts. Focuses on "*what*" is happening in the process, that is, action(s) and sequence(s) of events

2. *Correlative Knowledge*: It is based on the identification of the interrelationship between variables identified in the "description." Primarily focuses about "what output (Y) is correlated to which input (X)." Relationships are mostly univariate in nature; that is, interactions between two or more input variables and the effect on the output are not understood.

3. *Causal Knowledge*: It is based on what causes the interrelationships between variables identified in the "correlation." This knowledge is associated with an awareness about which variables are "critical" and which variables are not. By this stage, significant scientific knowledge has been established. Processes can be robust and optimization can begin.

4. *Mechanistic Knowledge*: It is based on the explanation of how "causal" relationships occur. Mechanistic knowledge is based on the development of models from basic physical, chemical, microbial, or biological mechanisms of observed phenomena underlying the ways in which the causal variables interact.

5. *First Principles Knowledge*: It is based on a theoretical understanding of prevailing mechanisms and why they occur. First principles knowledge enables prediction of behavior over a significantly wider range and situations. It expands mechanistic knowledge and as a result explains the fundamental reasons "*why*" an event takes place. With this level of knowledge, the inherent

process robustness can be further optimized and potentially generalized to other similar unit operations, processes, and products. An example of the use of mechanistic approach in pharmaceutical process development and scale-up in API synthesis is described by Hannon et.al. [59].

17.4.4 Product and Process Design and Development

The choice of desired level of understanding for each product and/or process is dependent upon a number of factors including safety, efficacy, quality, time, and cost. The minimum desired state is one where causal or empirical understanding is achieved for CQAs. This level is adequate for process control and can be quickly and economically achieved. Additional understanding is desired or required for ensuring product quality. To demonstrate adequate understanding of product performance, predictive models should be built. Each of these models has a residual error or uncertainty associated with performance of a product. This residual error can be used as to assess risk. Lowering the residual error can lead to risk mitigation through enhanced process understanding can lead to risk mitigation.

To design and develop a robust product that has the desirable TQPP, a scientist must apply the knowledge about the physicochemical and/or biological characteristics of the drug substance, excipients, and components to produce the dosage form that meets the patient needs. Product development and process development are interrelated and therefore the product design will incorporate knowledge from the process development. Figure 17.11 illustrates the interrelationship between these properties to design the product and the process for a dosage form. These properties impact the quality attributes in terms of patient safety and efficacy or the manufacturability of the product. The key objective is to design a product that delivers the performance requirements identified in the QTPP, at release and during the shelf life. It is important to ensure that both product and process are designed using science- and risk-based approaches to

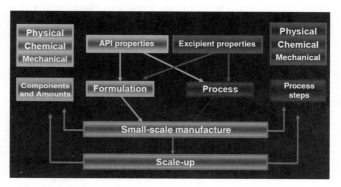

FIGURE 17.11 Relationship between product and process in drug product design.

understand and design strategies to manage variation from various sources. Good product design helps to mitigate risk of producing products that do not meet the acceptance criteria for the CQAs. It is desirable to design simple formulations and processes. Increased complexity in formulation and processes increase the challenges during the design/ development phase. Excipients (including processing aids, etc.) should be selected for their ability to provide the intended functionality throughout the shelf life of the product. The rationale and function of each excipient in the formulation should be justified. Formulation robustness studies should be conducted to identify the appropriate levels of ingredients and their functionality. These studies help to meet the performance criteria while tolerating some variability in composition, raw material characteristics, or process parameters. The appropriate use of suitable platform technologies (previously used and well-understood formulation approaches and manufacturing technologies frequently used in-house) may provide a resource efficient and rapid basis for product design and development.

Process and product design are closely associated with one another and are required for a successful product. The selection of process is dependent on the formulation and the inherent properties of the excipients and the API. It is also dictated by the QTPP that takes the patient profile (e.g., age and type of disease treatment) into consideration. The objective of a good process design is to develop a robust process in which all sources of variation are defined and controlled, end product variation is minimal, and the process is robust. The concept of process robustness in pharmaceutical processes is well documented [60]. Product and process understanding is the basis for effective process selection, implementation, and operation to establish control. There is no one size that fits all. S Changes in manufacturing process may result in the change in the quality attributes. For example, lactose is available in several grades including anhydrous direct tableting grade, spray dried form, and several particle size grades of lactose monohydrate. While lactose monohydrate may be useful for wet and dry granulation, the anhydrous type should be avoided in aqueous granulations. The conversion of anhydrous form to monohydrate form during aqueous granulation may contribute to about 5.3% weight gain, resulting in mass balance challenges in the formula. Polyethylene glycol (PEGs) are hygroscopic and therefore not suitable for wet granulation. Drug substances having poor aspect ratio (particle size, shape) may not be suitable for direct compression. Similarly, materials with poor mechanical properties may not be suitable for direct compression.

Quality risk management should be applied in each of the stages for process design and development. Ideally, a process should incorporate scaleable or scale-independent manufacturing operations or other strategies (e.g.,

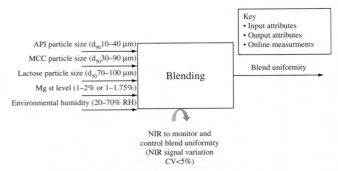

FIGURE 17.12 Example of an input, process, and output diagram for blending operation.

continuous processing) to mitigate the risk associated with scale-up.

In all cases, the pharmaceutical process has multiple-unit operations and an IPO diagram is a good tool to understand the variability of the inputs/processing parameters on the output(s). A sequence of IPOs helps in the understanding of the multivariate interactions between the individual unit operations and eventually on the product CQAs. Figure 17.12 shows an example of a blending unit operation with defined input parameters, process, and the output attribute that is measured.

The eventual goal of process and product understanding is to identify the process parameters that have an impact on the CQA/CMA and therefore influence the variability of the output if not controlled well. QRA in combination with the DOEs helps identify the true CPPs and therefore help in the mitigation of risks by developing adequate controls (increasing the detectability and also decreasing the probability of occurrence).

17.5 DESIGN SPACE DEVELOPMENT

Design Space: The design space is the practical expression of product and process understanding. As defined by the ICHQ 8(R1), design space is the multidimensional combination and interaction of input variables (e.g., material attributes) and process parameters that have been demonstrated to provide assurance of quality. Working within the design space is not considered as a change. Movement out of the design space is considered to be a change and would normally initiate a regulatory postapproval change process. Design space is proposed by the applicant and is subject to regulatory assessment and approval [13].

The design space is a subset of the entire *knowledge space*, that is, the knowledge about the product and the process and its performance. Knowledge from prior product and process experience, literature, first principles (e.g.,

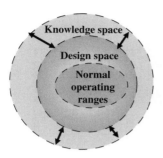

FIGURE 17.13 The link between the knowledge space, design space, and the normal operating range. Adapted from Ref. 61.

kinetics, thermodynamics, mass and heat transfer), theoretical models, process modeling, experiments (such as DoE), and empirical studies makes up the knowledge space, and the subset generating the right product quality is the design space. A process can be operated in the entire design space. The design space can be expressed or depicted in a number of ways. One of the approaches to depict a design space is to develop a typical set of normal operating ranges (NOR) based on the region where the product will always meet the QTPP irrespective of any interaction effects (a NOR that is developed and defined using a conservative approach)(Figure 17.13). ICH Q8(R2) provides a number of approaches to depict a design space in regulatory submission.

Developing the design space is a dynamic process and begins at drug conceptualization and continues to evolve over the entire life cycle of the process. It is imperative to have a robust change management system to capture the knowledge gained and assess impact if any to the design space.

A design space can be a time-dependent function (e.g., temperature and pressure cycle of a lyophilization cycle), or a combination of principal components of a multivariate model. Scaling factors can also be included if the design space is intended to span multiple operational scales.

A design space that spans multiple operations or the entire process can be developed or independent design spaces for one or more unit operations can be developed. The functional relationship between the CQAs and the related CPPs and MA, or how the outputs from one unit operation impact the process downstream dictate the strategy that will be undertaken for the design space development. Design spaces based on a holistic approach spanning the entire manufacturing process can provide more operational flexibility.

Typically the design space is being developed at small or pilot scale but is implemented and confirmed at larger /full scales. To avoid scale, equipment and site-specific issue variability from these sources should be included in the design space. A good example of this approach is using equipment/scale-independent parameters. Use of dimen-

sionless numbers and/or models for scaling would enable the scale-up process and minimize challenges during technology transfer and will enable a smoother implementation of help. If this is the case, technology transfer will be less complicated and continuous improvement activities over the product life cycle, will be more easily established. Several recent studies have reported different methods used to develop design space and the possible approaches to represent the same [61–67].

The following section discusses the various approaches/tools that enable the development of a design space.

17.5.1 Representation of Design Space

A combination of PARs does not constitute a design space. PARs based on univariate experimentation may provide some knowledge about the process. Design space is established when the PARs are studied in a multivariate fashion. In some instances, it may be helpful to know the edges of failure or the possible failure modes. However, it is not an essential part of the design space. The design space will serve as a set of internal or external specifications. In cases where the design space are quite simple, a set of NOR can be derived from the design space and the process can be operated within these NOR.

A design space can be presented in many different ways, but a pictorial fashion is often preferable as it summarizes both the process understanding and the experimental output (Figures 17.14 and 17.15). If the operation of a process is monitored online, the process can be documented in the *batch report* as a point or a trace in a representative design space plot (Figure 17.16), or as a signature in time-dependent process diagram representing the design space (Figure 17.17).

17.5.2 Design of Experiments (DoE)

Univariate experiments, that have traditionally been used during development, yield limited knowledge and interaction effects are not well understood. This is typically called one factor at a time approach or OFAT. DoE has the ability to study multiple interactions simultaneously and minimize the number of experiments needed to gain a high level of process/product understanding. DoE should be used during product and process development to understand, develop, and optimize processes.

The output from a QRA exercise serves as the starting point for the DOEs. Depending on the amount of knowledge that exists for a product/process, it is typical to perform screening DOEs are usually necessary to distinguish true from perceived risks and gain process/product knowledge. Once sufficient knowledge is gained from these DOEs, optimization DOEs are used to develop the design space

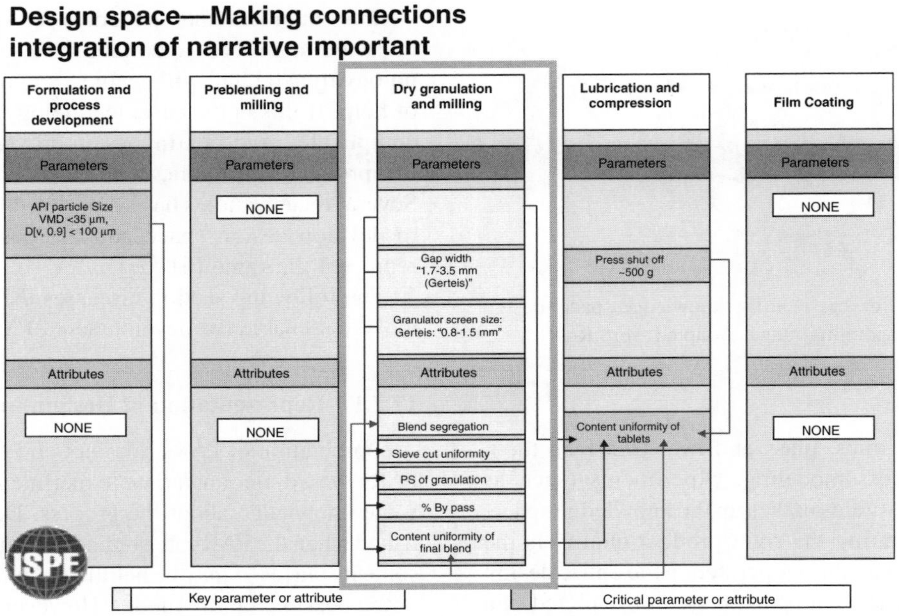

**Design space—Making connections
integration of narrative important**

Formulation and process development	Preblending and milling	Dry granulation and milling	Lubrication and compression	Film Coating
Parameters	**Parameters**	**Parameters**	**Parameters**	**Parameters**
API particle Size VMD <35 μm, D[v, 0.9] < 100 μm	NONE	Gap width "1.7-3.5 mm (Gerteis)" / Granulator screen size: Gerteis: "0.8-1.5 mm"	Press shut off ~500 g	NONE
Attributes	**Attributes**	**Attributes**	**Attributes**	**Attributes**
NONE	NONE	Blend segregation / Sieve cut uniformity / PS of granulation / % By pass / Content uniformity of final blend	Content uniformity of tablets	NONE

Key parameter or attribute	Critical parameter or attribute

FIGURE 17.14 Presentation of a design space as connections. From ISPE, Jim Spavins, Annual Meeting, November 2007.

on the parameters that are considered to be critical to the process and potentially contribute to the increased variability.

Screening studies enable us to focus on the key parameters that have a significant effect on a CQA when we have a large number of parameters to study. It helps in identifying the parameters of significant interest. When screening studies have shown that particular variables may affect a CQA, it will usually be necessary to study the relationship in more detail, for example, to determine whether the relationship is linear or nonlinear, whether the variables interact, and possibly to model the relationship mathematically. Ideally, further experimental investigation will lead to a mechanistic under-

standing of the phenomena, but this may not always be possible. Design of experiment approaches such as response surface studies, incorporating multivariate statistical analysis, enable the complexity of inter-relationships to be understood. Several examples of such multivariate approaches to study the robustness of formulation are available in the literature [68–70]. Response surface experiments map how the remaining high-impact parameters affect the experimental result. It helps in identifying both the main and the interaction effects. The next stage involves optimization of the best fit model that can best predict the understanding. Based on the prediction of the model, it is important to verify the fitness between the observed and the predicted to validate the model. Desirability limits help in choosing a set of desired attributes based on scientific and business decision (constraints). The use of statistical parameters [71] and different experimental designs including Bayesian approaches studied within the pharmaceutical industry have been extensively discussed in the literature [72–74].

17.5.3 First Principles Models

These models are based on theoretical understanding of prevailing mechanisms when they occur. This allows for prediction of behavior over a significantly wider range and situations. Typically, there is a higher degree of predictability. These include reaction mechanisms, kinetic models, correlations based on engineering principles, mass transfers, and so on.

Multiple design of experiments (DoE)

- Analyze data and results to determine
  - **establish key parameters**
  - **process variables**
  - **explore potential interactions**

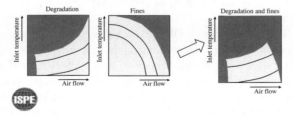

FIGURE 17.15 Presentation of a design space as contour plot. From ISPE, Jim Spavins, Annual Meeting, November 2007.

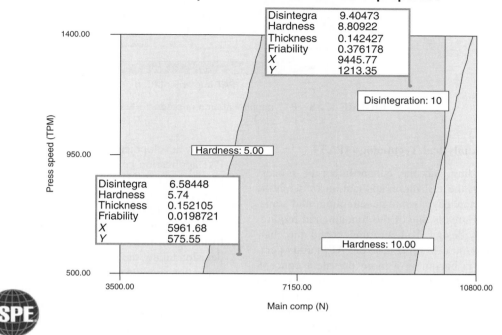

FIGURE 17.16   Presentation of a design space as overlay plot. From ISPE, Jim Spavins, Annual Meeting, November 2007.

17.5.4   Chemometrics

Process sensors and analyzers are important element of the PAT and QbD toolbox. Process sensors are univariate instruments measuring one variable or parameter and are typical used to control the process in terms of temperature, pressure, pH, TOC, and so on, or to control process equipment by measuring rotation speed, power consumption, time, frequency, and so on. Process analyzers are multivariate, that is, measuring many variables at the same time, and are typical being used to measure and control the process itself, for

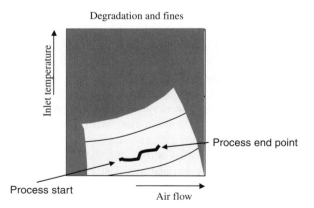

FIGURE 17.17   Presentation of a design space as overlay plot. From ISPE, Jim Spavins, Annual Meeting, November 2007.

example, a chemical reaction or process materials. Typical examples of process analyzers are spectrometers.

Chemometrics is the application of mathematical or statistical methods to chemical data. The International Chemometrics Society (ICS) offers the following definition: "Chemometrics is the science of relating measurements made on a chemical system or process to the state of the system via application of mathematical or statistical methods." Chemometric research spans a wide area of different methods that can be applied in chemistry. There are techniques for collecting good data (optimization of experimental parameters, DoE, multivariate analysis or multivariate statistics, calibration, signal processing) and for getting information from these data (statistics, pattern recognition, modeling, structure-property relationship estimations). Chemometrics tries to build a bridge between the methods and their application in chemistry.

In spectroscopy, the application of chemometrics is most often in the development of the calibration models. Calibration is achieved by using the spectra as multivariate descriptors to predict concentrations of constituents of interest using statistical approaches such as multiple linear regression, principal components analysis, and partial least squares. Other popular chemometry techniques include approaches for *ab initio* prediction of number of components, noise reduction, and multivariate curve resolution.

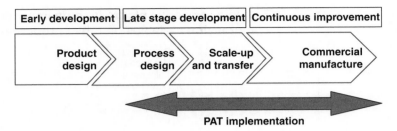

FIGURE 17.18 PAT implementation in product life cycle.

17.5.5 Process Analytical Technology (PAT)

Process understanding and the enhancement of process knowledge through the utilization/application of scientific and engineering knowledge, strategic sampling, and statistical control procedures is one of the fundamental requirements of quality by design. To better understand and characterize the product/process utilization of PAT in unit operations of high risk is becoming a more prevalent concept. Initially presented to the pharmaceutical industry as part of the "cGMPs for the 21st Century Initiative," PAT today is recognized as one of the key elements of quality by design that can be utilized during development to enhance process understanding and serve as an integral part of the control strategy in the commercial manufacturing process.

PAT [3] is defined as a system for designing, analyzing, and controlling manufacturing through timely measurements (i.e., during processing) of critical quality and performance attributes of raw and in-process materials and processes, with the goal of ensuring final product quality. PAT is often interpreted to be a measurement system rather than an integrated system that provides enhanced process understanding. The utilization of well-established concepts such as DoE, process analyzers, sampling theory, statistics/multivariate data analysis, and prior knowledge in combination with quality risk management facilitates the enhancement of process knowledge in high-risk unit operations. Increased process and product knowledge helps understand the sources of variability and therefore helps design a strategy for predicting and reacting to the variability to control the output. The appropriate use of PAT in conjunction with QbD will enable pharmaceutical manufacturers to supply products with an even higher level of quality to all customers, but most importantly, the patients. In short, PAT is an enabler to achieving the goals of QbD and deals with the "how" aspects of process understanding, while QbD defines the "what" aspects of process understanding [75].

PAT is often associated with in-process control testing in commercial manufacturing. However, to successfully implement PAT, sufficient knowledge of the process/product has to be gained in the early stages of development. As part of a QbD effort, PAT opportunities can be identified in early development. However, since the success rate for compounds

in early development (phase 1) is low, it is recommended that PAT application be initiated in the later phases of development (phase 2 and continued into phase 3). This approach is depicted in Figure 17.18 for a standard solid dosage form.

A key concept in the application of PAT analyzers as part of the drug product control strategy is that not all attributes that can be measured need to be controlled. A risk-based decision taking the process/product understanding into consideration should be utilized to select the process analyzer as part of the control strategy. Processing steps/attributes that may need to be monitored to increase process knowledge and mitigate risk are identified and evaluated utilizing appropriate process analyzers to enhance process understanding. Figure 17.19 demonstrates the output of a QRA exercise for an immediate release solid dose product using a wet granulation technique. In the case of this product, the granulation unit operations were identified as the area of the highest risk and process analyzer will be used to monitor the attributes of the granulation and enhance the process knowledge of this unit operation. In development, it is typical to investigate the feasibility of multiple process analyzers (e.g., a particle size analyzer, acoustic measurements, or the monitoring of process parameters such as torque or power in the wet granulation unit operation) to study an attribute (or a process parameter). Identifying and utilizing analyzers is just one aspect of a PAT system. The FDA guidance identifies PAT as a system that comprises four distinct tools: (1) multivariate tools for design, data acquisition, and analysis; (2) process analyzers; (3) process control tools; and (4) continuous improvement and knowledge management tools. As process analyzers are being applied either online, at-line, or in-line to gain process knowledge, it is equally important to focus on the other tools of the PAT system. Multivariate data analysis systems when used in an appropriate manner help identify complex interactions that may typically not be noticed in univariate studies. In combination with the knowledge gained during development and appropriate design and construction of the process equipment, the analyzer enables the definition of process signatures. These signatures can be used for process monitoring, control, and end point determination in instances where such signatures may impact product quality. A simple example of such a unit operation would be the monitoring of a blending unit operation using

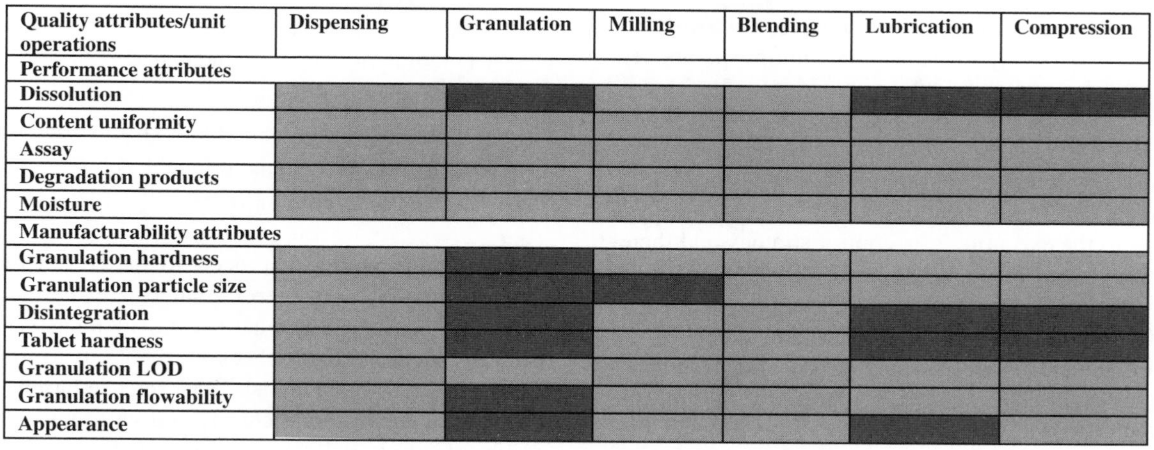

Quality attributes/unit operations	Dispensing	Granulation	Milling	Blending	Lubrication	Compression
Performance attributes						
Dissolution						
Content uniformity						
Assay						
Degradation products						
Moisture						
Manufacturability attributes						
Granulation hardness						
Granulation particle size						
Disintegration						
Tablet hardness						
Granulation LOD						
Granulation flowability						
Appearance						

FIGURE 17.19 Utilization of quality risk management to identify areas for PAT utilization.

a spectroscopic technique. In this instance, the end point for blending will be based on the process signature of the quality attribute (uniformity of the blend) compared to the traditional method of using an end point based on time.

Once risk-based decisions are made to identify the appropriate analyzers to monitor product/process variables, it is necessary to ensure that the methods are fit for purpose. This includes an appropriate design of studies to understand the impact of variations to the methods (from both an analytical measurement and a multivariate methodology perspective). Typical variations that need to be considered include variation of inputs (API, raw material)—physical and chemical variations, personnel variations, environmental variations, instrument variations, and so on. In the case where critical product attributes have an existing reference method, a comparison of the reference method to that of the PAT method must be completed. This will further demonstrate that the PAT system is fit for use and is a suitable alternative to measure the critical quality attribute(s) of the product. In addition to the variations, sufficient understanding regarding the placement of the sensor in the manufacturing equipment and the ability of the device to measure a representative sample should also be demonstrated. In addition, the development of a scientific and statistically sound risk-based sampling plan should also be one of the goals during development. It is very important to utilize the knowledge gained during atypical manufacturing runs during the development of PAT as they provide a very good opportunity to assess the capabilities of the system.

Several literature examples about the use of different PAT for control or monitoring have been illustrated [76–78]. It is up to the individual scientist to explore the use based on the defined risk and use it as an opportunity to use it within the defined guidelines to improve on the detectability that may eventually lead to risk reduction or mitigation. Real-time release testing (RTRT) [79, 80], is an added benefit of a well defined control strategy that may include the use of PAT.

17.6 CONTROL STRATEGY

ICH Q10 defines control strategy as a planned set of controls derived from current product and process understanding that assures process performance and product quality. The controls can include the following:

- Parameters and attributes related to drug substance and drug product materials and components
- Facility and equipment operating conditions
- In-process controls
- Finished product specifications
- The associated methods and the frequency of monitoring and control

The implementation of a control strategy inherently addresses the implementation of the design space. The control strategy for a product should be holistic and developed based on the impact of the changing characteristics of the product (e.g., powder blend to granulation to tablet) across the entire manufacturing process. The development of a robust control strategy facilitates greater process robustness and leads to lower variability in the process and the product. This in turn provides greater assurance of product quality and facilitates opportunities for real-time release of the product. It is important to understand that regulatory flexibility is the outcome of the enhanced process understanding and is not the goal of QbD.

Controls are developed on the basis of the process understanding and can include

- *EPC (Engineering Process Control) [81]*: Simple control loops that control simple parameters with well-documented effects of final product quality.
- *SPC (Statistical Process Control)*: Allows monitoring of primary parameters to improve understanding and control over time.

- *MSPC (Multivariate Statistical Process Control)*: Advanced control taking into account multiple parameters and correlations.

ISPE has described a level-based concept to design a control strategy based on different levels of control [82]. Development and deployment of quality systems contribute immensely to the execution of a control strategy and therefore form a crucial element in the implementation of the design space. The deployment of a science and a risk-based approach to quality as opposed to the checkbox mentality to ensure compliance is one of the fundamental differences in this approach. This requires a change in culture, and a proactive approach to quality needs to be adopted. For instance, the implementation of robust quality risk management systems within the manufacturing site and the development of a proactive reaction to the failure modes that may be potentially observed during day-to-day operations would facilitate the smoother transition to the implementation of a control strategy. An example of one such instance when PAT is potentially part of the control strategy is provided in Figure 17.20.

Some other factors to consider while utilizing PAT as part of the control strategy may include the following:

- Disaster recovery plans
- Chemometric model maintenance
- Handling of outliers
- Batch release process in the RTR environment
- Quality risk management (enabler)
- Tracking and trending of data

The other factors that need to be considered in the implementation of a control strategy are the cultural and personnel aspects. The pharmaceutical industry has long been entrenched with the checkbox mentality to compliance. The embracement of a risk- and science-based approach quality is a fundamental change and will require a considerable effort. It will also require quality personnel who understand the fundamental aspects of QbD. This would include scientific and regulatory aspects of design space development and presentation that in turn would enable them to understand the variabilities and uncertainties in the tools used to develop design space. This would enable them to assess changes appropriately. In summary, control strategy development and implementation is a fundamental aspect of QbD and is the journey that occurs throughout development and is not an afterthought.

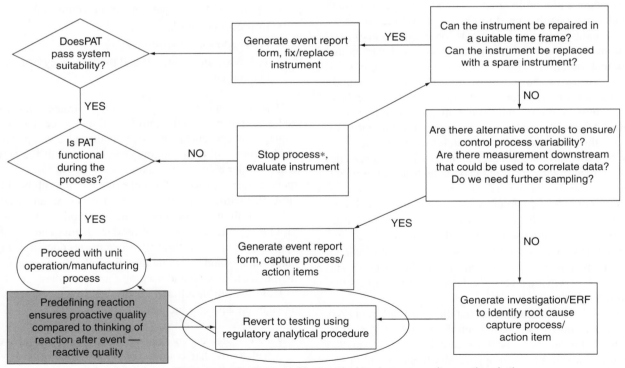

Please note that it may not be practical to stop some unit operations in the manufacturing process during the middle of the run; for example, blending

FIGURE 17.20 Decision tree for failure modes—PAT failure during the manufacturing process.

17.7 CONTINUAL IMPROVEMENT

This is a recurring activity to increase product and process understanding based on development / commercial experience and occurs throughout the life cycle of the product. The knowledge gained can be utilized to further reduce variability, eliminate deviations and defects, increase manufacturing efficiency and reduce costs. Continual improvement is particularly important as many of the deviations and rejects may be associated with variability in materials. During the development of products, this knowledge may be limited. Once the product is in routine manufacturing, this understanding is enhanced due to greater knowledge of the raw material (lot to lot variability and supplier to supplier variability). The design space at the time of registration, represents the process understanding at that snapshot of time. During commercialization, more process understanding will be gained and the design space can be updated as an activity within the continual improvement program. A good amount of time series data is collected during manufacturing operations. This can be particularly useful for gaining additional process understanding, especially to study the effect of raw material, environment, and process conditions [83].

Robust product and processes are the foundations for reliable and consistent supply of products. ICH Q8 describes process robustness as the ability of a process to tolerate variability of materials and changes of the process and equipment without negative impact on quality.

Historically, this opportunity to apply continual improvements was not available due to rigid and fixed processes that were locked during registration of the product. This systematic approach has created an opportunity to facilitate continuous improvement that has allowed using flexible regulatory approaches. Continuous improvement is facilitated by the elements of pharmaceutical quality systems as described in ICH Q10.

17.8 CONCLUSIONS

QbD is a systematic, risk-based approach to pharmaceutical development. The goal is to enhance product and process understanding. If a suitable control strategy is enabled, then it can lead to regulatory flexibility and possibly reduction in end product testing. Although it may require more resources, time, and effort initially, these should be offset by reduced time and resources to scale-up, transfer, and commercialization of products. Pharmaceutical companies are in the process of estimating the business value of QbD and more data should emerge in the next few years. It can also be applied to legacy products that were not previously filed by this approach. Current applicants are in the process of estimating the value

of this approach, but more data should emerge in the next few years. The key benefits include structured development, improved manufacturing efficiency, regulatory flexibility, and better knowledge management. QbD begins with predefined objectives and places an emphasis on product and process understanding and control. It enables the development of process that are more amenable to scale-up and transfer and facilitates innovative concepts such as real-time release testing, continuous processing, and continuous quality verification.

REFERENCES

1. The metamorphosis of manufacturing, IBM Business Consulting Services Executive Briefing, http://www.ibm.com/services/us/imc/pdf/ge510-4034-metamorphosis-of-manufacturing.pdf.

2. Innovation and continuous improvement in pharmaceutical manufacturing pharmaceutical CGMPs for the 21st Century, U.S. Food and Drug Administration, September 2004, www.fda.gov/cder/gmp/gmp2004/manufSciWP.pdf.

3. PAT — A Framework for Innovative Pharmaceutical Development, Manufacturing, and Quality Assurance, September 2004, http://www.fda.gov/cder/guidance/6419fnl.pdf.

4. ICH Q8(R2) Pharmaceutical Development, Step 4, August 2009, http://www.ich.org/cache/compo/276-254-1.html.

5. ICH Q9 – Quality Risk Management, Step 4, November 2005, http://www.ich.org/cache/compo/276-254-1.html.

6. ICH Q10 – Pharmaceutical Quality System, Step 4, June 2008, http://www.ich.org/cache/compo/276-254-1.html.

7. Chemical, pharmaceutical and biological information to be included in dossiers when process analytical technology (PAT) is employed, Reflection Paper, EMEA/INS/277260/2005, http://www.emea.europa.eu/Inspections/docs/PATGuidance.pdf.

8. Juran JM. *Juran on Quality by Design – The New Steps for Planning Quality into Goods and Services*, Free Press, 1992.

9. Yu LX. Pharmaceutical quality by design: product and process development, understanding, and control. *Pharm. Res.* 2007; 25(4):781–791.

10. ICH Harmonized Tripartite Guideline on Pharmaceutical Quality Systems Q10, step 4 version, dated June 4, 2008.

11. Hulbert MH, Fely LC, Inman EL, Johnson AD, Kearney AS, Michaels J, Mitchell M, Zour E. Risk management in pharmaceutical product development process. *J. Pharm. Innov.* 2008; 3:227–248.

12. ICH. Draft consensus guideline: pharmaceutical development annex to Q8. Available at http://www.ich.org/LOB/media/MEDIA4349.pdf (accessed 11/21/2007).

13. ICH Q8 (R1) Pharmaceutical Development: Quality by Design, May 2006.

14. FDA CDER. Draft guidance for industry and review staff. Target product profile-strategic development process tool, March 2007.

15. Yu LX, Raw A, Lionberger R. U.S. FDA question-based review of generic drugs: a new pharmaceutical quality assessment system. *J. Generic Med.* 2007;4:239–248.

16. ISPE PQLI. Draft PQLI summary update report. http://www.ispe.org/cs/pqli_product_quality_lifecycle_implementation_/draft_pqli_summary_update_report (accessed 11/21/2007).

17. Lionberger RA, Lee SL, Lee L, Raw A, Yu LX. Quality by design: concept for ANDAs. *AAPS J.* 2008;10(2):268–276.

18. FDA Office of Generic Drugs. Model quality overall summary for an extended release capsule. Available at http://www.fda.gov/cder/OGD/QbR/OGD_Model_QOS_ER_Capsule.pdf (accessed 11/21/2007).

19. FDA Office of Generic Drugs. Model quality overall summary for an immediate release tablet. http://www.fda.gov/cder/OGD/QbR/OGD_Model_QOS_IR_Tablet.pdf (accessed 11/21/2007).

20. Potter C, Beerbohm R, Coups A. A guide to EFPIA's Mock P.2 document. *Pharm. Technol. Eur.* 2006;18:39–44.

21. Qureshi SA. Developing discriminatory drug dissolution tests and profiles: Some thoughts for consideration on the concept and its interpretation. *Dissolution Technologies*, 2006; November, 18–23.

22. Gray V, Kelly G, Xia M, Butler C, Thomas S, Maycock S. The science of USP 1 and 2 dissolution: present challenges and future relevance. *Pharm. Res.* 2009;6:9822–9835.

23. FDA. Guidance for Industry: bioavailability and bioequivalence studies for orally administered drug products-general considerations (revised), CDER, Rockville, MD, 2003.

24. Klien S, Rudolph M, Dressman JB. Drug release characteristics of different meslazine products using USP apparatus 3 to simulate passage through the GI tract. *Dissolution Technologies* 2002;9:6–12.

25. Wilding IR. Evolution of the biopharmaceutics classification system to oral modified release formulations: what do we need to consider? *Eur. J. Pharm. Sci.* 1999;8:157–159.

26. Dutta S, Qiu Y, Samara E. Once-a-day extended release dosage form of divalproex sodium III: development and validation of a level A IVIVC. *J. Pharm. Sci.* 2005;94:1949–1956.

27. FDA. *Guidance for Industry: SUPAC-MR*, CDER, Rockville, MD, 1995.

28. Selen A. QbD and biopharmaceutics for clinically relevant dissolution testing, quality by design: defining clinically relevant design goals, November 20, 2008, Atlanta, GA, American Association of Pharmaceutical Scientist.

29. Dickinson PA, Lee WW, Stott PW, Townsend AI, Smart JP, Ghahramani P, Hammett T, Billett L, Behn S, Gibb YC, Abrahamsson B. Clinical relevance of dissolution testing in quality by design. *AAPS J.* 2008;10(2):1603–1607.

30. Cogdill RP, Drennen JK. Risk-based quality by design (QbD): A Taguchi perspective on the assessment of product quality, and the quantitative linkage of drug product parameters and clinical performance. *J. Pharm. Innov.* 2008;3:23–29.

31. Tong C, D'Souza SS, Parker JE, Mirza T. Dissolution testing for the twenty-first century: linking critical quality attributes and critical process parameters to clinically relevant dissolution. *Pharm Res.* 2007;24(9):1603–1607.

32. Amidon GE, He X, Hageman MJ.In Abraham DJ, editor. *Burgers Medicinal Chemistry and Drug Discovery*, Vol. 2, Chapter 18, Wiley–Interscience, New York, 2004.

33. Yu LX, Furness MS, Raw A, Woodland KP, Nashed NE, Ramos E, Miller SPF, Adams RC, Fang F, Patel RM, Holcombe FO, Chiu Y, Hussain AS. Scientific considerations of pharmaceutical solid polymorphism in abbreviated new drug applications. *Pharm. Res.* 2003;20:531–536.

34. Sun D, Yu LX, Hussain MA, Wall MA, Smith RL, Amidon GL. *In vitro* testing of drug absorption for drug " developability" assessment: forming an interface between *in vitro* pre-clinical data and clinical outcomes. *Curr. Opin. Drug Discov. Dev.* 2004;7:75–85.

35. Yu LX, Ellison CD, Hussain AS. Predicting human oral bioavailability using in-silico models. In Krishna R,editor. *Applications of Pharmacokinetics Principles in Drug Development*, Kluwer, New York, 2004.

36. Amidon GL, Lennernas H, Shah VP, Crison JR. A theoretical basis for a biopharmaceutical drug classification: the correlation of *in vitro* drug product dissolution and *in vivo* bioavailability. *Pharm. Res.* 1995;12:413–420.

37. Kesisoglou F, Wu Y. Understanding the effect of API properties on bioavailability through absorption modeling. *AAPS J.* 2008;10(4):516–525.

38. Liu DJ. QbD—application of simulation and modeling for designing optimum dosage forms. 43rd AAPS Annual Arden House Conference, West Point, New York, February 2008.

39. ICHQ6A. Q6A Specifications: Test Procedures and Acceptance Criteria for New Drug Substance and New Drug Products: Chemical Substances, 2000.

40. ASTM–WK13538-Standard practice for Identification of Critical Attributes of Raw Materials in Pharmaceutical Industry.

41. Mroz C.A changing paradigm: regulatory and supply chain considerations when selecting and qualifying excipients for your drug products. Excipient Fest Europe, 2008. http://www.excipientfest.com/europe/presentations/EFE185CarlMroz-Colorcon.PPT.

42. Guy A.Role of Excipients & Excipient suppliers in Quality by Design, Balchem, Eastern Pharmaceutical Technology Meeting, 2008.

43. Hlinak AJ, Kuriyan K, Morris KR, Reklaitis GV, Basu PK. Understanding critical material properties for solid dosage form design. *J. Pharm. Innov.* 2006;1:12–17.

44. Moreton C. Functionality and Performance of Excipients. *Pharm. Technol. Suppl.* 2006;s4–s14.

45. Weins RE. Pharmaceutical excipient testing and control strategies, PQRI Excipient Working Group, 2007. http://ipecamericas.org/newsletters/Wiens.pdf.

46. Mitchell S, Balwinski KM. A framework to investigate drug release variability arising from hypromellose specifications in controlled release matrix tablets. *J. Pharm. Sci.* 2007;1–9.

47. Soh JLP, Boersen N, Carvajal MT, Morris KN, Peck GN, Pinal R. Importance of raw material attributes for modeling ribbon and granule properties in roller compaction: multivariate analysis on roll gap and NIR spectral slope as process critical control parameters. *J. Pharm. Innov.* 2007;2:106–124.

48. Amidon GE. Physical and mechanical property characterization of powders. In Brittain HG, editor. *Physical Characterization of Pharmaceutical Solid*, Marcel Dekker, New York, 1995, pp. 281–320.

49. Hendeles L, Wubbena P, Weinberger M. Food-Induced dose dumping of once-a-daily theophylline. *Lancet* 1984;22:1471.

50. Guidance for Industry, Food Effect Guidance Food-Effect Bioavailability and Fed Bioequivalence Studies. http://www.fda.gov/cder/guidance/5194fnl.pdf.

51. FDA Alert for Healthcare Professionals (July 2005): Hydromorphone Hydrochloride Extended-Release Capsules (marketed as Palladone™). http://www.fda.gov/cder/drug/InfoSheets/HCP/hydromorphoneHCP.pdf.

52. FDA. Draft Guidance, 2007. http://www.fda.gov/cder/guidance/bioequivalence.

53. Fadda HM, Mohammed MAM, Basit A. Impairment of the *in vivo* drug release behavior of oral modified release preparations in the presence of alcohol. *Int. J. Pharm.* 2008;360:171–176.

54. Henderson LS, Tenero DM, Campanile AM, Baidoo CA, Danoff TM. Ethanol does not alter the pharmacokinetic profile of the controlled release formulation of carvedilol. *J. Clin. Pharmacol.* 2007;47:1358–1365.

55. Nosal R, Schwartz T. PQLI definition of criticality. *J. Pharm. Innov.* 2008;3:69–78.

56. Seibert KD, Sethuraman S, Mitchell JD, Griffiths KL, McGarvey B. The use of routine process capability for determination of process parameter criticality in small molecules. *J. Pharm. Innov.* 2008;3:105–112.

57. E2363-06a, ASTM E55 Committee Terminology Standard for Definitions.

58. ASTM E55- WK 5935. Standard Guide for Process Understanding Related to Pharmaceutical Manufacture and Control.

59. Hannon J, Clark P.Application of mechanistic thinking in pharmaceutical process development and scale-up. Presentation to FDA CDER/ONDQA, February 28, 2008.

60. Liebowitz S, McCarthy R, Glodek M, McNally G, Oksanen C, Schultz T, Sundarajan M, Vorkapich R, Vukovinsky K, Watts C, Millili G. Process robustness: a PQRI white paper. *Pharm. Eng.* 2006;26(6):1–11.

61. Garcia T, Cook G, Nosal R. PQLI key topics: criticality, design space, and control strategy. *J. Pharm. Innov.* 2008;3:60–68.

62. Lipsanen T, Antikainen O, Raikkonen H, Airaksinen S, Ylirussi J. Novel description of a design space for fluidized bed granulation. *Int. J. Pharm.* 2007;345:101–107.

63. Xie L, Wu H, Shen M, Augsburger LL, Lyon RC, Khan MA, Hussain AS, Hoag SW. Quality by design: effect of testing parameters and formulation variables on the segregation tendency of pharmaceutical powders measured by the ASTM D 6940-04 segregation tester. *J. Pharm. Sci.* 2008;97(10):4485–4497.

64. Ende DAM, Bronk KS, Mustakis J, O'Connor G, Santa Maria CL, Nosal R, Watson TJN. API quality by design example from the torcetrapib manufacturing process. *J. Pharm. Innov.* 2007;2:71–86.

65. Fields T. From proven acceptable ranges to design space. *J. Validation Technol.* 2007;14(1):50–53.

66. Lepore J, Spavins J. PQLI design space. *J. Pharm Innov.* 2008; 3:79–87.

67. MacGregor JF, Bruwer MJ. A framework for the development of design and control spaces. *J. Pharm. Innov.* 2008;3:15–22.

68. Gabrielsson J, Nystrom A, Lundstedt T. Multivariate methods in developing an evolutionary strategy for tablet formulation. *Drug Dev. Ind. Pharm.* 2000;26(3):275–296.

69. Gabrielsson J, Lindberg NO, Palsson M, Nicklasson F, Sjostrom M, Lundstedt T. Multivariate methods in the development of a new tablet formulation. *Drug Dev. Ind. Pharm.* 2003; 29(10):1053–1075.

70. Gabrielsson J, Lindberg NO, Pihl AC, Sjostrom M, Lundstedt T. Robustness testing of a tablet formulation using multivariate design. *Drug Dev. Ind. Pharm.* 2006;32:297–307.

71. Gao Z, Moore T, Smith AJ, Doub W, Westenberger B, Buhse L. Gauge repeatability and reproducibility for accessing variability during dissolution testing. *AAPS Pharm. Sci. Tech.* 2007; 8(4):E1–E5.

72. Lunney PD, Cogdill RP, Drennen JK. Innovation in pharmaceutical experimentation part 1: review of experimental designs used in industrial pharmaceutics research and introduction to Bayesian d-optimal experimental design. *J. Pharm. Innov.* 2008;3:188–203.

73. Peterson JJ. A Bayesian approach to the ICH Q8 definition of design space. *J. Biopharm. Stat.* 2008;18:959–975.

74. Peterson JJ, Snee RD, McAllsiter PR, Schofield TL, Carella AJ. Statistics in pharmaceutical development and manufacturing. *J. Qual. Technol.* 2009;41(2):111–134.

75. Sekulic SS. Is PAT Changing Product/Process Development? *Am. Pharm. Rev.*, 2007, July/August, 30–34.

76. Papp MK, Pujara CP, Pinal R. Monitoring of high shear granulation using acoustic emission: predicting granule properties. *J. Pharm Innov.* 2008;3:113–122.

77. Akseli I, Libordi C, Cetinkaya C. Real-time acoustic elastic property monitoring of compacts during compaction. *J. Pharm. Innov.* 2008;3:134–140.

78. Shah RB, Tawakkul MA, Khan MA. Process analytical technology: chemometric analysis of Raman and near infra-red spectroscopic data for predicting physical properties of extended release matrix tablets. *J. Pharm. Sci.* 2007;96(5):1356–1365.

79. Walter K. Introduction to real time process determination. *Pharm Eng.*, 2007, March/April, 1–9.

80. Radspinner D.Implementing PAT-Industry example, RPS/FDA meeting, London, UK, December 14, 2004.

81. Bolton R, Tyler S. PQLI engineering controls and automation strategy. *J. Pharm. Innov.* 2008;3:88–94.

82. Davis S, Lundsberg L, Cook G. PQLI control strategy model and concepts. *J. Pharm. Innov.* 2008;3:95–104.

83. Stryczek K, Horacek P, Klema J, Castells X, Stewart B, Geoffroy J-M. Capitalizing on aggregate data for gaining process understanding: effect of raw material, environmental and process conditions on the dissolution rate of a sustained release product. *J. Pharm. Innov.* 2007;2:6–17.

18

ORAL CONTROLLED RELEASE-BASED PRODUCTS FOR LIFE CYCLE MANAGEMENT

Nipun Davar and Sangita Ghosh

Transcept Pharmaceuticals, Inc., Point Richmond, CA, USA

18.1 INTRODUCTION: LIFE CYCLE MANAGEMENT IS A NECESSARY TREND

There are a few new products in the pipeline and the pharmaceutical industry is realizing the necessity to reposition or reformulate the existing drugs in order to extend their life cycles and sustain their value. Of the 94 new drugs approved by the FDA in 2007, only 16 (17%) were new molecular entities. Thirty-two (33%) approvals were for new formulations of which two were new salts. Approval of new indications for existing drugs is a potent strategy to expand a well-established franchise. Ten (11%) known drugs gained approval for new indications in 2007. However, several drugs are rapidly losing their patent protection. Between 2007 and 2009, 35 drug patents would expire leaving them vulnerable to generics [1]. In 2008 alone, it is estimated that the U.S. patent expiry of branded drug products may affect over $64 billion in global sales due to the entry of generic versions. This threat of generic competition is also driving the innovator companies to place a stronger emphasis on life cycle management.

The business case for life cycle management of existing drugs is reasonably straightforward. The plethora of safety information for the existing drugs reduces not only the cost and time of development but also offers lower risk to the sponsors. In many cases, the companies could utilize their sales force to leverage the existing relationships that they have developed with the prescribers. These benefits need to be balanced with the reduced level of regulatory exclusivity for the existing drugs compared to those of the new chemical entities. Controlled release (CR) technologies often play a pivotal role in enhancing the value of products based on off-patent drugs. The sponsors often rely on patent protection offered from the controlled release technologies to extend the period of market protection beyond regulatory exclusivity. Also, CR technologies could often lead to clinical differentiation resulting in better efficacy or superior safety. In terms of 2008 dollars, the estimated cost of developing a product using off-patent molecule and an established oral CR technology is $50–70 MM, of which approximately $10 MM is required for technical pharmaceutical development. The time required for developing a product is 4–7 years.

For the products based on off-patent molecules to succeed, the commercial potential comes from one of the three areas (Figure 18.1). First, the new product should offer enhanced efficacy or safety profile compared to the existing products. For instance, oral controlled release technology could result in reduced frequency of clinically meaningful adverse events (see Procardia XL®). Second, the product should be able to extend the patent life of the molecule (see Concerta®). Finally, the product must provide an increased source of revenue by targeting toward a new patient population by seeking a new indication (see Dilacor® XR) [2].

Pharmaceutical companies have employed several strategies to differentiate the follow-on products and to extend the patent life. These strategies are not necessarily mutually exclusive and may be pursued in combination.

Controlled Release or reformulation is perhaps the most common strategy that has been successfully employed so far. Several products having been developed through oral CR technology, with an improved performance over existing drugs, have been approved (Table 18.1), while many are

Oral Controlled Release Formulation Design and Drug Delivery: Theory to Practice, Edited by Hong Wen and Kinam Park
Copyright © 2010 John Wiley & Sons, Inc.

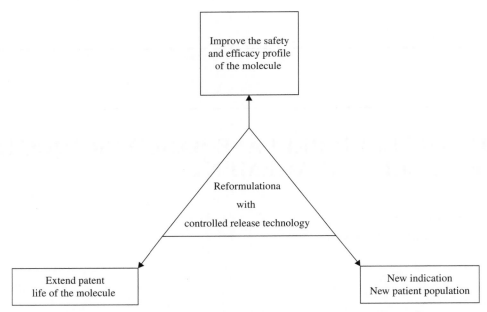

FIGURE 18.1   Life cycle management of existing products can be achieved through reformulation and further expansion of the scope of the molecule.

under development (Table 18.2). Controlled release dosage forms could offer several advantages and provide means for clinical differentiation and patient benefit. Table 18.3 provides examples of various CR-based reformulated products and their clinical benefits. The pharmacokinetic (PK) and clinical improvements form the basis of patent application allowing the protection of these products in the marketplace.

New indications for existing drugs are a means to achieving life cycle management. This is done through identifying applications of the disease severity extensions (e.g., Aricept®), new indications, or new patient populations (geriatric or pediatric use).

Active enantiomer: Replacement of an already approved racemate by a single active enantiomer has been used to create new products. Esomeprazole is the s-isomer of omeprazole (Prilosec) and was launched in 2000, 2 years before the expiry of omeprazole patent. Racemic omeprazole and esomeprazole are equally absorbed and esomeprazole is less susceptible to small intestinal and hepatic metabolism [3–5]. Therefore, the same degree of acid suppression is obtained with lower doses of esomeprazole. The composition of matter patent for esomeprazole will expire in 2020.

Prodrugs and metabolites of the existing drugs could be developed into new products. For example, the approved antihistamine terfenadine is a prodrug used for the treatment of allergic rhinitis and has been replaced by the active metabolite fexofenadine that shows less cardiac side effects [6]. Fexofenadine has become the allergy drug of choice due to its vastly improved side effect profile.

Combination of existing drugs is often used as a life cycle management strategy. The two drugs selected could be

delivered immediately, or one of these could be delivered in a controlled manner. For example, Avandamet XR is in phase 3 clinical trials and is a combination of extended release rosiglitazone and metformin hydrochloride by GlaxoSmithKline for the treatment of type 2 diabetes. Another example of a combination product under development by GlaxoSmithKline is a combination of carvedilol controlled release and lisinopril for the treatment of hypertension [7].

Abuse deterrence of opioid products is a societal benefit and could be leveraged as a differentiating feature to improve existing products. Several pharmaceutical companies are using innovative tamper-resistant dosage forms to reduce the abuse potential of these drugs. Remoxy™ (Pain Therapeutics Inc. and King Pharmaceuticals) is an extended release oxycodone in a viscous gel cap designed to be resistant to extraction using high-proof alcohol as well as crushing. The new drug application (NDA) for Remoxy™ has been submitted in June 2008. Another example is Acurox™ (Acura Pharmaceuticals Inc.), which is oxycodone with a subtherapeutic amount of immediate release (IR) niacin. Acura claims that the small amount of niacin would not affect patients taking the recommended dosage, but those trying to take excessive amounts of the product will feel the unpleasant therapeutic effects of niacin such as flushing, itching, sweating, and other forms of discomfort. Embeda® (Alpharma) is a combination of extended release morphine and uses sequestered naltrexone core, intended to deter crushing to release the active ingredient. Embeda® was approved by the FDA in August 2009.

TABLE 18.1 Oral Controlled Release Product Approvals Since 2006

Product	Drug	Technology	Indication	Company
Ranexa tablets	Ranolazine	Matrix tablet	Angina pectoris	CV Therapeutics
Clarinex – D12 hour tablets	Desloratidine + pseudo-ephedrine sulfate	Bilayer tablets with desloratidine IR and pseudo-ephedrine ER	Allergic rhinitis	Schering-Plough
Opana ER tablets	Oxymorphone HCl	TIMERx-N, matrix tablet	Moderate to severe pain	Endo Pharmaceuticals Inc.
Coreg CR capsules	Carvedilol phosphate	Multiparticulates	Heart failure	SB Pharmaco
Invega tablets	Paliperidone	OROS	Schizophrenia	Janssen
Lialda	Mesalamine	MMX delayed release	Ulcerative colitis	Shire PLC
Amrix ER tablets	Cyclobenzaprine hydrochloride	Multiparticulates	Relief of muscle spasm	ECR Pharmaceuticals
Seroquel XR tablets	Quetiapine fumarate	Matrix tablets	Schizophrenia	AstraZeneca
Zyflo CR tablets	Zileuton	Matrix tablets	Prophylaxis and chronic asthma	Critical Therapeutics Inc.
Sanctura XR capsules	Trospium chloride		Overactive bladder	Indevus Pharmaceuticals Inc.
Protonix delayed release granules	Pantoprazole sodium	Multiparticulates	Erosive gastroesophageal reflux disease	Wyeth Pharmaceuticals Inc.
Morphine sulphate	Morphine sulphate	Matrix tablets	Management of moderate to severe pain	KV Pharmceutical
Moxatag extended release tablets	Amoxicillin		Treatment of pharyngitis and tonsillitis	Middlebrook Pharmaceuticals Inc.
Luvox CR capsules	Fluvoxamine maleate	SODAS (Spheroidal Oral Drug Absorption System)	Social anxiety disorder and obsessive compulsive disorder	Jazz Pharmaceuticals Inc.
Simcor extended release tablets	Niacin and simvastatin	Hydrogel drug delivery	Comprehensive cholesterol management	Abbott
Esomeprazole delayed release for suspension	Esomeprazole magnesium		Dyspepsia, GERD	AstraZeneca
Pristiq tablets	Desvenlafaxine succinate		Major depressive disorder and vasomotor symptoms associated with menopause	Wyeth Pharms Inc.
Omeprazole delayed release for suspension	Omeprazole magnesium		Dydpepsia and GERD	AstraZeneca
Aplenzin	Bupropion hydrobromide		Depression	Biovail Laboratories
Venlafaxine hydrochloride extended release tablets	Venlafaxine hydrochloride	Osmodex® controlled release technology	Major depressive disorder and social anxiety disorder	Osmotica Pharmaceutical Corp.
Requip XL	Ropinirole	Geometrix	Parkinson's disorder	GlaxoSmithKline

TABLE 18.2   Representative Oral Controlled Release Products under Development Since 2006

Product	Drug	Company	Indication	Phase of Development	Technology
Jurnista	Hydromorphone	J&J	Severe pain	Approved in EU; outlicensed in the United States	OROS
Lamictal XR	Lamotrigine	GSK	epilepsy	Approvable	
Prograf MR	Tacrolimus	Astellas	Liver transplant rejection and several new indications	Approvable	
Remoxy	Oxycodone	Pain Therapeutics/ Durect	Pain	Post PIII	ORADUR
Rosiglitazone XR	Rosiglitazone	GSK	Alzheimer's disease	PIII	
Gepirone ER	Gepirone	GSK	Major depressive disorder	PIII	Matrix
Avandamet XR	Rosiglitazone + metformin HCl	GSK	Type 2 diabetes	PIII	
Coreg CR + ACE inhibitor	Carvedilol + ACE	GSK	Hypertension	PIII	Micropump
Gabapentin GR	Gabapentin	Depomed	Diabetic peripheral neuropathy	PIII	AcuForm™

TABLE 18.3   Patient Benefits of Drug Delivery in Life Cycle Management Products

Benefit	Examples
Maintain or optimize efficacy	Concerta®, Covera-HS®, Ditropan XL®, Jurnista™
On-demand or patterned delivery	Ionsys®, Nicoderm CQ®
Reduce side effects	Procardia XL®, Ditropan XL®, Doxil®
Reduce drug interactions	Ditropan XL®
Minimize abuse	Concerta®
Targeted delivery	Doxil®
Overcome high first-pass metabolism	Duragesic®, Ditropan XL®

18.2   DEVELOPMENT OF ORAL CR-BASED PRODUCTS

This section intends to provide an overview of the development plan for products based on oral controlled release technologies using off-patent molecules. A detailed description of the development of oral CR-based products is presented elsewhere [8, 9]. Several functional areas representing multiple disciplines are responsible for the development of the product. The pivotal input comes from the drug delivery or formulation scientists, clinical pharmacologists, clinicians, and regulatory, marketing, and project management associates. An overview of the various activities during the development is depicted in Figure 18.2.

Formulation Design and Development: After assessing the key physicochemical and pharmacokinetic properties of

the drug, in the simplest case, the estimated dose for the formulation to be used to assess the feasibility of the controlled release dosage form is determined using Equation 18.1, where C_{ss} is the concentration of drug in plasma at steady state, CL_o is the oral clearance of drug based on solution dosage form and is calculated using the product of volume of distribution and elimination rate constant and divided by oral bioavailability (BA), and k_0 is the *in vitro* release rate from the CR tablet. The dose is calculated using *in vitro* release rate and dosing interval, which is usually either 12 or 24 h.

$$k_0 = C_{ss} \times CL_o \qquad (18.1)$$

The equation above assumes constant release rate and absorption of drug throughout the dosing interval. Since the colonic absorption could be different from the absorption in the upper gastrointestinal tract, the dose could be underestimated and may need adjustment. Often tablets with two release rates are included in the first pharmacokinetic study in healthy volunteers to estimate the final dose and release rate of the product.

Advanced Formulation Development Studies: After proving that the drug is absorbed well in the lower parts of the gastrointestinal tract, additional optimization of the formulation is performed. These *in vitro* studies are conducted to understand the effect of the levels of each rate-controlling excipient on the *in vitro* release rate. These studies also help in finalizing the amounts of various rate-controlling excipients in the formulation with a goal to optimize the release rate and reduce batch-to-batch variability. These formulations are packaged and placed on informal stability studies to evaluate

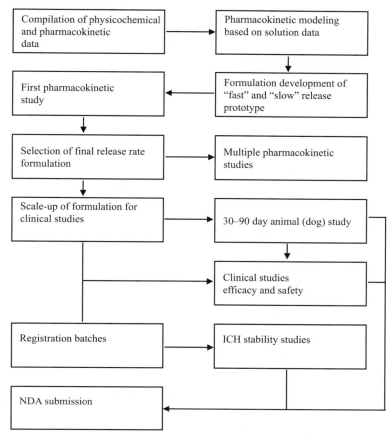

FIGURE 18.2 Development plan for oral controlled release (CR) products.

the impact of temperature and humidity on release rate and other quality parameters.

Scale-up and Registration Stability Studies: At this stage the formulations are scaled up to meet the needs of batch sizes required for registration stability batches. If possible, several studies to understand the effect of processing conditions, such as coating parameters, and granulation conditions on the *in vitro* release rate are performed. Based on the ICH guidance, three batches of all dosage strengths in the to-be-marketed formulation and commercial packaging are required to be placed on stability at 40°C/75% RH for 6 months and 25°C/60% RH over the shelf life of the product.

Preclinical Animal Toxicology Studies: The premise of the life cycle management programs is that the there is an adequate toxicology information on the drug. Assuming that the polymeric excipients in the controlled release product are included in the inactive ingredient list of approved products by the FDA, the toxicological program is relatively short. Sometimes if the choice of the drug is such that the carcinogenicity information is not demonstrated, these additional data may be required. For a typical program, prior to initiating a large clinical trial in humans, studies include oral dosing study of 30–90 days in dogs to evaluate gastrointes-

tinal tolerability and systemic toxicity. Additional studies to demonstrate *in vitro/in vivo* drug release rate to compare the *in vivo* release profile in the dog with the *in vitro* release profile may be useful. These studies also demonstrate that the controlled release product does not have dose dumping effect in animals [10].

Pharmacokinetic Studies: After the first pharmacokinetic study in healthy volunteers to select the dose and release rate, additional proof-of-concept studies are often conducted to assess the impact of new plasma profile—pulsatile, zero order, or ascending—on pharmacodynamic (PD) measurements. For instance, several studies were conducted to determine the effect of rate of delivery of methylphenidate (MPH) (Concerta®) on concentration–effect relationships [11, 12]. Additional studies could include multiple-dose pharmacokinetic studies on the CR dosage form and comparison with the immediate release form, food effect study, interaction with alcohol, and dose proportionality study. Often an *in vitro/in vivo* correlation (IVIVC) may be required to justify the release rate specifications of the CR dosage form. In the absence of IVIVC, the release profile specifications could be justified using bioequivalence (BE) of the target plasma profile compared to one with the "slow" or "fast" profile.

Clinical Studies: For an approved drug in a well-established CR form, one or two placebo-controlled phase 3 clinical trials are required to prove the efficacy. Additional long-term studies (1 year) may be required to determine the safety of the novel CR product based on ICH guidelines. In situations when estimating the dose from the initial PK studies is difficult, a phase 2 dose finding study could be included into the program.

Regulatory Aspects: The most common regulatory pathway for registration and approval by the U.S. FDA of the CR products is to submit a new drug application under Section 505 (b)(2) of the Federal Food, Drug, and Cosmetic Act. The details are presented in the FDA's regulation at 21 CFR 314.54. A 505(b)(2) application relies on information that was not conducted by or for the applicant and for which the applicant has not obtained a right of reference. Typically, prior information that is referenced could include toxicological studies conducted with the new chemical entity. The information could come from published literature or from the agency findings of safety and effectiveness for an approved drug. Unlike the full NDA, the filing or approval of a 505(b)(2) application may be delayed due to patent or exclusivity protections covering an approved product. If one or more of the clinical studies, other than the bioavailability/bioequivalence studies, were essential for approval of the CR product, the 505(b)(2) application may be granted 3 years of Hatch–Waxman exclusivity [13]. The review period at the FDA for these applications may vary but could be as short as 10 months.

In general, it is helpful to engage with the FDA throughout the development process. The first PK study often serves as the IND opening study. Face to face discussions with the FDA could take place during pre-IND meeting, end of phase 2 meeting, and pre-NDA meeting.

18.3   IMPORTANCE OF PATENTS FOR PRODUCTS INTRODUCED TO MANAGE LIFE CYCLE

The number of prescriptions filled by generic drugs increased from 19 to 47% between 1984 (the year Hatch–Waxman Act was introduced) and 2002. The proportion of generics as a percentage of total prescriptions is reported to be close to 60% [14]. Patents are pivotal part of the strategy in developing products using existing drugs as they provide the necessary market protection beyond the 3 years of regulatory exclusivity. Technically, the U.S. Patent and Trademark Office issues protection for 20 years after the first filing date [15].

The compositions of matter patents cover the drug substance and offer the strongest protection from competition. The method-of-use patents are directed toward the use of the invention, in this case, it would be the medical condition. These offer second level of protection after the composition of matter. The formulation patents cover the delivery system. The protection offered by the formulation patents is moderate. Most debate and litigation has focused on the patentability or the degree of protection offered by the formulation patents.

The CR product patents broadly cover the unique and novel aspects of the formulations. Many CR-based patents claim the shape of the plasma profile. Some claim a method of treating a disease using certain drug formulated in a novel and unique delivery system. Several examples of patent claims utilized to protect CR patents are presented later in this chapter.

18.4   PRODUCT CASE STUDIES

The use of oral controlled release technologies as the basis for managing life cycle of off-patent molecules has been exemplified. The case studies that are listed in Table 18.4 and explained in the following sections demonstrate that this strategy has been successfully employed across various therapeutic areas such as central nervous system, cardiovascular, and urology. More than one oral controlled release technology has been employed to develop these products making the technology both its core and its Achilles heel.

18.4.1   Paliperidone ER (INVEGA®)

Paliperidone is a psychotropic drug that belongs to the chemical class of benzisoxazole derivatives. It is indicated

TABLE 18.4   Examples of Oral Controlled Release Products

Product	Drug	Therapeutic Area	Technology
Invega	Paliperidone	CNS	OROS
Concerta	Methylphenidate	CNS	OROS
Focalin XR	Dexmethylphenidate hydrochloride	CNS	SODAS
Metadate CD	Methylphenidate	CNS	Multiparticulates
Cardizem CR and LA	Diltiazem	Cardiovascular	SODAS
Dilacor XR	Diltiazem	Cardiovascular	Geomatrix
Procardia XL	Nifedipine	Cardiovascular	OROS
Coreg CR	Carvedilol	Cardiovascular	Multiparticulates
Seroquel XR	Quetiapine fumarate	CNS	Matrix tablets
Ditropan XL	Oxybutynin	Urology	OROS

for the treatment of schizophrenia, which is characterized by a range of symptoms, including delusions, hallucinations, agitation, hostility, emotional withdrawal, incoherent thoughts and speech, lack of spontaneity, and affective flattening.

From the start, the drug delivery and pharmacokinetic scientists collaborated to design the product that would demonstrate an improved tolerability profile. Paliperidone ER is designed using OROS® osmotic release technology. It is a capsule-shaped tablet comprised of three layers as the core. The core tablet is enclosed within a subcoat and a semipermeable membrane. Two laser-drilled orifices are present on the drug-layer side of the tablet. The drug is released in gradually increasing concentration over a period of 24 h. The system is designed to match the maximum plasma concentration available from an immediate release tablet of the parent drug. The gradual rise in the plasma concentration (ascending profile) allows reaching the minimum effective concentration with a single dose, thus eliminating the need to titrate the drug over several days, and more important, overcoming the adverse effect of orthostatic hypotension. The rapid rise in plasma concentration of the parent drug from the immediate release tablet results in orthostatic hypotension that could lead to falling in elderly patients. By overcoming the need to titrate an effective dose, paliperidone ER patients can better handle the situations of drug holidays and potentially remain compliant over the course of the treatment.

A pharmacokinetic–pharmacodynamic study comparing ascending (5.5 mg dose), flat (4.5 mg dose), and immediate (4 mg dose) drug release profiles was conducted with paliperidone. The study concluded that the flat and ascend treatments provided lower incidences of orthostatic hypotension compared to the immediate release profile. Also, ascend treatment resulted in less prolactin elevation compared to other treatments [16].

The inventors of paliperidone ER have attempted to patent [16] the oral dosage form that offer a unique *in vitro* ascending drug release profile and characteristics of the plasma profile *in vivo*. The PK attributes included maximum plasma concentration-to-dose ratio, time to maximum plasma concentration, and area under the plasma concentration curve (AUC). The claims also include numerical limits for ratios of various partial areas under the plasma curve, for instance, AUC 0-6/AUC 0-24.

18.4.2 OROS® Methylphenidate (Concerta®)

Concerta® is a once-daily form of methylphenidate that is indicated for attention deficit/hyperactivity disorder (ADHD). It addressed an unmet patient need by overcoming the need to distribute a scheduled substance to children at school. More important, from a clinical perspective, it differentiated from Ritalin SR® by its improved efficacy and by overcoming tolerance to the drug during the course of the

day. Methylphenidate, a central nervous system stimulant, has been mostly used in immediate release and conventional sustained release (SR) formulations for the treatment of ADHD. Concerta® was designed and developed using OROS® technology from ALZA and launched by Johnson & Johnson as the once-daily version of methylphenidate to enter the ADHD market, providing significant improvements over existing treatments. The annual sales of Concerta® in 2007 reached over a billion dollar [17].

Methylphenidate has a short half-life of 1–3 h [18, 19]. Concerta® is designed to provide an ascending plasma profile that allows a fast onset of action followed by a sustained effect for an additional 12 h and is designed for a single morning dose that allows controlled release of drug over school day and late afternoon. The system is trilayer, membrane-coated tablet core enclosed within a drug overcoat layer. The overcoat contains approximately 20% of the dose, which is for immediate release, and the core and middle layers contain increasing concentrations of drug that are released over time. The OROS® design successfully overcomes the deficiencies of the existing IR and SR methylphenidate products. IR formulations need to be taken several times a day for effective treatment and the SR product (Ritalin SR®, Novartis, E. Hanover, NJ) lacks a fast onset of action and yields a flat plasma profile that results in tolerance to the drug [20].

A study of the pharmacodynamic effects of methylphenidate delivered by Concerta® showed that the drug plasma concentrations increased over the first 2 h following release from the drug overcoat layer, and a further increase was achieved as drug from the first layer of the tablet core was released. Peak plasma concentrations were achieved 6–8 h after administration following release from the second layer of the tablet core.

In clinical trials in children between 6 and 12 years of age, Concerta® was significantly more effective than placebo. The efficacy of Concerta® was similar to IR methylphenidate administered thrice a day and demonstrated similar onset of action [21]. Concerta® also minimizes the fluctuations between peak and trough concentrations associated with immediate release methylphenidate (Figure 18.3) [22].

Methylphenidate is a controlled substance and is subjected to abuse. Because of the tablet's nondeformable shell, OROS® methylphenidate cannot be cut and is less prone to be abused. The Drug Abuse and Warning Network mentions OROS® methylphenidate 50 times compared to 588 mentions of all other methylphenidate brands between 2000 and 2002; of the 548 citations for abuse of oral methylphenidate, only 49 concerned the OROS® formulation. There was no mention of any abuse of the OROS® formulation via other routes (sniffing or snorting and injecting). The OROS® market share is estimated at 5–48% of the total, and abuse of OROS® formulation has been significantly less than that of the other methylphenidate products [23].

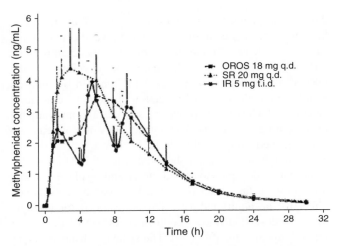

FIGURE 18.3 Mean methylphenidate plasma concentrations in 36 adults, following a single dose of Concerta® 18 mg q.d. and immediate release methylphenidate 5 mg t.i.d. administered every 4 h.

ALZA has been successful in protecting the concept of drug release in an ascending manner over longer periods in its patents. There are two issued U.S. patents that specifically protect Concerta® [24, 25]. These patents claim a method of treatment of ADHD using methylphenidate-containing composition that achieves substantially ascending plasma concentrations over various time periods ranging from about 4–9.5 h.

There are a few other patent applications pending that are not specific to methylphenidate but include various classes of drugs, such as CNS drugs, anti-infectives, and others. It also claims a method of treatment for a patient in a clinical situation where therapeutic effectiveness decreases at time periods before the end of intended therapy using a composition that achieves ascending plasma concentration over a period of 8 h. Another patent application narrowly claims the osmotic system for a list of drugs.

18.4.3 Dexmethylphenidate Extended Release (Focalin® XR)

The conventional commercial preparations of methylphenidate contain the racemate, DL-methylphenidate. The *d-threo* enantiomer, dexmethylphenidate hydrochloride in Focalin® XR, has been approved for the ADHD indication in both adults and children [26]. Novartis first launched the immediate release version of dexmethylphenidate (Focalin®) and later developed an extended release version that was approved in 2005 for the same indication. Focalin® XR has demonstrated efficacy in children [27], uses a dose that is half of other methylphenidate products that are racemates, and allows less frequent dosing than immediate release preparations [28].

Focalin® XR (dexmethylphenidate hydrochloride) capsule is an extended release formulation of dexmethylphenidate

with a bimodal release profile. It uses the proprietary SODAS® (Spheroidal Oral Drug Absorption System, Elan Corporation, plc) technology. Each bead-filled Focalin® XR capsule contains half the dose as immediate release beads and half as enteric-coated, delayed release beads, thus providing an immediate release of dexmethylphenidate that starts acting within the first hour of dosing and a second delayed release of dexmethylphenidate for control of symptoms throughout the work day.

The FDA's Orange book lists several drug delivery-based patents [29–31] that may offer market protection for Focalin® XR until 2019. The inventors claim a dosage form for oral administration of methylphenidate of two groups of particles—first group for immediate release and second group of particles for delay in drug release by 2–7 h. The formulation comprises 2–17% by weight of drug with binders and coating of ammonio methacrylate [29]. In another invention, the authors claim a method of treating a disease amenable to treatment with a phenidate drug using a once-daily dosage form comprising two groups of particles each containing methylphenidate. The first group consists of 2–99% by weight of methylphenidate for immediate release. The second group consists of 2–75% by weight of methylphenidate along with one or more binders and coating consisting of ammonio methacrylate copolymer. The second group delays the release of methylphenidate by 2–7 h after administration. More important, this invention claims dosage form that provides *in vivo* plasma concentrations of methylphenidate comprising two maxima separated temporally by 2–7 h and these two plasma concentrations do not differ by more than 30% [32]. An additional invention that does not seem specific to methylphenidate claims a multiparticulate modified release composition that includes two or more sets of drug-containing particles. The first set includes a CNS drug. Subsequent sets of particles comprise of modified release coating or matrix material such that the dosage form delivers the drug from these particles in a pulsatile manner [31].

Focalin® XR is administered once daily and results in a lower second peak concentration, higher interpeak minimum concentrations, and less peak and trough fluctuations than Focalin tablets given in two doses 4 h apart. This is due to an earlier onset and more prolonged absorption from the delayed release beads.

The AUC (exposure) after administration of Focalin® XR given once daily is equivalent to the same total dose of Focalin® tablets given in two doses 4 h apart [33].

18.4.4 Methylphenidate Controlled Release (Metadate® CD)

There is a clear need to overcome the burden of dispensing methylphenidate, which is a schedule II controlled substance, to children in the middle of a school day. A drug delivery-based product, Metadate® CD became available in 2003 based on Eurand's Diffucaps™ technology to address this need. It consists of multiparticulates or beads acting as drug reservoir. Each Metadate® CD capsule consists of 30% methylphenidate as immediate release beads and 70% as delayed release beads from which the drug is released over 6–10 h. In one study of healthy adults, in the first 6 h after administration of medication, those taking Metadate® CD had higher plasma levels of methylphenidate than those taking equivalent Concerta® [34]. The second pulse of methylphenidate in Metadate® CD may have reduced bioavailability and therefore reduced efficacy.

Based on the U.S. Patent 6,344,215, the inventors Betmann et al. have claimed the delivery system in a very specific manner. They have claimed a modified release methylphenidate capsule comprising of immediate release and extended release MPH-containing beads in a specific ratio (30:70) to deliver certain dose of MPH. The claim adds further specification by describing the formulation of the two types of beads. The immediate release beads include a core particle coated with MPH-containing water-soluble film. The extended release beads are made up of a core particle coated with MPH-containing water-soluble film, which is further coated with a dissolution rate-controlling polymer in an amount up to 20%. The claim includes *in vitro* cumulative release profiles of MPH using various proportions of the immediate and extended release beads. Additional claims specify ethylcellulose as the dissolution rate-controlling polymer and polyvinylpyrrolidone as the film forming excipient.

18.4.5 Controlled Release Diltiazem (Cardizem® CD, Cardizem® LA, and Dilacor® XR)

Cardizem® (diltiazem, a calcium channel blocker) provides an excellent example of how incremental therapeutic improvements due to reformulation can extend the life cycle of a drug through extension of patent protection. Cardizem® was first developed by Marion Laboratories (now Aventis) as a thrice-daily oral tablet formulation whose original patent expired in 1988. A sustained release twice-daily formulation, Cardizem® SR was later developed and further reformulated to once-daily formulation Cardizem® CD. Biovail purchased the patent in 2002 and incorporated a new drug delivery technology through reformulation to develop a very successful long-acting version of the drug, Cardizem® LA, 14 years after the original patent ended.

18.4.5.1 Cardizem® CD Cardizem® CD is formulated as capsules and consists of a dual microbead system using SODAS technology from Elan [35, 36]. The beads that are in the range of 1–2 mm in diameter are made using core seeds to which is applied a first coating containing the diltiazem hydrochloride. Over the first coating, further coatings of

either a thin copolymer (rapid release diltiazem beads) or a thick copolymer (delayed release diltiazem beads) are applied. Within the GI tract, the soluble polymers dissolve, leaving pores within the outer membrane. Fluid then enters the core of the beads and dissolves the drug. The resultant solution diffuses out in a controlled, predetermined manner allowing the prolongation of the *in vivo* dissolution and absorption phases [37, 38].

The U.S. Patent 5,002,776 describing the Cardizem CD formulation expired in March 2008. The other U.S. patents describing the formulation will expire in May 2011 [39, 40].

18.4.5.2 Cardizem® LA Cardizem® LA is a recent commercially available form of chronotherapeutic diltiazem hydrochloride that provides 24 h blood pressure control with a once-daily dose and offers physicians a flexible dosing range from 120 to 540 mg. Cardizem® LA was approved by the FDA in 2003 for hypertension and angina. The formulation of Cardizem® LA employs immediate release diltiazem (uncoated) and extended release diltiazem beads (coated) that are mixed with a hydrophobic or wax material compressed to a tablet and control the release of diltiazem from the dosage form [41]. This polymer creates a lag time in tablet dissolution, allowing maximal concentrations within 11–18 h post dose [42]. This delayed peak in plasma concentration of the drug allows patients to take the medication at bedtime and have better coverage during the morning hours. Cardizem® LA has a different pharmacokinetic profile and offers compliance advantages compared to Cardizem® CD. Bedtime dosing of Cardizem® LA provides peak diltiazem concentrations within 11–18 h, whereas bedtime dosing of Cardizem® CD provides peak diltiazem within 10–14 h (Figure 18.4) [43, 44]. To obtain possible chronotherapeutic benefits with Cardizem® CD during the 6:00 a.m. to noon time period, the patient would have to administer the capsule at about 2:00 a.m., while with Cardizem® LA the patient has the advantage of administering the capsule before bedtime. When dosed at bedtime, Cardizem® LA demonstrates a unimodal peak plasma concentration profile, as opposed to the CD formulation, which exhibits a bimodal profile, with decreasing plasma concentrations during the early morning hours.

Evening administration of Cardizem® LA 360 mg correlated to an additional 3.3 mmHg diastolic blood pressure reduction, compared with morning administration of 360 mg [45]. With evening administration, clinical trials have shown Cardizem® LA increased reduction in blood pressure in the early morning hours, which is when patients may be at the greatest risk of significant cardiac events.

For chronic stable angina, evening doses of 180, 360, or 420 mg of Cardizem® LA statistically improved exercise tolerance compared to morning doses of 360 mg or placebo. For all evening doses, particularly during 7–11 a.m., time to onset of angina and myocardial ischemia was significantly

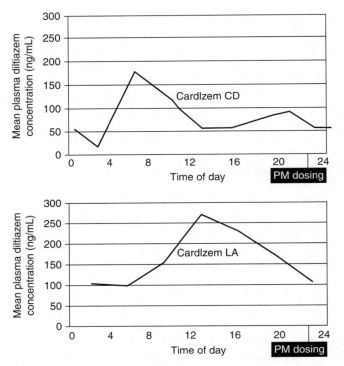

FIGURE 18.4 Comparison of Cardizem LA and Cardizem CD plasma concentrations when administered at bedtime.

delayed compared to placebo. Overall, adverse events in all the Cardizem LA treatment groups were lower than those of placebo.

Two U.S. patents protect the Cardizem LA product [46, 47]. The U.S. Patent 7,108,866 claims the treatment of hypertension and angina with chronotherapeutic dosing of Cardizem LA and will expire in 2019, while the U.S. Patent 6,923,984 covers drug product claims and offers patent protection till 2021.

Cardizem achieved revenues of $260 million in 1988. Hoechst Marion Roussel collaborated with Elan to develop the twice-daily formulation, Cardizem SR, and the revenues peaked in 1989 at $400 million and remained steady until 1991, when Cardizem CD was introduced. Twelve years after the original product was marketed, these line extensions of that same molecule bring in about $800 million yearly despite generic alternatives.

18.4.5.3 Dilacor® XR

Another example of successful life cycle management of diltiazem is Dilacor® XR. Rhone-Poulenc Rorer extended the life of the product by new formulation and a new indication. The product was launched in 1992 and is available as 120, 180, and 240 mg capsules. Unlike Cardizem that was mainly used for the treatment of angina, it was developed for the treatment of hypertension. The product also benefited from a new indication, when later

in the products' life cycle, Dilacor® XR received approval for the treatment of angina. This added three more years of patent protection and gave Rhone-Poulenc the opportunity to penetrate the angina market [48].

A unique formulation technology was used to create a patent-protected brand of Dilacor® XR, which is produced as two-piece hard gelatin capsules, with each capsule containing a multitude of tablets [49]. Dilacor® XR uses Geomatrix™ delivery system and contains a degradable controlled release tablet formulation designed to predictably release diltiazem at a constant rate over a 24 h period. Each tablet comprises a cylindrical core containing 60 mg diltiazem hydrochloride mixed with inactive ingredients, which include a polymer that swells and forms a gel upon contact with aqueous fluids. Because the gel has high viscosity, it swells and dissolves slowly in the gastrointestinal fluids and thereby retards the rate of release of the diltiazem hydrochloride. To further retard the release, insoluble polymeric barriers are affixed to the top and bottom of the cylindrical core, thus leaving only the periphery exposed to the gastrointestinal fluid. Figures 18.5 and 18.6 schematically represent Dilacor® XR formulation and mechanism of drug release from the delivery system [50]. The formulation of Dilacor® XR capsules successfully accomplishes gradual release to enable once-daily dosing, but requires complex and expensive procedures to produce. In particular, production of the tablets contained in Dilacor® XR capsules requires production of cores containing the diltiazem hydrochloride and the affixing of the insoluble platforms to it. Another difficulty with the Dilacor XR formulation is that the tablets are larger than desirable, so the capsules containing 4 tablets (240 mg strength) are of size 00, having an external diameter of about 8.5 mm and length of about 23.7 mm, making them difficult to swallow. Controlled absorption of diltiazem begins within 1 h, with maximum plasma concentrations being achieved 4–6 h after administration.

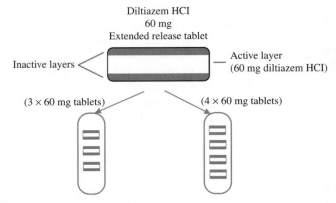

FIGURE 18.5 Depiction of the new, extended release formulation of Dilacor XR as produced by the patented Geomatrix system.

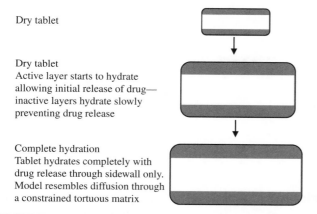

Dry tablet

Dry tablet
Active layer starts to hydrate
allowing initial release of drug—
inactive layers hydrate slowly
preventing drug release

Complete hydration
Tablet hydrates completely with
drug release through sidewall only.
Model resembles diffusion through
a constrained tortuous matrix

FIGURE 18.6 Schematic representation of Geomatrix release mechanism for Dilacor XR.

In vivo release of diltiazem occurs throughout the gastrointestinal tract, with controlled release still occurring for up to 24 h after administration, as determined by radiolabeled methods. The presence of food did not affect the ability of Dilacor® XR to maintain a controlled release of the drug and did not impact its sustained release properties over 24 h after administration.

The tablets used in Dilacor® XR are made in accordance with the invention of U.S. Patents 4,839,177 and 5,422,123. The U.S. Patent 4,839,177 describes a system for the controlled release of active substances and claims the delivery device and the process and expired in December 2006. The U.S. Patent 5,422,123 describes tablets with controlled-rate release of active substances and claims the formulation as well as the process and protects the product through June 2012.

18.4.6 OROS® Nifedipine (Procardia XL®)

The collaborative efforts of Pfizer and ALZA led to the development of the cardiovascular drug nifedipine as Procardia XL® [51] for the treatment of hypertension and angina pectoris. This product was introduced in 1989 to replace Pfizer's existing product Procardia®, which was an immediate release formulation given thrice-a-day. Procardia XL® successfully used ALZA's controlled release oral osmotic (OROS) delivery system that enabled a consistent drug delivery over time and, therefore, extended efficacy, optimized blood pressure control, and reduced side effects. In the OROS nifedipine tablets, as fluid from the GI tract entered the outer semipermeable membrane of the tablet, it dissolved nifedipine, caused the "push" layer to swell, thereby expelling the drug slowly into the GI tract through a laser-drilled hole in the drug side of the membrane. The constant flow provided circadian control of drug against

heart attacks compared to sometimes dangerous variations in plasma levels in drugs taken in the immediate release dosage form.

The success of Procardia XL® is due to multiple advantages that it offers. Clinical trials conducted to compare efficacy, safety, compliance, and relative costs of treatment between Procardia XL® and the immediate release formulation demonstrated fewer side effects (32% versus 58%), higher compliance rates (93% versus 76%), and lower cost of treatment for Procardia XL® [52]. In another study, that switched patients from IR to OROS treatment, the latter resulted in reduction in angina and nitroglycerin usage. Also, based on questionnaires that asked about frequency of symptoms, activity work performance, and energy level, 87% of the patients reported stable or improved quality of life after switching to OROS nifedipine [53].

Calcium channel blockers administered in IR dosage forms may be associated with vasodilatory side effects and reflex activation of the sympathetic nervous system. In a study that exemplifies the effect of rate of delivery on clinical effects, Kleinbloessem et al. administered two regimens of IV infusion to evaluate hemodynamic effects in six healthy volunteers [54]. The first regimen resulted in steady-state blood plasma concentrations over 5–7 h, and the second regimen achieved the same results within 3 min; the concentrations were similar. During the gradual-rise infusion, heart rate was unchanged and diastolic blood pressure fell slowly by 10 mmHg. With a faster infusion rate, heart rate increased immediately and remained elevated for the duration of the infusion. At the end of the gradual-rise regimen, a sudden increase in the infusion rate for 10 min produced tachycardia and an unexpected increase in blood pressure. The authors hypothesized that these unexpected effects could be due to baroreceptor activation. Overall, the study established a clear role of rate of drug input on its pharmacologic effect.

18.4.7 Carvedilol Phosphate Extended Release Capsules (Coreg CR™)

Coreg CR™ is a once-daily controlled release formulation of Coreg (carvedilol phosphate) indicated for the treatment of mild-to-severe heart failure, left ventricular dysfunction following myocardial infarction, and hypertension [55]. Coreg CR™ has been developed by GSK in collaboration with Flamel and uses its Micropump platform. Flamel's Micropump delivery system utilizes a micron-scale technology for encapsulating small molecules for oral administration and allows a more convenient once-daily dosing schedule. Each microparticle consists of an active ingredient core coated with a diffusion controlled layer and releases the drug at a controlled rate, extending the duration of action and limiting its concentration, which can limit side effects (Figure 18.7).

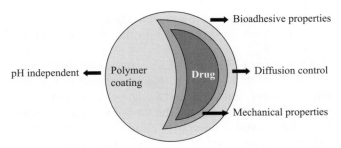

FIGURE 18.7 Schematic representation of a micropump. Each microparticle of micropump consists of a core of active ingredient and a polymer coating that may be soluble in a pH-dependent or a pH-independent manner. Micropump acts as a delivery system that releases the drug at a controlled rate extending the duration of action of the drug and limiting its concentration that can limit side effects.

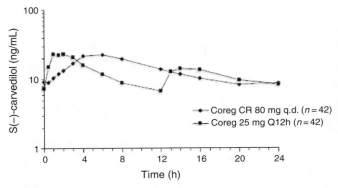

FIGURE 18.8 Mean steady-state plasma concentration–time profiles of Coreg and Coreg CR in patients with left ventricular dysfunction.

Coreg CR™ consists of hard gelatin capsules filled with carvedilol phosphate immediate release and controlled release microparticles that are drug layered and then coated with methacrylic acid copolymers. The capsule contains particles of sizes between 200 and 500 μm that release about 12.5% of the dose as immediate release component, 37.5% of the dose is released by micropumps at pH 5.5, while 50% of the dose is released at pH 6.4–6.8.

The development costs for a CR-based follow-on product could be lower compared to the costs to introduce a new chemical entity. Clinical trials necessary to support efficacy and safety requirements for the immediate release carvedilol as new chemical entity for the treatment of heart failure and myocardial infarction required close to 8900 patients. Approval of Coreg CR™ in these conditions was simply based on equivalence of pharmacokinetic and pharmacodynamic (β-blockade) parameters between Coreg CR™ and immediate release carvedilol.

The once-daily dosing of controlled release formulation replicated the antihypertensive efficacy of the twice-daily dosing of the immediate release formulation. In a 122 patient crossover study, the pharmacokinetic and pharmacodynamic profiles were compared between patients receiving carvedilol CR and carvedilol IR. The PK profiles for R(+)- and S(−)-carvedilol for the two formulations were equivalent (based on area under the curve, maximum plasma concentration [C_{max}], and trough drug concentration). Carvedilol CR delayed C_{max} by 3.5 h compared to carvedilol IR. In addition, patients receiving Coreg CR™ reported 24% fewer adverse events than those receiving immediate release Coreg [56]. Figure 18.8 shows the pharmacokinetic equivalence of the controlled release and immediate formulations of Coreg.

The controlled release formulation Coreg CR™ simplifies the treatment regimen and offers potential for improved compliance for patients with heart disease. The combination of this benefit with the proven effectiveness of Coreg offers a treatment advantage for the growing number of heart patients who take multiple medications every day [57]. The first patent for Coreg expired in September 2007. Additional patents were granted on carvedilol controlled release formulation that extends the patent life till 2023.

Among the patents covering Coreg CR™, the delivery-based patent claims microcapsules of reservoir kind containing drug for oral use. The microcapsules are characterized further in that these consist of drug-coated particles with at least one coating film that is composed of (i) 50–80% by weight of film forming polymer that is insoluble in digestive tract, such as ethyl cellulose or cellulose acetate, (ii) 5–15% of nitrogen-containing polymer, such as PVP, (iii) 4–15% of plasticizer, such as hydrogenated castor oil, and (iv) at least one surface-active agent or lubricating agent, such as magnesium stearate. The claim also specifies the particle size preferably in the range of 100–500 μm. These microcapsules are expected to remain in the small intestine for at least 5 h to increase drug absorption [58].

Coreg CR™ was launched in March 2007 and represents 21% of new prescriptions and 13% of existing prescriptions of Coreg. Currently, 90% of cardiologists have prescribed Coreg CR and expect to increase their usage. Coreg CR was developed to replace Coreg IR, a twice-daily beta-blocker with sales of $1.4 billion in 2006. Coreg IR hit generic status in September 2007 and Glaxo has been switching patients to Coreg CR™.

Another life cycle management strategy that GSK has employed to further retain market share is to develop a combination product of Coreg CR with an ace inhibitor, lisinopril. Several clinical trials are being conducted to examine the benefits of combining Coreg CR™ and lisinopril for the treatment of hypertension [59].

18.4.8 Quetiapine Fumarate Extended Release Tablets (Seroquel XR™)

Seroquel XR™ is the follow-on product for AstraZeneca's Seroquel® (quetiapine fumarate) that was approved in the United States in May 2007 [60] and is currently indicated for the treatment of schizophrenia. Seroquel XR™ is a once-daily dosage form and incorporates extended release technology by film-coated matrix tablets comprising hypromellose as a release controlling copolymer with diffusion and erosion controlled release [61].

The original version of Seroquel® was launched in 1997 and is approved in 88 countries for the treatment of schizophrenia, in 79 countries for the treatment of bipolar mania, and in 11 countries, including the United States, for the treatment of bipolar depression. Although the global sales of Seroquel® for 2007 passed the $4 billion mark, generic drugmakers are challenging patents protecting its exclusivity. In addition, Seroquel® will lose patent exclusivity in most major markets by 2012. Patents have been obtained on the composition of extended release dosage form of quetiapine claiming gelling agents in combination with quetiapine. These patents for Seroquel XR™ extend till 2017 [62].

Additional life cycle management has been recently (January 2008) sought by AstraZeneca by seeking approval from the U.S. regulators to sell an extended release version of its Seroquel® drug to treat both manic and depressive episodes associated with bipolar disorder as monotherapy, adjunct therapy, and maintenance therapy in adult patients. The company based its submission to the U.S. Food and Drug Administration on clinical studies demonstrating the effectiveness of the once-daily pill in treating depression, opening up a potential new market for the product. Clinical data showed that patients with major depressive disorder and generalized anxiety disorder who received Seroquel XR™ once daily experienced significant reductions in symptom severity compared to those on placebo in each of three trials. The effect was seen with all of the doses of the drug tested [63].

18.4.9 OROS® Oxybutynin Hydrochloride (Ditropan XL®)

In the 1990s, a similar OROS system was used to reformulate oxybutynin chloride, an anticholinergic and antispasmodic agent for the treatment of overactive bladder and urge-urinary incontinence by Ortho-McNeil. Oxybutynin works by binding to the muscarinic receptors and blocking the attachment of acetylcholine to those sites on the bladder muscles. The attachment of neurotransmitters (acetylcholine) to the receptors site (called cholinergic) sets in motion a sequence of changes that result in muscle contractions. Blocking this combination prevents the contraction.

The incidence of systemic anticholinergic side effects such as dry mouth, dry eyes, constipation, and headache are associated with the use of the immediate release formulation and may limit compliance with dosing regimens. Oxybutynin is extensively metabolized in the liver and small intestine resulting in low oral bioavailability (6%). Therefore, immediate release formulations of oxybutynin have to be administered two to three times a day, making it a good candidate for controlled release formulations.

Ditropan XL®, a controlled release formulation of oxybutynin chloride, is a once-daily dosage form that incorporates the OROS® technology for drug delivery to the gastrointestinal tract. OROS® oxybutynin delivers the drug by an osmotic process in a constant zero-order manner over approximately 24 h. It was designed such that no appreciable quantity of the drug is delivered in the first 2–3 h after the patient ingests the system. Most of the drug is released when the system reaches the colonic portion of the gastrointestinal tract, 3–5 h following dosing.

The extended release formulation of oxybutynin hydrochloride offers several benefits over the immediate release formulation. In an open-label, two-way crossover, multiple-dose study, the pharmacokinetics of a once-daily controlled release formulation, OROS® oxybutynin chloride (Ditropan XL®), was compared with that of immediate release oxybutynin. It was demonstrated that the OROS® system provides relatively constant oxybutynin concentrations throughout the day and night, minimizing the peak/trough fluctuations associated with medications given several times a day by 66–81%. With OROS® oxybutynin chloride, mean relative bioavailability was higher (153%) for oxybutynin [64, 65]. Fewer subjects reported adverse event with OROS® oxybutynin chloride than with IR oxybutynin. In particular, symptoms of dry mouth were reduced from 46 to 25% [66]. The lower incidence of anticholinergic effect like dry mouth has been attributed to the reduced formation of a primary metabolite, N-desethyloxybutynin, with the OROS® formulation [67–69]. This increased parent drug bioavailability and reduced bioavailability of N-desethyloxybutynin may be due to reduced gut-wall first-pass metabolism. Within 3–5 h after dosing, OROS® oxybutynin chloride systems are thought to reach the colon, where cytochrome P450-mediated oxidation, the primary metabolic pathway, appears to be less extensive than in the small intestine.

18.5 SUMMARY

High proportions of the new drug applications approved by the FDA are utilizing existing drugs in combinations with oral controlled release technologies. The CR technology plays a pivotal role in defining the success of these products by adding patent or market protection beyond the regulatory exclusivity. Clinical differentiation, better tolerability, or higher efficacy is often enabled by the technology as well. There are a number of examples of CR-based products, but the successful ones—Concerta® and Procardia XL®, provided superior clinical differentiation and patient benefit and were well protected with strong patent positions.

REFERENCES

1. Available at http://drugtopics.modernmedicine.com/drugtopics/ Supplements/Drug-patent-expirations-2007-2009/ArticleStandard/Article/detail/414709.

2. Reformulation strategies: comparisons of past and future reformulation strategies. Available at http://www.datamonitor. com. Accessed Sept 2006.

3. Andersson T, Hassan-Alin M, Hasselgren G, Rohss K, Weidolf L. Pharmacokinetic studies with esomeprazole, the (S)-isomer of omeprazole. *Clin. Pharmacokinet.* 2001;40:411–426.

4. Andersson T, Rohss K, Bredberg E, Hassan-Alin M. Pharmacokinetics and pharmacodynamics of esomeprazole, the S-isomer of omeprazole. *Aliment. Pharmacol. Ther.* 2001;15: 1563–1569.

5. Lind T, Rydberg L, Kyleback A, Jonsson T, Hasselgren G, Holmberg J, Rohss K. Esomeprazole provides improved acid control vs. omeprazole in patients with symptoms of gastro-oesophageal reflux disease. *Aliment. Pharmacol. Ther.* 2000;14:861–867.

6. Grant JA, Danielson L, Rihoux J-P, DeVos C. A comparison of cetirizine, ebastine, epinastine, fexofenadine, terfenadine, and loratadine versus placebo in suppressing the cutaneous response to histamine. *Int. Arch. Allergy Immunol.* 1999; 118:339–340.

7. Available at http://www.gsk.com/investors/product_pipeline/ docs/gsk-pipeline-feb08.pdf.

8. Modi N, Gupta SK. Development of controlled release products. In Bonate PL, Howard DR, editors. *Pharmacokinetics in Drug Development: Regulatory and Development Paradigms*, Vol. 2, AAPS Press, Arlington, VA, 2004, pp. 343–358.

9. Yacobi A, Halperin-Walega E. *Oral Sustained Release Formulations, Design and Evaluation*, Pergamon, New York, 1988.

10. Wong PSL, Gupta SK, Stewart BE. Osmotically controlled tablets. In Rathbone MJ, Hadgraft J, Roberts MS, editors. *Modified Release Drug Delivery Technology*, Marcel Dekker, New York, 2003, pp. 101–114.

11. Swanson JM, Wigal SB, Udrea D, Lerner M, Agler D, Flynn D, Fineberg E, Davies M, Kardatzke D, Ram A, Gupta S. Evaluation of individual subjects in the analog classroom setting: I. Examples of graphical and statistical procedures for within-subject ranking of responses to different delivery patterns of methylphenidate. *Psychopharmacol. Bull.* 1998; 34:825–832.

12. Swanson J, Gupta S, Guinta D, Flynn D, Agler D, Lerner M, Williams L, Shoulson I, Wigal S. Acute tolerance to methylphenidate in the treatment of attention deficit hyperactivity disorder in children. *Clin. Pharmacol. Ther.* 1999;66: 295–305.

13. Guidance for Industry. Application covered by Section 505(b)(2), Food and Drug Administration, CDER, October 1999.

14. Furrow ME. Pharmaceutical patent life-cycle management after KSR v. Teleflex. *Food Drug Law J.* 2008;63:275–320.

15. Nicolaou S. Regulatory and intellectual property issues in drug delivery research. In Wang B, Siahaan TJ, Soltero RA, editors. *Drug Delivery Principles and Application*, John Wiley & Sons, Inc., New Jersey, 2005, pp. 435–442.

16. US Patent Application 2005/0208132.

17. Available at http://www.drugpatentwatch.com.

18. Wargin W, Patrick K, Kilts C, Gualtieri CT, Ellington K, Mueller RA, Kraeme G, Breese GR. Pharmacokinetics of methylphenidate in man, rat and monkey. *J. Pharmacol. Exp. Ther.* 1983;226:382–386.

19. Patrick KS, Mueller RA, Gualtieri CT, Breese GR. *Psychopharmacology: The Third Generation of Progress*, Raven, New York, 1987, pp. 1387–1395.

20. Modi NB, Lindemulder B, Gupta SK. Single and multiple-dose pharmacokinetics of an oral once-a-day osmotic controlled-release OROS (methylphenidate HCl) formulation. *J. Clin. Pharmacol.* 2000;40:379–388.

21. Pelham WE, Gnagy EM, Burrows-Mclean L, Williams A, Fabiano GA, Morrisey SM, Chronis AM, Forehand GL, Nguyen CA, Hoffman MT, Lock TM, Fiebelkorn K, Coles EK, Panahon CJ, Steiner RL, Meichenbaum DL, Onyango AN, Morse GD. Once-a-day Concerta methylphenidate versus three-times-daily methylphenidate in laboratory and natural settings. *Pediatrics* 2001;107–105.

22. Concerta® [package insert], ALZA Corporation, Mountain View, CA, 2007.

23. Spencer TJ, Biederman J, Ciccone PE, Madras BK, Dougherty DD, Bonab AA, Livni E, Parasrampuria DA, Fischman AJ. PET study examining pharmacokinetics, detection and likeability, and dopamine transporter receptor occupancy of short- and long-acting oral methylphenidate. *Am. J. Psychiatry* 2006; 163:387–395.

24. Lam AC, Shivanand P, Ayer AD, Hatamkhany Z, Gupta SK, Guinta DR, Christopher CA, Saks SR, Hamel LG, Wright JD, Weyers RG,inventors, Alza Corporation, assignee. Methods and devices for providing prolonged drug therapy. US Patent 6,930,129, 2005 Aug 16.

25. Lam AC, Shivanand P, Ayer AD, Weyers RG, Gupta SK, Guinta DR, Christopher CA, Saks SR, Hamel LG, Wright JD, Hatamkhany Z,inventors, Alza Corporation, assignee. Methods and devices for providing prolonged drug therapy. US Patent 6,919,373, 2005 July 19.

26. Doyle BB, *Understanding and Treating Adults with Attention Deficit Hyperactivity Disorder*, American Psychiatric Publishing, 2006.

27. Srinivas NR, Hubbard JW, Quinn D, Midha KK. Enantioselective pharmacokinetics and pharmacodynamics of DL-threo-methylphenidate in children with attention deficit hyperactivity disorder. *Clin. Pharmacol. Ther.* 1992;52:561–568.

28. Spencer T, Biederman J, Wilens T, Doyle R, Surman C, Prince J, Mick E, Aleardi M, Herzig K, Faraone S. A large, double-blind randomized clinical trial of methylphenidate in the treatment of adults with attention-deficit/hyperactivity disorder. *Biol. Psychiatry* 2005;57:156–163.

29. Mehta AM, Zeitlin AL, Dariani MM,inventors, Celegene Corporation, assignee. Delivery of multiple doses of medications. US Patent 5,837,284, 1998 Nov 17.

30. Mehta AM, Zeitlin AL, Dariani MM,inventors, Celegene Corporation, assignee. Delivery of multiple doses of medications. US Patent 6,635,284, 2003 Oct 21.

31. Devane JG, Stark P, Fanning NMM,inventors, Elan Corporation, plc, assignee. Multiparticulate modified release composition. US Patent 6,730,325, 2004.

32. Mehta AM, Zeitlin AL, Dariani MM,inventors, Celegene Corporation, assignee. Delivery of multiple doses of medications. US Patent 6,635,284, 2003 Oct 21.

33. Focalin® XR [package insert], Novartis Pharmaceuticals Corporation, East Hanovar, NJ, 2006 Label information.

34. Gonzalez MA, Pentikis HS, Anderi N, Benedict MF, DeCory HH, Hirshey Dirksen SJ, Hatch SJ. Methylphenidate bioavailability from two extended-release formulations. *Int. J. Clin. Pharmacol. Ther.* 2002;40(4):175–184.

35. US Patent 5,968,552, 1999 Oct 19.

36. Verma R, Garg S. Current status of drug delivery technologies and future directions. *Pharma. Technol.* 2001;25(2):1–14.

37. Prisant LM, Elliott WJ. Drug delivery systems for treatment of systemic hypertension. *Clin. Pharmacokinet.* 2003;42(11):931–940.

38. Felicetta JV, Serfer HM, Cutler NR, Comstock TJ, Huber GL, Weir MR, Hafner K, Park GD. A dose-response trial of once-daily diltiazem. *Am. Heart J.* 1992;123(4 Pt 1):1022–1026.

39. Hendrickson DL, Dimmitt DC, Williams MS, Skultety PF, Baltezor MJ,inventors, Carderm Capital L.P., assignee. Diltiazem formulation. US Patent 5,286,497, 1994 Feb 15.

40. Hendrickson DL, Dimmitt DC, Williams MS, Skultety PF, Baltezor MJ,inventors, Carderm Capital L.P., assignee. Diltiazem formulation. US Patent 5,470,584, 1995 Nov 28.

41. Page RL. Evolving concepts in chronotherapy with CCBs: implications for pharmacists. Available at https://secure.pharmacytimes.com/lessons/200503-03.asp. Accessed 2005.

42. Cardizem LA [package insert], Biovail Pharmaceuticals, Bridgewater, NJ, April 2004.

43. Sista S, Lai JC, Eradiri O, Albert KS. Pharmacokinetics of a novel diltiazem HCl extended-release tablet formulation for evening administration. *J. Clin. Pharmacol.* 2003;43:1149–1157.

44. Thiffault J, Landriault H, Gossard D, Raymond M, Caille G, Spenard J. The influence of time of administration on the pharmacokinetics of a once-a-day diltiazem formulation: morning against bedtime. *Biopharm. Drug Dispos.* 1996;17:107–115.

45. Glasser SP, Neutel JM, Gana TJ, Albert KS. Efficacy and safety of a once daily graded-release diltiazem formulation in essential hypertension. *Am. J. Hypertens.* 2003;16:51–58.

46. Albert KS, Maes PJ,inventors, Biovail Laboratories International SRL, assignee. Chronotherapeutic diltiazem formulations and the administration thereof. US Patent 7,108,866, 2006 Sept 19.

47. Remon JP, inventor, Universiteit Gent, assignee. Cushioning wax beads for making solid shaped articles. US Patent 6,923,984, 2005 Aug 2.

48. Agarwal N, Thakkar N. Surviving patent expiration: strategies for marketing pharmaceutical products. *J. Product Brand Manage.* 1997;6(5):305–314.

49. Conte U, La Manna A, Colombo P,inventors, Jagotec AG, assignee. Tablets with controlled-rate release of active substances. US Patent 5,422,123, 1995 June 6.

50. Frishman WH. A new extended-release formulation of diltiazem HCl for the treatment of mild-to-moderate hypertension. *J. Clin. Pharmacol.* 1993;33:612–622.

51. Procardia® XL [package insert], Pfizer Inc., New York, NY, 2003 Label information.

52. Mohiuddin SM, Lucas BD Jr, Shinn B, Elsasser GN. Conversion from sustained-release to immediate-release calcium entry blockers: outcome in patients with mild-to-moderate hypertension. *Clin. Ther.* 1993;15(6):1002–1010.

53. Brodgen RN, McTavish D. Nifedipine gastrointestinal therapeutic system (GITS): a review of its pharmacodynamic and pharmacokinetic properties and therapeutic efficacy in hypertension and angina pectoris. *Drugs* 1995;50:495–512.

54. Kleinbloesem CH, Van Brummelen P. Rate of increase in the plasma concentration of nifedipine as a major determinant of its hemodynamic effects in humans. *Clin. Pharmacol. Ther.* 1987;41:26–30.

55. Available at http://www.gsk.com/media/pressreleases/2006/2006_05_17_GSK833.htm.

56. Henderson LS, Tenero DM, Baidoo CA, Campanile AM, Hatter AH, Boyle D, Danoff TM. Pharmacokinetic and pharmacodynamic comparison of controlled-release carvedilol and immediate-release carvedilol at steady state in patients with hypertension. *Am. J. Cardiol.* 2006;98(7A):17L–26L.

57. Weber MA, Sica DA, Tarka EA, Iyengar M, Fleck R, Bakris GL, Coreg 367 Investigator Group. Once-daily carvedilol is effective in treating hypertension. Presented at the American College of Cardiology Annual Meeting, 2006.

58. Autant P, Selles JP, Soula G,inventors, Flamel Technologies, assignee. Medicinal and/or nutritional microcapsules for oral administration. US Patent 6,022,562, 2000 Feb 8.

59. Available at http://clinicaltrials.gov.

60. Mcguire S. AstraZeneca seeks new Seroquel XR indications. Available at http://www.mmm-online.com/AstraZeneca-seeks-new-Seroquel-XR-indications/article/100315. Accessed 2008.

61. Seroquel® XR [package insert], AstraZeneca Pharmaceuticals LP, Wilmington, DE, 2007 Label information.

62. Parikh BV, Timco RJ, Addiks WJ,inventors, Zeneca Limited, assignee. Pharmaceutical compositions using thiazepine. US Patent 5,948,437, 1999 Sept 7.

63. Available at http://www.astrazeneca.com/pressrelease/5379. aspx.

64. Gupta SK, Sathyan G. Pharmacokinetics of an oral once-a-day controlled-release oxybutynin formulation compared with immediate-release oxybutynin. *J. Clin. Pharm.* 1999;39:22–29.

65. Sathyan G, Chancellor MB, Gupta SK. Effect of OROS controlled release delivery on the pharmacokinetics and pharmacodynamics of oxybutynin chloride. *Br. J. Clin. Pharmacol.* 2001;52(4):409–417.

66. Anderson RU, Mobley D, Blank D, Saltzstein D, Susset J, Brown JS. Once daily controlled versus immediate release oxybutynin chloride for urge urinary incontinence. *J. Urol.* 1999;161:1809–1812.

67. Lindeke B, Hallström G, Johansson C, Ericsson K, Olsson LI, Strömberg S. Metabolism of oxybutnin: establishment of desethyloxybutynin and oxybutynin *N*-oxide formation in rat liver preparations using deuterium substitution and gas chromatographic mass spectrometric analysis. *Biomed. Mass Spec.* 1981;8:506–513.

68. Aaltonen L, Allonen H, Lisalo E, Juhakoski A, Kleimola T, Sellman R. Antimuscarinic activity of oxybutynin in the human plasma quantitated by a radioreceptor assay. *Acta Pharmacol. Toxicol.* 1998;55:100–103.

69. Madersbacher H, Jilig G. Control of detrusor hyperflexia by the intravesical instillation of oxybutynin hydrochloride. *Paraplegia* 1991;29:84–90.

19

GENERIC ORAL CONTROLLED RELEASE PRODUCT DEVELOPMENT: FORMULATION AND PROCESS CONSIDERATIONS

SALAH U. AHMED

Abon Pharmaceuticals, Northvale, NJ, USA

VENKATESH NAINI

Teva Pharmaceuticals, Pomona, NY, USA

19.1 INTRODUCTION: GENERIC DRUG INDUSTRY DYNAMICS

It has been about a quarter century since the Drug Price Competition and Patent Term Restoration Act of 1984 (Hatch–Waxman) was passed [1–3]. Since then the generic drug industry has seen tremendous growth accounting for more than $29 billion in annual sales in 2008 [4]. At present, generic drugs make up for 63% of prescriptions dispensed in the United States [5]. Over the next 7 years, additional $77 billion worth of brand-name products will be open for generic competition [6]. Increasingly, the brand pharmaceutical companies are facing an uphill task maintaining a robust pipeline of new drugs. In addition to losing patent protection, increasing cost and risk associated with new drug development [7] have resulted in eroding profits and consolidation in the pharmaceutical industry [8]. In light of these challenges, pharmaceutical companies are constantly on the lookout for extending patent life on existing products or formulating new delivery systems. In many instances, such new dosage forms offer genuine therapeutic benefits to the patient [9], but in other instances could be viewed as "patent-evergreening" or "layered innovation" [10, 11]. An increasing proportion of these complex drug delivery systems are oral controlled

release preparations. A classic example of such life cycle management, successfully using controlled release technology, was diltiazem hydrochloride (Cardizem) originally sold as three-times-a-day regimen with annual revenues of $260 million in 1988. The brand company successfully launched a twice-daily Cardizem SR with peak annual revenues of $400 million until 1991. In a further iteration, a once-daily Cardizem CD was introduced, which enjoyed peak market of $900 million by 1996 [12].

While brand companies utilize drug delivery technologies, predominantly oral controlled release (CR) dosage forms, to extend and expand their product franchises, generic companies are also becoming more sophisticated. They are increasingly using such technologies and well thought out legal strategies for reasonable success with their own oral CR portfolios. Generic companies themselves are dealing with additional competition and pressures to cut costs of ingredients and manufacturing, partly due to new players from India, China, and Europe. This has led to tremendous price erosion, especially for immediate release (IR) and commodity products, when multiple generics enter the market compared to a single generic [13]. Many specialty and established generic firms are turning to value-added niche products with higher "barrier to entry," especially with regard to formulation and processing technologies. More often, generic companies are using sophisticated business development deals involving API sourcing and specialized product development licensing and alliances to stay ahead in the game [14]. One aspect of this strategy is to develop generic versions of extended or controlled release branded formulations [15, 16]. Typically, these formulations are more difficult to develop and are protected by patents and other exclusivities as discussed below. This Chapter will provide an overview of formulation and process development of generic oral controlled release products.

19.2 PATENTS AND EXCLUSIVITY HURDLES FACED BY THE GENERIC INDUSTRY

Generic CR product development encompasses most of the challenges faced by branded development programs. In addition, they also face constraints due to patents covering formulation composition, processing, and in-process and finished product quality specifications. The patents themselves are granted by the patent office (USPTO), and FDA has no role in issuing them [17]. When a branded company files the new drug application (NDA), it is required to provide to the FDA a list of all patents covering the product and its use. This information is included in the *Approved Drug Products with Therapeutic Equivalence Evaluations* (Orange Book) [18]. The ANDA applicant is required to certify at the time of filing that its generic drug will not infringe upon these patents or the patents are invalid. There are four types of certification described as follows [19]:

Paragraph I: No patents referencing the listed product are in Orange Book.
Paragraph II: Patent filed in Orange Book has expired.
Paragraph III: Generic will be marketed after patents expire.
Paragraph IV: "Patent Challenge," the listed patents are either invalid or noninfringed.

The Paragraph IV certification is the most demanding and yet financially beneficial to generic companies if they eventually prevail in the courts. The intent of generic companies here is to market the product before relevant patents expire. The generic company is granted a 180-day exclusivity to market its product if successful in the court [20]. Under Paragraph IV certifications, the firm is also required to notify the NDA and/or patent owner. The patent holder has 45 days to initiate a patent infringement suit against the ANDA certifier. This process then leads to a 30-month stay, unless it is resolved in favor of the generic applicant before that. The Paragraph IV process is complicated with various other competing factors and is briefly summarized in Figure 19.1 [21–23].

Generic manufacturers are required to certify using one of the following approaches when filing an ANDA under Paragraph IV:

1. *Noninfringement* (Generic product avoids the patent claims)
2. *Invalidity* (RLD patent claim scope is too broad)

In noninfringement cases, the branded company tries to interpret the claims much more broadly compared to the generic defendant who wants a very narrow interpretation. This would eventually exclude the generic product from any such claims. The burden of proof to establish noninfringement generally falls on the branded company (e.g., Ditropan® XL; *Alza Corp vs. Mylan Laboratories*). In cases of invalidity, the claims are deemed invalid when they do not meet the standards of patentability (novelty, nonobviousness, etc). The burden to establish invalidity generally lies with the generic company (e.g., Effexor® XR; *Wyeth vs.Teva Pharmaceuticals*). In the case of extended release patents, the generic company would prefer noninfringement compared to invalidity since it is easier to prevail in the courts, it is quicker to litigate, and it also prevents other generic manufacturers from coming to the market.

Extended release branded products may be covered by patents that deal with one or more of the following aspects: (a) drug substance (b) dosage form (c) dosing range and frequency (d) composition and process (e) release profile (*in vitro/in vivo*) (f) pharmacokinetic characteristics. Formulation- and process-related patent claims include the particular type and viscosity grade of polymer (hypromellose, hydroxypropyl cellulose, polyethylene oxide, carbopol, etc.) and their levels used. Other aspects could cover design of the drug delivery system: matrix-controlled, reservoir system, coated pellets, use of bi-, tri-, or multilayered tablets, osmotic pumps, etc. Both *in vitro* dissolution profiles and *in vivo* drug release rates of CR dosage forms are also the subject matter of many patents. Furthermore, the plasma profile or a specific pharmacokinetic parameter such as C_{max}, T_{max}, or AUC of the delivery system may be claimed in the patent. The goal of branded companies is to obtain ER product patents with claims as broad as possible, but at the same time sufficiently narrow, so as not to be considered "prior art." Table 19.1 lists some examples of extended release branded products on the market along with their associated patent claims [15]. In addition to patent exclusivities, branded products may also be granted nonpatent (regulatory) exclusivities as described below. During these exclusivity periods, generic versions may not be approved.

1. *New Chemical Entity (NCE) – 5 years*: New drug that has not been approved before.

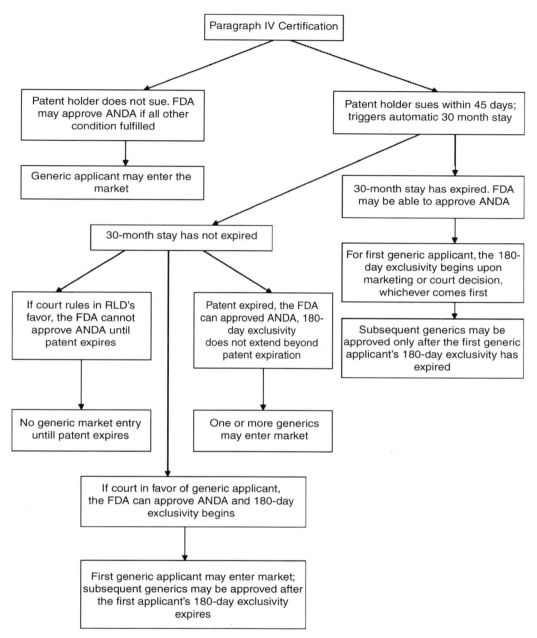

FIGURE 19.1   Paragraph IV certifications and approval of generic products (adapted from Ref. 20).

2. *New Clinical Investigations (NCIs) – 3 years*: New formulations or indications of existing drugs. New clinical studies are essential for approval.

3. *Orphan Drug Exclusivity – 7 years*: For drugs affecting less than 200,000 persons.

4. *Pediatric Exclusivity – 6 Months*: The drug is tested in pediatric population.

As described above, generic development of controlled release forms encompasses several legal and regulatory hurdles. Furthermore, the generic product is required to be pharmaceutically equivalent and bioequivalent to the branded product as per the FDA guidelines. More recently, generic companies are facing additional hurdles with the launch of authorized generics [24], citizen's petitions challenging validity of generic approvals, free trade agreements (CAFTA, NAFTA, etc), and ongoing patent reforms that may favor branded companies [4]. In spite of these challenges, generic CR product development promises higher rewards to the business.

TABLE 19.1   Examples of Branded Extended Release Products with Associated Patent Claims

Brand Product	Generic Name	Patent(s)	Claim Types
Adderall XR®	Amphetamine salts	US 6,322,819	FPK
		US 6,605300	FPK
Biaxin® XL	Clarithromycin	US 6,010,718	FPK
		US 6,551,616	F
		US 6,872,407	PK
Concerta®	Methylphenidate	US 6,919,373	PK
		US 6,930,129	PK
Depakote® ER	Divalproex	US 6,419,953	F
		US 6,511,678	FPK
		US 6,528,090	FPK
Ditropan® XL	Oxybutynin	US 6,124,355	PK
Effexor® XR	Venlafaxine	US 6,274,171	F, FPK
		US 6,403,120	PK, FPK
		US 6,419,958	PK
Glucophage® XR	Metformin	US 6,475,521	F, FPK
Niaspan®	Niacin	US 6,406,715	PK
		US 6,676,967	PK
Toprol® XL	Metoprolol	US 5,001,161	F
Wellbutrin® XL	Bupropion	US 6,096,341	FPK

Adapted from Ref. 15.

F, formulation; PK, pharmacokinetic; FPK, combination of formulation and pharmacokinetic characteristics

19.3   DOSAGE FORM DESIGN AND FORMULATION CONSIDERATIONS

19.3.1   Characterization of RLD

The reference product's formulation and patent listings are thoroughly evaluated after consultation with patent attorney. Exclusivity and patent data are obtained from the FDA's Orange Book [25]. The formulation group, working closely with legal and regulatory groups, usually comes up with a strategy for developing noninfringing generic CR formulation and process. The process starts with comprehensive evaluation of the RLD. Qualitative composition of RLD is usually obtained from its package insert and Physician's Desk Reference (PDR) [26]. Additional information about the RLD can be gathered from its *Summary Basis of Approval,* obtained through the Freedom of Information (FOI) services [27]. Although confidential sections in these documents are redacted, they still provide valuable insight into formulation, process development, and biopharmaceutics of the RLD.

Generic CR dosage forms are expected to be pharmaceutically equivalent and bioequivalent to the RLD. Two products may be considered pharmaceutically equivalent if they have the same active ingredient same dosage form and

strength and are bioequivalent [28]. The generic product is not expected to contain the same inactive ingredients as RLD and is generally not required to have the same drug release mechanism, as long as it meets the bioequivalence (BE) criteria [28]. In addition, several patent constraints may exist for using certain ingredients or technologies for manufacturing the generic CR dosage form. Probability of formulating a bioequivalent CR product dramatically increases if its composition is qualitatively and quantitatively similar to the RLD. It is especially important to closely match rate-controlling or other functional excipients, such as stabilizers and buffering agents used in RLD's formulation.

On the basis of one's experience and vast body of excipient literature [29], these ingredients can be classified as (a) release controlling and (b) nonrelease controlling excipients. In addition, several physical and chemical evaluation tools can be used to decode RLD's formulation and identify grades of rate controlling excipients. Preformulation tools and their use in evaluating RLD's formulation, composition, and process are summarized in literature reviews [30, 31]. Importance of such tools in expediting generic CR product development is well emphasized.

Optical evaluation significantly helps in understanding of the physical attributes of the innovator product. Low magnification microscopy is a valuable tool for obtaining information about the RLD's finished product and excipient grades used and their morphology. In some instances, video microscopy can be employed to track swelling, erosion, and dissolution phenomena in CR systems. Microscopic chemical imaging with Raman and near-infra red (NIR) spectroscopy has been used to estimate drug ingredient-specific particle size and its distribution in matrix systems [32] and estimate the grade and level of polymer in coated particles and beads [33, 34]. Scanning electron microscopy (SEM) is one of the most useful techniques for evaluating RLD composition and gain insight into its fabrication technique. Observing the cross section of a matrix tablet can provide information about the use of wet granulation or direct compression in the process. A cross section of coated bead provides information on the numbers of layers used to control drug release and their thickness [35]. SEM pictures of empty "hulls" following dissolution of a matrix system or reservoir bead can give the formulator further clues for designing the generic CR product [36, 37]. Figure 19.2 shows examples of SEM applications in CR systems.

The physical integrity of the oral CR product plays a critical role in drug release process under a variable physiological environment of the GIT. Texture analysis is used for evaluating hydrated controlled release matrices and mechanical strength of coated beads and particulate reservoir systems. Dürig and Fassihi used texture analyzer for characterizing hydrated gaur-gum matrices and the effect of other added excipients on gel dynamics [38]. Figure 19.3 shows the force–displacement curve obtained

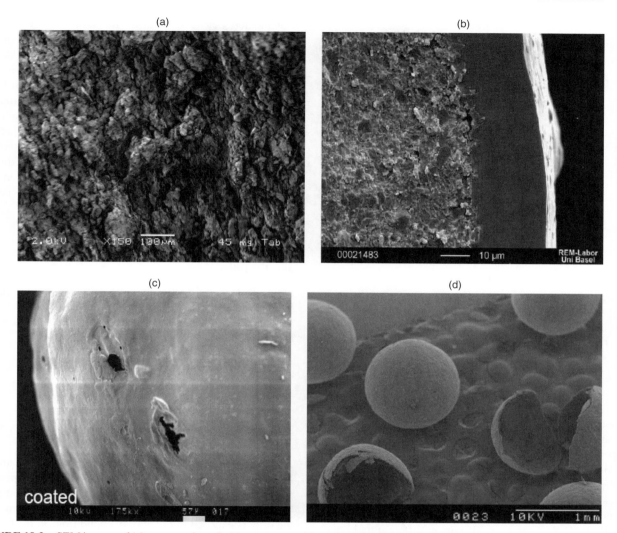

FIGURE 19.2   SEM images of (a) cross section of tablet compressed from granulated matrix; (b) cross section of a mutilayered bead; (c) and (d) dried empty "hulls" following dissolution of reservoir coated beads (b–d adapted, respectively, from Refs [25, 35, 37]).

using a texture analyzer, indicating various regions of hydration in the matrix, which correlated well with microscopic observation of these layers. Such information collected during preformulation evaluation of RLD can be used in developing a generic product with similar hydration profile. Dispersing RLD matrix in small volume of water and analyzing rheological profile of the resulting dispersion can also yield valuable information on viscosity and grade of rate-controlling polymer used.

The probability of a generic CR product to be bioequivalent to a branded product greatly depends on comparable drug release *in vitro*. Dissolution testing methods for CR dosage forms must capture multiple pH environments, changes in local GI fluid contents, residence times, and exposure to varying hydrodynamics and shear along the GIT [39, 40]. It is imperative that dissolution test methods should be predictive of the *in vivo* performance to draw

meaningful conclusions from comparative profiles of RLD and generic product, especially during product development phase. USP Apparatus I (Basket) and II (Paddle), which are routinely used for testing immediate release dosage forms, are also useful for CR dosage forms. However, due to their inability to dynamically change pH and induce additional hydrodynamic forces, USP Apparatus III (reciprocating cylinder, "BioDis") and USP Apparatus IV (flow-through cell) have become increasingly popular. The Apparatus III is especially suited for testing enteric coated tablets and multiparticulate beads. Recently, Dressman and coworkers used BioDis apparatus and changing pH media representing the entire GIT (simulated fluids ranging from gastric and intestinal to colonic pH) to differentiate two extended release bead formulations of mesalazine for colon delivery [41]. The method successfully predicted *in vivo* performance of two bead formulations coated with either Eudragit FS alone

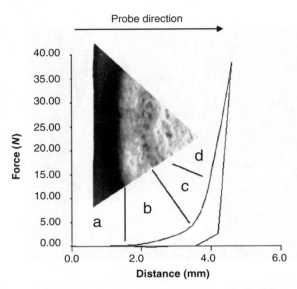

FIGURE 19.3 Texture analysis showing force–displacement profile and photomicrograph of hydrated matrix CR tablet following 3 h swelling. X-axis shows distance of penetration of the 2 mm analyzer probe. (a) Fully hydrated region; (b) partially hydrated region; (c) infiltrated region; and (d) glassy core region (adapted from Ref. 38).

(dissolves at pH >7) or in combination with Eudragit RL/RS. Apparatus IV is useful for evaluating CR dosage forms of very poorly soluble drugs. Other unconventional dissolution methodologies for evaluating CR dosage forms have been presented in the literature [42]. Before initiating any generic formulation development, two–three lots of RLD product are evaluated using multiple dissolution apparatus in biorelevant media (SGF, FaSSIF, and FeSSIF) under different hydrodynamic conditions to simulate shear forces in GIT under fasted and fed states [43]. If the FOI or literature includes IVIVC studies performed on the RLD, such information should be incorporated into the generic CR development program.

19.3.2 Pharmacokinetics of CR Dosage Form

Basic understanding of pharmacokinetics of CR dosage form is critical to designing a bioequivalent generic product. The "Biopharmaceutics" section of RLD's Summary Basis of Approval (FOI) provides valuable information regarding pharmacokinetic studies conducted by the innovator concerned (e.g., BE studies between clinical and commercial formulations), food effect, expected intrasubject variability, and IVIVC. In case of drugs with an absorption window, the RLD formulation may be designed to target specific regions of the GIT [44]. Generic versions of such drugs should very closely mimic RLD's *in vivo* drug release profile in order to be bioequivalent. Generic formulation design should also incorporate the effect of food, mechanical or shear forces

encountered during GI transit, and other coingested substances such as alcohol on bioavailability. The concern with "dose dumping" of CR formulations when exposed to alcohol has recently prompted regulatory agencies to require comparative profiles of generic and innovator products in dissolution media containing varying levels of alcohol ranging from 5–40% [45]. In these instances, the generic product is generally expected to perform on par with innovator product and should not be better or worse. There are several examples in the literature when such *in vitro* release profiles indicated *in vivo* dose dumping, such as hydromorphone (Pallodone® SR) and other opioids [46], and in other instances such as carvedilol CR, there was no such correlation [47].

19.3.3 Physicochemical Characterization of Drug and Excipients

Drug substance properties that play critical role in designing bioequivalent generic CR product are summarized below [48, 49]. Other preformulation aspects of CR drug development are covered in a separate chapter in this book.

1. Ionization, pK_a, solubility, and permeability
2. Polymorphism, salt form, and crystallinity
3. Particle size and surface area
4. Drug–excipient compatibility and stability

19.3.3.1 Ionization, pK_a, Solubility, and Permeability
Most drugs are weak acids or bases, with pH-dependent solubility. Drug substance pK_a and its pH-dependent solubility data are essential for designing CR dosage form since the drug is exposed to range of pH conditions in the GIT. In addition, maintaining pH-independent drug release from the dosage form may require inclusion of pH-modifying excipients or buffering agents in the formulation [50]. Literature information for ionization and solubility is available only for older drugs; hence the generic formulator is expected to generate this data for many drugs during the preformulation stage. For poorly soluble drugs, release from the CR dosage form is determined by its solubility and partitioning into the surrounding polymer matrix. Permeability assessments are usually made from published bioavailability data (>90% absorbed can be considered as highly permeable).

19.3.3.2 Polymorphism, Salt Form, and Crystallinity
As per regulatory requirements, a generic product is not required to have the same polymorphic form as RLD, except in instances where dramatic differences in solubility and bioavailability have been established for different polymorphic forms [51]. X-ray powder diffraction (XRPD) is routinely used for screening polymorphic forms of API, quantify certain formulation components, and detect drug polymor-

phic form in the finished dosage form, especially with high drug loads and no significant matrix interference [52]. XRPD is also very useful in monitoring phase transformations upon processing (wet granulation, milling, drying, etc.) and storage, which may adversely impact finished product performance [53]. Recently, Raman microscopic imaging has been used to detect polymorphic form of drug dispersed within a CR dosage form at very low levels. In some instances, polymers and manufacturing process used for preparing the CR system may convert the drug to a metastable form or even render it amorphous, as in the case of carbamazepine CR based on a hypromellose matrix [54]. Salt forms are generally used to enhance solubility of drugs or optimize its other physicochemical properties such as stability, compressibility, hygroscopicity, and so on. In some instances, the innovator may choose to use a less soluble salt form to achieve extended release, as in the case of trazodone tosylate that is less soluble compared to its hydrochloride salt. Typically, the generic product is required to use the same salt form as RLD due to issues of solubility and stability along the GIT and stability concerns with alternative salt forms [55].

19.3.3.3 Particle Size and Surface Area

For drugs with poor solubility, particle size reduction is the most common way to increase solubility and bioavailability [48]. For highly soluble drugs, particle size may not be important for dissolution. A CR dosage form is often designed using a highly water-soluble drug dispersed in poorly soluble excipients and polymers. The size of resulting granules, particles, or beads is often manipulated to achieve the desired drug release profile. For example, achieving granules of right consistency and size with a highly soluble drug is critical when used in wet granulated matrix systems. This in turn may affect drug release from monolithic matrix systems. Particle size plays a key role in determining process efficiency when drug is layered onto beads. More uniform and smooth surfaces have been reported with beads layered using drug solutions or micronized drug suspensions, compared to coarser particles [36]. In case of directly compressed matrix systems, polymer particle size plays an important role in the rate of hydration and subsequent erosion. Particle size and shape of materials can be assessed using microscopy/image analysis, light obscuration, and laser light scattering. For some highly insoluble drugs, surface area of incoming API raw material may be monitored and controlled in addition to its particle size.

19.3.3.4 Drug–Excipient Compatibility and Stability

Drug substance is selected after comparative impurity evaluation of RLD and multiple vendor API lots under ambient and stressed conditions. Basak et al. have reviewed regulatory aspects of impurities in generic drug substances and finished products [56]. Selection of excipients for CR dosage form is coupled with drug–excipient compatibility studies.

Excipient selection should also consider their hygroscopicity and heat sensitivity due to their potential degradation under accelerated storage conditions. Several drug–excipient mixtures are screened in ratios typically observed in finished dosage form using thermal techniques such as isothermal microcalorimetry (IMC) [57] or stressed under high temperature and humidity conditions followed by HPLC analysis [58].

19.3.4 Formulation Development

On the basis of characterization of RLD and preformulation studies, drug substance and excipients of appropriate quality are selected after receiving input from patent counsel, for potentially infringing compositions. Drug release mechanism of generic CR product may be different from RLD, to circumvent patents (e.g., matrix design to match *in vivo* release profile of an osmotic delivery system). Appearance of the generic dosage from is based on innovator product's market presentations and input from the marketing department. Using this information, the tablet or capsule tooling is sourced. Selected inactive ingredients in the formulation are thoroughly reviewed for listing in the *Inactive Ingredient Guide* (*IIG*) for that specific route of administration [59]. Ingredients not listed in *IIG* should not be used under all circumstances to avoid delays in filing and acceptance of the ANDA. If alternative ingredients must be used, a correspondence should be initiated with the Office of Generic Drugs (OGD), very early in the product development cycle.

Several terminologies are synonymously used to describe oral CR products, such as modified release (MR), extended release (ER), sustained release (SR), delayed release (DR), and long acting (LA). Malinowski and Marroum have reviewed regulatory aspects of oral CR products with definitions of various types of CR dosage forms [18]. For practical purposes, ER and DR could be considered as subsets of the MR dosage forms. ER dosage forms allow at least a twofold reduction in dosing frequency compared to the IR product. DR products, on the other hand, enable a significant time lag in drug release from the time of administration. Enteric coated products fall into this last category. Controlled release technologies and their application for oral drug delivery are well reviewed in the literature [60, 61] and in other chapters of this book. The majority of oral CR innovator products on market today fall into the following categories:

1. Matrix controlled systems
2. Reservoir controlled systems
3. Osmotic pump systems

Matrix controlled systems are usually monolithic tablets containing hydrophilic water-soluble polymers, hydrophobic polymers or waxes, or a combination of all these ingredients [62]. A review of polymers and other ingredients used in these systems is provided elsewhere in this book. Hypro-mellose-based matrix systems are the most common used and extensively studied. Briefly, the following factors influence drug release from these dosage forms:

1. Drug dose, particle size, solubility, ionization, parti-tioning, and stability [63, 64].
2. Polymer level, viscosity grade, and particle size are major determinants in modulating drug release [65].
3. Fine particle grades of hypromellose hydrate faster and hence result in a better controlled release [66].
4. Other excipients in the matrix such as fillers (water soluble versus insoluble), release modifiers (buffering agents and stabilizers), salts, or electrolytes [67].
5. Polymers of different viscosity and hydrophobicity can be combined for desired drug release.
6. Higher surface/volume ratio of smaller size tablets may require higher polymer amounts compared to larger tablets to achieve a similar drug release [68].
7. Method of manufacture – direct compression, wet granulation (high shear and fluid bed), and compres-sion forces [69].

A variety of proprietary technologies based on matrix control have been patented, such as TIMERx® (Penwest) that uses a unique blend of polymers (xanthan and locust bean gum) [70] and Geomatrix® (Skyepharma) that uses multi-layered tablets with restricted swellable drug layer sand-wiched between insoluble layers [71]. Usually, alternative combinations of hydrophilic or water insoluble polymers and waxes can be used to tailor drug release to match the innovator [72]. When using a combination of hydrophilic polymers, the Philipof equation can be used to predict drug release on the basis of polymer concentration and viscosity [73]:

$$\eta = (1 + KC)^8 \qquad (19.1)$$

where η is viscosity in mPa·s; K is constant for each indi-vidual polymer; and C is concentration in percentage.

A further modification can be used for changing the level of single high molecular weight (MW) polymer with a combination of high and low molecular weight polymers at a higher total polymer concentration. This approach could be taken when the use of certain polymers is protected by patents:

$$F_1 = [\{(K_1 C_{org}/C_{blend}) - K_1\}]/[K_1 - K_2] \qquad (19.2)$$

where F_1 is weight fraction of higher MW polymer in blend; K_1 is constant for higher MW polymer (use 2% solution viscosity and Equation 9.1); K_2 is constant for lower MW polymer (use 2% solution viscosity and Equation 9.1); C_{org} is

iriginal polymer level in formulation; and C_{blend} is desired polymer level in formulation.

The excipient supplier Colorcon has introduced a model-ing system (HyperStart®) for hydrophilic matrices based on drug solubility and filler characteristics [74]. This can be very useful in designing a generic CR matrix system as shown in Figure 19.4, whereby a matrix formulation of highly soluble drug metformin, predicted using HyperStart®, was closely comparable in drug release to innovator product Glucophage® XR 500 mg. In some cases, innovator product design (bilayer matrix) may release drug in a biphasic manner—immediate release layer followed by a sustained release component or a combination of two drugs may be used one with IR and the other with SR. This biphasic release can be mimicked using a single monolithic matrix, whereby the immediate-release portion is added as-is and the delayed-release portion is granulated using hydrophilic or hydropho-bic rate-controlling polymers. Using this approach and hy-drophilic polymers, HPC for IR and HPMC for DR, it was possible to formulate acetaminophen matrix tablet matching drug release profile of the innovator's biphasic release [75]. Liu and Fassihi used HPC/HPMC in a bilayer composite matrix dosage form of highly soluble, low-dose drug alfu-sozin hydrochloride to match the zero-order release of marketed product based on trilayer Geomatrix® system [76].

Reservoir systems typically consist of monolithic (tablets) or multiparticulate units coated with water-insoluble but permeable membranes. A drug may diffuse through the membrane for extended release or the coating may dissolve above certain pH, as in enteric coated systems. Multiparti-culate units may be beads or pellets formed by the extrusion spheronization process, by direct pelletization in rotor fluid bed, or by coating drug onto nonpareil sugar or microcrys-talline cellulose beads. Extrusion is performed on wet gran-ulated masses or hot melt wax matrices. Multiparticulates

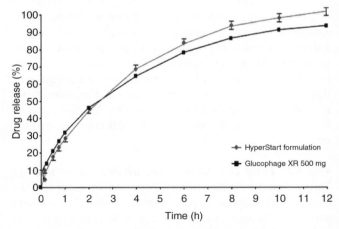

FIGURE 19.4 Comparison of metformin hydrochloride release from HyperStart® matrix formulation and innovator product Glu-cophage XR (adapted from Ref. 74).

offer several advantages over single-unit systems, such as less *in vivo* variability in gastric emptying and absorption and less risk of "dose dumping" [77]. Polymers used for reservoir coating systems are discussed in other chapters of this book. Product is coated at 10–20°C above the polymer's minimum film-forming temperature (MFFT), which is reduced by using plasticizers. Plasticizers help reduce coating temperatures required for the process and also improve a polymer's mechanical properties. Several formulation factors influence drug release from reservoir systems and these can be summarized as [78, 79]:

1. Substrate/reservoir composition for extruded/pelletized beads.
2. Bead morphology, density, porosity, and drug loading.
3. Starting bead particle size and composition (sugar versus MCC).
4. Drug layering onto beads from solution versus suspension.
5. Drug solubility, ionization, and diffusion through membrane.
6. Buffering agents and solubility enhancers in the reservoir.
7. Polymer type and system (aqueous dispersion versus organic solution).
8. Coating weight gain and porosity of film.
9. Release modifiers in the film such as pore formers and swelling agents.
10. Levels and type of the plasticizer used in the membrane.

Generic equivalents of coated reservoir tablets are easier to develop compared to multiparticulate formulations. For DR products, generics can use alternative enteric polymers with matching pH sensitivity, if certain polymers are excluded by patents [80]. Microscopic evaluation of RLD's bead product will provide insight into its formulation strategy. The desired drug release in RLD may be achieved by a mixture of IR and CR beads. In another instance, multilayered beads consisting of membrane-enclosed core coated with additional IR layer may be used. Depending on the innovator's formulation and process patent position, the generic product can be designed accordingly. Siepmann et al. has provided a good overview of modulating drug release from coated reservoir systems using a combination of polymers and other release modifiers [81]. These formulation strategies could be useful in tailoring generic product's drug release from a multiparticulate system to match the innovator's *in vivo* drug release. In another example, Ho et al. used Pluronic® F-68 with poorly soluble drug nifedipine in multilayered beads to match a branded product using combination of IR and CR beads filled inside a capsule [82]. The optimized formulation

consisted of 1.5 times more drug in CR core compared to IR outer layer and ratio of nifedipine:pluronic was kept at 1:1, in both layers. Texture analysis is a valuable tool for comparing bead strength of generic and innovator products, especially under circumstances when they undergo further compression. Additional cushioning agents such as microcrystalline cellulose or cushion coatings may be added to the beads to provide adequate acid resistance to enteric coated beads comparable to the innovator. Some examples are present in the literature when the generic alternative prepared from simple matrix could match drug release profile from a beaded product. In a study by Pillay and Fassihi, a polyethylene oxide (POE), sodium bicarbonate, metoprolol tartrate matrix system was able to mimic pH-independent drug release from the innovator product Toprol® XL that consisted of CR beads in a disintegrating tablet (see Figure 19.5) [83].

Osmotic system involves a semipermeable membrane with a laser-drilled hole enclosing a formulation of drug, an osmotic agent, and other excipients [84]. The membrane allows water permeation into the dosage form, resulting in increased osmotic pressure that pushes the dissolved drug through the hole. The first marketed products using this technology provided a zero-order release for highly soluble drugs. Later innovations at Alza and other companies provided osmotic forms containing poorly soluble drugs (Push-Pull® and L-OROS®/Alza, Ensotrol®/Shire, and Zero-Os®/ADD). Marketed innovator products using Alza's Push-Pull technology include Ditropan® XL (oxybutynin chloride), Procardia® XL (nifedipine), Concerta® (methylphenidate hydrochloride), Tegretol® XL (carbamazepine), and DynaCirc® CR (isradipine). Due to strong patent protection of these technologies, generic products are usually formulated to circumvent these patents. In one example, a capsule product containing a combination of IR, ER, and enteric coated beads of carbamazepine was shown to be bioequivalent to the innovator Tegretol® XL osmotic tablet [85].

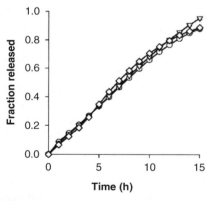

FIGURE 19.5 Comparison of release profiles of compressed multiple unit commercial product, Toprol® XL, in buffer media pH 2.6 (■) and pH 6.8 (▽) to a new directly compressed pH-independent monolithic system (◇) (adapted from Ref. 83).

In other cases, simple nonproprietary technology such as indented core tablet made with modified punches has been used to substitute expensive laser drilling of hole in the osmotic system [86]

19.3.5 Manufacturing Process Considerations

Batch size in generic industry is determined taking into account several factors including anticipated market capture at the time of launch. This is related to the status of the ANDA filing (e.g., Paragraph IV – "patent challenge"), number of anticipated first filers, existing and forecasted market for a particular product, expected number of generic launches at the time of launch (market share), and timing and/or success of patent litigation. Other factors influencing submission batch size for generic manufacturers include the following:

1. Available production capacity
2. Investment in new technology
3. Raw material cost
4. Stability of the finished product
5. Analytical testing issues

The formulation and process developed at the pilot scale is scaled up to production sizes to produce a minimum of 100,000 units for bio- and submission batches. At the time of launch, this may be scaled up, depending on the above-mentioned factors. Details of manufacturing process considerations for CR products are covered in other chapters and a review of scale-up considerations focused on generic products is available [87].

19.4 BIOPHARMACEUTICAL CONSIDERATIONS

Bioequivalence assessments are typically made using the average bioequivalence approach. Other approaches such as population BE, which assesses total variability in the population, and individual BE, which assesses subject-by-formulation interaction, have been proposed and can be considered [88]. According to the FDA's guidance, controlled release products filed under an ANDA require the following studies: (a) a single-dose nonreplicate BE fasting study comparing highest strength of the test and RLD and (b) a single-dose food-effect, nonreplicate BE study comparing highest strength of the test and RLD products [89]. Single-dose studies are preferred for CR products since they are more sensitive to differences in drug release and absorption from different dosage forms compared to multiple dose studies. In addition, due to tremendous implications for "dose dumping" with CR products when administered with food, and also formulation-dependent

food interactions may not be predicted by *in vitro* dissolution, the agency always requires a food study [90]. Typically, the crossover design is recommended for BE studies. In some instances, a parallel design may be considered for drugs with long half-life, where crossover design is difficult to conduct, and for cytotoxic drugs, where it is considered unethical to expose individuals to the drug twice, as with the crossover design [91].

One of the most important considerations for designing bioequivalence studies is the sample size (number of subjects) selection. The number of subjects chosen depends on a number of factors including the probability of concluding bioequivalence, the true difference between the test and the RLD products and expected intrasubject coefficient of variation (CV) for AUC and C_{max} (BE metrics measured in a study). The probability factor is the power of the study and is usually pegged at 80 or 90%. An approximate formula for estimating sample size has been proposed as follows [92]:

$$N = 2 \times (t_{\alpha,2N-2} + t_{\beta,2N-2})^2 [CV/(\nabla - \delta)]^2 \qquad (19.3)$$

where N is total number of subjects required for the study; t is appropriate value from the t-distribution; α is significance level (usually 0.1); $1-\beta$ is power (usually 80 or 90%); CV is intrasubject coefficient of variation; ∇ denotes the bioequivalence limit [$\ln(1.25) = 0.223$]; and δ is the difference between test and RLD (e.g., for T/R ratio of 1.1, $\delta = \ln(1.1) = 0.0953$; for T/R ratio of 0.90, $\delta =$ absolute value of $\ln(0.90) = 0.1053$].

For CR products δ can be assumed to be the maximum expected difference between formulations based on *in vitro* performance, for example, a 10% difference in drug release using a biorelevant dissolution method. One issue with above equation is that, in order to calculate N, one has to know t-values from the t-distribution for $2N-2$ degrees of freedom (in other words, N should be known). To avoid this problem and to obtain an approximate number of subjects, the following rules can be followed: (a) if the difference between products (δ) is other than zero, use approximately $t_{\alpha,2N-2} = 1.70$ and $t_{\beta,2N-2} = 0.85$; (b) if difference between products (δ) is zero (no difference between test and RLD), use approximately $t_{\alpha,2N-2} = 1.75$ and $t_{\beta,2N-2} = 1.34$. Diletti et al. have published tables with more accurately determined sample sizes, based on log-transformed data used in a bioequivalence study [93]. Table 19.2 is adapted from their work and shows sample sizes for various values of CV, power of 80 and 90%, and T/R ratios for log-transformed data [93].

One of the main issues in bioequivalence is that of highly variable drugs (CV > 30%) and has been extensively discussed in several conferences. Drugs with high CV present challenges to generic developers because of its impact on sample size and the associated high cost of BE study. The

TABLE 19.2 Sample Size for Given CV, Power, and Ratio (T/R) for Log-Transformed Data

CV (%)	Power (%)	μ_T/μ_R							
		0.85	0.90	0.95	1.00	1.05	1.10	1.15	1.20
5.0	80	12	6	4	4	4	6	8	22
7.5		22	8	6	6	6	8	12	24
10.0		36	12	8	6	8	10	20	76
12.5		54	16	10	8	10	14	30	118
15.0		78	22	12	10	12	20	42	168
17.5		104	30	16	14	16	26	56	226
20.0		134	38	20	16	18	32	72	294
22.5		168	46	24	20	24	40	90	368
25.0		206	56	28	24	28	48	110	452
27.5		248	68	34	28	34	58	132	544
30.0		292	80	40	32	38	68	156	642
5.0	90	14	6	4	4	4	6	8	28
7.5		28	10	6	6	6	8	16	60
10.0		48	14	8	8	8	14	26	104
12.5		74	22	12	10	12	18	40	162
15.0		106	30	16	12	16	26	58	232
17.5		142	40	20	16	20	34	76	312
20.0		186	50	26	20	24	44	100	406
22.5		232	64	32	24	30	54	124	510
25.0		284	78	38	28	36	66	152	626
27.5		342	92	44	34	44	78	182	752
30.0		404	108	52	40	52	92	214	888

Adapted from Ref. 93.

FDA has considered several proposals for reducing the number of subjects for highly variable drugs without increasing the risk to patients [94]. Of all the proposals, the approach using "Scaled Average BE" has gained popularity whereby the BE limits are widened based on intrasubject variability of the RLD [95]. These studies are performed using a replicate or partial replicate crossover design, under which RLD is given twice to the same subject in different periods. In these studies, in addition to meeting the 90% confidence interval, the FDA also requires that the point estimates (T/R mean ratio) should not exceed predetermined limits. Because of their ability to assess subjects by treatment interaction, replicate crossover studies have been suggested for bioequivalence of CR products with different drug release mechanisms (e.g., osmotic delivery versus matrix controlled) [91].

According to FDA's guidance, a biowaiver for non-RLD listed strengths of CR products is granted under the following circumstances [89]:

1. The CR product is a beaded capsule whereby the strength differs only by the number of beads containing active ingredient inside the capsule. Biowaivers for non-RLD strengths are granted based on dissolution similarity (f_2 factor > 50) to the biostrength.

2. The non-RLD strengths have to be *proportionally similar* to biostrength. The waiver is granted based on dissolution similarity (f_2 factor > 50) under at least three pH conditions. The FDA guidance defines the term proportionally similar in several ways:

   a. Active and inactive ingredients are in exactly the same proportion between all strengths.

   b. Ratio of inactive ingredients to the total weight of dosage form is within the defined limits of SUPAC-MR Level II changes [96].

   c. With highly potent drugs, the weight of dosage form remains the same across all strengths. The same inactive ingredients should be used in all strengths and the drug differences are compensated using one or more of the excipients. Changes to these excipients should be within Level II changes of the SUPAC-MR.

Drug and/or metabolite concentrations in plasma should be determined according to good laboratory practices (GLPs). The bioanalytical methods must be well characterized, fully validated, and proper documentation should be followed [97]. The method validation usually includes (a) stability of the analyte in a biological matrix and under storage, (b) specificity, (c) precision, (d) limit of quantitation, (e) linearity, and (f) reproducibility.

19.5 REGULATORY CONSIDERATIONS FOR GENERIC CR

The Office of Generic Drugs has implemented Question-Based Review (QbR) for the chemistry, manufacturing, and controls section of an ANDA [98]. This is with the overall goal of incorporating concepts of pharmaceutical GMPs for the twenty-first century [99] and process analytical technology [100] into its review process. This facilitates assurance of product quality, product performance-based specifications, continuous process improvement, enhancement and standardization of the review process, and reduction in the CMC review time. One of the main requirements for generic companies under this initiative is to submit a Quality Overall Summary (QOS) that addresses the QbR. A model QOS for CR product is available on the OGD's web site [101]. The document details information required by the QbR format for selection of raw materials, formulation development, process development, scale-up and in-process controls, and finished product specifications.

19.6 CONCLUSIONS

The hurdles for generic product development are significantly higher for oral CR products compared to their IR counterparts. The CR products introduced in the market by branded companies are either first-time entry of a new chemical entity or life-cycle management of existing products. While new drug pipelines are declining for branded companies, they are increasingly launching CR products as improved versions of existing drugs. Except in rare cases, these products have multiple patents covering formulation composition, process, dissolution, and pharmacokinetic parameters. While the validity of most of these patents is challenged in courts, the development scientist is often encountered with enormous challenges in formulating a CR product bioequivalent to RLD, with limited options presented outside the patents.

It is imperative for the product development scientist in generic arena to utilize the knowledge and expertise in varied disciplines including physicochemical attributes of drugs, excipients and dosage form, process engineering, pharmacokinetics of drug and dosage form, and regulatory and legal issues. The coordination of multidisciplinary expertise in an efficient manner for executing product development steps with aggressive time lines helps businesses succeed in an increasingly competitive world.

REFERENCES

1. Kibbe AH. Generic drugs and generic equivalency. In: Swarbrick J, editor. *Encyclopedia of Pharmaceutical Technology*, Vol. 3, 3rd edition, Informa Healthcare, New York, 2007, pp. 1891–1896.

2. Lopez G, Hoxie T. Generics. In: Edwards LD, Fletcher AJ, Fox AW, Stonier PD, editors. *Principles and Practice of Pharmaceutical Medicine*, 2nd edition, John Wiley & Sons Inc., New York, 2007, pp. 381–385.

3. NICHM. *A Primer: Generic Drugs, Patents and the Pharmaceutical Marketplace*. 2002, National Institute for Healthcare Management (NICHM), Washington. Accessed at http://www.nihcm.org/~nihcmor/pdf/GenericsPrimer.pdf.

4. Jaeger K. America's generic pharmaceutical industry: opportunities and challenges in 2006 and beyond. *J. Generic Med.* 2006;4(1):15–22.

5. Frank RG. The ongoing regulation of generic drugs. *N Engl. J. Med.* 2007;357(20):1993–1996.

6. Snow DB. Maximizing generic utilization: the power of pharmacy benefit management. *J. Generic Med.* 2007;5(1): 27–38.

7. DiMasi JA, Hansen RW, Grabowski HG. The price of innovation: new estimates of drug development costs. *J. Health Economics* 2003;22:151–185.

8. Congressional Budget Office. CBO Study: How increased competition from generic drugs has affected drug prices and returns in the pharmaceutical industry. July 1998, Congressional Budget Office, Washington DC. Accessed at http://www.cbo.gov/publications/.

9. Speers M, Bonnano C. Economic aspects of controlled drug delivery. In Mathiowitz E, editor. *Encyclopedia of Controlled Drug Delivery*, Vol. 1, John Wiley & Sons Inc., New York, 1999, pp. 341–347.

10. Kesselheim AS. Intellectual property policy in the pharmaceutical sciences: the effect of inappropriate patents and market exclusivity extensions on the health care system. *AAPS J.* 2007;9(3):E306–E311.

11. Pearce JA. How companies can preserve market dominance after patents expire. *Long Range Plan.* 2006;39(1):71–87.

12. Baichwal A, Neville DA. Adding value to product's life-cycle management: product enhancement through drug delivery systems. *Drug Del. Tech.* 2001;1(1):1–3

13. FDA. *Critical Path Opportunities for Generic Drugs.* Office of Generic Drugs (OGD), Center for Drug Evaluation and Research, Food and Drug Administration, 2007. Accessed at http://www.fda.gov/oc/initiatives/criticalpath/reports/generic.html.

14. Hoffman J. Global generics profitability: increasing need for business development and licensing. *J. Gen. Med.* 2004; 2(1):9–17.

15. Sklar SH. Extended-release drug patents: can they save big pharma's blockbuster medicines from the generic scrap heap?. *Pharm. Law Ind. Rep.* 2006;4(6):1–8.

16. Bates AK. Generic threat to XR/XL drugs? January 2007. Accessed at www.eularis.com, August 2008.

17. Melithil S. Patent issues in drug development: perspectives of a pharmaceutical scientist-attorney. *AAPS J.* 2005;7(3): E273–E727.

18. Malinowski HJ, Marroum PJ. Food and Drug Administration requirements for controlled release products. In Mathiowitz E, editors. *Encyclopedia of Controlled Drug Delivery*, Vol. 1, John Wiley & Sons Inc., New York, 1999, pp. 381–395.

19. Glover GJ. The influence of market exclusivity on drug availability and medical innovations. *AAPS J.* 2007;9(3): E312–E316.

20. Guidance for Industry. *180-Day Generic Drug Exclusivity Under the Hatch–Waxman Amendments to the Federal Food, Drug and Cosmetic Act.* Office of Generic Drugs, Food and Drug Administration, 1998.

21. Guidance for Industry. *180-Day Exclusivity When Multiple ANDAs are Submitted on the Same Day.* Office of Generic Drugs, CDER, Food and Drug Administration, 2003.

22. Federal Trade Commission. *Generic Drug Entry Prior to Patent Expiration: An FTC Study.* Federal Trade Commission, 2002. Accessed at http://www.ftc.gov/os/2002/07/genericdrugstudy.pdf.

23. Guidance for Industry. *Court Decisions, ANDA Approvals, and 180-Day Exclusivity Under the Hatch–Waxman Amendments to the Federal Food, Drug, and Cosmetic Act.* Center for Drug Evaluation and Research, Food and Drug Administration, 2000.

24. Thomas JR. Authorized generic pharmaceuticals: effects on innovation. CRS Report for Congress, 2008, Congressional Research Services, Washington DC. Accessed at http://www.fas.org/sgp/crs/index.html.

25. Electronic Orange Book: Approved Drug Products with Therapeutic Equivalence Evaluations. Accessed at http://www.fda.gov/cder/ob/.

26. *Physicians' Desk Reference 2009*, 63rd edition, Montvale, NJ, Thomson Reuters, 2008.

27. *CDER Freedom of Information: Handbook for Requesting Information and Records from FDA.* Accessed at http://www.fda.gov/cder/foi/index.htm.

28. Guidance for Industry. *Oral Extended (Controlled) Release Dosage Forms: In Vivo Bioequivalence and In Vitro Dissolution Testing.* Office of Generic Drugs, CDER, Food and Drug Administration. 1992.

29. *Handbook of Pharmaceutical Excipients.* 5th edition, Rowe RC, Sheskey PJ, Owen SC editors.), 2005. Pharmaceutical Press, UK.

30. Bansal AK, Koradia V. The role of reverse engineering in the development of generic formulations. *Pharm. Tech.* 2005;8:50–55.

31. Koradia VS, Chawla G, Bansal AK. Comprehensive characterization of the innovator product: targeting bioequivalent generics. *J. Generic Med.* 2005;2(4):335–346.

32. Maribel R. New dimensions in tablet imaging. *Pharm. Tech.* 2008;32(3):10–15.

33. Application Note: Confocal Raman Microscopy Imaging in the Pharmaceutical Industry. WiTec, Germany. Accessed at http://www.witec.de/en/download/Raman/.

34. Nelson MP, Panza J, Maier J, Treado PJ. Raman chemical imaging for controlled release systems. ChemImage Corporation, 2008, Excipient Fest, San Juan Puerto Rico.

35. Pollinger N. Optimization of pharmaceutical fluid-bed layering and pellet coating processes following software supported design of experiments. In: *Industry Trends in Controlled Release Pharmaceutical Dosage Form Development.* Glatt-Air Open House Meeting, Ramsey, NJ, 2007.

36. Jones D. The use of (common and) uncommon tools to solve uncommon problems for controlled release products. In: *Industry Trends in Controlled Release Pharmaceutical Dosage Form Development.* Glatt-Air Open House Meeting, Ramsey, NJ, 2007.

37. Heinicke G, Schwartz JB. Ammonio polymethacrylate coated diltiazem: drug release from single pellets, media dependence and swelling behavior. *Pharm. Dev. Tech.* 2007;12:285–296.

38. Dürig T, Fassihi R. Guar-based monolithic matrix systems: effect of ionizable and non-ionizable substances and excipients on gel dynamics and release kinetics. *J. Control. Rel.* 2002;80:45–56.

39. Lee SL, Raw AS, Yu L. Dissolution Testing. In Krishna R, Yu L, editors. *Biopharmaceutics Application in Drug Development*, New York, Springer, 2008, pp. 47–74.

40. Tong C, D'Souza SS, Parker JE, Mirza T. Dissolution testing for the twenty-first century: linking critical quality attributes and critical process parameters to clinically relevant dissolution. *Pharm. Res.* 2007;24(9):1603–1607.

41. Klein S, Rudolph MW, Skalsky B, Petereit H-U, Dressman JB. Use of BioDis to generate a physiologically relevant IVIVC. *J. Control. Rel.* 2008;130(3):216–219.

42. Pillay V, Fassihi R. Unconventional dissolution methodologies. *J. Pharm. Sci.* 1999;88(9):843–851.

43. Klein S, Wunderlich M, Dressman JB, Stippler E. Development of dissolution tests on the basis of gastrointestinal physiology. In: Dressman J, Kramer J, editors. *Pharmaceutical Dissolution Testing*, New York, Taylor & Francis, 2005, pp. 193–227.

44. Davies SS. Formulation strategies for absorption windows. *Drug Disc. Today.* 2005;10(4):249–257.

45. Levina M, Vuong H, Rajabi-Siahboomi AR. The influence of hydroalcoholic media on hypromellose matrix systems. *Drug Dev. Ind. Pharm.* 2007;33:1125–1134.

46. Walden M, Nicholls FA, Smith KJ, Tucker GT. The effect of ethanol on release of opioids from oral prolonged-release preparations. *Drug Dev. Ind. Pharm.* 2007;33:1101–1111.

47. Henderson LS, Tenero DM, Campanile AM, Baidoo CA, Danoff TM. Ethanol does not alter the pharmacokinetic profile of the controlled release formulation of carvedilol. *J. Clin. Pharmacol.* 2007;47:1358–1365.

48. Lee SL, Raw AS, Yu L. Significance of drug substance physicochemical properties in regulatory quality by design. In: Adeyeye MC, Brittain HG, editors. *Preformulation in Solid Dosage Form Development.* New York: Informa Healthcare, 2008, pp. 571–583.

49. Physicochemical factors affecting bioequivalence. In: Sarfaraz N, editor. *Handbook of Bioequivalence Testing*, New York, Informa Healthcare, 2007, pp. 163–196.

50. Varna MVS, Kaushal AM, Garg S. Influence of microenvironmental pH on the gel layer behavior and release of basic drug from various hydrophilic matrices. *J.Control. Rel.* 2005;103:499–510.

51. Raw AS, Furness MS, Gill DS, Adams RC, Holcombe RC, Yu LX. Regulatory considerations of pharmaceutical solid polymorphism in Abbreviated New Drug Applications (ANDAs). *Adv. Drug Del. Rev.* 2004;56:397–414.

52. Stephenson GA. Applications of X-ray powder diffraction in the pharmaceutical industry. *Rigaku J.* 2005;22(1):2–15.

53. Zhang GGZ, Law D, Schmitt EA, Qiu Y. Phase transformation considerations during process development and manufacture of solid dosage forms. *Adv. Drug Del. Rev.* 2004;56:371–390.

54. Katzhendler I, Azoury R, Friedman M. Crystalline properties of carbamazepine in sustained release hydrophilic matrix tablets based on hydroxypropyl methylcellulose. *J. Control. Rel.* 1998;54:69–85.

55. Verbeeck RK, Kanfer I, Walker RB. Generic substitution: the use of medicinal products containing different salts and implications for safety and efficacy. *Eur. J. Pharm. Sci.* 2006;28:1–6.

56. Basak AK, Raw AS, Al Hakim AH, Furness S, Samaan NI, Gill DS, Patel HB, Powers RF, Yu L. Pharmaceutical impurities: regulatory perspective for Abbreviated New Drug Applications. *Adv. Drug Del. Rev.* 2007;59:64–72.

57. Schmitt EA, Peck K, Sun Y, Geoffroy J-M. Rapid, practical and predictive excipient compatibility screening using isothermal microcalorimetry. *Themochim. Acta.* 2001;380: 175–183.

58. Serajuddin ATM, Thakur AB, Ghosal RN, Fakes MG, Ranadive SA, Morris KR, Varia SA. Selection of solid dosage from composition through drug–excipient compatibility testing. *J. Pharm. Sci.* 1999;88(7):696–704.

59. Inactive ingredient search for approved products. CDER, Food and Drug Administration. Accessed at http://www. accessdata.fda.gov/scripts/cder/iig/.

60. Qiu Y, Zhang G. Research and development aspects of oral controlled-release dosage forms. In: Wise DL, editor. *Handbook of Pharmaceutical Controlled Release Technology*, New York: Marcel Dekker, 2000, pp. 465–503.

61. Larsson A, Abrahmsen-Alami S, Juppo A. Oral extended-release formulations. In: Gad SC, editor. *Pharmaceutical Manufacturing Handbook: Production and Processes*, New York: John Wiley & Sons Inc., 2008, pp. 1191–1222.

62. Tiwari SB, Rajabi-Siahboomi AR. Extended-release oral drug delivery technologies: monolithic matrix systems. In: Jain KK, editor. *Drug Delivery Systems*, Totowa: Humana Press, 2008, pp. 217–242.

63. Fu XC, Liang WQ, Ma XW. Relationship between the release of soluble drugs from HPMC matrices and the physicochemical properties of drugs. *Pharmazie* 2003;58:221–222.

64. Rao KV, Devi PK, Buri P. Influence of molecular size and water solubility of the solute on its release from swelling and erosion controlled polymeric matrices. *J. Control Rel.* 1990;12:133–141.

65. Velasco MV, Ford JL, Rowe P, Rajabi-Siahbhoomi AR. Influence of drug: hydroxypropylmethylcellulose ratio, drug and polymer particles size and compression force on the release of diclofenac sodium from HPMC tablets. *J. Control Rel.* 1999; 5775–5785.

66. Mitchell SA, Balwinski KM. Investigation of hypromellose particles size effects on drug release from sustained release from hydrophilic matrix tablets. *Drug Dev. Ind. Pharm.* 2007;33:952–958.

67. Levina M, Rajabi-Siahbhoomi AR. The influence of excipients on drug release from hydroxypropyl methylcellulose matrices. *J. Pharm. Sci.* 2004;93:2746–2754.

68. Reynolds TD, Mitchelle SA, Balwinski KM. Investigation of the effect of tablet surface area/volume on drug release from hydroxypropylemethylcelluose controlled release matrix tablets. *Drug Dev. Ind. Pharm.* 2002;28:457–477.

69. Sheskey PJ, Williams DM. Comparison of low-shear and high-shear wet granulation techniques and the influence of percent water addition on preparation of a controlled relapse matrix tablet containing HPMC and a high-dose, highly water soluble drug. *Pharm. Tech.* 1996;20:80–92.

70. McCall TW, Baichwal AR, Staniforth JN. TIMERx oral controlled release drug delivery system. *Drug Pharm. Sci.* 2003;126:11–19.

71. Colombo P, Conte U, Gazzaniga A, Maggi L, Sangalli ME, Peppas NA, Manna AL. Drug release modulation by physical restrictions of matrix swelling. *Int. J. Pharm.* 1990;63:43–48.

72. Tiwari SB, Rajabi-Siahboomi AR. Modulation of drug release from hydrophilic matrices. *Pharm. Tech. Eur.* 2008, Accessed at http://www.ptemag.com/.

73. Using dow excipients for controlled release of drugs in hydrophilic matrix systems. September 2006. Accessed at www.dowexcipients.com.

74. Levina M, Gathoskar A, Rajabi-Siahboomi AR. Application of a modeling system in the formulation of extended release hydrophilic matrices. *Pharm. Tech. Eur.* 2006, Accessed at http://www.ptemag.com/.

75. Tiwari D, Lewis RK, During T, Harcum WW. Development of single layer acetaminophen extended release tablet with biphasic release. 2006, Annual Meeting of the Controlled Release Society, Vienna, 2006. Accessed at http://www. herc.com/aqualon/pharm/index.html.

76. Liu Q, Fassihi R. Zero-order delivery of a highly-soluble, low dose drug alfusozin hydrochloride via gastro-retentive system. *Int. J. Pharm.* 2007;348(1–2):27–34.

77. Abrahamsson B, Alpsten M, Jonsson UE, Lundberg PJ, Sanberg A, Sundgren M, Svenheden A, Toelli J. Gastro-intestinal transit of a multiple-unit formulation (metoprolol CR/ZOK) and non-disintegrating tablet with the emphasis on colon. *Int. J. Pharm.* 1996;140:229–235.

78. Siepmann J, Siepmann F, Paeratakul O, Bodmeier R. Process and formulation factors affecting drug release from pellets coated with ethylcellulose pseudolatex aquacoat. In: McGinity

JW, Felton LA, editors. *Aqueous Polymeric Coatings for Pharmaceutical Dosage Forms*, 3rd edition, New York: Informa Healthcare, 2008, pp. 203–236.

79. Bodmeier R, Siepmann J. Nondegradable polymers for drug delivery. In: Mathiowitz E, editor. *Encyclopedia of Controlled Drug Delivery*, Vol. 2, New York, John Wiley & Sons Inc., 1999, pp. 664–689.

80. Agyilirah GA, Banker GS. Polymers for enteric coating applications. In Tarcha PJ, editor. *Polymers for Controlled Drug Delivery*, Boca Raton, CRC Press, 1991, pp. 39–66.

81. Siepmann F, Siepmann J, Walther M, MacRae RJ, Bodmeier R. Polymer blends for controlled release coatings. *J. Control. Rel.* 2008;125:1–15.

82. Ho H-O, Chen C-N, Sheu M-T. Influence of Pluronic F-68 on dissolution and bioavailability characteristics of multi-layer pellets of nifedipine for controlled release delivery. *J. Control. Rel.* 2000;68:433–440.

83. Pillay V, Fassihi R. A novel approach for constant rate delivery of highly soluble bioactives from a simple monolithic system. *J. Control. Rel.* 2000;67:67–78.

84. Verma RK, Krishna DM, Garg S. Formulation aspects in the development of osmotically controlled oral drug delivery systems. *J. Control. Rel.* 2002;79:7–27.

85. Stevens RE, Limsakun T, Evans G, Mason DH. Controlled, multidose, pharmacokinetic evaluation of two extended-release carbamazepine formulations (Carbatrol and Tegretol XR). *J. Pharm. Sci.* 1998;87(12):1531–1534.

86. Liu L, Xu X. Preparation of bilayer-core osmotic pump tablet by coating indented core tablet. *Int. J. Pharm.* 2008;352:225–230.

87. Ahmed SU, Naini V, Wadgaonkar D. Scale-up, process validation, and technology transfer. In Shargel L, Kanfer I, editors. *Generic Drug Development: Solid Oral Dosage Forms*, New York, Marcel Dekker, 2005, pp. 95–136.

88. Guidance for Industry. *Statistical Approaches to Establishing Bioequivalence*. Center for Drug Evaluation and Research, Food and Drug Administration, 2001.

89. Guidance for Industry. *Bioavailability and Bioequivalence Studies for Orally Administered Drug Products: General Considerations*. Center for Drug Evaluation and Research, Food and Drug Administration, 2003.

90. Davit BM, Conner DP. Issues in bioequivalence and development of generic drug products. In: Sahajwalla CG, editor. *New Drug Development: Regulatory Paradigms for Clinical Pharmacology and Biopharmaceutics*, New York, Marcel Dekker, 2004, pp. 399–416.

91. Patnaik RN. Bioequivalence assessment: approaches, designs, and statistical considerations. In: Sahajwalla CG, editor. *New Drug Development: Regulatory Paradigms for Clinical Pharmacology and Biopharmaceutics*, New York, Marcel Dekker, 2004, pp. 561–586.

92. Bolton S, Bon C. *Pharmaceutical Statistics: Practical and Clinical Applications*, 4th edition, New York, Marcel Dekker, 2004, pp. 168–170 and 325–360.

93. Diletti E, Hauschke D, Stenijans VW. Sample size determination for bioequivalence assessment by means of confidence intervals. *Int. J. Clin. Pharmacol. Ther. Toxicol.* 30 (Suppl.): 1992; S51–S58.

94. Haidar SH, Davit B, Chen M-L, Conner D, Lee L, Li QH, Lionberger R, Makhlouf F, Patel D, Schurimann DJ, Yu LX. Bioequivalence approaches for highly variable drugs and drug products. *Pharm. Res.* 2008;25(1):237–241.

95. Haidar SH, Kwon HH, Lionberger R, Yu LX. Bioavailability and bioequivalence. In: Krishna R, Yu L, editors. *Biopharmaceutics Application in Drug Development*, New York, Springer, 2008, pp. 262–289.

96. Guidance for Industry. *SUPAC-MR: Modified Release Solid Oral Dosage Forms. Scaleup and Postapproval Changes: Chemistry, Manufacturing, and Controls; In vitro Dissolution Testing and In vivo Bioequivalence Documentation*. Center for Drug Evaluation and Research, Food and Drug Administration, 1997.

97. Sarfaraz N. Bioanalytical method validation. *Handbook of Bioequivalence Testing*. Informa Healthcare, New York, 2007, pp. 237–264.

98. Question Based Review for CMC Evaluation of ANDAs. Office of Generic Drugs, Center for Drug Evaluation and Research, Food and Drug Administration. Accessed at http://www.fda.gov/cder/ogd/QbR.htm.

99. A Risk Based Approach to Pharmaceutical CGMPs for the 21st Century. Center for Drug Evaluation and Research, Food and Drug Administration. Accessed at http://www.fda.gov/cder/gmp/.

100. Guidance for Industry. *PAT: A Framework for Innovative Pharmaceutical Development, Manufacturing and Quality Assurance*. Center for Drug Evaluation and Research, Food and Drug Administration, 2004.

101. Example Quality Overall Summary for Controlled Release Capsules. Office of Generic Drugs, Center for Drug Evaluation and Research, Food and Drug Administration. Accessed at http://www.fda.gov/cder/ogd/QbR/OGD_Model_QOS_ER_Capsule.pdf.

20

THE SCIENCE AND REGULATORY PERSPECTIVES OF EMERGING CONTROLLED RELEASE DOSAGE FORMS*

RAKHI B. SHAH AND MANSOOR A. KHAN
Division of Product Quality Research, Office of Testing and Research, Office of Pharmaceutical Sciences, Center for Drug Evaluation and Research, Food and Drug Administration, Silver Spring, MD, USA

20.1 BACKGROUND

Controlled release (CR) drug delivery systems ensure improved therapeutic response by providing more consistent blood levels compared to immediate release formulations. Commonly used terminologies include extended release, prolonged release, sustained release, slow release, controlled release, and so on. The term controlled release is now associated with those systems from which therapeutic agents may be automatically delivered at predefined rates over a long period of time [1]. Reduced fluctuation in plasma makes it possible to administer higher doses, thereby reducing the frequency of administration. This simple dosing regimen results in higher patient compliance and convenience. Also, a drug can be targeted to hard-to-reach areas such as tumors that enhances its bioavailability and thus can be used for targeted delivery. The release of drug can be modified from few hours to even a year in some cases. However, such systems could be very complex. Drug withdrawal might be very difficult and there is a potential for dose dumping since a very large dose of drug is present in the body at a given time. This could prove lethal in certain cases. There is a possibility of immune response with certain carriers used to deliver the drugs. Finally, many of the test methods are still evolving to study the safety and efficacy of such delivery systems. Despite all these drawbacks, pharmaceutical firms are increasingly using CR delivery systems as a means of drug delivery due to some of the merits discussed earlier.

*The views expressed in this chapter are only of the authors and do not necessarily reflect the policy of the agency.

Controlled drug delivery can be obtained from a variety of dosage forms and delivery systems including suspensions, emulsions, liposomes, microspheres, gels, implants, tablets, capsules, transdermal patches, gastrointestinal therapeutic systems, drug eluting stents, pulmonary products, and so on. Numerous technologies such as matrix systems, reservoir systems, pellets, and so on have been used to formulate CR dosage forms that are outlined in previous chapters. This chapter deals with regulatory overview, drug approval process in the United States, and regulatory definitions and requirements of CR dosage forms.

20.2 REGULATORY OVERVIEW

The first regulation of drugs in the United States was to curb the import of adulterated drugs under the provision of Drug Importation Act of 1848. It was not until 1906 that Pure Food and Drugs Act gave an authority to the Food and Drug Administration (FDA) to regulate misbranding. The Act was later amended by Shirley Amendment to regulate therapeutic claims by patent medicine firms. However, there were many shortcomings in the Act and in 1938 Federal Food, Drug, and Cosmetic Act or FD&C Act gave FDA authority to regulate drugs and other medical products. It required manufacturers to file a New Drug Application (NDA) with the FDA and seek FDA's approval before a drug could be marketed in the United States. Safety was the main focus of the NDA. Efficacy was included after the FD&C Act was amended by Kefauver–Harris Amendment. These statutes or legislations

Oral Controlled Release Formulation Design and Drug Delivery: Theory to Practice, Edited by Hong Wen and Kinam Park
Copyright © 2010 John Wiley & Sons, Inc.

are laws passed by the Congress. The FDA is mandated to implement the statutes and does so by making regulations that are published in Code of Federal Registrar (CFR). The CFR Title 21 pertains to drugs with Sections 312, 314, and 601 dealing with Investigational New Drugs (INDs), NDA, and Biological License Application (BLA), respectively. Section 211 deals with the current Good Manufacturing Practices (cGMPs).

20.2.1 FDA Review of CR Formulations

Any new or generic drug product application is reviewed by the FDA, with appropriate modifications, to consider various types of dosage forms. This section explains general principles governing the regulation of the drug products.

A preclinical candidate is identified in the pharmaceutical firm on the basis of its analytical, toxicological, biopharmaceutical profile in animal models. At this stage, the sponsor of the firm submits an IND to the FDA for review. The process from drug discovery to approval for marketing is schematically shown in Figure 20.1.

Under present regulations in the United States, the use of a human drug product not previously authorized for marketing in the United States requires the submission of an IND to the agency. The FDA regulations of 21 CFR 312.22 and 312.23, respectively, contain the general principles underlying the IND submission and the general requirements for content and format. Section 312.23(a)(7)(i) requires an IND for each phase of investigation to include sufficient chemistry, manufacturing, and control (CMC) information to ensure proper identity, strength or potency, quality, and purity of the drug substance and drug product. The type of information submitted will

TABLE 20.1   Schematic of Clinical Trial Phases

Phase	Approximate Number of Human Subjects	Approximate Length	Purpose
1	20–200	Months	Safety
2	100s	Months to 2 years	Short-term safety, dosage, efficacy
3	100s to 1000s	1 to 4 years	Safety, dosage, efficacy

depend on the phase of the investigation, the extent of the human study, the duration of the investigation, the nature and source of the drug substance, and the drug product dosage form. Pre-IND meeting between the sponsor and the FDA takes place before IND is submitted for review. An IND review is mainly to determine if the intended drug is safe enough to the volunteers and to determine if the drug is efficacious to justify further development. INDs are not approved but they are called open INDs once they become effective. At this stage, a pharmaceutical firm can begin clinical trials.

If the drug demonstrates adequate safety in initial human studies, termed phase 1, progressive human clinical trials through phases 2 and 3 are undertaken to assess safety and efficacy. Phases are depicted schematically in Table 20.1.

As the clinical trials progress, design, development, scale-up, process controls, labeling, packaging, and so on are carried out by the pharmaceutical firms. Sponsors file NDA with the FDA seeking an approval to market the new drug product. NDA is a registration document submitted by a sponsor to the FDA to review and decide whether to approve marketing of a new drug. NDA contains CMC data, pharmacology, toxicology, metabolism, clinical safety, and efficacy data, as well as proposed labeling. The NDA review process is schematically depicted in Figure 20.2.

A review is mainly performed to assess whether the proposed drug formulation is safe and effective and whether the benefits outweigh the risks involved. Various components of drug application are reviewed on the basis of the type of data (Figure 20.2). Information on drug substance and drug product is detailed in the NDA.

For a drug substance, the following information should be presented [2]: general information (chemical structure, name, IUPAC name, solubility, etc.), manufacture (manufacturer, manufacturing process and process controls, structure of the proposed starting materials, control of critical steps, and intermediates), and characterization of chemical structure, particle size data, physical properties (i.e. polymorphic or solid-state form), if applicable in case of CR drug products, critical quality attributes including, but not limited to, identity, purity, quality, potency or strength, and impurities, container closure system, and stability data are all detailed in the NDA.

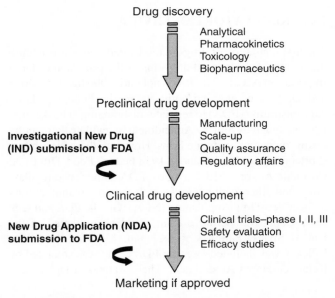

FIGURE 20.1 Schematic representation of the new drug approval process from discovery to marketing.

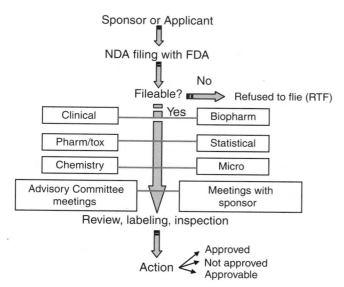

Sponsor or Applicant

NDA filing with FDA

Fileable? — No → Refused to flie (RTF)

Yes

Clinical		Biopharm
Pharm/tox		Statistical
Chemistry		Micro
Advisory Committee meetings		Meetings with sponsor

Review, labeling, inspection

Action — Approved / Not approved / Approvable

FIGURE 20.2 Schematic representation of the New Drug Application review by the FDA.

For drug products, the following information is included in the NDA: description of the composition of drug product—quantitative composition of all the ingredients in the drug products should be provided; manufacture (manufacturers, batch formula, description of manufacturing process, and controls); physicochemical tests (e.g., identity, assay, content uniformity, degradants, impurities, dissolution, viscosity, and particle size); and biological, pharmacological, toxicological, and microbiological data. In addition, data on the particle size distribution and/or polymorphic form of the drug substance used should be included for CR solid dosage forms, if applicable, so that relevant correlations can be established between data generated during early and late drug development and *in vivo* product performance. The analytical procedure, degradation profile, container closure system, and stability data for drug products are all part of NDA.

CMC review assesses the critical quality attributes and manufacturing processes of new drugs, establishes quality standards to assure safety and efficacy, and facilitates new drug development. Impurity profile and stability data are also reviewed. CMC review for generic applications (ANDA, Abbreviated New Drug Application) has recently been modified to include question-based reviews (QBRs) to transform the review into a modern, science and risk-based quality assessment.

Pharmacology and toxicology review evaluate nonclinical data to aid in selecting safe doses for "first in man" clinical trials and point attention to potential toxicities that should be monitored in the clinic. It also assesses toxicities that cannot be addressed in clinical trials such as carcinogenicity, teratogenicity, mutagenicity, and chronic toxicity. Various screening test methods are used to identify whether DNA

damage is a possible outcome from exposure to a certain chemical. Tests are also conducted to find out the distribution of the chemical in various tissues of the body. If the molecule does not cause DNA damage during *in vitro* testing, if it is metabolized quickly and does not reach sensitive organs, or if it is not absorbed, then it is less likely to present certain kinds of health hazards.

Clinical safety and efficacy review evaluates whether the proposed drug is safe and effective in humans on the basis of clinical studies data. Statistical significance of the data is evaluated under statistical review. Biopharmaceutical and microbiological reviews are all very important in terms of drug approval process. Drug release profiles for CR dosage forms are a key quality attribute and are reviewed in conjunction with the CMC review. *In vitro in vivo* correlation (IVIVC), if developed for CR formulations, is also reviewed and is briefly outlined in the chapter later.

During its review of newer dosage forms such as CR dosage forms from emerging technologies, the FDA may call for additional data from the sponsor needed to support the applications, if not supplied in the original applications. The FDA requests information when the agency considers such information relevant to determining whether a particular CR drug product is safe and effective. If the FDA determines such data are needed for a class of drugs, the agency may issue guidance to applicants recommending that they may be submitted in the original application.

20.2.2 Regulatory Definitions of CR Formulations

CR formulations are defined in CDER (Center for Drug Evaluation and Research) data standards manual. The definitions exist for different types of extended release dosage forms and are given below [3]. These definitions are important because of the fact that a product has to be pharmaceutically equivalent and bioequivalent to be therapeutically equivalent. For example, if a firm markets CR capsules, a generic firm also has to develop CR capsule as an equivalent product for 505(j) applications.

Tablet, Multilayer, Extended Release: A solid dosage form containing medicinal substances that have been compressed to form a multiple layered tablet or a tablet within a tablet, the inner tablet being the core and the outer portion being the shell, which, additionally, is covered in a designated coating; the tablet is formulated in such manner as to allow at least a reduction in dosing frequency as compared to that drug presented as a conventional dosage form.

Tablet, Film Coated, Extended Release: A solid dosage form that contains medicinal substances with or without suitable diluents and is coated with a thin layer of a water-insoluble or water-soluble polymer; the tablet is formulated in such manner as to make the contained medicament available over an extended period of time following ingestion.

Tablet, Extended Release: A solid dosage form containing a drug that allows at least a reduction in dosing frequency compared to that drug presented in conventional dosage form.

Capsule, Extended Release: A solid dosage form in which the drug is enclosed within either a hard or a soft soluble container made from a suitable form of gelatin, and which releases a drug (or drugs) in such a manner as to allow a reduction in dosing frequency compared to that drug (or drugs) presented as a conventional dosage form.

Capsule, Film Coated, Extended Release: A solid dosage form in which the drug is enclosed within either a hard or a soft soluble container or "shell" made from a suitable form of gelatin; additionally, the capsule is covered in a designated film coating, and which releases a drug (or drugs) in such a manner as to allow at least a reduction in dosing frequency compared to that drug (or drugs) presented as a conventional dosage form.

Capsule, Coated, Extended Release: A solid dosage form in which the drug is enclosed within either a hard or a soft soluble container or "shell" made from a suitable form of gelatin; additionally, the capsule is covered in a designated coating, and which releases a drug (or drugs) in such a manner as to allow at least a reduction in dosing frequency compared to that drug (or drugs) presented as a conventional dosage form.

Suspension, Extended Release: A liquid preparation consisting of solid particles dispersed throughout a liquid phase in which the particles are not soluble; the suspension has been formulated in a manner as to allow at least a reduction in dosing frequency compared to that drug presented as a conventional dosage form (e.g., as a solution or a prompt drug-releasing, conventional solid dosage form).

Suppository, Extended Release: A drug delivery system in the form of a suppository that allows a reduction in dosing frequency.

Solution, Gel Forming, Extended Release: A solution that forms a gel when it comes in contact with ocular fluid, and which allows at least a reduction in dosing frequency.

Pellets, Coated, Extended Release: A solid dosage form in which the drug itself is in the form of granules to which varying amounts of coating have been applied, and which releases a drug (or drugs) in such a manner as to allow a reduction in dosing frequency compared to that drug (or drugs) presented as a conventional dosage form.

Patch, Extended Release: A drug delivery system in the form of a patch that releases the drug in such a manner as to allow a reduction in dosing frequency compared to that drug presented as a conventional dosage form (e.g., a solution or a prompt drug-releasing, conventional solid dosage form).

Patch, Extended Release, Electrically Controlled: A drug delivery system in the form of a patch that is controlled by an electric current that releases the drug in such a manner as to allow a reduction in dosing frequency compared to that drug presented as a conventional dosage form (e.g., a solution or a prompt drug-releasing, conventional solid dosage form).

Liquid, Extended Release: A liquid that delivers a drug in such a manner as to allow a reduction in dosing frequency compared to that drug (or drugs) presented as a conventional dosage form.

Insert, Extended Release: A specially formulated and shaped nonencapsulated solid preparation intended to be placed into a nonrectal orifice of the body, where the medication is released, generally for localized effects; the extended release preparation is designed to allow a reduction in dosing frequency.

Injection, Suspension, Extended Release: A sterile preparation intended for parenteral use that has been formulated in a manner as to allow at least a reduction in dosing frequency compared to that drug presented as a conventional dosage form (e.g., as a solution or a prompt drug-releasing, conventional solid dosage form).

Injection, Powder, Lyophilized, For Suspension, Extended Release: A sterile freeze-dried preparation intended for reconstitution for parenteral use that has been formulated in a manner as to allow at least a reduction in dosing frequency compared to that drug presented as a conventional dosage form (e.g., as a solution).

Injection, Powder, For Suspension, Extended Release: A dried preparation intended for reconstitution to form a suspension for parenteral use that has been formulated in a manner as to allow at least a reduction in dosing frequency compared to that drug presented as a conventional dosage form (e.g., as a solution).

Granule, For Suspension, Extended Release: A small medicinal particle or grain made available in its more stable dry form, to be reconstituted with solvent just before dispensing to form a suspension; the extended release system achieves slow release of the drug over an extended period of time and maintains constant drug levels in the blood or target tissue.

Granule, Delayed Release: A small medicinal particle or grain to which an enteric or other coating has been applied, thus delaying release of the drug until its passage into the intestines.

For Suspension, Extended Release: A product, usually a solid, intended for suspension prior to administration; once the suspension is administered, the drug will be released at a constant rate over a specified period.

Film, Extended Release: A drug delivery system in the form of a film that releases the drug over an extended period in such a way as to maintain constant drug levels in the blood or target tissue.

Fiber, Extended Release: A slender and elongated solid thread-like substance that delivers drug in such a manner as to allow a reduction in dosing frequency compared to that drug (or drugs) presented as a conventional dosage form.

Core, Extended Release: An ocular system placed in the eye from which the drug diffuses through a membrane at a constant rate over a specified period.

Bead, Implant, Extended Release: A small sterile solid mass consisting of a highly purified drug intended for implantation in the body that would allow at least a reduction in dosing frequency compared to that drug presented as a conventional dosage form.

If a product of emerging technologies yields a controlled release dosage form other than these products, appropriate review divisions need to be consulted for defining a new dosage form.

20.2.3 Excipients in CR Dosage Forms: Issues and Guidance from FDA

Excipients are ingredients that are intentionally added to therapeutic and diagnostic products to provide a function and are not intended to exert a therapeutic effect. According to 21 CFR 210.3(b)(8), an inactive ingredient is any component of a drug product other than the active ingredient. Excipients are used to formulate the active drug substance into a pharmaceutical dosage form.

Excipients must meet the requirements established in 21 CFR 330.1(e) for over-the-counter (OTC) drug products and 21 CFR 314.94(a)(9) for generic drug products. For generic drug products, the excipients for parenteral, ophthalmic, or otic routes should be the same as the Reference Listed Drug (RLD) products. For oral, dermal, or topical drug products, selection of excipients is flexible, with the only criterion being it should be safe.

Well-known excipients are listed in Inactive Ingredient Guide (IIG) as published on the FDA web site [4]. IIG lists the maximum amount of excipient used per dosage unit. For multiple units, chronicity of dosing, maximum daily dose, and so on, the sponsors usually justify the use in the amounts used. For compendial excipients, references to quality standards (e.g., USP, NF) should be provided. There should be documented human use in the proposed level or they should be known excipients that should be generally recognized as safe (GRAS).

For noncompendial excipients, a specification sheet should be provided that identifies the tests and acceptance criteria and indicates the types of analytical procedure (e.g., HPLC) used. A complete description of the analytical procedures should be submitted upon request. A brief description of the manufacture and control of these components or an appropriate reference should be provided, for example, Drug Master File (DMF) or NDA. Information for excipients not used in previously approved drug products in the United States (e.g., novel excipients) should be equivalent to that submitted for drug substances. Section 21 CFR 314.94 (a)(9) ii states that "an applicant shall identify and characterize the inactive ingredients in the proposed drug products and provide information demonstrating that such inactive ingredients do not affect the safety or efficacy of the proposed drug product." A new excipient means any inactive ingredients that are intentionally added to therapeutic and diagnostic products but that (1) we believe are not intended to exert therapeutic effects at the intended dosage, although they may act to improve product delivery (e.g., enhance absorption or control release of the drug substance) and (2) are not fully qualified by existing safety data with respect to the currently proposed level of exposure, duration of exposure, or route of administration. New excipients should be evaluated for safety. A guidance for industry is available for performing nonclinical safety studies for pharmaceutical excipients [5]. There are numerous other guidances for determining safety that include potential pharmacological actions [6], genotoxicity [7], and reproductive toxicology [8, 9]. Acute toxicology studies should be performed in both a rodent species and a mammalian nonrodent species by the route of administration intended for clinical use [10]. In some cases, a dose-escalation study is considered an acceptable alternative to a single-dose design [11]. It is recommended that the absorption, distribution, metabolism, and excretion of the excipient be studied following administration by the clinically relevant routes to the same species as are used in the nonclinical safety studies [12, 13]. Novel excipients both intended for a short-term, intermediate, or long-term use and intended for nonoral routes of administration need to qualify for safety as outlined in the safety evaluation guidance published by the FDA [5].

Excipient functionality is defined as attributes of an excipient that can enhance/modulate product performance (e.g., solubilization, dissolution, and bioavailability), stability, and manufacturability of the drug product. For CR dosage forms, dissolution rate-controlling excipients serve an important role in regulating the drug release from the dosage form. Critical quality attributes for excipient functionality is important to identify and control for CR dosage forms. For example, if the particle size of an excipient is important that might be contributing to the controlling of the drug release, it should be specified. Variability in critical properties of an excipient can have a major impact on product performance and quality. Some sources of variability include lot-to-lot variability from the same manufacturer, different production sites for one manufacturer, different manufacturers, shipping and storage conditions, and so on. Tests in pharmacopoeial monographs may not fully address functionality of the excipient. Appropriate functionality tests for critical excipients should be developed and excipient characterization performed during development of dosage forms to ensure product quality and performance [14].

Drug–excipient compatibility studies are warranted as there might be stability issues. Degradation may be caused by interaction between functional groups of drug and excipients. Physical interaction between a drug and an excipient might compromise quality. Adsorption of drug by excipient might affect drug release if the excipient is not used for modifying release in CR dosage forms. Also, many drugs have amino groups and these might interact with certain excipients such as lactose [15]. Literature reports drug–excipient interaction or compatibility study approaches [15–19].

20.2.4 Specifications and Product Quality Attributes for CR Dosage Forms

A "specification" is defined as a list of tests, reference to analytical procedures, and appropriate acceptance criteria that are numerical limits, ranges, or other aspects [20]. Conformance to specifications means that the drug substance and/or drug product, when tested according to the listed analytical procedure, will meet the listed acceptance criteria. Specifications are critical quality standards that are proposed and justified by the manufacturer and approved by regulatory authorities as conditions of approval. Specifications should focus on characteristics found to be useful in ensuring the safety and efficacy of the drug substance and the drug product. The experience and data obtained during development of a new drug substance or product should form the basis for setting the specifications. It may be possible to propose excluding or replacing certain tests on this basis. Particle size test may be performed as an in-process test, or may be performed as a release test, depending on its relevance to product performance. Justification of specification should refer to relevant developmental data (linkage to performance, stability safety, efficacy, etc.), ICH or FDA guidances, pharmacopoeia standards, and test data. In addition, a reasonable range of expected analytical and manufacturing variability should be considered. Justification is also needed if an alternative approach is proposed.

Particle size, *in vitro* dissolution, residual solvents, osmolarity, sterility, stability, impurities, hardness, friability, aggregation, and so on are some of the key product quality attributes that may be relevant to dose, stability, or other characteristics significant to biological interactions or product quality for CR dosage forms. Particle size of the drug substance (as well as excipient) might be important in controlling the release. It might also affect targeting ability and reticuloendothelial uptake of parenteral CR dosage forms [21]. It might affect stability, and safety in certain cases. Residual solvents are also an important quality consideration for CR formulations where organic solvents are utilized during coating or manufacturing of the drug product. The safe levels of residual solvents are identified in ICHQ3C guidance [22]. However, it is not clear whether the total amount of residual solvent is important or the amount released *in vivo* on a daily basis is important [21]. Very recently, the United States Pharmacopeia (USP) issued new testing requirements and limits for residual solvents in the drug products, and the FDA has issued a draft guidance to assist manufacturers on this issue [23, 24]. Solvent recovery systems should comply with EPA (Environmental Protection Agency) or OSHA (Occupational Safety and Health Administration) regulations. Dissolution, stability, impurity, and potency are some of the important causes of product recalls for CR dosage forms.

A case study was presented at PQRI workshop for a two-component extended release tablets where drug product specification was based on four pilot studies [25]. Dissolution was not shown to be affected by particle size distribution and hence there was a wide range for acceptance criteria. However, failures in release profiles of one of the components were obtained in commercial batches. Milling and change in compression force was introduced in the drug product that resulted in failure of dissolution in another component of the drug product. Thus, this example serves to substantiate the importance of specification and acceptance criteria of a key quality attribute for CR dosage forms [25].

20.2.5 Drug Release and IVIVC

In vitro dissolution has been extensively used as a quality control tool for CR dosage forms. If judiciously used, it is a useful method for determining product quality and sometimes to evaluate the clinical performance of dosage forms. Some of the uses of *in vitro* drug release tests include (1) formulation development to include assessment of dose dumping and *in vivo* stability; (2) quality control to support batch release; (3) evaluation of the impact of manufacturing process changes on product performance; (4) substantiation of label claims; and (5) compendial testing [21]. The USP test for drug release for CR dosage forms is based on drug dissolution from the dosage unit against elapsed test time. Description of various test procedures and apparatus may be found in the USP [26]. The individual monographs contain specific criteria for compliance with the test and apparatus and test procedures to be used. However, for novel and emerging CR dosage forms, selection of media, time points based on the duration of the use, selection of apparatus, impeller speed, and sink condition all provide a good scientific basis for method selection. Evaporation of the drug from the media should also be considered. Drug release can be modeled and mechanistic understanding becomes important in cases when dose dumping becomes an issue [27–29]. Ideally, dissolution test should be physiologically based, that is, selection of media or apparatus can be made to mimic *in vivo* conditions. If one can predict the *in vivo* performance from *in vitro* dissolution data, it will minimize unnecessary human testing [30]. Therefore, investigations of *in vitro in vivo* correlations between *in vitro* dissolution and *in vivo* bioavailability are increasingly becoming an integral part of CR drug product development. General approaches for *in vitro* test method design can include (1) identification of release media and conditions that result in reproducible release rates, (2) preparation of formulation variants that are expected to have different biological profiles, (3) testing of formulation variants both *in vitro* and *in vivo*, and (4) modification of *in vitro* release methods to allow discrimination between formulation variants that have different *in vivo* release profiles [21].

IVIVC *In vitro in vivo* correlation refers to the relationship between *in vitro* dissolution and *in vivo* availability. For a successful development of an IVIVC, *in vitro* dissolution has to be a rate-limiting step for bioavailability. CR drug products present themselves as promising candidates for establishing IVIVC of their performance since they represent dosage forms that exhibit dissolution rate-controlled properties. CR formulations allow profiling of drug release over extended time periods *in vitro* and *in vivo*, which substantiates the value of the correlation. If the release from the dosage from is the rate-controlling step, the permeability of the drug must be sufficiently high to guarantee sink conditions in the gastrointestinal tract during the *in vivo* dissolution process [31]. Thus, CR formulations with potential for a successful IVIVC have to be BCS (Biopharmaceutical Classification System) class I or II drug products [32]. However, the dynamic environment along the entire gastrointestinal tract has to be considered in evaluating the potential for IVIVC for CR products due to their extended release time profile [33, 34]. Factors to be considered include (1) the impact of available fluid volume, the presence of surfactants, and motility pattern on drug product dissolution and drug solubility; (2) the regional differences in intestinal permeability; (3) the regional differences in intestinal metabolism and secretion; and (4) the transit time dependence of all the above factors [31].

In vitro dissolution test can also be used as a surrogate for bioequivalence (BE) in the presence of a meaningful IVIVC. The FDA guidance on IVIVC has been developed (1) to reduce regulatory burden by decreasing the number of bios-tudies needed to approve and maintain CR products on the market and (2) to set clinically more meaningful dissolution specifications [35]. It is anticipated that with a predictive IVIVC, the biostudies, which are generally required for major manufacturing changes as outlined in SPAC guidelines [36], are replaced by a simple *in vitro* dissolution test [30].

Four categories of IVIVCs have been described in the FDA guidance [35]:

1. Level A correlation represents a point-to-point relationship between the *in vitro* dissolution and the *in vivo* input rate (e.g., the *in vivo* dissolution of the drug from the dosage form). In general, these correlations are linear; however, nonlinear correlations are also acceptable. Level A correlation is considered most informative and very useful from a regulatory viewpoint.

2. Level B correlation uses the principles of statistical moment analysis. The mean *in vitro* dissolution time is compared either with the mean residence time or with the mean *in vivo* dissolution time. Although this type of correlation uses all the *in vitro* and *in vivo* data, it is not considered a point-to-point correlation. Furthermore, since it does not uniquely reflect the actual *in vivo* plasma level curve, this may not be very useful from the regulatory point of view.

3. Level C correlation establishes a single-point relationship between a dissolution parameter (e.g., t50% or percent dissolved in 4 h) and a pharmacokinetic parameter (e.g., AUC or C_{max}). Level C correlation does not reflect the complete shape of the plasma concentration–time curve, therefore is not the most useful correlation from a regulatory point of view. However, this type of correlation can be useful in early formulation development.

4. Multiple level C correlation relates one or several pharmacokinetic parameters of interest to the amount of drug dissolved at several time points of the dissolution profile. Multiple level C correlation can be useful as level A IVIVC from a regulatory point of view. However, if one can develop a multiple level C correlation, it is likely that a level A correlation can be developed as well.

The guidance outlines five categories of biowaivers:

1. Biowaivers without an IVIVC
2. Biowaivers using an IVIVC: nonnarrow therapeutic index drugs
3. Biowaivers using an IVIVC: narrow therapeutic index drugs
4. Biowaivers when *in vitro* dissolution is independent of dissolution test conditions

5. Situations for which an IVIVC is not recommended for biowaivers

If IVIVC is developed, this should be used to set dissolution specifications for the CR products in such a way that the fastest and slowest release rates allowed by the upper and lower dissolution specifications result in a maximum difference of 20% in the predicted C_{max} and AUC. The guidance describes procedures for setting dissolution specifications in cases where there is no IVIVC, where there is a level A IVIVC, and where there is a level C IVIVC. When there is no IVIVC, specification ranges at each time point of 20% or less (of the label claim) are recommended. If justified, deviations from this criterion can be acceptable up to a maximum range of 25%. Beyond this range, the specification should be supported by bioequivalence studies [30].

20.2.6 Bioavailability and Bioequivalence for CR Dosage Forms

Bioavailability is defined as the rate and extent to which the active ingredient or active moiety is absorbed from a drug product and becomes available at the site of action. For drug products that are not intended to be absorbed into the bloodstream, bioavailability may be assessed by measurements intended to reflect the rate and extent to which the active ingredient or active moiety becomes available at the site of action (CFR Section 320.1). Bioavailability for orally administered drug products can be documented by developing a systemic exposure profile. A profile can be obtained by measuring the concentration of active ingredients and/or active moieties and, when appropriate, its active metabolites over time in samples collected from the systemic circulation [37]. The regulatory objective of bioavailability studies is to assess the performance of the formulations used in the clinical trials that provide evidence of safety and efficacy (21 CFR 320.25(d)(1)).

Bioequivalence is defined as *the absence of a significant difference in the rate and extent to which the active ingredient or active moiety in pharmaceutical equivalents or pharmaceutical alternatives becomes available at the site of drug action when administered at the same molar dose under similar conditions in an appropriately designed study* (CFR Section 320.1). BE studies are a critical component of ANDA submissions. The purpose of these studies is to demonstrate BE between the pharmaceutically equivalent generic drug product and the corresponding reference listed drug (21 CFR 314.94 (a)(7)). Together with the determination of pharmaceutical equivalence, establishing bioequivalence allows a regulatory conclusion of therapeutic equivalence [37]. If any part of a drug product includes a controlled component, the following recommendations from the guidance apply [37]:

NDAs: Bioavailability and bioequivalence Studies: An NDA can be submitted for a previously unapproved new molecular entity, new salt, new ester, prodrug, or other noncovalent derivative of a previously approved new molecular entity formulated as a CR drug product. The first CR drug product for a previously approved immediate-release drug product should be submitted as an NDA. Subsequent CR products that are pharmaceutically equivalent and bioequivalent to the listed drug product be submitted as ANDAs. Bioavailability requirements for the NDA of a CR product are listed in Section 320.25(f). The purpose of an *in vivo* bioavailability study for which a controlled release claim is made is to determine if all of the conditions are met that include (1) the drug product meets the controlled release claims made for it, (2) the bioavailability profile established for the drug product rules out the occurrence of any dose dumping, (3) the drug product's steady-state performance is equivalent to a currently marketed non-CR or CR drug product that contains the same active drug ingredient or therapeutic moiety and that is subject to an approved full NDA, and (4) the drug product's formulation provides a consistent pharmacokinetic performance between individual dosage units.

As noted in Section 320.25(f)(2), "the reference material(s) for such a bioavailability study shall be chosen to permit an appropriate scientific evaluation of the controlled release claims made for the drug product" such as (1) a solution or suspension of the active drug ingredient or therapeutic moiety, (2) currently marketed non-CR drug product containing the same active drug ingredient or therapeutic moiety and administered according to the dosage recommendations in the labeling, or (3) currently marketed CR drug product subject to an approved full NDA containing the same active drug ingredient or therapeutic moiety and administered according to the dosage recommendations in the labeling.

The guidance also recommends bioavailability studies to be conducted for an extended release drug product submitted as an NDA by (1) a single-dose, fasting study on all strengths of tablets and capsules and highest strength of beaded capsules, (2) a single-dose, food-effect study on the highest strength, and (3) a steady-state study on the highest strength.

Bioequivalence studies are recommended when substantial changes in the components or composition and/or method of manufacture for a CR drug product occur between the to-be-marketed NDA dosage form and the clinical trial material.

ANDAs: Bioequivalence Studies: For CR products submitted as ANDAs, the following studies are recommended: (1) a single-dose, nonreplicate, fasting study comparing the highest strength of the test and reference listed drug product and (2) a food-effect, nonreplicate study comparing the highest strength of the test and reference product. Because single-dose studies are considered more sensitive in addressing the primary question of bioequivalence (i.e., release of the drug substance from the drug product into the systemic circulation), multiple dose studies are generally not recommended, even in instances where nonlinear kinetics are present.

20.2.7 Pharmaceutical Equivalence

The Drug Price Competition and Patent Term Restoration Act amendments or Hatch–Waxman amendments of 1984 to the FD&C Act gave the FDA statutory authority to accept and approve for marketing ANDA for generic substitute of a pioneer product. To gain approval, ANDA for a generic CR formulation must, among other things, be both pharmaceutically equivalent and bioequivalent to the pioneer CR product, which is also termed the RLD product as identified in FDA's Approved Drug Products with therapeutic equivalence ratings (The Orange Book).

To be pharmaceutically equivalent, the generic and pioneer formulations must (1) contain the same active ingredient; (2) contain the same strength of the active ingredient in the same dosage form; (3) be intended for the same route of administration; and (4) generally be labeled for the same conditions of use. The FDA does not require that the generic and reference listed CR products contain the same excipients or that the mechanism by which the release of the active drug substance from the formulation takes place be the same.

20.2.8 Biowaiver of CR Dosage Forms

Waiver of *in vivo* studies in certain cases of CR dosage forms may be applicable, although BCS guidance provides relief only for BCS class 1 (high solubility, high permeability) drugs. The following cases are identified in FDA's guidance [37] when the biowaiver for CR drug products is applicable that include (1) beaded capsules of lower strength and (2) tablets of lower strength. The criteria are that when the CR drug product is in the same dosage form, but of a different strength, when it is proportionally similar in its active and inactive ingredients, and when it has the same drug release mechanism, an *in vivo* BE determination of one or more lower strengths can be waived based on dissolution profile comparisons in at least three dissolution media (e.g., pH 1.2, 4.5, and 6.8) with an *in vivo* study only on the highest strength. The f2 test can be used to compare profiles from the different strengths of the product. An f2 value of more than 50 can be used to confirm that further *in vivo* studies are not needed. Certain postapproval changes can also get biowaiver as outlined in the next section.

However, Bialer et al. considered using current parameters (C_{max}, AUC, and t_{max}) to assess bioequivalence for CR dosage forms [38]. For a reliable assessment of bioequivalence in CR formulations, there might be a role for the use of additional criteria that are outlined in their work [38].

20.2.9 Scale-Up Considerations for CR Dosage Forms

In addition to the regulatory requirements of safety and efficacy, the ability of experimental formulation to be reproducibly manufactured on high-speed production equipment in a cost-effective manner is vital for a successful drug product. The product must be capable of being processed and packaged on a large scale, often with the equipment that only remotely resemble those used in laboratory. Scale-up studies should include a close examination of the formula to determine its ability to withstand process modification and should also include a review of a range of relevant processing equipment to determine the most compatible, economical, simple, reliable, and reproducible in manufacturing the product [39]. The availability of raw materials that consistently meet the specifications of the formulations should be determined. Process controls should also be evaluated and validated during scale-up operations. Appropriate records and reports are issued to support cGMP initiative [40]. A process using the same type of equipment performs quite differently when the size of the equipment and the amount of material involved are significantly increased. For example, loading a mixer can become a complicated operation using sophisticated equipment when large volumes are involved. Also, lengthy transfer lines might result in material loss that must be accounted for and compensated. If the system is used to transfer materials for more than one product, steps to prevent cross-contamination must be taken. In addition, validated cleaning procedures should be in place. All critical features of a process should be identified so that as the process is scaled-up, it can be adequately monitored to provide assurance that the process is under control and that the product produced after scale-up maintains the specified attributes originally intended.

Information on the types of *in vitro* dissolution and *in vivo* bioequivalence studies for CR drug products approved as either NDAs or ANDAs in the presence of specified postapproval changes are provided in an FDA guidance for industry entitled "SUPAC-MR: Modified Release Solid Oral Dosage Forms: Scale-Up and Post-Approval Changes: Chemistry, Manufacturing, and Controls, *In Vitro* Dissolution Testing, and *In Vivo* Bioequivalence Documentation." It is a resource for manufacturers who intend to change (1) components and composition, (2) manufacturing site, (3) scale-up or scale-down, and/or (4) manufacturing (process and equipment) of CR formulations during postapproval period [36]. A challenge that still needs to be addressed in the guidance is to evaluate the influence of change in a multivariate manner.

The guidance defines (1) levels of change, (2) recommended CMC tests for each level of change, (3) recommended *in vitro* dissolution tests and/or *in vivo* bioequivalence tests for each level of change; and (4) documentation that should support the change [36].

FDA regulations laid down in 21 CFR 314.70(a) provide that applicants may make changes to an approved application in accordance with a guidance, notice, or regulation published in the Federal Register that provides for a less burdensome notification of the change (annual report). For CR solid oral dosage forms, consideration should be given as to whether the excipient is critical to drug release. The sponsor should provide appropriate justifications for claiming any excipient(s) as a nonrelease controlling excipient in the formulation of the modified release solid oral dosage form. The functionality of each excipient should be identified. The guidance also defines the level of changes and type of preapproval or documentation needed to submit.

- Level 1 changes are those that are unlikely to have any detectable impact on formulation's quality and performance. Generally, these changes are reported in the annual report.
- Level 2 changes are those that could have a significant impact on formulation's quality and performance.
- Level 3 changes are those that are likely to have a significant impact on formulation's quality and performance.

Level 2 and 3 changes generally need prior approval from the regulatory agency (FDA).

If manufacturing equipment for extended release dosage form is changed, preapproval might be needed if there is a change in the mechanism. Examples of changes are detailed in the FDA guidance on manufacturing equipment addendum [41].

It is recommended that for postapproval changes, the *in vitro* comparison be made between the prechange and postchange products. In instances where dissolution profile comparisons are recommended, an f2 test can be used. An f2 value of more than 50 suggests a similar dissolution profile. A failure to demonstrate similar dissolution profiles may indicate an *in vivo* bioequivalence study be performed. When *in vivo* studies are conducted, the comparison should be made for NDAs between the prechange and the postchange products, and for ANDAs comparison between the postchange product and the reference listed drug should be made [37].

20.2.10 Clinical Considerations of CR Dosage Forms

Patients should be advised for the dose and dosing frequency of CR drug products and instructed to not use them interchangeably or concomitantly with immediate-release forms of the same drug. Plasma concentration in terms of bioavailability is measured for CR drug products. For targeted CR drug products, concentration at the site of action might be more meaningful; however, it is not practical to achieve. Plasma concentration might be less than tissue concentration in some cases. Also, in certain cases such as for locally acting CR drug products or where absorption of drug from the CR product is minimal, plasma concentration is not relevant. In such cases, clinical pharmacodynamic studies or *in vitro* studies are recommended [42]. However, this is on case-by-case basis and consultation with the FDA is recommended.

20.2.11 Stability Consideration of CR Dosage Forms

Stability testing provides evidence on how the quality of a drug substance or drug product varies with time under the influence of temperature, humidity, and light, and to establish a retest period for the drug substance or a shelf life for the drug product and recommend appropriate storage conditions [43].

Stability studies for drug substance should include attributes that are susceptible to change during storage and are likely to influence quality, safety, and/or efficacy. The testing might cover physical, chemical, biological, microbiological, thermal, and sensitivity to moisture. Detailed storage conditions and testing procedures are given in the ICH Q1A(R2) guidance [43]. Stress testing of the drug substance is useful in identifying the likely degradation products, which can help establish the degradation pathways and the intrinsic stability of the molecule. This enables one to develop and validate stability indicating analytical methods. Stress testing generally includes the effect of temperatures (in 10°C increments, for example, 50°C, 60°C, and so on, above that for accelerated testing), humidity (e.g., 75% or more relative humidity), oxidation, hydrolysis, and photolysis on the drug substance [43–45]. In general, minimum three batches are used to test the stability of drug substance except photostability where only one batch can be used [43, 45]. These studies form an integral part of the information provided to regulatory authorities.

Stability studies for a drug product should be based on knowledge of the behavior and properties of the drug substance, results from stability studies on the drug substance, and experience gained from clinical formulation studies. Selection of batches, storage conditions, specifications, and so on are given in detail in ICH guidance [43]. For a new dosage form, stability test conditions are identified in ICH guidance Q1C [46]. Generally, three batches are used to evaluate stability of drug products. New dosage form in the guidance is defined as *a drug product that is a different pharmaceutical product type, but contains the same active substance as included in the existing drug product approved by the pertinent regulatory authority*. A change from an immediate release form to a CR dosage form is considered a new dosage form and therefore the guidance becomes applicable in such case. For studying drug products, applicability of reduced design, when applicable, can be gleaned from ICH guidance Q1D [47]. For evaluation of stability data, ICH Q1E guidance is very useful [48].

20.2.12 Labeling of CR Dosage Forms

Label should include dosing frequency for CR dosage forms. It should also state with which *in vitro* drug release test the product complies. For example, for theophylline CR capsules, seven tests are described in the monograph, each with different drug release times and tolerances. Other requirements are similar for all the drug products as per FDA regulations for prescription and over-the-counter drug labels. With new physician's labeling rule, a highlight section of most pertinent and important information is summarized in half a page on the label for prescription drug products. For OTC, drug facts label should contain dosing recommendations since they are consumed by patients without medical supervision. There are several final and draft guidance documents available from the FDA on labeling and interested readers can refer to those from the FDA web site [49–51].

20.2.13 Nonoral CR Dosage Forms

Dosage forms delivered by nonoral routes of administration may be designed to provide controlled drug release. The nonoral routes for CR dosage forms include topical, transdermal, parenteral, vaginal, rectal, ophthalmic, nasal, pulmonary, otic, and so on. Most of the regulatory discussions above are applicable to all these types of dosage forms. There might be additional requirements depending upon the type of dosage forms. For example, sterility assurance is an important criterion for parenteral, ophthalmic, pulmonary, nasal, and otic CR drug products. Different sterilization techniques exist, and on the basis of the susceptibility of the drug substance and type of drug products, one method of sterilization is chosen. In addition to sterility assurance, product integrity should be demonstrated [21]. Also, foreign particulate limits for these products are recommended. Particle size, residual solvents, syringeability, injectability, resuspendability, and so on are some of the regulatory requirements based on the type of products.

20.2.14 Combination CR Products

A combination product includes (1) a product comprised of two or more regulated components, that is, drug/device, biologic/device, drug/biologic, or drug/device/biologic, that are physically, chemically, or otherwise combined or mixed and produced as a single entity; (2) two or more separate products packaged together in a single package or as a unit and comprised of drug and device products, device

and biological products, or biological and drug products; (3) drug, device, or biological product packaged separately that according to its investigational plan or proposed labeling is intended for use only with an approved individually specified drug, device, or biological product where both are required to achieve the intended use, indication, or effect and where upon approval of the proposed product the labeling of the approved product would need to be changed, for example, to reflect a change in intended use, dosage form, strength, route of administration, or significant change in dose; or (4) any investigational drug, device, or biological product packaged separately that according to its proposed labeling is for use only with another individually specified investigational drug, device, or biological product where both are required to achieve the intended use, indication, or effect (21 CFR Section 3.2(e)). The primary mode of action, which is the single mode of action, of a combination product that provides the most important therapeutic action of the combination product determines the review procedures within the FDA under Section 503(g)(1) of the Act. Drug eluting stents are example of combination CR products, where drug is coated on the stents and thus becomes a combination product. Regulatory requirements can be gleaned from the guidance documents available with the FDA [52, 53].

20.2.15 Inspection

Drugs including biological drugs are subject to cGMP requirements laid down in 21 CFR Parts 210 and 211. These requirements govern the methods to be used in, and the facilities or controls to be used for, the manufacture, processing, packaging, or holding of a drug. They are intended to ensure that the drug meets the safety requirements of the Federal Food Drug and Cosmetics Act, has the identity and strength, and meets the quality and purity characteristics that it purports or is represented to possess. Tests used for inspection and product surveillance for CR dosage forms should be adequate to address quality and safety of products in emerging technologies. Modifications are warranted to address issues that pose challenges, for example, relating to tests that assess product stability or development of potentially hazardous by-products. Furthermore, with any product, scaling up to full production rates may affect several factors including batch-to-batch consistency, and it may be necessary to evaluate the adequacy of existing testing methodologies to assess such consequences of scale-up for products of emerging CR delivery.

20.3 CONCLUSIONS

Several technologies and drug delivery systems are in use and new platforms continue to emerge for CR dosage forms. FDA's current review processes are elaborate and comprehensive for evaluating safety and efficacy of almost all products, both traditional and those of new emerging technologies. Many of the challenges of emerging technologies can be dealt with the existing procedures. Although the FDA authorities may be adequate to meet novel challenges, in some cases, evolving state of the science regarding emerging CR technologies may warrant a case-by-case approach to assess whether sufficient evidence exists to show that the products satisfy the applicable statutory and regulatory standards. When newer products require significant changes in the review procedures, appropriate guidance documents with a thorough public input can be issued. Affected manufacturers and other interested parties can get timely information about FDA's expectations. This approach fosters predictability in the agency's regulatory processes, and enables innovation with enhanced transparency, while protecting public health. Overall, the chapter provided a brief overview of regulatory procedure for drug approval process in the United States with regulatory definitions and requirements of various CR dosage forms. Further research is needed with respect to determining the critical quality attributes, test methods to evaluate those attributes, linking clinical relevance of product quality attributes to safety and efficacy, long-term stability, scale-up, and IVIVC. For an enhanced product and process understanding, readers are encouraged to consult certain new guidance documents that include ICH Q8, Q9, and Q10.

REFERENCES

1. Lordi NG. Sustained release dosage forms. In Lachman L, Lieberman HA, Kanig JL, editors. *The Theory and Practice of Industrial Pharmacy*, 3rd edition, Lea & Febiger, Philadelphia, PA, 1987, pp. 430–456.

2. Food and Drug Administration. FDA Guidance for Industry, INDs for Phase 2 and Phase 3 Studies: Chemistry, Manufacturing, and Controls Information, Food and Drug Administration, May 2003.

3. CDER data standards manual, December 2006. Available at http://www.fda.gov/cder/dsm/DRG/drg00201.htm. Accessed July 2008.

4. Inactive ingredient search for approved drug products. Available at http://www.accessdata.fda.gov/scripts/cder/iig/index.cfm. Accessed July 2008.

5. Food and Drug Administration. FDA Guidance for Industry, Nonclinical studies for the safety evaluation of pharmaceutical excipients, Food and Drug Administration, May 2005.

6. ICH guidance for industry S7A. Safety Pharmacology Studies for Human Pharmaceuticals, International Conference on Harmonisation, July 2001.

7. ICH guidance for industry S2B. A standard Battery for Genotoxicity Testing of Pharmaceuticals, International Conference on Harmonisation, November 1997.

8. ICH guidelines for industry S5A. Detection of Toxicity to Reproduction for Medicinal Products, International Conference on Harmonisation, September 1994.

9. ICH guidelines for industry S5B. Detection of Toxicity to Reproduction for Medicinal Products: Addendum on Toxicity to Male Fertility, International Conference on Harmonisation, April 1996.

10. FDA guidance for industry. Single Dose Acute Toxicity Testing for Pharmaceuticals, Food and Drug Administration, August 1996.

11. ICH guidelines for industry M3. Nonclinical Safety Studies for the Conduct of Human Clinical Trials for Pharmaceuticals, International Conference on Harmonisation, November 1997.

12. ICH guidelines for industry S3A. Toxicokinetics: The Assessment of Systemic Exposure in Toxicity Studies, International Conference on Harmonisation, March 1995.

13. ICH guidelines for industry S3B. Pharmacokinetics: Guidance for Repeated Dose Tissue Distribution Studies, International Conference on Harmonisation, March 1995.

14. Hamad ML, Gupta A, Shah RB, Lyon RC, Sayeed VA, Khan MA. Functionality of magnesium stearate derived from bovine and vegetable sources- I. Dry granulated tablets. *J. Pharm. Sci.* 2008. Available at www.interscience.wiley.com. DOI 10.1002/jps.21381.

15. Monkhouse DC. Excipient compatibility possibilities and limitations in stability prediction. In Grimm W, Krummen A, editors. *Stability Testing in the EC*, Japan and the USA: Scientific and Regulatory Requirements, Wissenschaftliche Verlagsgesellschaft, Stuttgart, 1993, pp. 67–74.

16. Serajuddin ATM, Thakur AB, Ghoshal RN, Fakes MG, Ranadive SA, Morris KR, Varia SA. Selection of solid dosage form composition through drug–excipient compatibility testing. *J. Pharm. Sci.* 1999;88(7):696–704.

17. van Dooren AA. Design for drug–excipient interaction studies. *Drug Dev. Ind. Pharm.* 1983;9:43–55.

18. Wirth DD, Baertschi SW, Johnson RA, Maple SR, Miller MS, Hallenbeck DK, Gregg SM. Maillard reaction of lactose and fluoxetine hydrochloride, a secondary amine. *J. Pharm. Sci.* 1998;87:31–39.

19. Marini A, Berbenni V, Moioli S, Bruni G, Cofrancesco P, Margheritis C, Villa M. Drug–excipient compatibility studies by physico-chemical techniques: The case of Indomethacin. *J. Thermal Anal. Calorimetry* 2003;73(2):529–545.

20. ICH guidelines for industry Q6A. Specifications, Test Procedures and Acceptance Criteria for New Drug Substances and New Drug Products: Chemical Substances, International Conference on Harmonisation, December 2000.

21. Burgess DJ, Hussain AS, Ingallinera T, Chen ML. Assuring quality and performance of sustained and controlled release parenterals: workshop report. *AAPS PharmSci.* 2002;4(2): article 7.

22. ICH guidelines for industry Q3C. Impurities: Residual Solvents, International Conference on Harmonisation, December 1997.

23. General Chapter (467). Residual Solvents. United States Pharmacopeia 31/National Formulary 26.

24. FDA guidance for industry. Residual Solvents in Drug Products Marketed in the United States, Food and Drug Administration, August 2008.

25. Nasr M. Setting Specifications in the New Paradigm: Scientific and Regulatory Challenges. PQRI Workshop on Setting Specifications in the 21st Century, North Bethesda, MD, March 16, 2005.

26. General Chapter (724). Drug Release. United States Pharmacopeia 31/National Formulary 26.

27. Hussain AS. Preventing alcohol induced dose dumping is a desired product design feature. FDA-ACPS Meeting, October 26, 2005.

28. Meyer RJ, Hussain AS. Awareness topic: mitigating the risks of ethanol induced dose dumping from oral sustained/controlled release dosage forms. FDA-ACPS Meeting, October 26, 2005.

29. Khan MA. QbD perspectives on dose-dumping with alcohol. AAPS workshop on the role of dissolution and drug product life cycle. AAPS, Arlington VA, April, 2008.

30. Uppoor R. Regulatory perspectives on *in vitro* (dissolution)/*in vivo* (bioavailability) correlations. *J. Contr. Rel.* 2001;72: 127–132.

31. Lipka E, Amidon GL. Setting bioequivalence requirements for drug development based on preclinical data: optimizing oral drug delivery systems. *J. Control. Rel.* 1999;62(1–2): 41–49.

32. FDA guidance for industry. Waiver of *In Vivo* Bioavailability and Bioequivalence Studies for Immediate-Release Solid Oral Dosage Forms Based on a Biopharmaceutics Classification System. Food and Drug Administration, August 2000.

33. Kararli TT. Comparison of the gastrointestinal anatomy, physiology, and biochemistry of humans and commonly used laboratory animals. *Biopharm. Drug Disp.* 1995;16: 351–380.

34. Davis B, Morris T. Physiological parameters in laboratory animals. *Pharm. Res.* 1993;10:1093–1095.

35. FDA guidance for industry. Extended Release Oral Dosage Forms: Development, Evaluation and Applications of *In Vitro/In Vivo* Correlations. Food and Drug Administration, September 1997.

36. FDA guidance for industry. SUPAC-MR: Modified Release Solid Oral Dosage Forms. Scale-Up and Postapproval Changes: Chemistry, Manufacturing, and Controls; *In Vitro* Dissolution Testing and *In Vivo* Bioequivalence Documentation, September 1997.

37. FDA guidance for industry. Bioavailability and Bioequivalence Studies for Orally Administered Drug Products: General Considerations. Food and Drug Administration, March 2003.

38. Bialer M, Arcavi L, Sussan S, Volosov A, Yacobi A, Moros D, Levitt B, Laor A. Existing and new criteria for bioequivalence evaluation of new controlled release (CR) products of carbamazepine. *Epilepsy Res.* 1998;32(3):371–378.

39. Harder S, Buskirk GV. Pilot plant scale-up techniques. In Lachman L, Lieberman HA, Kanig JL, editors. *The Theory and Practice of Industrial Pharmacy*, 3rd edition, Lea & Febiger, Philadelphia, PA, 1987, pp. 681–710.

40. FDA guidance for industry. Current Good Manufacturing Practice for Phase 1 Investigational Drugs. Food and Drug Administration, July 2008.

41. FDA guidance for industry. SUPAC-IR/MR: Immediate Release and Modified Release Solid Oral Dosage Forms: Manufacturing Equipment Addendum, January 1999.

42. FDA interim guidance. Cholestyramine Powder: *In Vitro* Bioequivalence. Food and Drug Administration, February 1993. Available at www.fda.gov/Cder/Guidance/cholesty.pdf. Accessed August 2008.

43. ICH guidance for industry Q1A(R2). Stability Testing of New Drug Substances and Products. International Conference on Harmonisation, November 2003.

44. Shah RB, Tawakkul MA, Prasanna HR, Faustino PJ, Nguyenpho A, Khan MA. Development of a validated stability indicating HPLC method for Ranitidine hydrochloride syrup. *Clin. Res. Regulat. Affairs* 2006;23:35–51.

45. ICH guidance for industry Q1B. Photostability Testing of New Drug Substances and Products. International Conference on Harmonisation, November 1996.

46. ICH guidance for industry Q1C. Stability Testing for New Dosage Forms. International Conference on Harmonisation, May 1997.

47. ICH guidance for industry Q1D. Bracketing and Matrixing Designs for Stability Testing of New Drug Substances and Products. International Conference on Harmonisation, January 2003.

48. ICH guidance for industryQ1E. Evaluation of Stability Data. International Conference on Harmonisation, June 2004.

49. FDA guidance for industry. Adverse Reactions Section of Labeling for Human Prescription Drug and Biological Products: Content and Format. Food and Drug Administration, January 2006.

50. FDA guidance for industry. Clinical Studies Section of Labeling for Human Prescription Drug and Biological Products: Content and Format. Food and Drug Administration, January 2006.

51. FDA guidance for industry. Content and Format of the Dosage and Administration Section of Labeling for Human Prescription Drug and Biological Products. Food and Drug Administration, April 2007.

52. FDA guidance for industry and FDA staff. Early Development Considerations for Innovative Combination Products. Food and Drug Administration, September 2006.

53. FDA guidance for industry and FDA staff. Current Good Manufacturing Practice for Combination Products, September 2004.

INDEX

Oral Controlled Release Formulation Design and Drug Delivery: Theory to Practice, Edited by Hong Wen and Kinam Park
Copyright © 2010 John Wiley & Sons, Inc.